The Engineering Record

American Steam and Hot-Water Heating Practice

The Engineering Record

American Steam and Hot-Water Heating Practice

ISBN/EAN: 9783743465220

Manufactured in Europe, USA, Canada, Australia, Japa

Cover: Foto ©berggeist007 / pixelio.de

Manufactured and distributed by brebook publishing software (www.brebook.com)

The Engineering Record

American Steam and Hot-Water Heating Practice

AMERICAN

STEAM AND HOT-WATER HEATING

PRACTICE.

FROM

THE ENGINEERING RECORD.

(Prior to 1887 THE SANITARY ENGINEER.)

BEING A SELECTED REPRINT OF DESCRIPTIVE ARTICLES, QUESTIONS, AND ANSWERS.

WITH FIVE HUNDRED AND EIGHTY-FIVE ILLUSTRATIONS.

NEW YORK:
THE ENGINEERING RECORD,
1895.

PREFACE.

THE ENGINEERING RECORD (prior to 1887 THE SANITARY ENGINEER) has for sixteen years made its department of Steam and Hot-Water Heating and Ventilation a prominent feature.

Besides the weekly illustrated descriptions of notable and interesting current work, a great variety of questions in this field have been answered.

In 1888 Steam-Heating Problems was published. This was a selection of questions, answers, and descriptions that had been published during the preceding nine years, and dealt mainly with steam-heating.

The present book is intended to supplement this former publication, and includes a selection of the descriptions of hot-water, steam-heating, and ventilating installations in the different classes of buildings in the United States, prepared by the staff of THE ENGINEERING RECORD, besides a collection of questions and answers on problems arising in this department of building engineering, covering the period since 1888, in which the heating of dwellings by hot water has become popular in the United States.

The favor with which Steam-Heating Problems has been received encourages the hope that American Steam and Hot-Water Practice may likewise prove useful to those who design, construct, and have charge of ventilating and heating apparatus.

TABLE OF CONTENTS.

STEAM-HEATING NOTES AND QUERIES.

HOT-WATER HEATING NOTES AND QUERIES.

HEATING OF RESIDENCES AND APARTMENT HOUSES.

HOT-WATER HEATING IN A CHICAGO RESIDENCE.

THE new residence of Mr. J. B. Earl, St. Louis Avenue and Adams Street, Chicago, Ill., has recently been equipped with a hot-water heating apparatus arranged as shown in accompanying plans. Direct radiation is used throughout except in the reception room, hall, and library, which are heated by indirect coils in the basement. A hot-water heater of a rated net capacity of 1,900 square feet operates the 1,192 feet of direct and 480 feet of indirect radiators, and is designed to maintain a temperature of 85° Fahr. in the bathroom and of 75° Fahr. everywhere else in the house. The general arrangement of pipes is of a 4-inch main riser (A, Figs. 1 and 2), which takes the hot water from the heater up through a vertical shaft or closets to the attic, where, in the triangular space (S, Fig. 2) between the attic ceiling and the ridge of the roof it distributes the flow through a horizontal main B B, from which branches C C, etc., for the different radiators are taken off, and following the pitch of the roof terminate in vertical drop risers D D, etc., which are carried through the partitions to the basement. As they descend the hot water is supplied to the radiators, and passing through them is received at a lower point of the same pipe and taken as return water to the basement, where the bottoms of the risers are connected with horizontal mains to the boiler.

Figure 1 is a diagram of the attic piping, showing the delivery of the hot water through riser A and its horizontal distribution through pipes B B and C C, etc. The expansion tank is of steel, of 30 gallons capacity, and all the branches C C, etc., have their connections made on the sides of main B B. All pipe in the attic is covered with mineral wool jacketing, as are all the risers in the outside walls. A sheet-iron shield is interposed between the pipes and the lathing, and tees are left in the attic mains to provide for possible future connections.

Figure 2 is a basement plan showing in heavy black lines the system of return mains from the bottoms of drop risers D D, etc., to the heater. The heater is a Gurney double crown No. 1, style C, provided with an altitude gauge to show the height of water in the expansion tank, a thermometer in contact with the circulation, and a Butz automatic thermostatic regulator. I I are each 20-section standard indirect radiators, which deliver hot air to the first-floor registers, which are here indicated in dotted lines at R R R R. The small circles indicate the positions of the drop risers D D, etc., from the distributing mains in the attic. The short branches terminating with a cross are connections for first-floor radiators.

Figures 3 and 4 are plans of the first and second floors, respectively, showing arrangement of rooms, position of drop risers D D, etc., and the location and connection to them of the radiators. There are

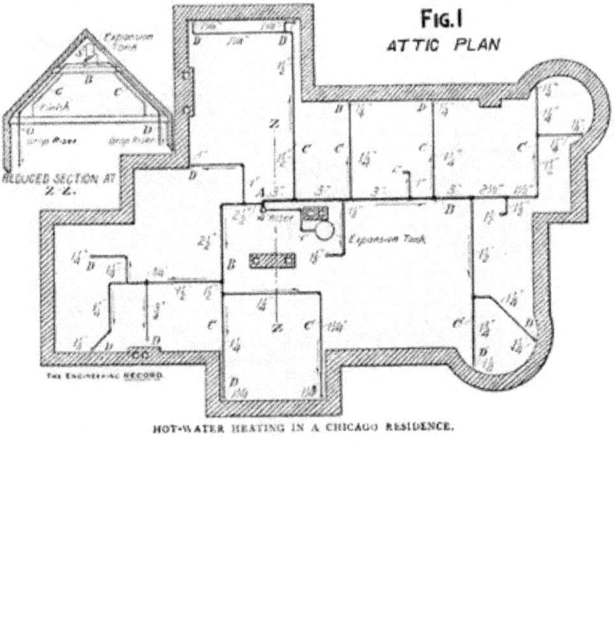

HOT-WATER HEATING IN A CHICAGO RESIDENCE.

in all 22 direct "Perfection" radiators of the Detroit pattern, each furnished with an air cock and a nickel-plated, wood wheel, Detroit quarter-turn finished valve on the return connection which drops through the floor before connecting to the return riser. P is a 1-inch pipe coil in the butler's pantry, having 10 feet of radiation for plate warming. The piping, indirect radiators and boiler satisfactorily endured a test of 50 pounds cold-water pressure applied by the Illinois Heating Company, Chicago, who installed the apparatus.

HOT-WATER HEATING OF A SUBURBAN RESIDENCE.

The recently built residence of W. W. Green, Esq., at Englewood, N. J., is a bluestone and frame structure about 36x65 feet in ground size, three stories high. It is built in accordance with the plans of Berg & Clark, architects, of New York City. The house, which stands upon high ground, and is exposed to the northwest winds, contains about 60,000 cubic feet to be heated. The heating is by a hot-water system designed and installed by the

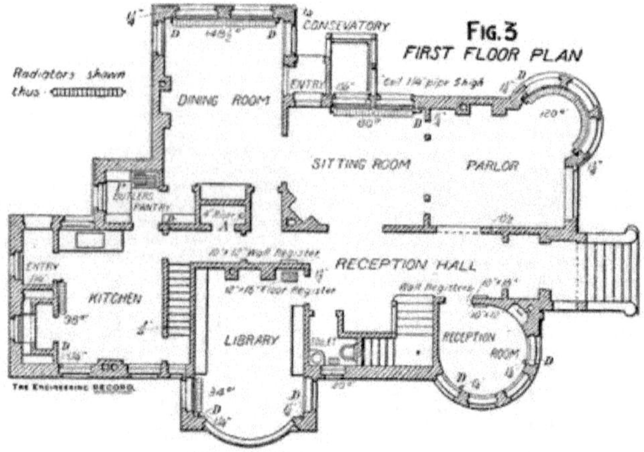

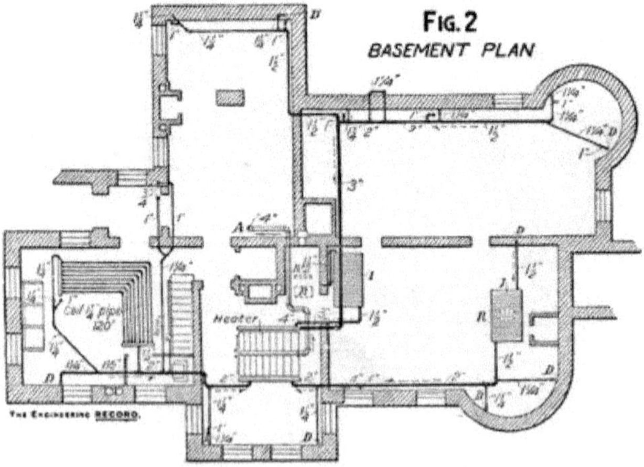

HOT-WATER HEATING IN A CHICAGO RESIDENCE.

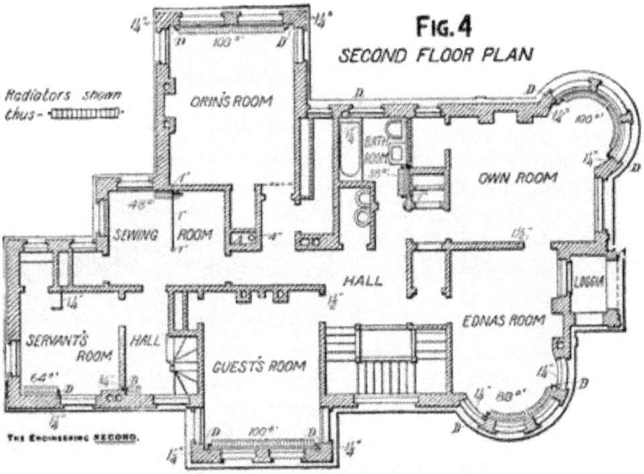

HOT-WATER HEATING IN A CHICAGO RESIDENCE.

Boynton Furnace Company, New York, the plans being the work of their Superintendent and heating engineer, James A. Harding. The heater is an exposed sectional cast-iron return flue No. 40 Boynton, set in the basement, as shown in Fig. 1. The hall, reception room, den, living room, and dining room on the first floor, Fig. 2, are heated by an independent indirect system, the location and sizes of floor registers, cold-air boxes and the heating surfaces of the heating stacks being given in their respective places on the plans. This system is supplied by a 3-inch main flow, which, rising gradually, may be traced on Fig. 1 to the point designated "syphon," where it rises perpendicularly into the closet room

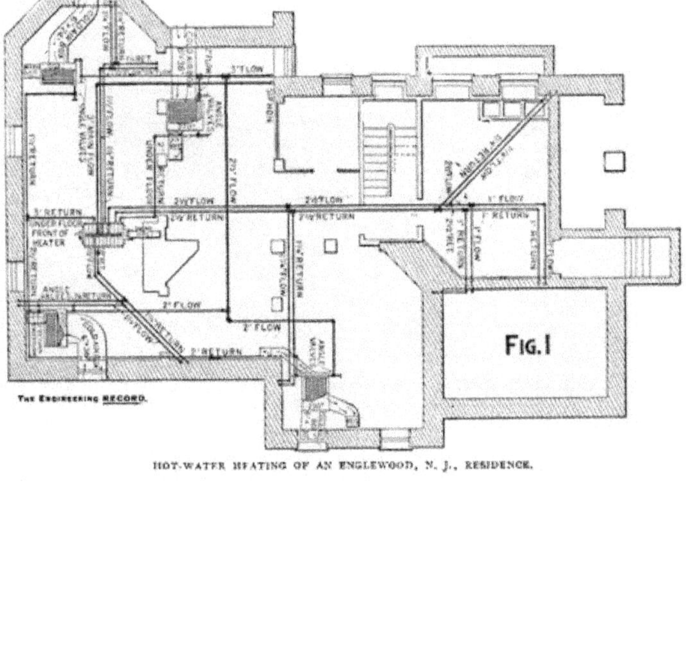

HOT-WATER HEATING OF AN ENGLEWOOD, N. J., RESIDENCE.

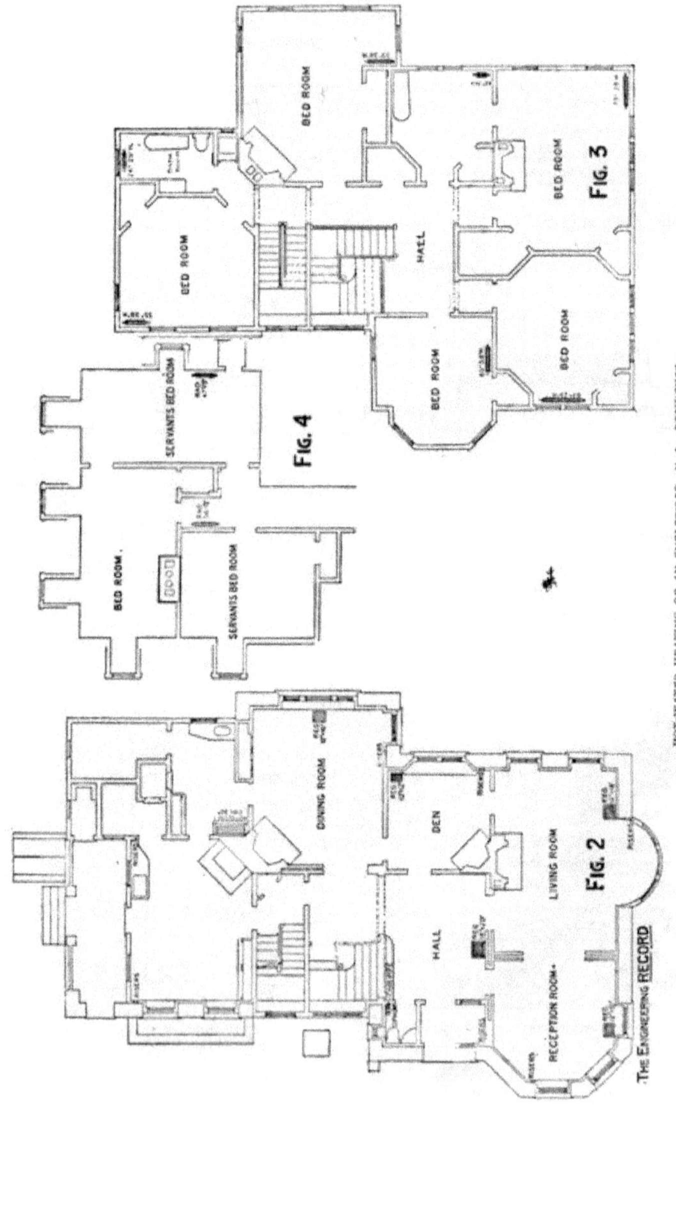

HOT-WATER HEATING OF AN ENGLEWOOD, N. J., RESIDENCE.

on the main floor, continuing close to the ceiling and then by a return bend passing back and downward to its rising point in the basement, and, full sized, to the point from which the 2½-inch branch is taken to supply the stacks heating the den, dining room and living room. Continuing, a 2-inch branch is taken off to the hall stack, the run terminating with a 1½-inch flow to the reception room stack. All of these flow pipes are on a gradual incline to the heaters, so that air or steam accumulating at any point in the system will travel to the highest point of the syphon in the closet on the main floor and thence by a ¾-inch relieving pipe emptying into the air space of the expansion tank in attic. The returns of all the indirect heaters are assembled in a 3-inch pipe which passes under the floor to the boiler. The den and dining rooms are heated from a joint stack partitioned proportionately to their sizes. All of the indirect stacks are of the American Radiator Company's "Perfection" pin style and are hung in two sections, an upper and a lower one, each of which has independent flow and return valves, so that either or both of the sections may be used as the season or service may require.

The size of the several rooms is: On first floor, 10 feet high, hall, 15x20 feet; reception room, 14x14 feet; living room, 15x21 feet; dining room, 15x18 feet; den, 12x12 feet; butler's pantry, 9x13 feet. On second floor, 9 feet 6 inches high, bedrooms over hall, 14x14 feet; reception room, 15x17 feet; living

FIG. 1

BASEMENT PLAN

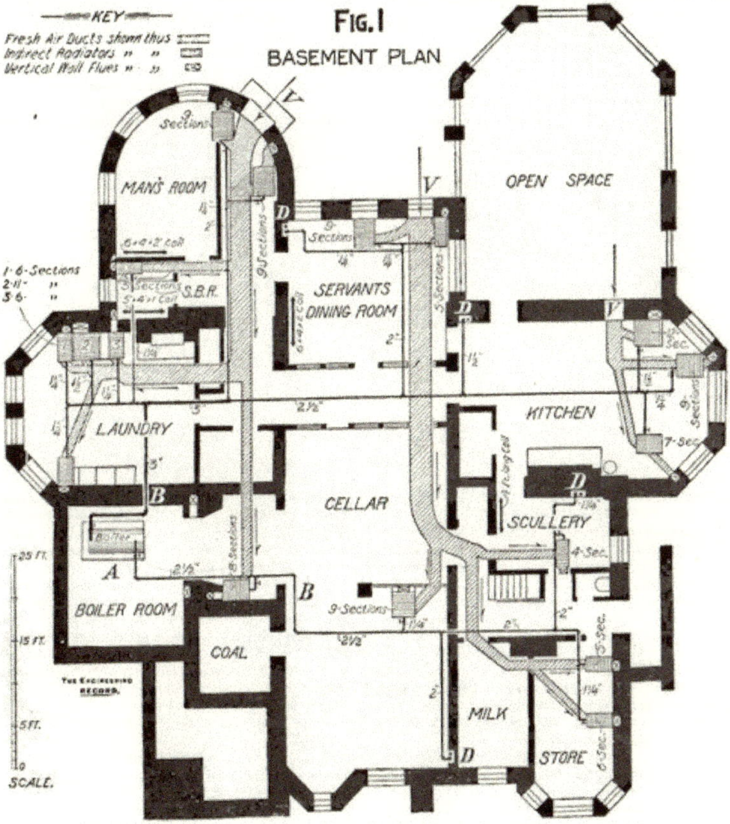

ALTERNATE STEAM OR HOT-WATER HEATING OF A YONKERS, N. Y., RESIDENCE. (See page 8.)

room, 15x20 feet; dining room, 16x20 feet; kitchen, 15x16 feet; front bathroom, 10x12 feet; rear bathroom, 7x11 feet. On third floor, 9 feet high, three servants' bedrooms, each 10x20 feet.

The direct heating is in three sections, the flow and return pipes for which and their sizes are shown on Fig. 1. The location of the radiators, which are of the American "Perfection" pattern, and their sizes are shown on the plans, Figs. 3 and 4. The expansion tank of this job is of the closed pattern, to be run open in mild weather, but furnished with a safety valve to be set at 10 pounds in severely cold weather, so that the water in the system may be heated to a higher temperature than can be done by the use of an open tank. The expansion pipe for the tank is connected to one of the return pipes, close to the boiler.

Although the house is in an exposed situation, as has been stated, the additional higher temperature of the water from the closed system, the double-decked indirect stacks, and the provision of a sufficiently large boiler, have served to furnish any range of heat required during the severe winter of 1892.

boiler A is a 12-section cast-iron Gold's pattern, with close-built brickwork, and having all necessary appliances for a steam-heating job. The heating mains B, the sizes and runs of which are marked on the plan, are all pitched up from the boiler, and corresponding return pipes C of the same size and lines are laid with a pitch down to the boiler. The foot of the risers D D D D, which supplies the direct radiators on the third floor, are dripped as shown in Fig. 2, each drip having the plug cock E. The safety valve F, water regulator G, steam gauge I, and damper regulator J have each a plug cock K, located as shown in Fig. 3. The stand-pipe L of the expansion pot, which is located in a closet on the third floor, has a plug cock marked M.

When required for hot-water heating the cock M on the stand-pipe L, Fig. 3, is opened, the cock E, Fig. 2, and all cocks marked K on Fig. 3 are closed. This puts all of the steam service appliances out of service, and prevents the circulation of hot water down the drip P, Fig. 2. Water is supplied to the system, and a pressure kept on it in the usual manner by a float valve in the expansion pot, at the

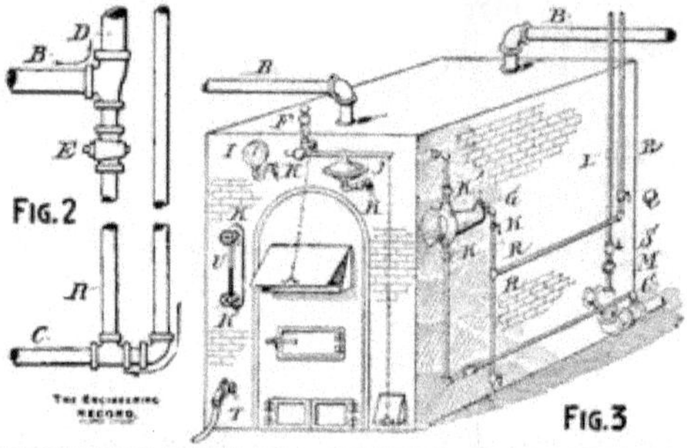

FIG. 2 FIG. 3

ALTERNATE STEAM OR HOT-WATER HEATING OF A YONKERS, N. Y., RESIDENCE.

ALTERNATE STEAM OR HOT-WATER HEATING OF A RESIDENCE.

The residence of Mr. J. E. Andrews, at Yonkers, N. Y., the plans of which were made by R. H. Robertson, architect, New York City, has a combination heating system which, during the milder weather at the beginning of the heating season, is used with hot water, and upon the advent of colder weather is changed to a steam heater, and changed back to hot water when the prevailing temperature is moderated in the spring. The house is of stone, three stories high with basement, covers a ground space of 92x84 feet, and has a large exposed surface on account of the bow, bays, and recesses which enter into its architectural composition.

The heating was done by Gillis & Geoghegan, New York, and is mainly of the indirect system. The location of the boiler, pipe lines, heating stacks, hot-air flues and fresh-air ducts, with sizes of each, is shown on the accompanying basement plan. The

top of stand-pipe L, the water supply to which is controlled by the cock Q on the cold-water pipe R. When required for steam heating the cocks M and Q are closed and the drip cock S is opened, which allows the water in the stand-pipe L to waste. The blow-off cock T is then opened, as are the air valves on the third floor, which allows the surplus water to pass off through the waste hose. The valves K on the water column U are then opened and the water allowed to waste until it has lowered to the desired level. The air valves on all the heaters having been opened and the water relieved from them, all the other cocks marked K, and the cock E are opened, and the system has been transformed to steam heating.

Fresh air is introduced to the several indirect stacks by galvanized-iron ducts starting from the points marked V on the plan, and which are cross-connected so that the supply may be taken from either point if necessary on account of high or con-

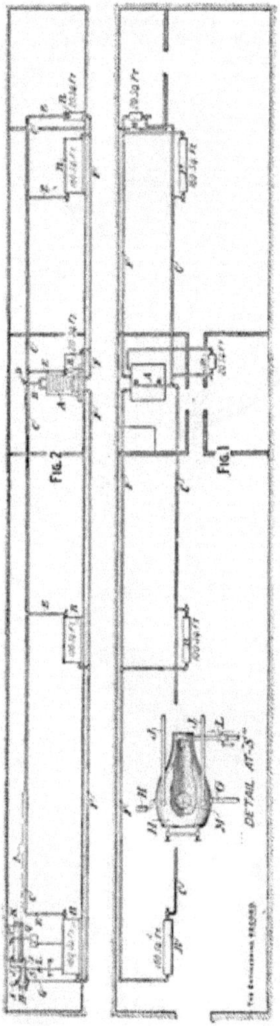

HOT-WATER HEATING OF A STORE.

trary winds. With but few exceptions the hot-air registers are set in the walls, the connection to them being built in as the walls went up. The use of this combined system during the past severe winter has justified the judgment of the designers in its ease of management and great range of heating.

HOT-WATER HEATING OF A STORE.

In the store of L. H. Biglow & Co., in Broad Street, New York City, there has recently been installed a compact hot-water heating plant, designed under conditions which make the job one of especial interest. These conditions were, in general, such that pipes could be run neither above nor below the store premises occupied by the firm. The size of the store, location and size of boiler, radiator and pipes are shown on the accompanying floor plan, Fig. 1. The elevation of the boiler, a No. 3 J. L. Mott Iron Works' "Sunray," the flow and return pipes, air valves, automatic water regulator, high-pressure and expansion tank are shown on Fig. 2.

To secure fireproofing, the boiler A was raised 1 foot above the store floor, upon a brick base, and placed about midway in the store. The 2½-inch riser B was connected into a branch tee supplying the two 2-inch flow pipes C C, which at this their highest point were 12 inches below the ceiling. The float air valve D was placed close upon top of the riser tee and served the purpose of an air escape. The flow pipes C C gradually descended to the down pipes E which entered the tops of the radiators, each having an angle valve to regulate the flow. The return pipes F were of the same size as the flow pipes, and were returned to the boiler between the floor joists.

Upon the top of the expansion pipe G was placed the automatic water regulator H, its water level I being but 3 inches above the highest point of C, or just enough to close the float valve D. The regulator H is connected by the pipes J J to the horizontal tank K. With the city water pressure turned on at the cock L, the float M closing against the inflow when the water level I is reached, the float air valve N on the top of the regulator allows the escape of air as the heated and expanded water rises in the pipe G. When the water has raised in the regulator and tank so as to float the valve N to a seat, the air is then confined and compressed above the water level O in the tank making it a closed system.

Upon the top of the tank K is set the safety valve P set to 10 pounds pressure. When this pressure has been passed the safety valve opens, releasing the air or water if it should rise to that point, passing off and down through the relief pipe Q to a closet tank. When the water has cooled and contracted sufficiently to allow the float valve N to drop, air enters, leaving the system an open one. Should water enough have been wasted to drop below the original water level I the regulator M performs its functions and supplies the deficiency. The work was planned and installed by the Blackmore Heating Company, New York City.

HOT-WATER RADIATORS BELOW THE BOILER LEVEL.

From sketches made during a visit to Washington, D. C., we herewith illustrate the arrangements in adjacent buildings on Eleventh Street, in that city, to secure hot-water circulation throughout the heating systems when some of the radiators were necessarily below the level of the boiler, or required the return pipes to fall below that level. Both buildings were three stories high above the basement.

Figure 1 shows the sytems in Johnson & Morris' office and shops on the basement floor. A is a Richardson & Boynton Co. No. 2 "Perfect" hot-water boiler, and B and C are 1-inch Box coils, and D is a Bundy radiator set on the same floor. Both the return pipes E and F are below the level of the bottom of the boiler, pipe F being *on* the floor and pipe E *under* the floor. G and H are radiators on the second floor, and I is the expansion tank. The overflow takes place through the horizontal branch J, which is connected to discharge pipe K by a tee L, open at the top. M and N are vent pipes and O is a supply branch to the city pressure. P is the attached thermometer.

Figure 2 is a diagram of the system next door to that shown in Fig. 1. A is a Boynton No. 4 "Perfect" boiler. F is its smoke flue, and G is its thermometer. B C D are radiators in the basement stove. E is a 2-foot single coil, warming a rear printing office at a slightly higher level, and I and J' are riser lines, each to a second and third floor radiator. V is a petcock. L is the expansion tank, connected by pipe M with the summit of line J, and overflowing through P. The latter is connected to discharge pipe Q by a tee R, whose upper branch is open. N is a cold-water supply.

Radiators B, C and D are set on the same floor with the boiler A, and their return pipe H is below this floor. Therefore, to promote the circulation, a 15-foot vertical loop S K was put on their supply main T, and provided at its summit K with an air-vent pipe O, connected to the overflow P.

Both of these systems have been in operation more than a year and are said to give good satisfaction and to keep all the radiators at the same temperature.

Both systems were put up by Johnson & Morris, of New York, and were described and exhibited to our representative by their Manager, Thomas Eagan.

HOT-WATER HEATING IN THE OFFICE AND SALESROOMS OF THE MURPHY & CO. VARNISH WORKS, NEWARK, N. J.

The accompanying illustrations, Figs. 1, 2, and 3, show the hot-water heating apparatus put into the office and salesrooms of Murphy & Co., in Newark, N. J., by the H. B. Smith Company, of New York, from plans by their engineer, Mr. Andrew G. Mercer, who has furnished us with the particulars of his work, giving the quantities of surface used, the size of pipes, both flow and return, the method of running them, the size of the boilers used and their connections. The illustrations made in our office from the working drawings furnished the foreman in charge of the work give all the data needed. We refer therefore, incidentally, to the general arrangement of the apparatus for the benefit of such of our readers as are not accustomed to the interpretation of plans.

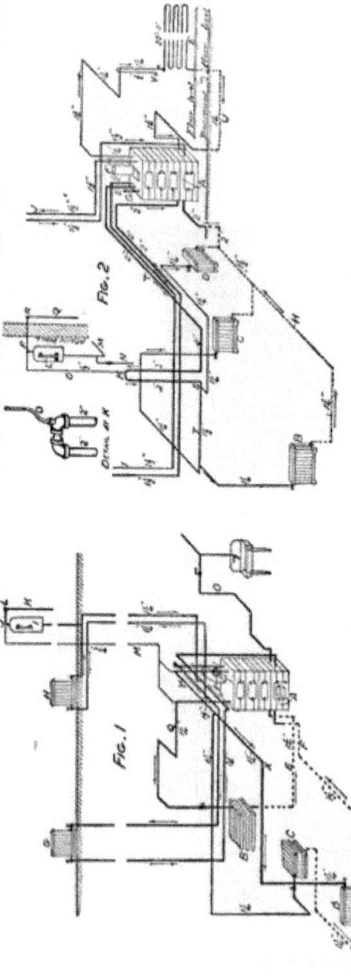

HOT-WATER RADIATORS BELOW THE BOILER LEVEL.

Figure 1 shows the ground-floor plan, there being no basement excepting the boiler room, which is about 6 feet below the common floor level. Two 10-section "Mercer" boilers are used, the total heating surface being 240 square feet and the grate surface 12 square feet.

A 3-inch flow-pipe leaves the top manifold of each boiler as seen in plan, Fig. 1, and also in the sketch, Fig. 3, which latter gives a general idea of the appearance of the boilers as set, and their connections as they appear when viewed from the platform at X, Fig. 1.

The two 3-inch boiler connections join with a 5-inch flow main, which rises to near the level of the ceiling, so that when it passes through the wall at C, it is close to the ceiling of the first story or tank rooms, where the varnish is stored. At *d* it is divided into two 3½-inch flow pipes, going right and left, so that at *e f* and *g* respectively they again divide into smaller branches, the sizes of which are plainly shown.

The pipes A B C and D are 1¼-inch wall-coils, *six* pipes high, and of the lengths shown. The supply pipes to these coils drop from the main flow pipe (as shown by the full black lines) and the return pipes from said coils return on the floor to the boiler room,

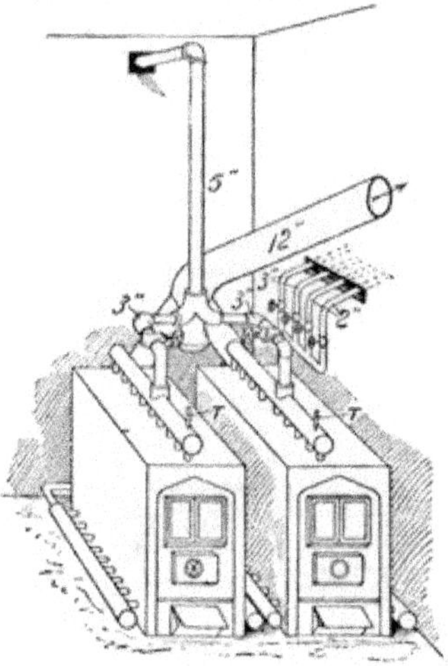

FIG. 3

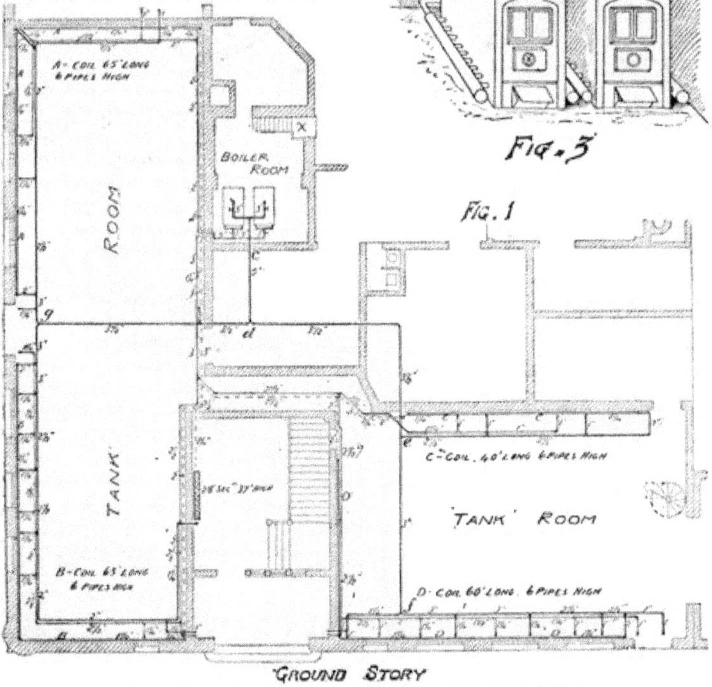

FIG. 1

GROUND STORY

HOT-WATER HEATING IN MURPHY & CO.'S VARNISH WORKS, NEWARK, N. J.

as shown by the broken lines. The detail of this is seen in Fig. 3.

The radiators on the second floor (Fig. 2) are supplied from the same flow main, as shown by the full black lines, and the return from these radiators is carried back to the boiler overhead, as shown by the double line, and on the same alignment as the main flow pipes.

Usually the flow and return pipes of a hot-water apparatus are of the same diameter; here it will be noticed, however, they are not, for the reason that each floor has its separate return pipe.

The points *g* and *f* on the main are the highest, all pipes having a uniform ascent as they approach them. From the top of each point an open vent is carried to the expansion tank on the second floor.

The second story, Fig. 2, which is the office floor, is warmed by "Union" radiators, the surface of each of which is shown in square feet on the plan. The cubic contents of the rooms is also shown, and an approximation can be made of the wall and window surface by those who desire to formulate data from these plans.

HOT-WATER HEATING PLANT IN A BROOKLYN RESIDENCE.

THE residence of Mr. I. S. Coffin, Remsen Street, Brooklyn, N. Y., has been recently remodeled under the plans and supervision of Mr. William B. Tubby, architect, New York City, and the new heating plant was designed by Mr. L. R. Blackmore, of the Blackmore Heating Company, New York, who installed the work. In the cellar is placed a No. 7 Sunray water heater, with a rated capacity of 2,000 square feet, and supplying 1,000 square feet of indirect radiation in six stacks, and 350 square feet of direct radiation in seven Ornate radiators, placed to heat the various rooms as shown by the plans and the accompanying schedule.

The dimensions of the building and heating plant are in accordance with the accompanying illustrations (page 16), which are prepared from data secured from the working drawings. The house is a typical city residence facing north, about 25 feet front, 46 feet deep, and four stories high above the basement and cellar, and having a two-story extension. Only two main lines of flow pipes are taken off from the heater. One goes immediately into a syphon 10 feet high, and returning to the cellar is divided into two branches that respectively supply the indirect stacks, which are arranged in two remote groups, one at the front and one in the rear end of the house. Each line is commanded by a valve that enables it to be cut out of circulation while the other is being operated alone, or any of its individual stacks may be separately turned on or off by their supply valves. All the supply pipes are hung from the cellar ceiling and are graded toward the stacks. The returns are run along the cellar wall near the floor and are graded toward the heater. As each

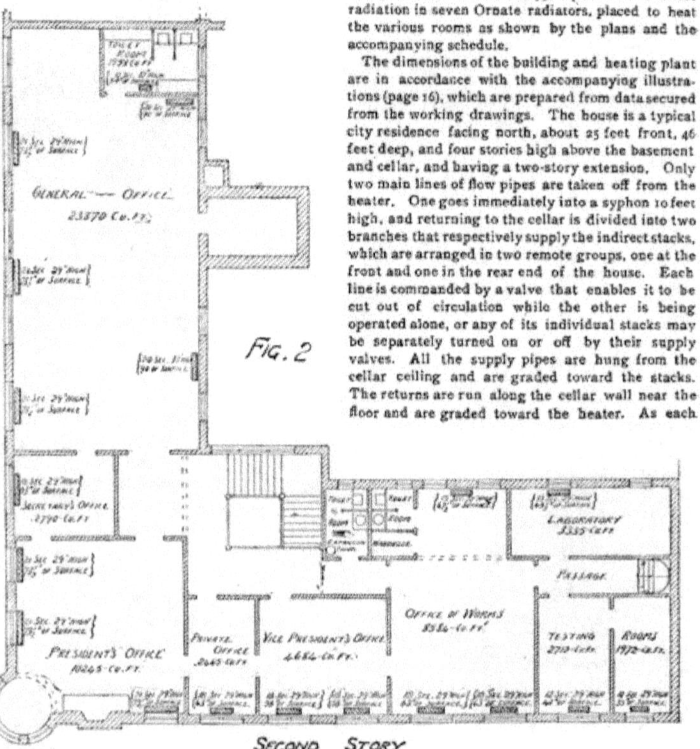

SECOND STORY

HOT-WATER HEATING IN MURPHY & CO.'S VARNISH WORKS, NEWARK, N. J.

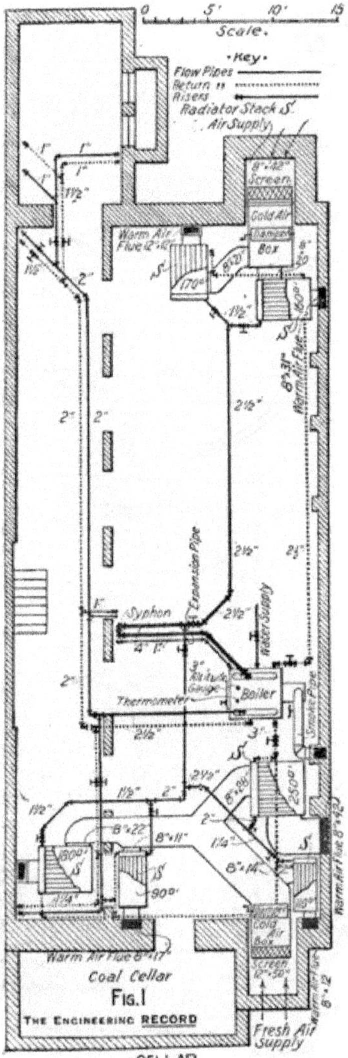

CELLAR

branch is taken off for the various risers, main and drip valves are provided so that any riser may be cut off and drained out independently.

The cold fresh air for the indirect stacks enters through two inlets at opposite ends of the house and is conducted to the various stacks by galvanized-iron ducts, No. 24 gauge, hung from the cellar ceiling. The entrances are provided with galvanized-iron wire screens of ¼-inch mesh and dampers of the full area of the ducts. Each stack is cased independently with galvanized iron, No. 24 gauge, with

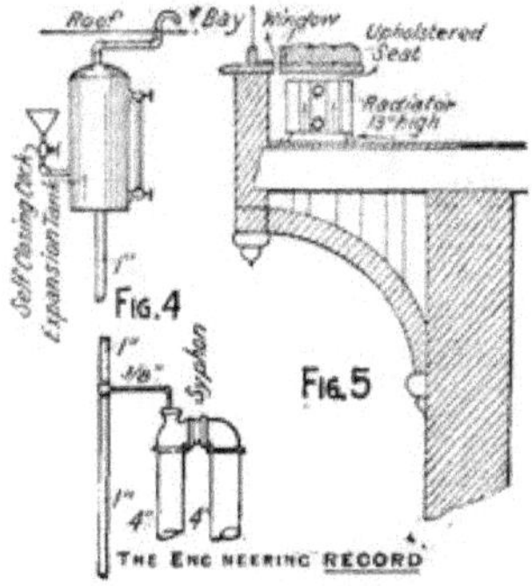

DETAILS, HEATING IN A RESIDENCE.

a 10-inch space under the stack for cold air and a 10-inch corresponding space for warm air over the stack. A mixing damper is provided at the end of each stack as shown in Fig. 6. All the hot-air flues are made of IX tin and are built into the walls, except flues for the hall and dining-room, which are carried outside the wall in basement. The size and dimensions of flues are shown on the plans.

The heater is furnished with a thermometer and altitude gauge. The plant is supplied with water by a ¾-inch lever handle stop-cock at the side of the boiler. The altitude gauge registers the exact height of water in the expansion tank, so that the engineer will not have to go to the expansion tank on the top floor in order to keep his plant filled with water. The altitude gauge has a fixed red index set to show the standard height required for the water (i. e., the tank about half-full) and a movable white index that indicates on the same dial the fluctuating height of water in the system.

Figure 4 shows the connections of the syphon (Fig. 1) to the expansion tank in the fourth story, which has an inverted overflow opening freely to the roof.

Figure 5 shows the special arrangement of a flue radiator under the divan seat in the bay window of the second-story boudoir, Fig. 2. Particular care was taken to leave a narrow opening between the seat and the wall, below the window-sill, for an upward current of hot air to pass through and warm the cold air that would otherwise descend from the window surface and fall upon the seat.

Figure 6 shows the arrangement of indirect stacks S S, etc., Fig. 1, in the cellar, and the control from the room heated of the mixing valves. The stacks are made with locked and bolted joints and their interiors are accessible by two slide doors in each, one to the hot-air and one to the cold-air chamber. Fresh cold air is always freely admitted when the main damper in the supply duct (Fig. 1) is open, and may pass through the wall duct A and be delivered from the register at any desired temperature up to the maximum power of the radiator by operating the mixing valve V, which can be set so as to close port B and open port F so as to have all the fresh air fully heated or so as to close F and open B so as to have none of the fresh air heated. Or it can be set at any intermediate position, as shown, so as to mix any required proportions of hot and cold air, the actual delivery of cold air in the room being of

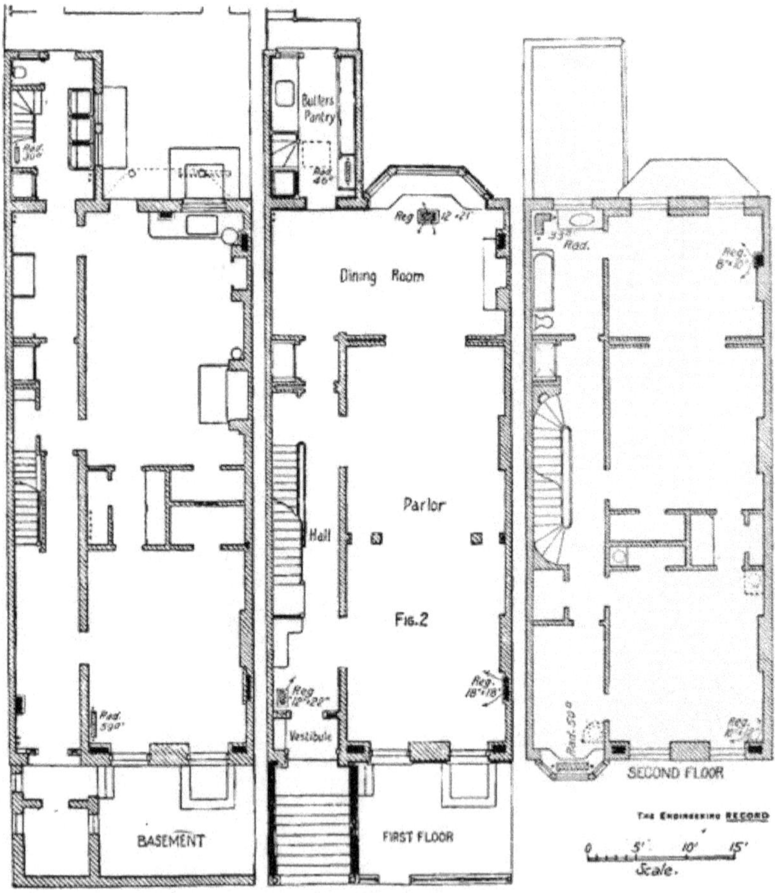

HOT-WATER HEATING PLANT IN A BROOKLYN, N. Y., RESIDENCE

Schedule.

Rooms.	Width.	Length.	Height.	Cubic Feet.	Ratio.	Square Feet. Direct.	Square Feet. Indirect.	Number of Settings.	State of Radiation.	Temperature.
BASEMENT.										
Billiard-room	16'	19'	9'	2,736	0.025	67.5	....	1	15 sections, 38 inches high, Ornate.	70
Laundry	8'	14'	9'	1,008	0.0345	41.5	....	1	7 sections, 38 inches high, Ornate.	70
FIRST FLOOR.										
Parlor	16'	38'6"	14	8,624	0.05	....	250	1	25 sections, Perfection Pin.	70
Dining-room	14'6"	23'6"	14'	4,776	0.036	....	172	1	17 sections, Perfection Pin.	70
Butler's pantry	8'	14'	14'	1,568	0.027	42	....	1	Detroit dining-room radiator.	70
Hall, first	6'6"	32'	14'	2,912 }						
Hall, second	7'	25'	11'	1,925 }	0.03	....	180	1	18 sections, Perfection Pin.	70
Hall, third	7'	20'	10'	1,400 }						
SECOND FLOOR.										
Parlor chamber	15'6"	17'	11'	2,530	0.043	....	110	1	11 sections, Perfection Pin.	70
Dining-room chamber	15'6"	12'	11'	2,046	0.034	....	70	1	16 sections, Perfection Pin.	70
Hall chamber	7'	14'	11'	1,078	0.043	47	....	1	13 sections, 13 inches high, Detroit flue.	70
Bathroom	7'	12'	11'	924	0.035	33	....	1	8 sections, Detroit corner radiator.	70
THIRD FLOOR.										
Parlor chamber	15'6"	16'6"	10'	2,560	0.04	....	100	1	9 sections, Perfection Pin.	70
Dining-room chamber	15'6"	17'	10'	2,640	0.0345	....	92	....	Included in second-floor stack.	70
Bathroom	7'	12'	10'	840	0.037	31	....	1	7 sections, 38 inches high, Ornate.	70
FOURTH FLOOR.										
Hall	16'	22'	9'	3,168	0.03	99	....	1	22 sections, 38 inches high, Ornate.	70

Total number of settings, 13.
Total number of square feet direct radiation 350
Total number of square feet indirect radiation 1,000
Fifty per cent. added for boiler power 500

Total 1,850
Boiler power of No. 7 Sunray, 2,000 feet.

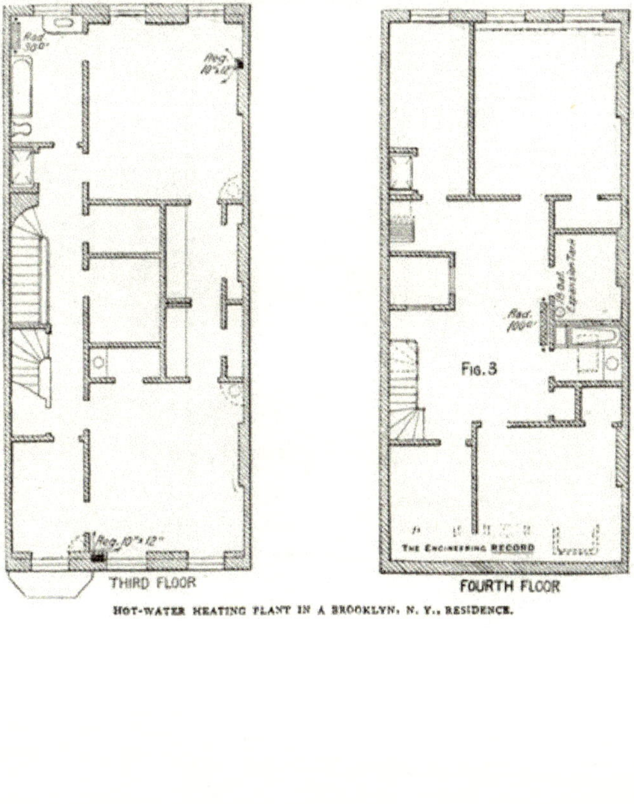

HOT-WATER HEATING PLANT IN A BROOKLYN, N. Y., RESIDENCE.

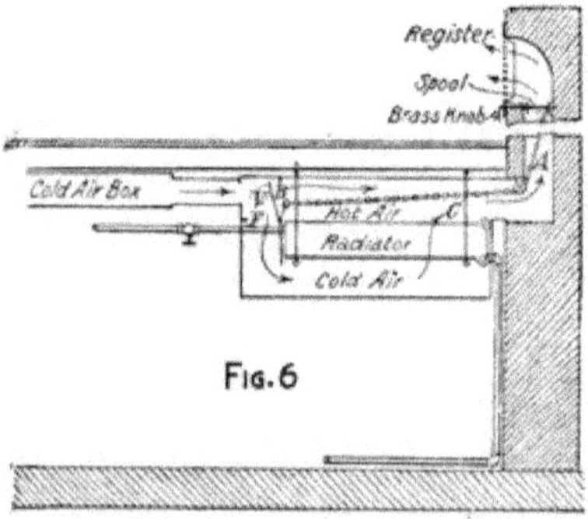

FIG. 6

course dependent on the available circulation created by the fireplace system of ventilation. The combined openings in ports B and F are always constant, it being impossible to close both at once. The valve V is conveniently operated by a bent arm, which is worked by a chain C, led over suitable sheaves and through the duct, up to a drum of small diameter, which is commanded by a crank at the register face. The damper closes port F by falling by gravity, but the diameter of the spool on which its chain is wound is so small that it cannot overhaul accidentally, but stays as left, and must be turned by hand to operate in either direction. The spool or drum was made of an ordinary piece of round iron, with a couple of washers slipped on for heads, and its small diameter enables its operation to easily and accurately control the mixing valve, since it takes several revolutions of the knob to completely reverse the damper.

INDIRECT STEAM OR HOT-WATER HEATING IN A MASSACHUSETTS RESIDENCE.

THE new residence of ex-Gov. Oliver Ames, 2d, at North Easton, Mass., is built in an exposed situation on high ground, and the severe requirements for heating it in the cold and windy climate of the locality required consideration, as did also the special architectural features of the house. The building is long and has a comparatively narrow area. The

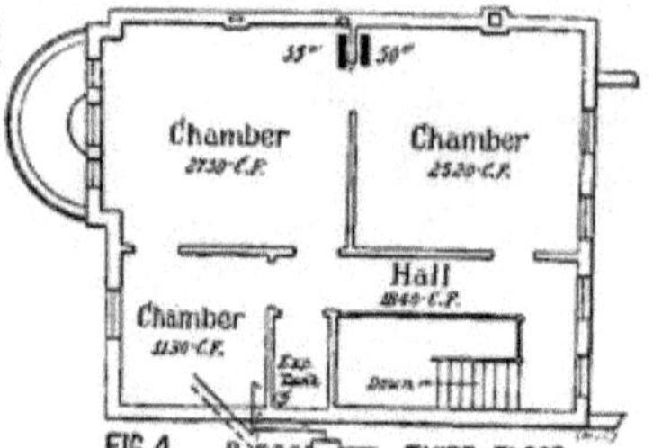

FIG. 4 THIRD FLOOR

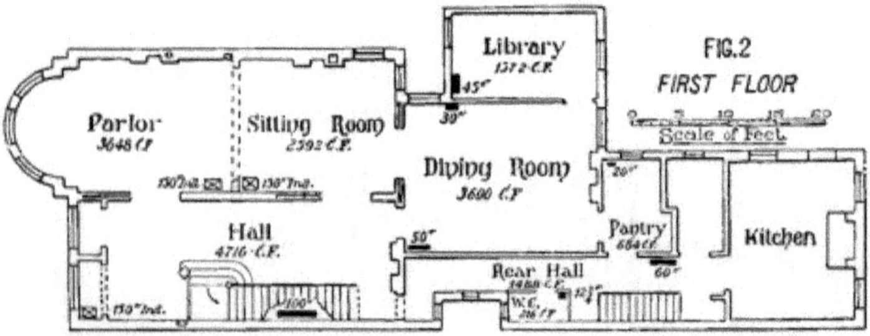

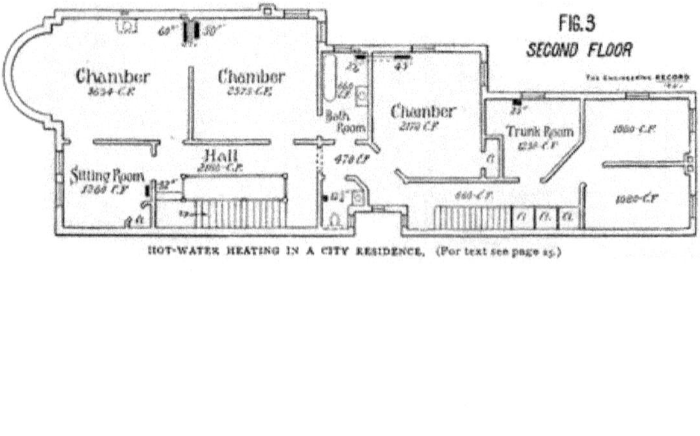

HOT-WATER HEATING IN A CITY RESIDENCE. (For text see page 15.)

house contains nearly 40 rooms beside the halls and has a very large volume of air to be warmed and renewed. Provision had also to be made for widely differing conditions to be met with in the family, private, guests' and servants' rooms in so large a household. A careful study of the conditions and requirements resulted in the adoption throughout of a system of indirect radiators, fireplaces and wall flues for ventilation. The radiators are all incased in galvanized-iron stacks, of which there are seven separate ones suspended from the basement ceiling and containing a total (rated) surface of 2,600 square feet of the Gold's pin type in 16-foot sections, and all of them connected up so that, in the fall, before the maximum efficiency of the apparatus is required, they can be operated by hot water, which can be drained off and the whole system put under steam pressure when colder weather demands a higher service. It can again be converted into a hot-water plant in the mild spring days when only a little heat is needed. All these operations are conveniently effected by two valves, and it is claimed that a considerable economy is attained by omitting steam for low duty.

Figure 1 is a basement plan showing the arrangement of stacks and flues to serve a floor area about 140 feet long by 46 feet in extreme width, and showing the size and location of the steam main, which is hung from the ceiling and connected with a parallel return pipe (not here shown) of corresponding size throughout. Fresh external air is received through three screened inlets I I I and drawn, as indicated by the full arrows, to the radiator stacks S S, which warm it and deliver it to the first, second, and third stories through cylindrical or rectangular galvanized iron ducts, as indicated by the dotted arrows. There is no direct radiation except that of the coil C in the laundry clothes-drying room. The boiler is of the tubular pattern of 34 horse-power, with 12 square feet of grate area, and is set in brick walls and fitted with a Locke safety valve and an automatic damper regulator.

Figures 2, 3, and 4 respectively, are plans of the first, second, and third floors, showing the location and size of registers. This heating and ventilating system was installed by A. A. Sanborn, of Boston, in accordance with the plans and specifications of Messrs. Rotch & Tilden, architects, also of Boston. The entire cost of the system was about $3,000. Provision has been made for the future control of the hot-air supply by electric thermostats.

UNUSUAL PIPING IN A HOT-WATER HEATING APPARATUS.

The plans accompanying this description (page 19) show the hot-water heating apparatus in the residence of Mr. S. F. Requa, at South Evanston, Ill., as it was installed by the Illinois Heating Company, of Chicago, Ill.

The building covers a rectangle 75x35 feet, and contains, beside the basement, a first and second story. The cubical contents of the building are about 30,000 cubic feet, and this is warmed by 1,153

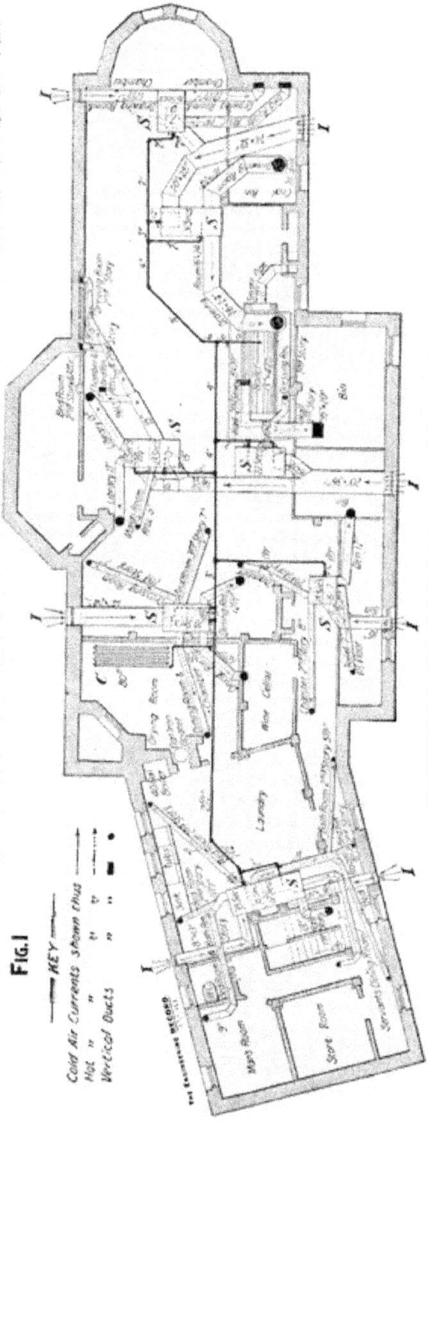

INDIRECT STEAM OR HOT-WATER HEATING IN A MASSACHUSETTS RESIDENCE.

square feet of hot-water radiating surface. With the exception of one indirect radiator of 180 square feet of surface warming the front hall, the building is warmed by direct radiation, there being 22 direct radiators located as shown on the plan.

The designer of this plant has departed from the usual method of running the flow and return pipe of a hot-water apparatus. The more common practice would locate the boiler in the central part of the basement and run lines of pipe in different directions and branches from these to the base of the risers supplying the radiators. The returns often parallel the flow pipes, so that the water for the radiators near the boiler would have to travel in such a system through a much shorter length of pipe than the water for radiators at a more distant point in the basement. Occasionally plants piped upon this system have given trouble, more generally, we believe, because of too small pipe sizes causing sufficient friction to prevent a proper circulation in the more distant radiators, rather than any inherent defect in the system.

Another point of advantage claimed for this system over the one more commonly used is that in the latter the water, when the apparatus is started, will pass through the nearest radiator and then return directly to the boiler through the vertical return pipe at the boiler at nearly the same temperature at which it leaves the boiler, and as a consequence reduces the weight of water in the return pipe and thus diminishes the motive power of the entire apparatus. In the system installed in Mr. Requa's residence the first radiator on the flow pipe, the designer claims, is the last to have its return water enter the boiler, thus keeping the vertical return pipe cool until the last radiator on the line is full.

In the plant under description the designer has sought to overcome any chance of failure in the more distant radiators by making the sum of the lengths of the flow and return pipes to and from a radiator a constant quantity and proportioning the diameters of the pipe to the quantity of water that is to pass through them. This is done in the following manner: A No. 1 Humbert heater is used. It will be noticed that the flow pipe from this, marked A, is carried entirely around one end of the basement, dropping as it leaves the boilers until it enters the boiler at the rear and the bottom. From the first radiator B on this main a return pipe runs into a return main C, beginning at that point. This runs parallel to the supply main, and is also carried around the basement, receiving the returns from various radiators and returning the water to the boiler.

INDIRECT STEAM OR HOT-WATER HEATING IN A MASSACHUSETTS RESIDENCE.

FIG. 1

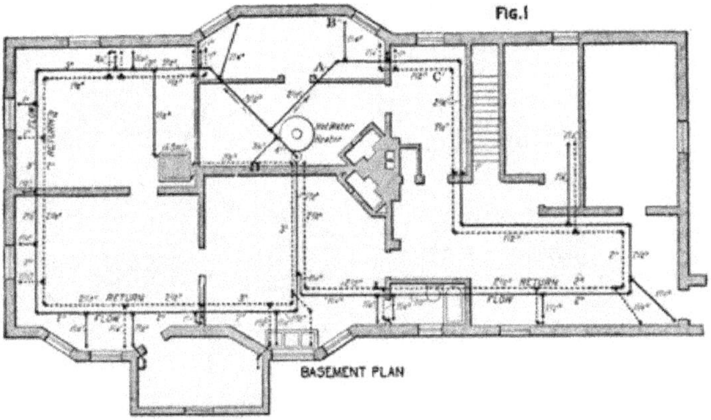

BASEMENT PLAN

FIG. 2

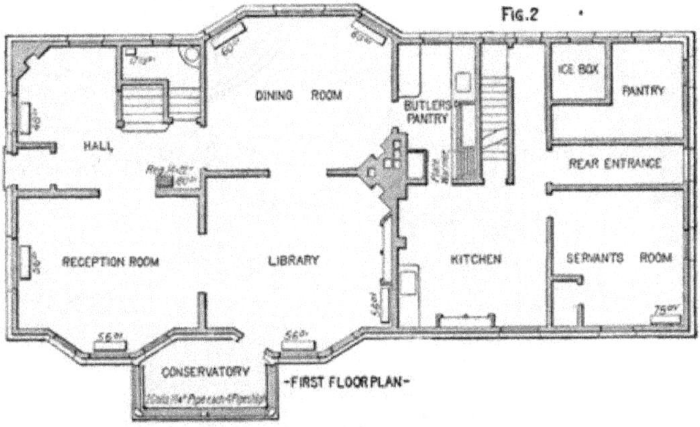

-FIRST FLOOR PLAN-

FIG. 3

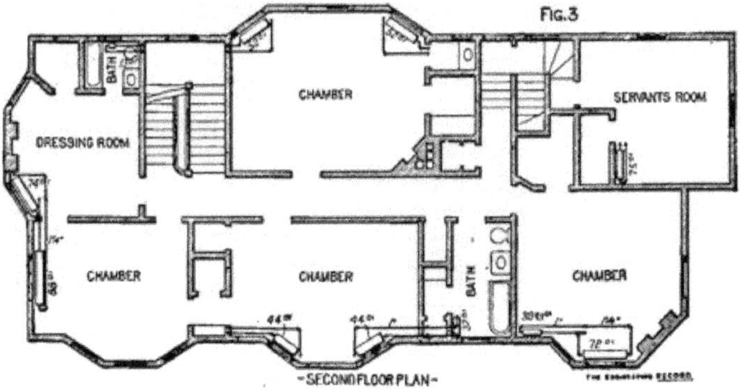

-SECOND FLOOR PLAN-

UNUSUAL PIPING IN A HOT-WATER HEATING APPARATUS.

In the case of the first radiator the water flows to the radiator and then passes entirely around the basement to the boiler. The same course is taken with the second radiator, etc., so that the water for each has to travel the same distance in making the circuit. The return main increases in size as the flow decreases. The other end of the building is supplied by a similar system.

VENTILATION AND HEATING OF THE RESIDENCE OF MR. CORNELIUS VANDERBILT.

PROMINENT even among the palatial private residences in New York City, the recently enlarged and remodeled residence of Mr. Cornelius Vanderbilt attracts much attention by its beautiful and imposing exterior. It is situated in Fifth Avenue occupying the block between Fifty-seventh and Fifty-eighth Streets. In its engineering service, providing for water supply, drainage, ventilation, heating, and lighting, the residence is notably complete. The water supply has been described and illustrated at length in THE ENGINEERING RECORD of January 12, 19, and 26, 1895, and the extensive and elaborate system of ventilation of the large ball-room, salon, banquet hall and living rooms is worthy the attention of all who have made a study of the ventilation and heating of large buildings. The architect of the

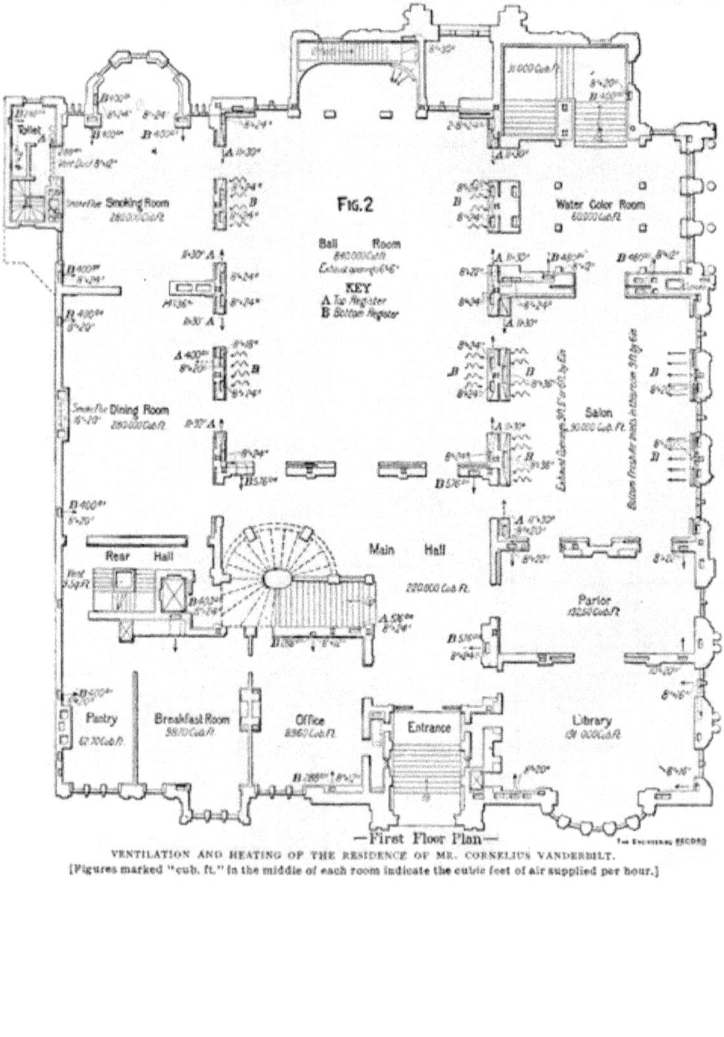

—First Floor Plan—

VENTILATION AND HEATING OF THE RESIDENCE OF MR. CORNELIUS VANDERBILT.

[Figures marked "cub. ft." in the middle of each room indicate the cubic feet of air supplied per hour.]

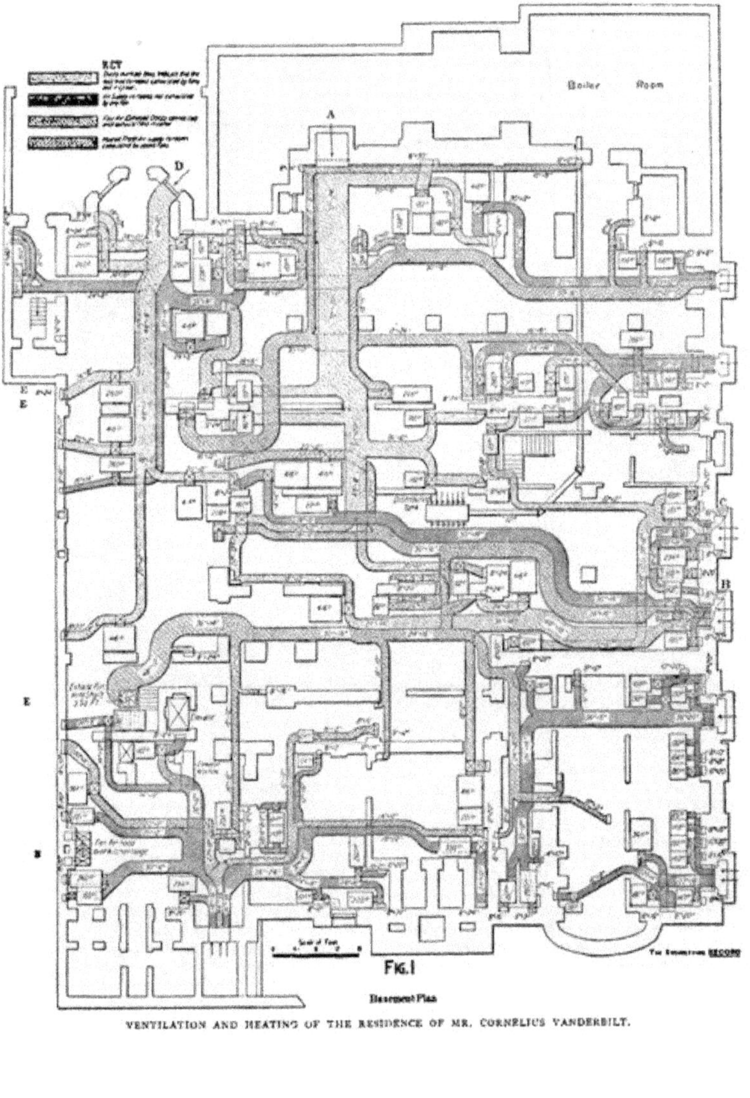

FIG. I

Basement Plan

VENTILATION AND HEATING OF THE RESIDENCE OF MR. CORNELIUS VANDERBILT.

building was Mr. George B. Post, of New York City, while Mr. Alfred R. Wolff, consulting engineer, was the designer of the heating and ventilating system. To these gentlemen we are indebted for facilities given us in the preparation of this article. In this part it will be endeavored to give as far as possible the details of the heating system, and in a subsequent issue the method of calculating the amount of heating surface, size of ducts, etc., to give the required amount of air will be shown, this data having been placed at our disposal by Mr. Wolff.

The building occupies the whole Fifth Avenue front of the block, and extends down the side streets for a distance of 150 feet, thus making the lot upon which it stands 125x150 feet in size. The residence consists of a cellar, basement, ground, first, second, third, and fourth floors. The cellar contains the boilers, heating coils, ducts, etc., of the heating system, the electric switchboard, and the filters, tanks, etc., for the water supply of the buildings. On the basement floor will be found the kitchen, laundry, storerooms, and servants' quarters. The residence has two entrances to the ground floor, the one on Fifty-seventh Street that is more usually used, and one on the Fifty-eighth Street side facing the entrance to the Central Park. The latter is used only upon state occasions, and leads from a carriage porch into a hall situated on the basement floor. On either side of this hall are located the dressing-rooms for the ladies and gentlemen, and from it a broad stairway leads to the reception hall on the first floor, large sliding doors lead from this to the salon and to the ball-room. The ball-room is 65 feet long and 50 feet wide, and connected with it are the dining and smoking room, as will be seen from Fig. 2, the first-floor plan. The dining-room contains the valuable collection of paintings owned by Mr. Vanderbilt.

On entering the Fifty-seventh Street entrance one finds the hall, enriched with elaborate carvings in Caen stone, from which doorways lead to the library, breakfast-room, parlor, etc., which are more generally used. The plan of the second floor shows, beside the arrangement of the bed chambers, boudoirs, etc., the location of the ventilating ducts that extend around the space above the hanging ceilings of the salon, ball-room, and smoking-room.

The building is warmed entirely by hot water and is ventilated partly by what is known as the indirect system, and in some cases the currents of air are stimulated by exhaust fans, but in no case is the air forced into any part of the building under pressure. Ninety-seven indirect stacks, of a total heating surface of 19,565 square feet, serve to warm the building. To supply the hot water three horizontal return-tubular boilers are provided, each containing 1,244 square feet of heating surface. Each boiler is 54 inches in diameter by 16 feet in length and contains 90 3-inch tubes 16 feet long. The boilers are located under the sidewalk at the northeast corner of the building, as it was inconvenient to put them in a central position in the cellar. As it was thought that there might be trouble in obtaining a proper circulation in the more distant stacks, the piping is run in a different manner from the ordinary practice. A large 12-inch main is carried overhead to a large distributing tank 3 feet in diameter and 8 feet long. This is suspended from the ceiling at a point near the center of the cellar and from it 14 heating mains radiate to the different parts of the basement to supply the indirect stacks.

The pipes are so pitched as they leave this tank as to cause any air that might find its way into the system to flow towards the stacks, and then leave the system through an air valve. The return pipes from each stack are carried as far as possible in conduits under the cellar floor, these being covered with iron plates. The return lines lead back to what might be called a collecting or receiving tank similar in size and placed under the distributing tank. The receiving tank is under the cellar floor, and a 12-inch return main leads from this to the boilers. Another advantage of this system is that each section of the heating system, as for instance the coils for the ball-room, those for the dining-room, smoking-room, etc., is supplied by one supply and return main independent of the others, and controlled by valves close to the distributing and collecting tank, so that each system can be controlled from a central point.

Turning now to the ventilation of the building it will be noticed from Fig. 1 that the ducts for supplying fresh air to the indirect stacks are divided into four general classes by a difference in the shading. The four classes are as follows: First, ducts supplying air to rooms which are exhausted by fans not in the cellar; second, ducts leading to rooms not exhausted by any fans; third, ducts exhausting air from rooms by means of fans located in the cellar; fourth, ducts supplying air to rooms exhausted by the last-named fans. Probably the most important, and certainly the most interesting, is the part of the plant that ventilates the ball-room, salon, and those parts of the building that are used on state occasions. The ducts for this system come under the first head, and they can readily be found on Fig. 1 by means of the key in the corner. The cubic contents of the ball-room is about 103,700 cubic feet, and it was supposed that it would at times contain 400 persons, and hence provision was made for supplying 14,000 cubic feet of air per minute, or 35 cubic feet per capita. The cold air for the ball-room enters the building on the north side through a duct, which for convenience of reference we have marked A. Branches from A convey the air to the nine indirect stacks, each of which contains 182 square feet of heating surface. An 8x24-inch duct leads from each to a 11x30-inch register discharging air into the ball-room. The registers are located at a point about 11 feet above the floor.

On the Fifth Avenue side of the building there will be found a duct marked C, shaded in a similar manner to A. This supplies four indirect stacks of 169 square feet of surface each, that warm the air for the salon. Two of the registers for the salon are slightly different from those in the ball-room, as they are placed behind divans within 6 inches of the floor. Each of the registers so placed is 9 feet wide and 6 inches in height, distributing the air at a low velocity and over a large area. An 18x48-inch duct B, starting close to C, supplies three stacks of 416 square feet of surface each for the main hall. Again, on the north side of the house will be noticed a duct D, supplying three stacks, each also of 416 square feet of surface, and which furnishes air for three registers in the dining-room. The registers contain 400 square inches each. Four stacks of 260 square feet warm the air for the smoking-room, this being also supplied with air from the duct D.

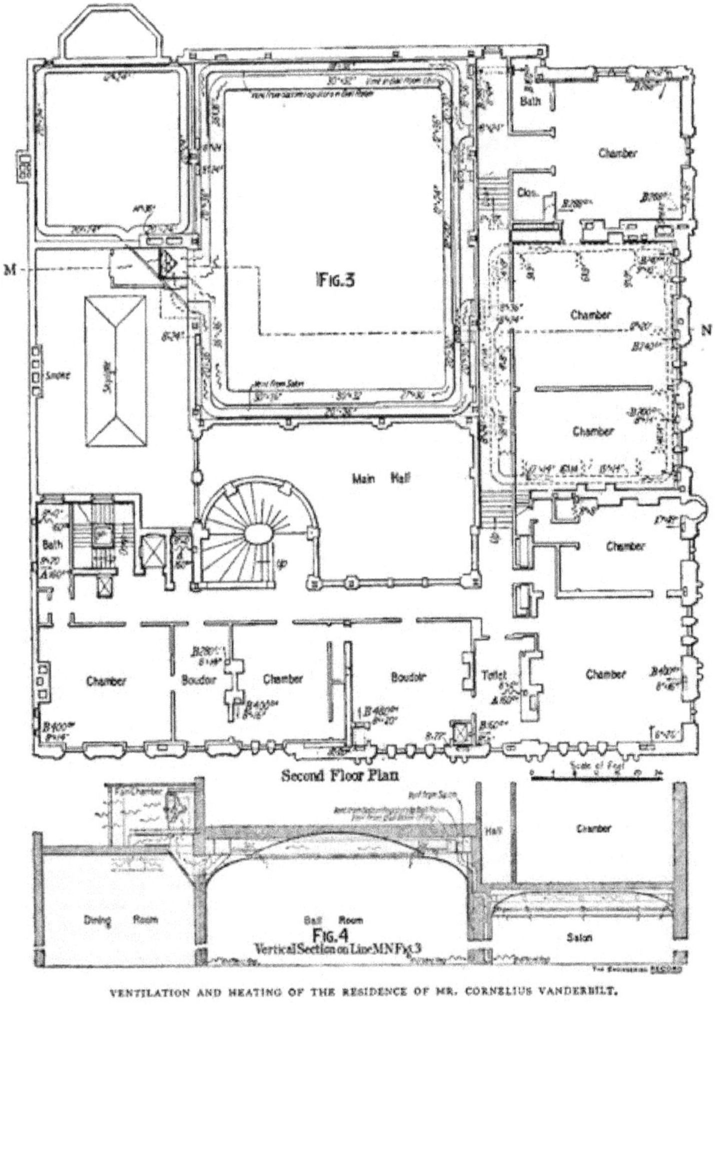

VENTILATION AND HEATING OF THE RESIDENCE OF MR. CORNELIUS VANDERBILT.

From what has been stated it is evident that each of the rooms mentioned is supplied by an entirely separate system. One 5-foot fan, however, located in a specially constructed fan chamber on the roof over the dining-room, serves to ventilate all of these rooms. If any single one of these larger rooms be not in use, the main vent duct leading from it to the the fan may be closed by a damper controlled from the basement and the speed of the fan reduced, as it is driven by an electric motor.

Beginning now with the salon on Fig. 2, the plan, it will be noticed that the room in question contains two vent registers, each 9 feet long by 6 inches in height. On Fig. 3 will be noticed a dotted duct which is carried around over the salon between the arched cornice and the floor above, as will perhaps be more clearly seen by Fig. 4, which is a vertical section through the building. The latter drawing shows the manner in which the two bottom vent registers are connected to the duct, as well as the manner in which the vitiated air leaves the salon through the opening between the cornice and the hanging ceiling and into the duct through openings provided for the purpose. The ball-room is ventilated by a similar duct passing around over the ceiling of the ball-room, but in this instance the air enters the duct through a slot in the bottom of the duct. At a point farthest from the fan this slot is 3½ inches broad, and it decreases to 1 inch in size at a point where the duct leads to the fan. Still another duct, which is carried around in the space over the hanging ceiling, is provided to draw the air from the four bottom vent registers, each of these being 6'x0" and located close to the floor. The smoking-room is vented by a similar system. The dining-room is vented by means of the glass diffuser under the skylight, which is raised several inches. The space between the diffuser and the skylight is connected to and exhausted by the exhaust fan.

Turning again to the basement, the ducts which supply air to the flues leading to the bedrooms, boudoirs, etc., on the upper floors of the building will be recognized by their being sectioned as shown in the second convention of the key in the corner of the drawing. The rooms supplied by these ducts generally contain fireplaces and are vented by them.

There remain but two other systems of ducts in the basement—those exhausting air from rooms by a fan and the ducts which supply air to these rooms. There are four vent fans in the basement. E E E E, and to aid in finding them the letter used to designate each fan is marked opposite the fan along the left-hand border of the drawing. These fans are used solely to ventilate the kitchen, laundry, and servants' quarters in the basement, and the ducts which supply air to these rooms can be easily determined by means of the key.

There will be noticed on Fig. 1 on each fresh-air duct, at a point near the indirect stack it supplies, a rectangle with two diagonal lines. These indicate the location of the switch dampers which regulate the temperature of the air. Figure 5 shows a sketch of a typical indirect stack with the sheet-iron connections, switch, etc. The dampers are shown as being down in the sketch so that the air flows under the coils, and turning, comes up through them to the short duct leading to the base of the flue. The switch or by-pass allows a constant flow of pure air. The dampers are operated by a chain connected to levers, the other end of the chain being connected to a specially designed nickel-plated regulator (Fig 6),

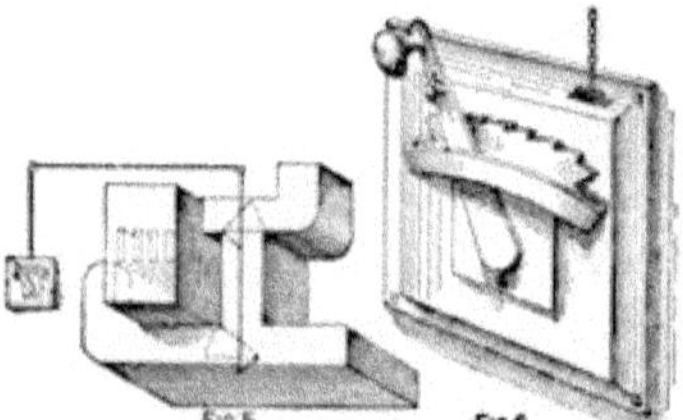

Fig. 5 Fig. 6
The Engineering Record.

either placed at some central point where it can be operated by the engineer or else it is located in the room to which the air supply it controls leads.

Baker, Smith & Co., of New York City, were the contractors for the plant.

HOT-WATER HEATING IN A MELROSE, MASS., RESIDENCE.

The method of installing the hot-water heating apparatus in the residence of Mr. John A. Fish, at Melrose, Mass., is shown by the accompanying drawings. The system of piping is simple. But one main circuit pipe is used in the basement, which not only supplies the radiators, but receives the return pipes also. The main circulating pipe in the basement is carried as closely as possible to the radiators in order to shorten the horizontal branches to the several radiators and risers.

It will be observed that the flow pipe connected to the radiator is the first pipe taken off the circuit, the tee being turned upwards, while the return pipe is taken into a tee on the circuit main, entering on its side. These connections are shown on the sketch. It is desirable in erecting piping on this principle that the flow and return pipe enter the circuit as shown—that is, the flow out of the top and the return pipe entering the side. The main circuit pipe as it leaves the heater is supposed to be perfectly level, with the exception of that part of it running back and dropping down into the return header at the bottom. This pipe has a slight incline towards the heater for the purpose of draining. All the piping in the cellar is covered with asbestos material.

A No. 303 Gurney hot-water heater and 330 square feet of radiation form the apparatus for the house, the conservatory being heated by a coil containing 100 feet of 1¼-inch pipe. The heating plant was laid out by Mr. John A. Fish, of the Gurney Hot-Water Heating Company, of Boston, Mass., and has worked satisfactorily and warmed the house perfectly in very cold weather.

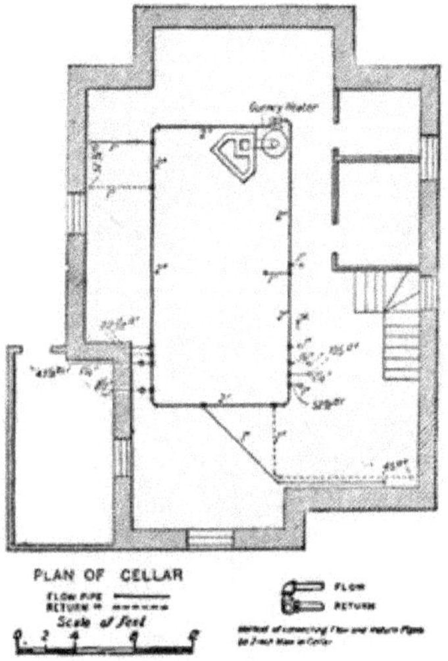

HOT-WATER HEATING IN A CITY RESIDENCE.*

THE accompanying drawings show the hot-water heating apparatus in the residence of Mr. M. Thalkeimer, of Richmond, Va. A No. 60 "Spence" heater supplies hot water for about 1,020 square feet of surface in radiators. Three of these are of the indirect system, while all the others heat by direct radiation. The air for the indirect radiators enters the building through a 36x12-inch duct in the basement. This divides into three branches, each of 150 square inches in section, which supply the registers on the first floor with air. Each indirect coil contains 150 square feet. As the indirect coils are on the same floor as the heater a syphon is introduced to maintain a circulation of water. In the case of the coils A and B the supply is run in a vertical direction and then dropped to the radiating coils, an air valve being placed at the highest point. In the case of the coil C the method is somewhat different. The pipe supplying that coil is carried to the third floor and supplies radiators on the second and third floors. A detail of their connections is shown by Fig. 5. It will be seen that in order to have a constant circulation through the indirect coil in the basement it would be necessary to have either one of the radiators on the riser open, and to permit both of them being closed at the same time a by-pass was introduced as shown in the figure. The piping is 1½ inches up to the by-pass and 1¼ inches beyond, so that at all times a part of the hot water rising in the supply will go through the by-pass and meet the other part which has been cooled by passing through the radiator, and thus give a mixture that will con-

* See also page 16 for Figs. 2, 3, and 4.

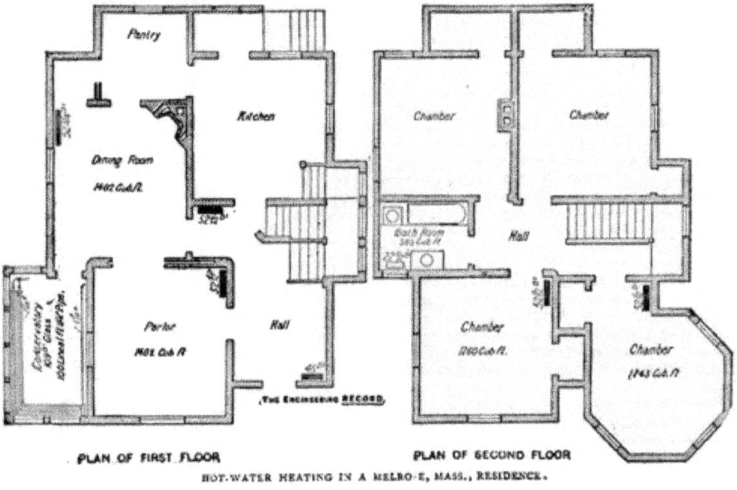

PLAN OF FIRST FLOOR

PLAN OF SECOND FLOOR

HOT-WATER HEATING IN A MELROSE, MASS., RESIDENCE.

tain a sufficient amount of heat for the needs of the indirect coil C in the basement.

At D, third-floor plan, is shown the method of connecting the expansion tank to the riser and return marked E in Fig. 1, the foundation plan.

The plant was laid out by Mr. Percival H. Seward, of the American Boiler Company, of New York, while Messrs. West & Branch, of Richmond, Va., were the contractors for the work.

REMODELED HEATING PLANT IN A CITY RESIDENCE.

The accompanying plans show the heating apparatus as it now stands in the residence of Mr. T. J. Hayward, of Baltimore, Md. The building is located upon a double lot facing west and is exposed on the south and east. The lot upon which the building stands being about 60 feet wide, an open space is left between it and the next house south; this forms an open passage for the circulation of air coming from any quarter. A strong wind from either the east or north beats against the walls of the adjoining house at the south and rebounds against the south wall of Mr. Hayward's residence. On account of these conditions the supply of air for the indirect radiation is taken from the south with nearly the same results as though taken from the north and western directions, as would be the proper practice if the house stood exposed on all sides.

At the time of Mr. Hayward's purchase the house was fitted with two sets of indirect steam-heating apparatus, the front one located at B with an auxiliary stack of radiators at D intended to heat the front hall, the parlor and back parlor, and the rooms over the same. The rear apparatus was located at A, with an auxiliary stack at E, the stack at A being intended to heat the dining and smoking rooms and the rooms over them, while the auxiliary stack at E was to heat through one flue the rear chamber, conservatory, and bath on the second floor and two rooms on the third floor. As the amount of radiation was insufficient to properly heat the house, the arrangement of the rear part being especially unsatisfactory, and heat was required in the two small rooms over the front hall, the rear heater was removed and a hot-water apparatus installed in its place, and at the same time adding to it the two unheated rooms and also the indirect radiator supplying heat to the front hall, this being done to relieve somewhat the front apparatus. The following season the steam apparatus at the front was removed, and the entire apparatus made over to warm the building by hot water. A Chesapeake boiler was installed at B, and the hot-water boiler in the rear removed, its supply and return pipes, however, connected with the new mains in the front part of the building.

The building contains three stories besides the basement, but only the basement, first, and second floors are shown, these answering the purpose of this description. The building contains approximately 77,000 cubic feet of space, and this is warmed

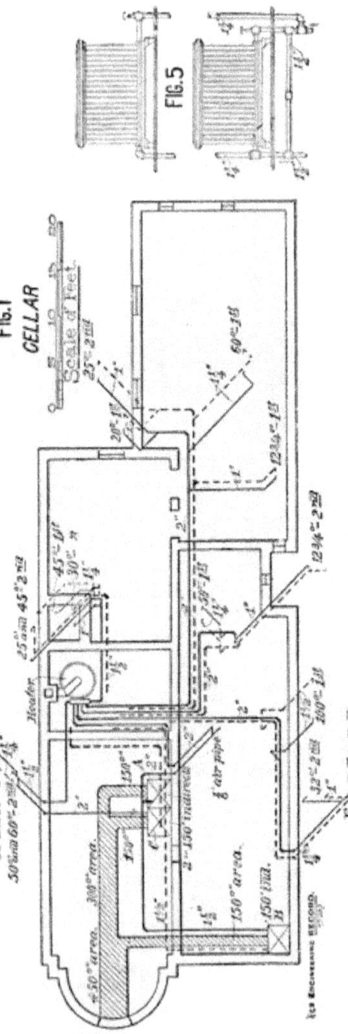

HOT-WATER HEATING IN A RICHMOND, VA., RESIDENCE.

by 1,450 square feet of indirect and 928 square feet of direct radiation. The distribution is shown on the plans. The indirect stacks at B are located in the space above the boiler, they being inclosed in brickwork and connected to the outer air by the duct C, which also supplies air to another cluster of stacks as shown.

Sectional views are given (page 28) of the Chesapeake boiler, which was designed by Mr. Charles W. Newton, of Bartlett, Hayward & Co., that firm installing the heating apparatus. The boiler consists, as will be seen, of two cast-iron manifolds connected by 1½-inch pipe as shown. The lower manifold, which is cast with a right or left return pipe connection, is rectangular in shape, the larger faces being bored to receive the 1½-inch pipes which form the greater part of the heating surface of the boiler. The sides of the lower manifold contain a web or fins that extend out as far as the brick setting. The manifold also serves as a baffle-plate, compelling the products of combustion to pass up between the tubes and then drop in the rear of the smoke connection.

The boiler is fitted with a rocking grate, a rib on each grate bar being connected to the rocking bar by pins as shown. The general method of construction of the boiler gives, the makers claim, a boiler that is easily cleaned, economical as to cost of constructing and operation, and one that will not be liable to injury by expansion. The boiler can be made of less or greater capacity by increasing or decreasing the length of pipes and grate surface. Its small height makes it desirable for use in cellars with low ceilings.

HEATING AND VENTILATION OF A PHILADELPHIA SUBURBAN RESIDENCE.

The accompanying illustrations show details of the heating and ventilating system designed by the Onderdonk Heating and Ventilating Company for Charles S. Onderdonk, and installed in his house at Wyncote, Pa., of which Wilson Brothers & Co., of Philadelphia, are the architects. The system is one of indirect hot water, with ventilation of every room into a central stack 25 inches in diameter in the clear, the draft in which is induced by a 10-inch smoke pipe from the boiler. All rooms in the front part of the house are connected to this central stack either directly where it passes through such rooms or adjacent to them, or by means of flues, which are located in the partitions, proceed to the cellar and are then led by means of horizontal ducts into the base of the stack.

The kitchen or frame part of the building receives its ventilation by means of a brick stack J, Fig. 2, 18 inches in the clear, in which an 8-inch cast-iron pipe is placed, which induces the ventilation in that stack. An opening is made immediately over the kitchen

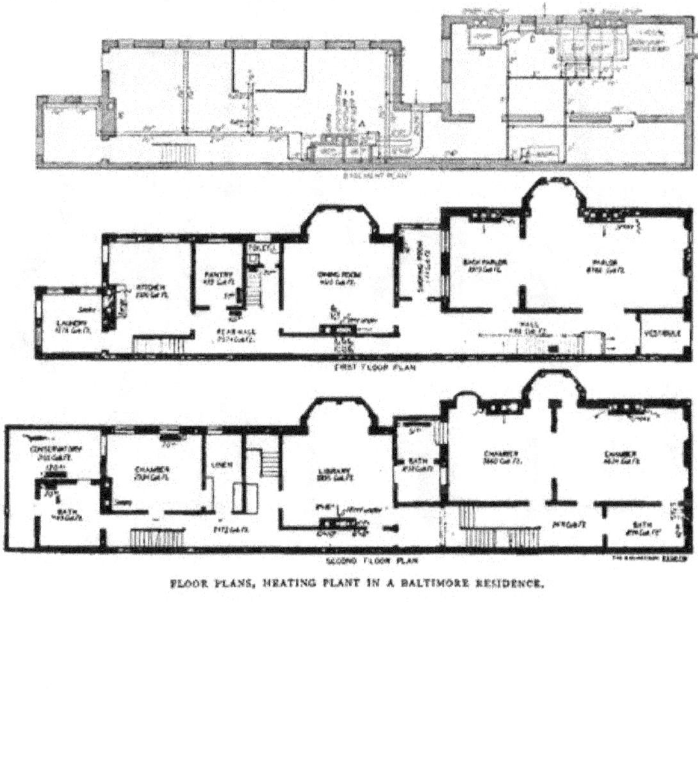

FLOOR PLANS, HEATING PLANT IN A BALTIMORE RESIDENCE.

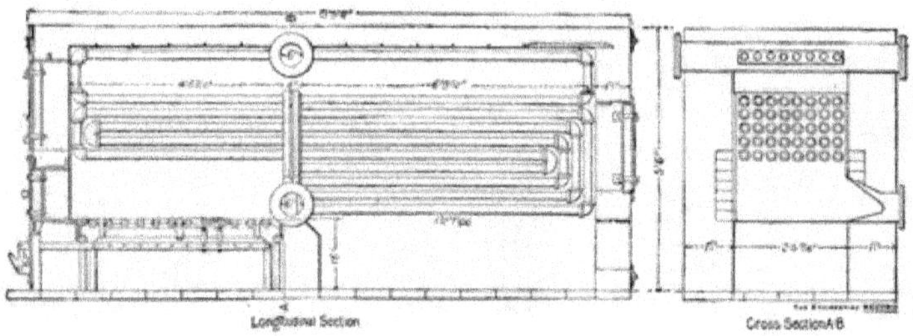

THE CHESAPEAKE BOILER. (See page 26.)

range and one at the ceiling of the kitchen, and the rooms above the kitchen are ventilated into this stack.

A, B, C, D, E, F, G, and I are downtake flues exhausting the foul air from the owner's bedroom, parlor, sitting-room, den, etc. The water-closet on the second floor is ventilated by means of a duct containing 12 square inches of area, which rises from the adjacent partition and is connected to the central ventilating stack, a connection also being made to the kitchen stack for use in the summer months when the main ventilating stack is not heated. The pipe in the main flue is 10 inches in diameter.

The valves on the flow pipes to each radiator stack are automatically controlled by a Johnson electric thermostat in the rooms respectively served by them. The radiators have no air valves, but are fitted with an air pipe connected to an open riser extending in the main ventilating stack to above the level of the expansion tank. All registers for the admission of warm air to the rooms are located 6 inches below the ceiling, and those for ventilation are set on the

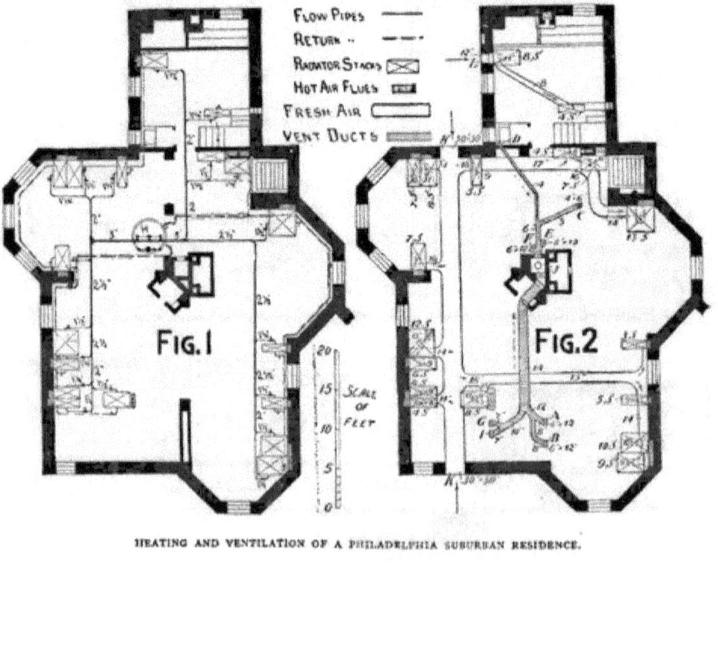

HEATING AND VENTILATION OF A PHILADELPHIA SUBURBAN RESIDENCE.

opposite side of the room just above the wash board. Each stack of radiators is controlled in addition on both flow and return by Pratt & Cady brass gate valves.

The whole system of piping is covered with magnesia sectional covering. In addition thereto the water in the boiler is prevented from reaching the boiling point by means of a Powers limiting device, which closes off the draft just before the water reaches the boiling point. The water-closet in the laundry in the basement is ventilated into the kitchen stack. The galvanized-iron flues for heating when erected were thoroughly wrapped with asbestos paper, the openings in the wall being thoroughly parged for the reception of the flue.

Fresh cold air is admitted from outdoors through windows K K and L, the two former of which are opposite each other, and are so arranged on opposite sides of the house that either one may be closed and the supply be drawn through the other one according to conditions of sunshine, shadow, and prevailing winds. The branches from the supply ducts to the radiator stacks underneath the latter and so hidden by them, are here shown dotted.

In Figs. 3, 4, and 5, the plans of the first, second, and third floors respectively, the fresh and foul-air flues are marked M, N, and H respectively, to indicate which floors they serve. All registers are marked R, with the size.

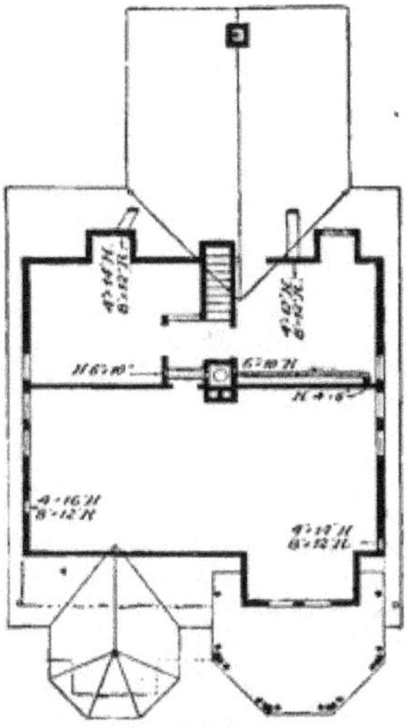

FIG. 5.

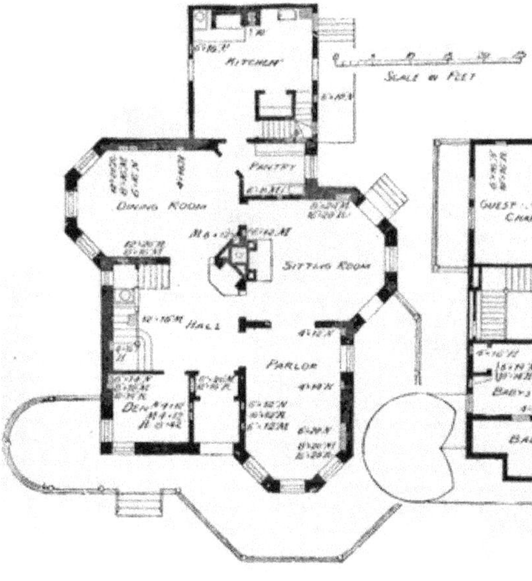

FIG. 3.

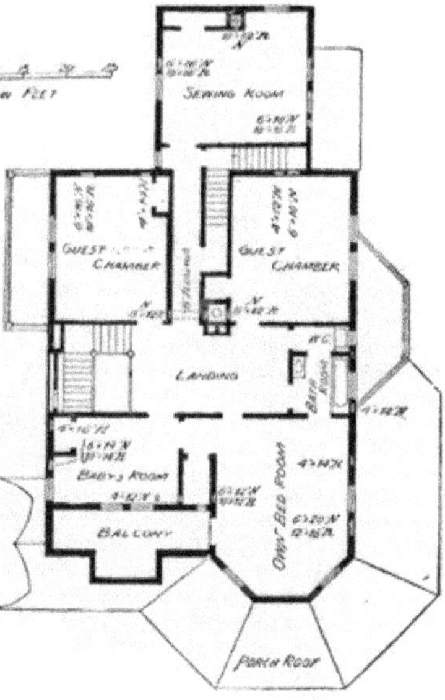

FIG. 4.

HEATING AND VENTILATION OF A PHILADELPHIA SUBURBAN RESIDENCE.

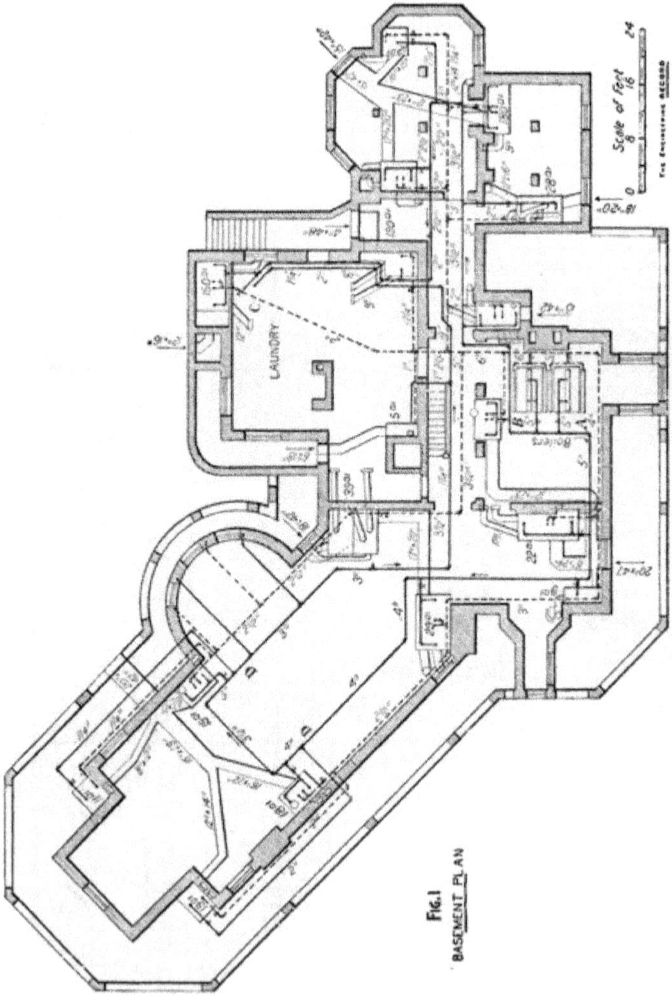

FIG. 1 BASEMENT PLAN

HOT-WATER HEATING IN AN ORANGE, N. J., RESIDENCE.

HOT-WATER HEATING IN A COUNTRY RESIDENCE.

The residence of Mr. John Pettit, of Orange, N. J., was designed by Mr. Alfred H. Thorp, of New York City, while the heating plant was installed by the H. B. Smith Company, New York City. The building, which contains about 113,500 cubic feet, is warmed and ventilated by the indirect hot-water system, 17 indirect stacks containing in all 3,070 square feet of radiating surface in Gold pin radiators. Two coils, containing in all 105 square feet of surface, are placed in the conservatory, warming that by direct radiation.

Figures 1, 2, and 3 show the basement, first and second floors of the residence. There is beside these a third story, the plan of which is not shown in this description. One of the principal features of this plant is the manner of running the pipe mains in the basement, the building being of a somewhat peculiar shape. The boiler plant consists of two Mercer boilers, and from these a 5-inch main A, Fig. 1, is carried around through the greater part of the cellar to the point C. Another 5-inch branch B leaves the boilers to supply indirect stacks in the remainder of the basement, this branch likewise terminating at the point C. At C both of the mains are connected to a 16x60-inch expansion tank located on the floor above. As each pipe is pitched so as to rise slightly as it departs from the boiler and each connection to the indirect stacks drops slightly as it leaves the

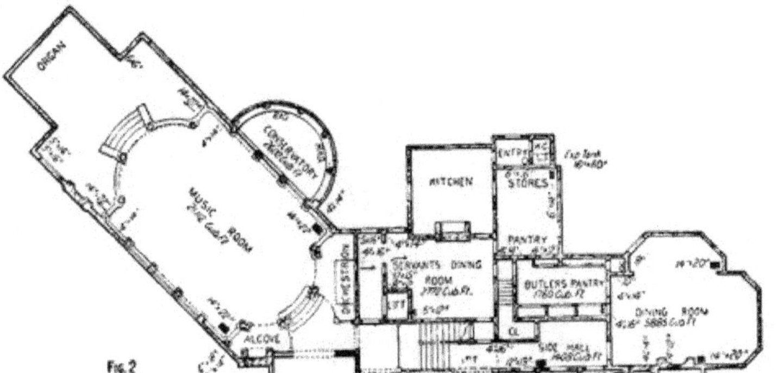

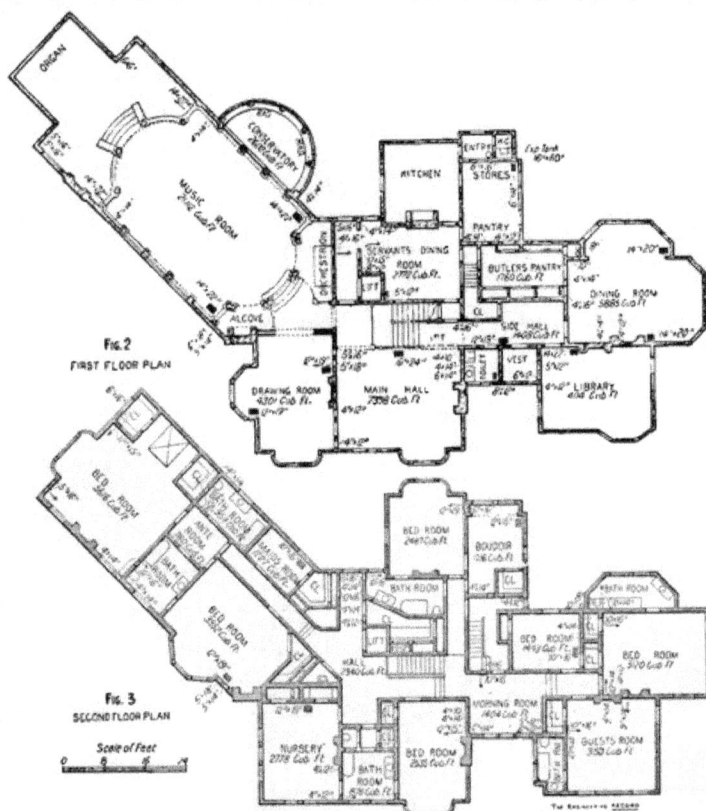

FIG. 2 FIRST FLOOR PLAN

FIG. 3 SECOND FLOOR PLAN

HOT-WATER HEATING OF AN ORANGE, N. J., RESIDENCE.

main, any air in the system will flow toward C, the expansion tank, and in this manner it is freed from air without the use of a single air valve.

Still another interesting point may be noted by following the main B around until the points D D are reached, when a 2-inch pipe leaves the main for supplying an indirect stack for heating the rooms on the second and third floors. The stack is placed underneath the piazza of the house, this location being necessitated by the low level of the floor of the music-room. An examination of the end of the building, Fig. 2, will show that the music-room is several feet lower than the remainder of the first floor, the organ being placed in the extreme end of the music-room at the lowest level. As it would be inconvenient to place the stack under the floor and be sure of a good circulation it was placed under the piazza as shown. The expansion tank is closed. A vent pipe leads to the roof, thus avoiding a waste of hot water over the ordinary which must occur in the common open-tank systems with overflow.

INDIRECT HEATING IN A RESIDENCE.

In a residence at Chestnut Hill, Philadelphia, for Mr. J. Levering Jones, there are one or two details in the heating system which will probably be of interest. They are shown in the accompanying plans of the building, which were made by Mr. George F. Pearson, architect, of Philadelphia, Pa. The heating plans were laid out by the Onderdonk Heating and Ventilating Company, of the same city.

All of the floor plans of the building are shown, and from these it will be noticed that the building is heated entirely by the indirect system. Forty-two indirect stacks situated in the basement and containing in all 2,232 square feet of Gold pin radiators warm the air for the building. No system of ventilation is provided other than that obtained through numerous fireplaces, windows, etc. The heating apparatus is an ordinary low-pressure, gravity return steam-heating system. The radiators are inclosed in galvanized-iron boxes as usual, and the heat is conducted from these through metal flues placed in the walls and delivered through registers to the

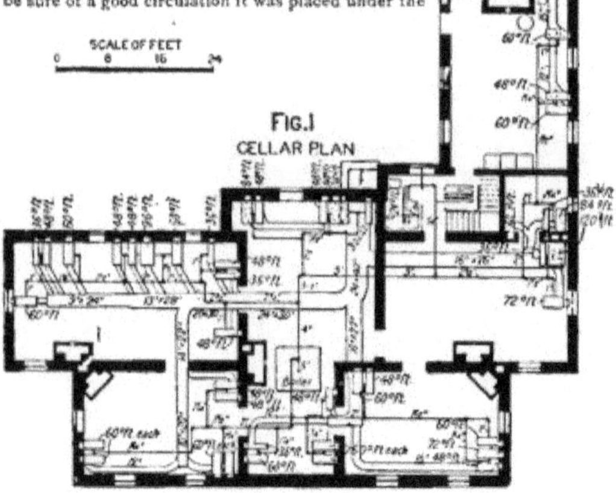

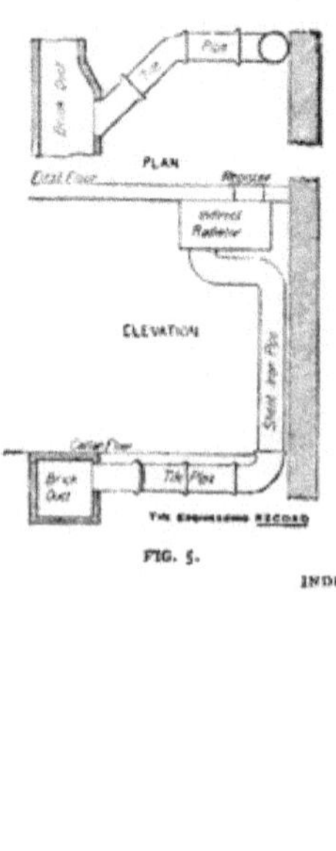

FIG. 5.

INDIRECT HEATING IN A PHILADELPHIA RESIDENCE.

rooms to which they lead. The sizes of the flues are shown on Fig. 2, the first-story plan and the register sizes in the rooms in which they are located.

An interesting part of the plant lies in the manner of supplying air to the radiators. Because of a desire to save head room in the basement, the cold air is supplied to the different stacks by means of a brick duct with terra-cotta branches, both run under the floor of the basement. These terra-cotta branches extend out from the brick duct to a point directly beneath the indirect stack and connections are made from them to the stack casing by means of galvanized iron pipes. Figure 5 is a rough sketch showing the detail of a section of brick cold-air ducts and connections to the indirect stacks. The joints of the terra-cotta pipe are made with cement in the ordinary way, and the terra-cotta elbow which occurs directly beneath the indirect stack is so placed that the bell comes over the cement floor of the cellar. From the elbow the galvanized-iron pipe rises vertically to the radiators.

FIG. 3

SECOND STORY

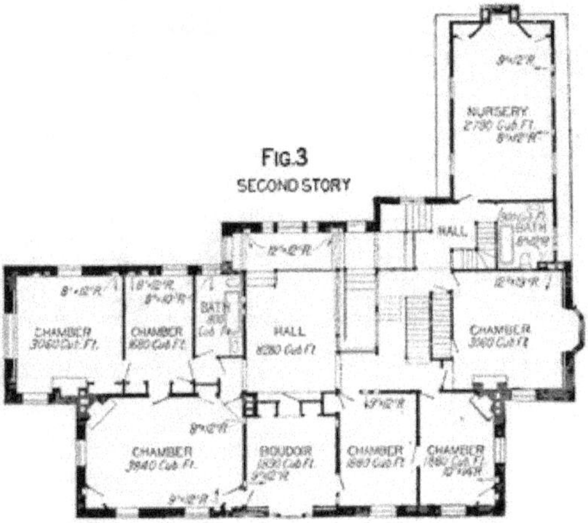

FIG. 4

THIRD STORY

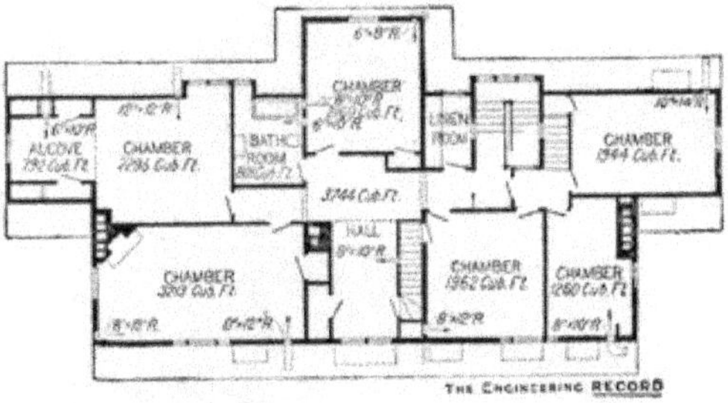

THE ENGINEERING RECORD

INDIRECT HEATING IN A PHILADELPHIA RESIDENCE.

HEATING OF CHURCHES.

ONE-PIPE HOT-WATER HEATING OF A CHURCH.

The operation of a hot-water job which was executed recently led to the unrestricted employment of the contractors to install in a new church edifice a hot-water system which should heat it throughout to a temperature of 70° Fahr. during any weather. The conditions appearing favorable, it was decided to install there a one-pipe system consisting of a single horizontal main leaving the top of the heater, encircling the room, and returning to the bottom of the heater, so that the water should circulate continuously through it, from top to bottom, and be diverted to the various radiators through short vertical branches, and after circulating through them return by parallel vertical branches to the main pipe a few feet beyond where it was withdrawn. The system embraced two separate mains or circuits independently connected to the heater, one to warm the auditorium and one to warm the adjacent Sunday-school room, parlors, and small rooms.

Figure 1 shows the arrangement of the rooms to be warmed and the location of the radiators, all of which were direct except one. The total volume of air to be heated was about 120,500 cubic feet, of which 32,856 cubic feet, including the Sunday-school hall and the adjacent rooms, was designed to be served by one 3-inch main, Fig. 2. The 4-inch main served only the auditorium radiators, and each main was separately controlled by a valve at the heater, so that in case it is desired to heat either the Sunday-school or auditorium alone, the fire may be proportioned for the work and the heat may be concentrated entirely on the single branch.

Figure 2 is a plan of the basement showing the location of heater and radiator branches and arrangement and size of mains. The original Sunday-school main S, which was afterwards removed, is shown by a broken double line, while the auditorium main M, which is still in service, is shown by a full double line. The branches to the radiators are shown by full single lines and the direction of flow of the water is indicated by the arrows.

Figure 3 shows the manner in which the radiator connections were made, the flow being taken from the top of the main and the return brought to its side. The branches were $1\frac{1}{4}$ inches, but the single-column radiators were tapped for 1-inch connections,

ONE-PIPE HOT-WATER HEATING OF A CHURCH.

and they were made as shown by reducing elbows. The expansion tank was set very high so as to cause a considerable pressure at the heater. When the system was operated the radiators at the beginning of the circuits, as at A and B, were much hotter than the most remote ones, as C and D respectively, the difference being as much as 20 degrees—i. e., the

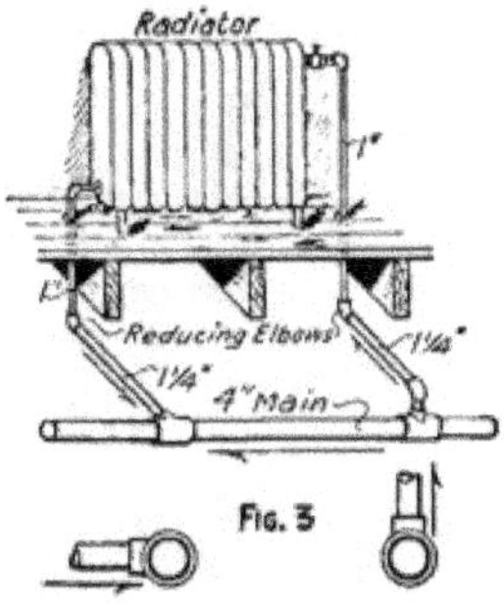

Fig. 3

water at A was 200° Fahr. when it was 180° Fahr. at C. The results secured in the auditorium were, however, satisfactory, and the general temperature of the atmosphere there was high enough. This was not the case in the Sunday-school room, however, and especially in one of the adjacent parlors which it was desired to keep warmed most of the time, and in which it was found impossible to secure a sufficiently high temperature even when the water at the heater was 225° or 230° Fahr., and the valve two-thirds closed on the main M so as to divert most of the circulation to the system S. Repeated trials were made and every promising device suggested was unsuccessfully tried to make it work satisfactorily. The main was given an increased pitch downwards in the direction of the arrows, some of the hottest radiators were choked so as to diminish their circulation, but nothing sufficiently raised the temperature of the remote radiators. Still the auditorium would be the hottest when its valve was two-thirds closed, and finally the pipe S was taken out, and the radiator branches connected onto the two separate flow and return mains F and R, shown by the single broken black lines. This arrangement secures a uniform operation of the radiators, but has not yet been tested by extremely cold weather The heater is rated for 2,200 square feet of radiating surface, and the mains are hung from the floor joists and are all jacketed.

The duty required of the heater is not greater than is generally allowed by these contractors in numerous satisfactory jobs, and they are somewhat perplexed by the difference of the results secured in the auditorium and in the Sunday-school part. In the latter room it has been suggested that there might be some specially unfavorable conditions, but no unusual exposure is mentioned, although it is suggested that the recesses for the large vertical sliding partitions may furnish a conduit for the withdrawal of large quantities of hot air from the top of the room.

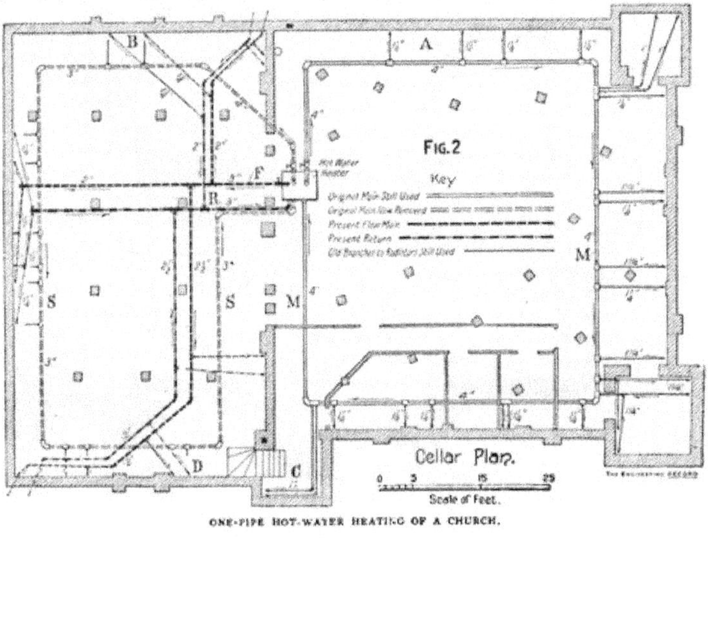

ONE-PIPE HOT-WATER HEATING OF A CHURCH.

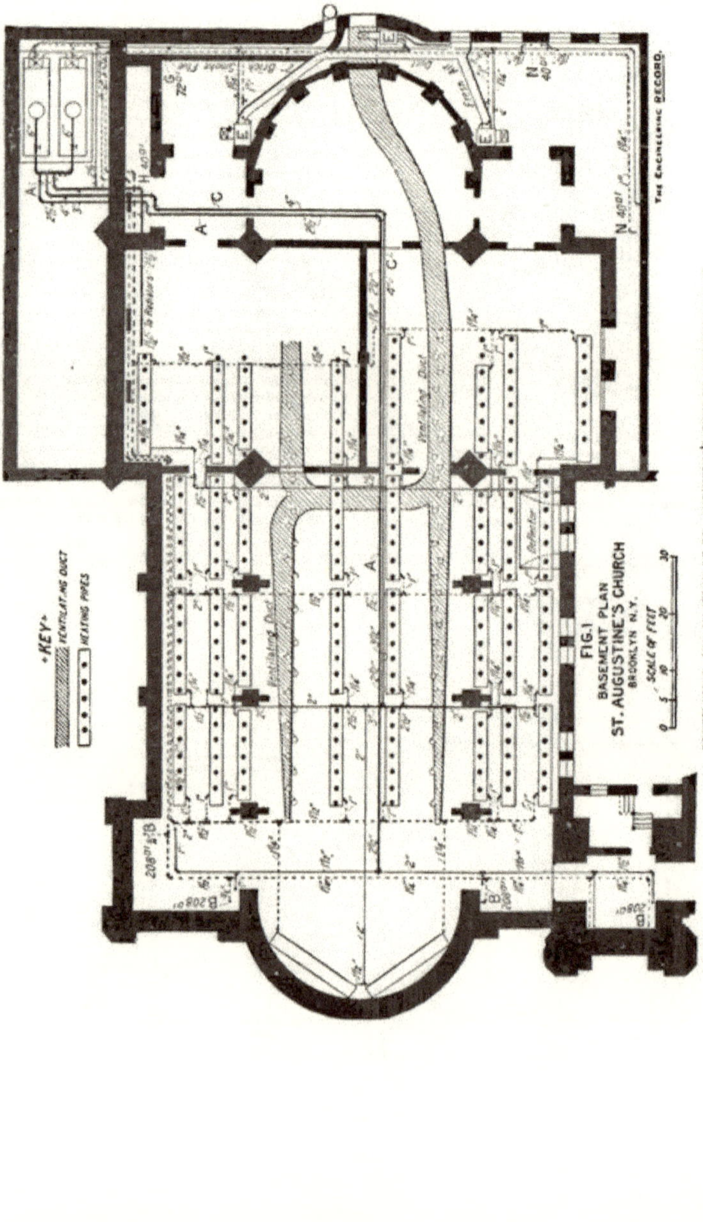

VENTILATION AND HEATING OF ST. AUGUSTINE'S CHURCH, BROOKLYN, N. Y.

THE VENTILATION AND HEATING OF ST. AUGUSTINE'S CHURCH, BROOKLYN, N. Y.

THE heating and ventilating of St. Augustine's Roman Catholic Church, at the corner of Sixth Avenue and Sterling Place, Brooklyn, N.Y., presents some features of interest. The architects of the building were Messrs. Parfitt Brothers, of Brooklyn, N. Y., while the heating and ventilating plant was designed by Mr. E. Rutzler, of New York City.

The church is about 176 feet in length by 135 feet in width. Figure 1 shows a plan of the basement of the building showing the building walls and columns in solid black, while the duct for carrying off the impure air is shaded.

Two return-tubular boilers, each 54 inches in diameter and 16 feet long, supply steam for the heating system. The boilers are set beneath the sidewalk in Stirling Place while the fuel-room is immediately in front of the boilers, this being a convenient location for receiving the coal from the sidewalk as well as being near the furnace doors. Each boiler is provided with 70 3-inch tubes, a Rutzler automatic damper regulator, and a lever and a spring safety valve.

Steam leaves each boiler through a 6-inch pipe which runs into an 8-inch cross-drum, and from the latter 2½, 3, and 4-inch mains lead to the different divisions of the heating system. Each feeder pipe from the boilers is supplied with a stop valve, as is each heating main. The 2½-inch main A supplies five direct radiators B B in the vestibule. Each has

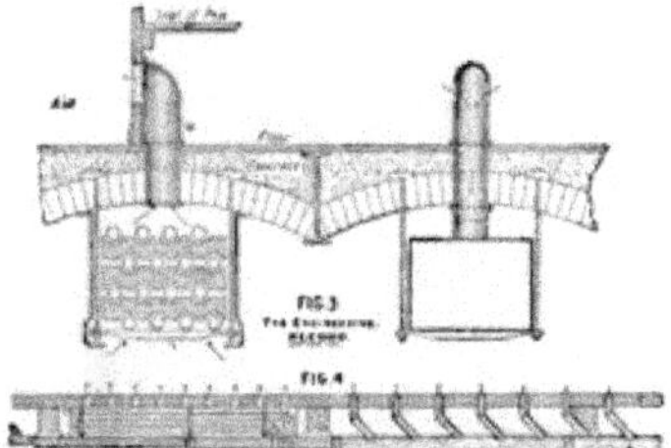

FIG. 3

FIG. 4

208 square feet of surface and the whole warms 9,814 cubic feet of air. A radiator of 40 square feet of surface, also supplied by this line of pipe, heats 2,700 square feet of surface in the tower-room.

The 4-inch main C supplies 3,500 square feet of surface in indirect coils and stacks distributed about the basement. This surface is to warm the air entering the church, the latter containing about 502,200 cubic feet of air. The 3-inch main D supplies the three direct stacks E E, the two 40 foot radiators N in the north sacristy heating 13,104 cubic feet of air and one 72-foot radiator G in the south sacristy heating 5,600 cubic feet of air, one 40-foot radiator H warming the 2,772 cubic feet of the south passage, two 48-foot and one 84-foot radiators in the basement.

All of the steam mains are suspended by flexible hangers, which are fastened to the I beams of the main floor. The return pipes, which are shown by the dotted lines, are all gathered into a 4-inch return drum at the boilers. Each radiator or coil is provided with a steam and return valve and an automatic air valve.

The entire basement beneath the main part of the church is used as an air chamber to supply the in-

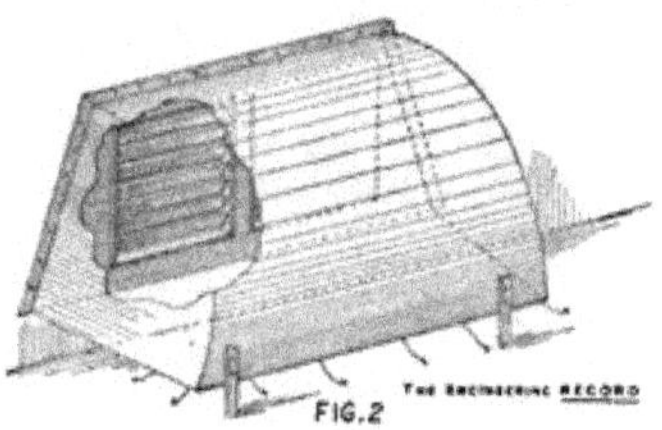

FIG. 2

direct coils. The air enters the basement through two windows in the north wall of the church. The window contains movable louver boards to regulate or cut off entirely the entrance of the air. Within the basement a galvanized hood spans the windows as shown by Fig. 2, so as to diffuse the entering air over the floor of the basement.

The heating stacks or coils are constructed in the method shown by Figs. 3 and 4. Two of these stacks were placed under the bow front of the auditorium and the other 32 were put up under the brick arches between the I beams which support the main floor of the church.

Each stack is composed of four horizontal flat-spring coils of 1 inch pipe, put together with branch tees and staggered as shown in Fig. 3. Cast-iron saddle pieces are introduced at the proper distances to hold the pipes in position and to keep their centers from sagging. These stacks vary in length from 15 to 23 feet. They rest upon horizontal bars which are suspended by ¾-inch bolts which pass up through the brick arch, a cast-iron plate being introduced between the bolt head and the brickwork of the arch. The lower end of the bolt is threaded for a considerable distance so as to adjust the pitch of the coils to insure their proper drainage. The upper edges of a galvanized-iron apron is made draft-tight with the brick arches so that the air passing into the church has to come in contact with the coils and be thoroughly warmed. The air enters the church through 4½-inch openings in the manner shown by the sectional view, Fig. 3. The location of these openings is shown by the solid black circles in Fig. 1. Figure 4 is a longitudinal section through the heating coils, the ventilating duct being shown in elevation beyond. The registers which finally deliver the heat to the church are shown in section by Fig. 3.

The registers for drawing off the vitiated air from the church are placed under the middle of the pews. Each has an opening of 4x8 inches. These registers

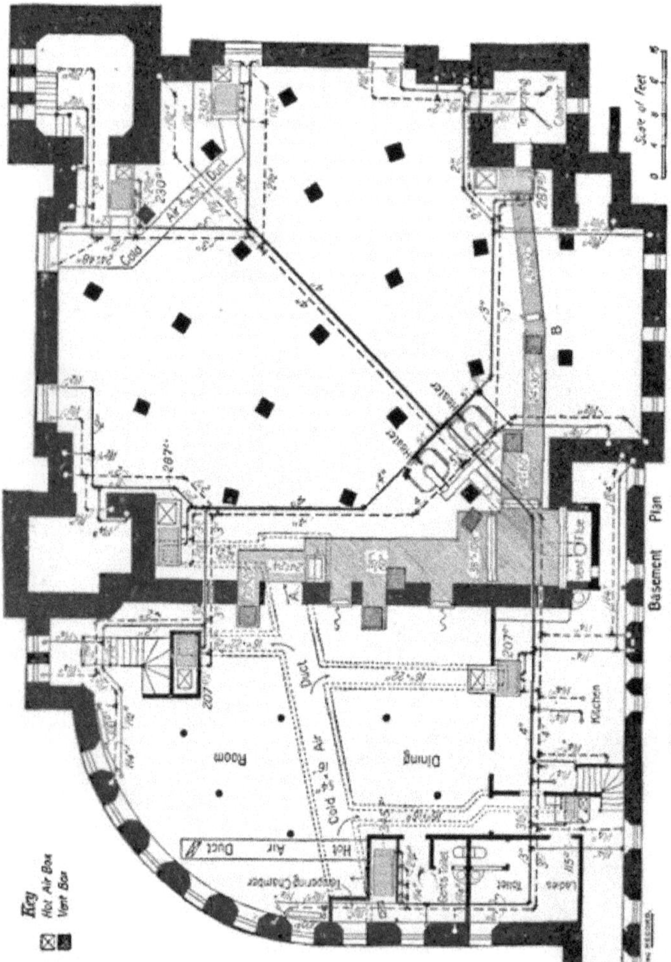

HOT-WATER HEATING APPARATUS IN A DANBURY, CONN., CHURCH.

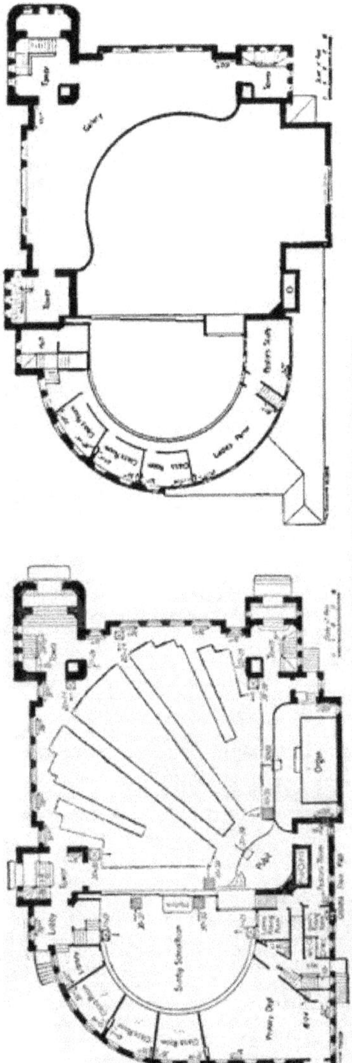

HOT-WATER HEATING APPARATUS IN A DANBURY, CONN., CHURCH.

are connected to the horizontal ducts by the smaller inclined ducts shown in Fig. 4. The exhaust ducts have a sectional area of 19 square feet as they run into a vertical ventilating shaft. The ventilating shaft is 150 feet in height and has a sectional area of 26 square feet. It is intended at some later date to serve also as a ventilating shaft for an adjacent school. This shaft contains a sheet-iron stack 30 inches in diameter for carrying off the smoke from the boilers. It was originally intended to exhaust the air from the church with a fan, but it was found out that the heat radiated from the sheet-iron chimney to the air in the surrounding ventilating stack was sufficient to cause a draft, so that a fan was not necessary.

HOT-WATER HEATING APPARATUS IN A DANBURY, CONN., CHURCH.

THE accompanying cuts show the plans of the basement, first floor and gallery of the Second Baptist Church of Danbury, Conn., the heating plant of which was designed by Anson W. Burchard, M. E., and installed by the J. M. Ives Company, of Danbury, under the direction of the designer, to whom we are indebted for the main part of the following description.

In designing a system of heating and ventilation for this building it was considered necessary to provide an apparatus of the simplest character, and which could be operated by an unskilled attendant. While it would have been possible to obtain more uniform and reliable results by the use of fans for the purpose of creating a circulation of air through the ventilating system, it was decided that an apparatus in which they were employed would require too much attention on the part of the janitor; and a heated flue was adopted as the simplest arrangement for this purpose.

A hot-water system was considered more economical of fuel than steam, particularly during those periods when, there being no service in the building, it would be unnecessary to maintain its temperature above 50° Fahr.; and during the mild weather of the spring and fall seasons the temperature of the air admitted for ventilation could be regulated by lowering the temperature of the stacks, without using mixing dampers, and overheating avoided thereby.

To simplify construction it was decided to carry out all the vitiated air through a single flue, located as near the center of the church as possible, and the fresh air was introduced near the outer walls, so that the direction of circulation is from the outside towards the center. The vitiated air is drawn from the building by numerous registers connected by ducts to the base of the large ventilating flue. The latter was built in the walls of the building and was carried 2 feet above the ridge, where it was covered by a stone cap, through a hole in which the smoke pipe passed. The covering stone was raised 16 inches above the top of the flue. The smoke pipe was made of No. 14 iron, in 10-foot sections, which were "Barffed" after being put together, and these were put in place as the walls of the flue were carried up, being held in position by frames of 1-inch pipe, built into the walls. A ladder, made by laying pieces of 1-inch pipe across the flue at intervals of 16 inches, was provided to enable the pipe to be inspected or renewed if necessary.

To stimulate the draft a furnace of the brick-set type, having a firepot 32 inches diameter, with a large cast-iron radiator, was set in the bottom of the flue. The fire and ash doors were accessible from cellar, and manhole and cleaning doors were provided to give access to the flue and stack. The smoke pipe from furnace was connected with boiler stack.

The main outlet ducts were built of wood, lined with tin, with seams locked to make them as nearly air-tight as possible. A large tight fitting damper was provided in the main vent duct near its connection with the flue, to enable the current to be regulated and to prevent the warm air from being drawn out of the building at those times when no service is being held.

The air inlet ducts in the main body of the building are made of galvanized iron, and those in the rear part, under the floor of the dining-room, are built of brick, with 8 inch side walls, with a 4-inch arched top and cemented bottom. Dampers were provided where these ducts connect with the outer air. A connection is provided (A and B in the basement plan) between the main ventilating duct and the main fresh-air duct, with a damper, so that air may be drawn through the radiators from the church, and an internal circulation established, making use of the indirect radiators to assist in warming up the building before service. From the large stack at the rear separate ducts were carried to each of the several classrooms, though not so shown on the plan.

The indirect radiators used are of the cast-iron pin type, the sections being 36 inches long and 12 inches deep, and were figured as containing 11½ feet of surface per section. The distribution of this surface is shown on the basement plan. The total amount of indirect radiation is 2,565 square feet. About 2,839 square feet of direct radiation is also provided. It has been estimated that the building contains about 223,360 cubic feet, and that the air supply will amount to 537,200 cubic feet of air per hour. This is sufficient to change the air about once in 25 minutes, but on extraordinary occasions the capacity of the apparatus may be increased beyond this figure.

In practice it is usual to bring the temperature of the building up to about 75° Fahr. half an hour before service. The dampers in the air inlet ducts and the main damper in the ventilating duct are kept closed until this time, a brisk fire having been started in the furnace in the shaft in the meantime.

The by-pass damper is then closed and the outer air and main ventilating duct dampers opened, which insures a fresh supply of air when the congregation assembles. The apparatus is operated in this manner for about 30 minutes after the close of service to insure the removal of all the vitiated air.

It was intended to provide sufficient direct radiation to maintain a temperature of 68° Fahr. in the building with an outside temperature of zero, and this was figured on a basis of 0.8 square foot of radiating surface, commercial rating, per square foot of exposed glass or its equivalent of exposed wall surface. The walls of the building being thick and lined with sheathing and ceiling, with paper between, 12 square feet of exposed wall surface was taken as equivalent to one of glass. The indirect radiation was calculated to warm the air admitted, although practically it aids considerably in maintaining the temperature of the building. The air ducts were calculated for a velocity of air through them of 7 feet per second, with an allowance for long or crooked pipes, this data having been obtained by trial in buildings provided with similar apparatus. In the classrooms the air is introduced at the ceilings.

The radiation in the entrance towers was as great as could be conveniently located, without regard to exposed surface. The study was provided with a proportionately larger radiator than other parts of the building, to insure a comfortable temperature there when the remainder of the building might be comparatively cold. The mains were considered ample to warm the small rooms in the basement and were left uncovered for this purpose.

Provision was made for ventilating the dining-room in the basement by drawing air from the large stack at the rear of the building, it being expected that the registers in the classrooms would be closed when it was desired to occupy this room.

No valves are provided for the stacks, but gate valves are inserted in the flow and return connections to each heater, enabling the apparatus to be run with one heater during the week. The boilers have 9 square of grate surface and 230 square feet of heating surface each. The expansion pipes are run separately from each heater to the expansion tank. There is a thermometer in each main flow pipe near the heater. The apparatus was erected in the year 1892.

STEAM HEATING OF A BROOKLYN, N. Y., CHURCH.

All Saints' Protestant Episcopal Church, situated at the corner of Seventh Avenue and Seventh Street, Brooklyn, N. Y., has a simple and effective steam-heating system which was installed by the E. T. Weymouth Company, New York City. Its main features are clearly set forth in the accompanying cuts. In a lower basement adjoining the church edifice are placed the two No. 30 "Allright" boilers A, Fig. 1, cross-connected by steam and return mains and flue connections so that either or both may be used as the service may require. Steam leaves the boilers through the 3-inch valves B which are close connected to the 5-inch horizontal cross pipe C. Pipe C passes through a rear well into the basement chapel and rises 30 inches at D. It is then run near the wall in a segment of a circle E made to conform to the apsidal termination of the chancel and at F rises to within 10 inches of the chapel ceiling. From this point by the bend G it is carried to the front of the edifice. Crossing to the other side return is made to point D by bends, drops, and a segment which are counterparts of those already described. The circuit from the boilers to point D is 350 feet in length and all is of 5-inch wrought-iron pipe. There is a fall of 5 inches in the main from the top of riser F to the point H from which there is an easy flow to the boiler. The foot of the steam risers at D and F, to each of which water runs with the steam, are dripped by the pipe I and below the water line into the main return pipe J, which has been reduced to 3 inches at D.

With the exception of the single heaters in the basement, library K, and the pastor's study P on the main floor, which have independent steam and return connections, so that they may be used while the remainder of the radiators are shut off, all connections M to radiators are taken from the top of the main steam pipe C and are upon the "one-pipe system." Each pair of radiators which are located at the base of the double windows is controlled by one valve and a long right and left connection, as shown in Fig. 2.

HEATING OF THE TEMPORARY CHAPEL, CATHEDRAL OF ST. JOHN THE DIVINE.

THE ceremonies attending the laying of the cornerstone ot the Cathedral of St. John the Divine in New York City on December 27, 1892, were notable not only on account of the gathering of prelates and dignitaries, but by reason of the amplitude of the material preparations for the comfort of the participants and attendants, which were complete to such a degree as to be remarkable, if not indeed singular in the ordering of what was practically an outdoor function at a most trying season of the year. The cathedral will be the foremost ecclesiastical structure in this country, and perhaps the first in the Western Hemisphere, and its sightly location on the plateau at Morningside Park is one of the highest points on Manhattan Island. The exposure of the location and the probability of cold or disagreeable weather led Bishop Potter and the Board of Trustees of the cathedral to decide to build a temporary structure to shelter the participants in the ceremonial. The plans were accordingly drawn by Messrs. Heins & LaFarge, of New York, the architects of the cathedral, and the matter of carrying out the arrangements for comfort and safety of those attending was placed in charge of Sexton Thomas P. Browne, of St. Agnes' Chapel, New York, with the result that even the most trivial matters were made to work almost mechanically.

Typifying the form of a cathedral a pavilion was constructed cruciform in plan around the cornerstone, 106 feet in one direction and 54 feet in the other. The foundation for the cornerstone had been laid with great expedition by David H. King, Jr., builder, of New York. A raised platf.rm about 15 feet square was erected in the center of the pavilion around the cornerstone for the use of those officiating in the ceremony proper and the Bishop of Albany who made the occasional address. On all sides of the square platform were wide spaces facilitating passage to the aisles in the nave and transepts. In nave and transepts were 12 tiers of raised platforms upon

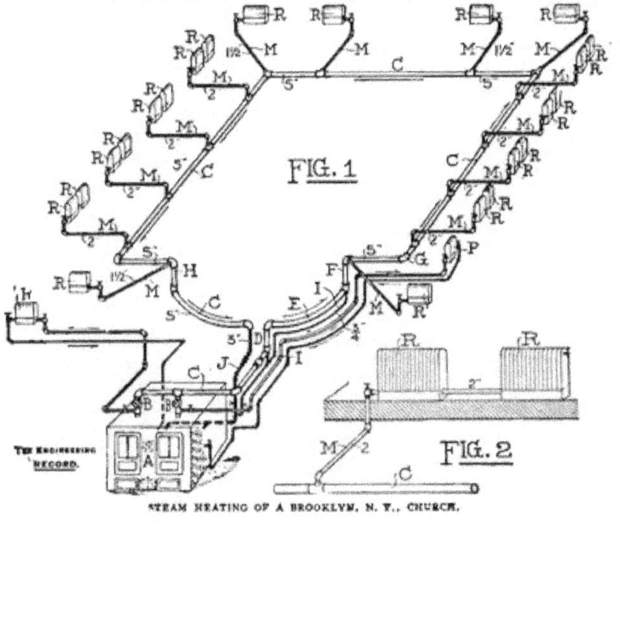

STEAM HEATING OF A BROOKLYN, N. Y., CHURCH.

which were placed 1,100 camp chairs. At each end of these platforms and in their rear were 4-foot aisles.

The walls of the pavilion, shown at D in the accompanying cut, were built up to a height of 10 feet above the rear seats with heavy timber and sheathed on the outside with matched stuff, painted a lead color, and covered with cloth on the inside. The roof C was of canvas, supported at four points by poles, giving a clear space of 50 feet from the floor in the center and was securely fastened to the top of the wooden walls. This arrangement was found sufficient to prevent drafts while securing the needed ventilation. Nine electric arc lamps, furnished by the Mount Morris Electric Light Company, were suspended from the poles and roof and brilliantly illuminated the entire space. The derrick B, securely fastened by guys to heavy ground stays, was used for hoisting and setting the stone. The floor of the pavilion and platform was covered with body Brussels carpet.

The heating arrangements were placed in the hands of Baker, Smith & Co., of New York, and comprehended a complete steam-heating plant. By working a large force of men continuously for three days and nights the installation was completed on time, and although it was exposed during erection to the chilling midwinter blasts, it was found on trial to work perfectly, so that during the ceremony, while the temperature without was near zero, it was kept at 70° Fahr. within the pavilion, and that without noise.

The locomotive boiler F, 54 inches in diameter and 21 feet long, was housed in an old barn E, its smoke end passing through the rear wall and the stack being braced from the roof, as shown. The exposed end of the boiler was protected by felting. Radiators R were placed about the center platform and along the walls at the ends and in the rear of the tiers of seats. There were 2,400 square feet of heating surface in the radiators, and the highest of them was below the water line in the boiler. The 4 inch main steam pipe G left the dome with a stop valve, passed across the space from E to the outside of and down D to a point below the inside upper aisle, where it entered, connecting with the reducing valve H, which cut the 40 pounds pressure on the boiler to 10 pounds for heating. The heating main branched into two 2½-inch pipes I I, each following the general form of the pavilion under the seating floor, preserving a uniform fall and meeting again at the opposite side, dropping after connection through J into the main return K, through which all the waters of condensation passed to the pump governors. The main return followed the contour and declination of the main steam pipe to the governor L, which was of the Kieley pattern, and controlled automatically the duplex Worthington pump M, which forced back to the boiler through N all the waters of condensation. Steam was laid on to the governor and pump through P at the full boiler pressure, pipe Q acting as a steam equalizer.

There were used in the mains and risers of this job over 3,000 feet of pipe, the greater part of which, being in the well inclosed space under the seats and uncovered, gave off sufficient heat to keep the floors

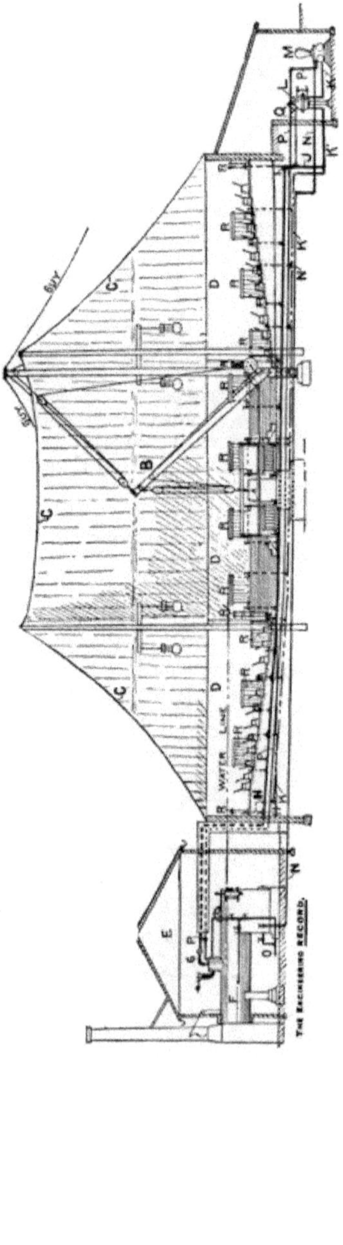

HEATING OF THE TEMPORARY CHAPEL, CATHEDRAL OF ST. JOHN THE DIVINE, NEW YORK CITY.

warmed. The exposed pipes between the pavilion and boiler-house were covered with felt, as was the end of the boiler through the wall. Fresh water was laid on to the injector through O.

HEATING AND VENTILATION OF A BALTIMORE CHURCH.

THE First Methodist Episcopal Church, Baltimore, Md., with connecting parsonage, occupying the northwest corner of St. Paul and Third Streets in that city, was recently erected according to the plans of McKim, Mead & White, of New York. It is a handsome edifice of stone, 120x170 feet in size, with a square tower on the street corner 187 feet high. The auditorium has a seating capacity of 1,400. The sittings are arranged in tiers and so spaced as to give all occupants a full view of the altar, pulpit, and organ. The domed ceiling rises 65 feet above the main floor and is frescoed from a scale drawing furnished by Prof. Simon Newcomb and representing the heavens as they appeared at 1 o'clock the night after the church was dedicated. The lighting is by stained-glass windows just below the base of the dome and by incandescent electric lamps. A chapel and Sunday-school room adjoins the auditorium on the rear, this part of the edifice containing also a parlor and reading room, a kitchen, toilet rooms, etc. There are six exits from the auditorium and three from the chapel. A parsonage is located to the right of the main entrance to the church and forms a part of the architectural composition. It has, however, no direct communication with the church edifice.

The edifice had been practically completed when Bartlett, Hayward & Co., heating engineers, of Baltimore, were called upon to provide the building with a plant adequate to the work to meet the conditions as found. There were no flues in the walls other than those for smoke and fireplace flues. Air could only be taken from one side of the cellar, as the side next the street was against an embankment. Two heating boilers were located in the basement, and were connected to the several indirect radiators, as shown in the accompanying figures. Fresh air is taken in at the points U U U, etc., Fig. 1. The brick-inclosed heating stacks G were uniformly located, and in order to insure to each stack an abundant supply of air the passageway H, following the contour of the foundation wall, was built from floor to ceiling with ducts leading to the several stacks from duct H, Fig. 2. The air when heated passes up and into the chambers B, which are formed by the sub ceiling V extending around the outer wall and lengthwise the center walls of the basement from 1 to 2 feet below the ceiling proper. Both ceilings are made tight, so that air could not be passed

FIG. 2

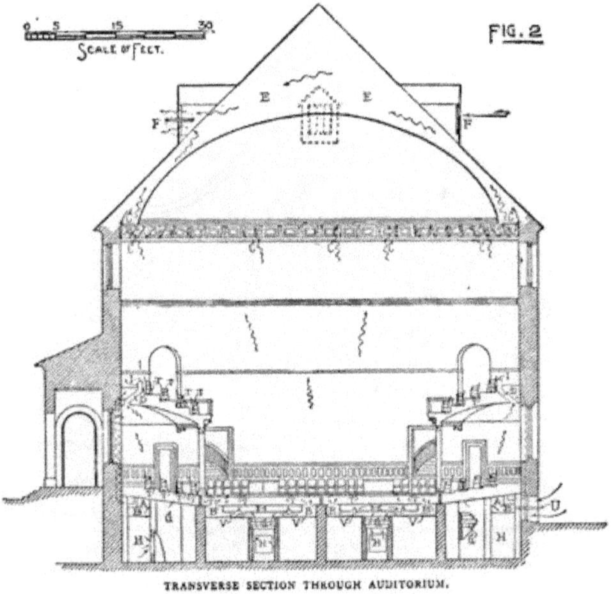

TRANSVERSE SECTION THROUGH AUDITORIUM.

through either of them excepting by the openings provided. Over each duct leading to the chambers B is the cold-air duct H, connected without dampers to the main fresh air chamber H. In the top of each of these ducts are the hinged dampers A. When these dampers are down or open the cold air is allowed to pass down from H and into B, at the same time shutting off the flow of hot air from the heaters G into B. When these dampers are up or closed the hot air from G is allowed to pass into B, while the cold air from H is entirely shut off. Any intermediate position of these dampers will allow of a corresponding proportionate flow of hot or cold air, which, meeting and mixing in the duct leading to chamber B, is then passed into the ducts formed tinues so long as the contact is maintained. The battery (two cells of Leclanche) acts only when the damper A is in motion and is thrown out of circuit when the damper A is completely opened or closed, the result in ordinary winter weather being that the damper is partially opened and is only fully closed in the most extreme cold. By this arrangement the entire chamber B is filled with warmed air, and the auditorium floor is kept at a comfortable temperature. The warmed air ascending from the numerous and uniformly distributed inlets T stimulates a steady current of the vitiated air of respiration through the paneled openings C in the ceiling cornice to the space E between the ceiling and the roof, and thence to the external air

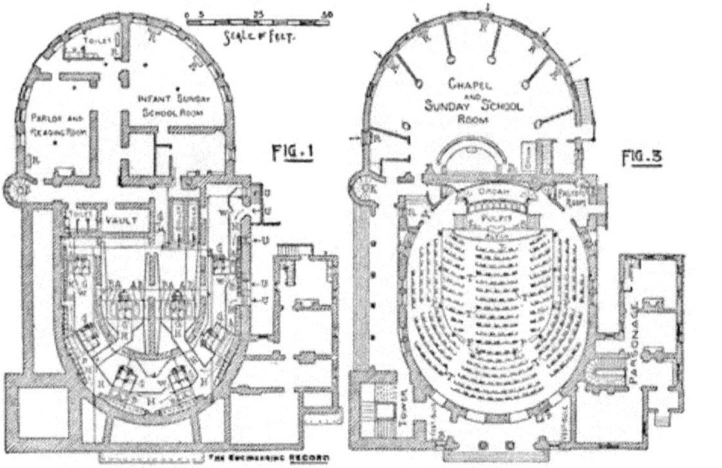

HEATING AND VENTILATION OF THE FIRST M. E. CHURCH, BALTIMORE, MD.

by the main floor joists and the original basement ceiling, through the openings M, then through the several small floor openings T on the main floor and gallery, graduated to give an equal flow under each seat. The chamber below the gallery receives its proportion through the wall ducts, shown by the dotted lines, Fig. 2.

The mixing dampers A are automatically regulated by a thermo-electric device, the invention of Maj. George M. Sternberg, Surgeon U. S. A. The opening A and the passage from G to B being properly graduated, the damper A is set at such a point that the proper amount of hot or cold air can be mixed and passed to B. Each of the eight thermostats located on the gallery columns at P, Fig. 3, is arranged to make an electrical contact on one side at 70 degrees and on the other at 69 degrees. Contact with either side slowly moves the damper A toward the opening or closing point. This movement continues through the dormers F, the center-hung dampers of which are controlled by chains descending to the gallery floor in the box L, Fig. 4, where the sexton can at any time reach them by the circular stairs K. The foul-air lock under the gallery is broken by the ducts D, Fig. 2, the currents being deflected rearwards by the pedestal registers I, Fig. 5. The movement of the entire body of air is gradually upwards. The registers S S on each side of the pulpit, Fig. 3, are under the individual control of the occupant of the pulpit, who can at any time open or close them by a lever X, Fig. 6, hidden beneath the pulpit top, the dampers being so arranged that he can have hot, cold or tempered air at will. This system of air inflow, regulation and distribution is designed to allow of a uniform current of cooled, fresh air from basement to auditorium during the summer season. The doors W allow entrance to the air passageway H for collecting accumulated dust, for whitewashing or repairs.

The operation of ventilation is in use only from the time the congregation has assembled, during service and until the auditorium has been emptied, when the sexton closes the tight dampers at U, opens the door Y, Fig. 3, and the doors W, Fig. 1, and closes the dampers F in the roof dormers. The air of the auditorium then passes through Y down the circular passageway K and by the basement hall into H by the doors W, into and through the heating stacks G,

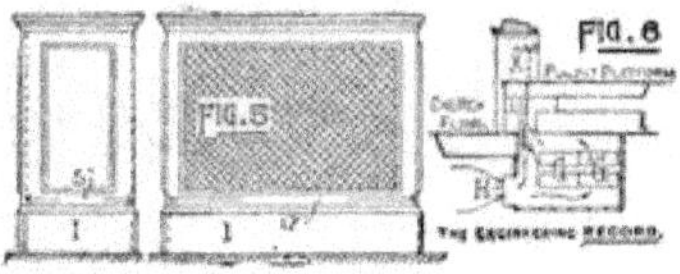

making the circuit again into the auditorium. By this arrangement the auditorium is at no time left to be chilled, and a very small amount of fuel suffices to keep it in a condition to be comfortably heated on short notice, when the operation is reversed and the air is taken from the outside.

The parlor and reading room, women's toilet and infants' Sunday-school rooms in the basement of the chapel building are heated by direct radiators. The chapel and Sunday-school room above, the pastor's

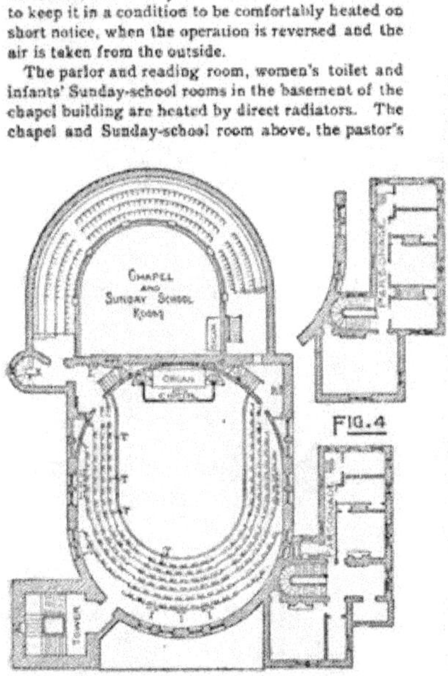

room and the vestibule on the main floor are heated by direct radiators, those in the chapel and Sunday-school room having individual fresh-air ducts from outside, which are regulated to suit by close-fitting dampers, making a combination of direct or indirect heating at will. The pastor's room is also supplied with a fireplace. The kitchen is heated by the range, which is part of a complete cooking outfit for use in entertainments. The plumbing is of approved form. The men's toilet room and the vault are ventilated into flues adjoining the fireplace in the parlor. The fuel room is ample for coal storage, and a fireproof vault is provided for storing the church records, etc.

Figure 1 shows a plan of the basement, Fig. 2 a transverse section through the auditorium showing radiator chambers, ducts, tempering apparatus and dormer ventilating dampers, Fig. 3 a plan of the main floor showing the location of hot-air flues, fresh-air inlets to radiators, warm and cold air inlets, thermostats on gallery columns, Fig. 4 a plan of the gallery, showing air inlets and location of pedestal registers, Fig. 5 end and front elevations of pedestal register for vent flues from under the gallery, and Fig. 6 the arrangement for individual control, from the pastor's desk, of the air supply to the pulpit platform.

HEATING AND VENTILATION OF A ROCKFORD CHURCH.

The Congregational Church at Rockford, Ill., has recently been completed according to plans of Architect D. S. Schureman. The heating system is steam of the low-pressure gravity type. The ventilation is accomplished by the plenum movement. Both are clearly shown by the accompanying illustrations. Figure 1 is the basement plan, Fig. 2 is the first-floor plan, and Fig. 3 is a sectional elevation through the church auditorium. Referring to the basement plan, it will be seen that the boiler is located at a lower level than any other part of the basement, so as to permit the return of condensation from the indirect radiators and other parts of the system by gravity. The basement plan also shows location of the large massed stack of indirect radiators, the fan and the fan chamber, the electric motor and the motor-room, and the sizes of the main supply pipes to the different direct radiators of the several portions of the house. The low space under the whole church auditorium is used as a plenum chamber into which air is forced, and thence finds its way into the auditorium through numerous small round register openings placed under the seats.

The Sunday-school section is supplied with fresh air by means of a separate system of galvanized-iron pipes leading directly from the fan, discharging air into the several rooms through registers, as shown. The indirect radiator chamber is provided with a tempering damper, so that air may be forced into the plenum space, or Sunday-school section, at any desired temperature. It is also provided with an indoor connection damper A, Fig. 1, so that air can be circulated over and over again in the church auditorium, so as to more rapidly warm the building in extremely cold weather. The choir platform is so arranged that a room underneath it is formed with an exceptionally high ceiling from which the large massed stack of indirect radiators is suspended. The first-floor plan shows the location of the direct radi-

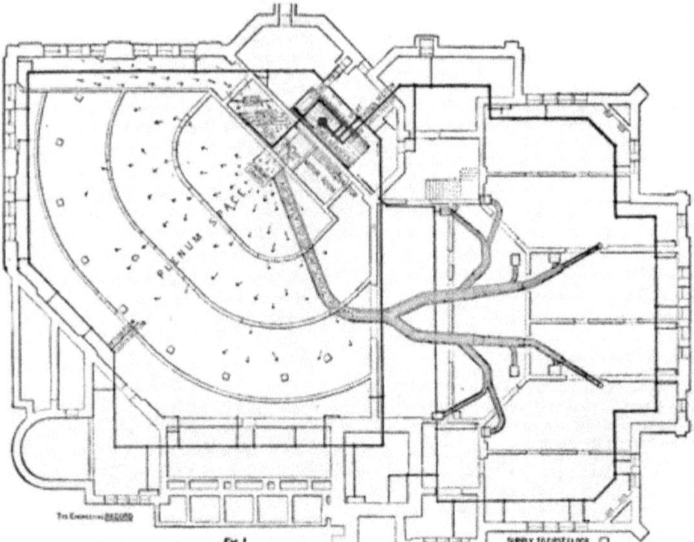

FIG. 1

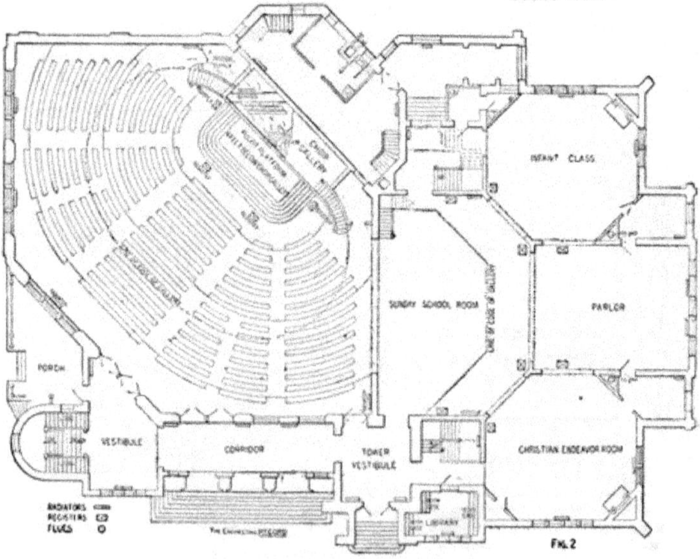

FIG. 2

HEATING AND VENTILATION OF A ROCKFORD, ILL., CHURCH.

ators and a few large register openings. Small registers under seats are not shown. In the second story of the church there are radiators in the gallery and also in small Sunday-school rooms. The apparatus is laid out with the Sunday-school direct radiators in one independent section, the direct radiators of the church in another section, and the indirect radiators of the ventilation in another section, so that the apparatus can be operated to suit the requirements of a church building.

In practice it is found that the direct radiators in the Sunday-school section are kept going during the greater part of the winter season. When an audience is assembled the fan is started up for the purpose of furnishing ventilation. Should the rooms then become overheated, the tempering damper is set to reduce the temperature of the incoming air to a point between 65 and 70 degrees. Should the rooms still remain overheated, the direct radiators are shut off by means of the main controlling valves in the boiler-room. When the assembly has dispersed the fan is stopped and the valves in the boiler-room controlling the direct-radiator section are then turned on again. The same method also obtains in the larger part of the building, excepting that the direct-radiator section is not turned on again until a few hours before the auditorium is used. Then all the direct radiators and the indoor connections are brought into use to rapidly accomplish the desired results. After the air has been introduced into the auditorium through the small round registers under each seat it is carried directly upward and finds its exit through skylight ventilators in the roof, which are closed when the church is not in use. It will be thus seen that the entire auditorium becomes essentially one ventilating shaft in which the current is always upward. This method of ventilation can be used in summer as well as in winter, regardless of the open windows. The fireplaces and other vertical vent flues furnish ventilation for several rooms in the Sunday-school section of the building. It is claimed that considering relative first cost, this scheme is better adapted to the conditions and requirements of ventilation for the different seasons than other systems requiring exhaust fans in addition to the plenum fans, and complicated system of exhaust ducts and flues.

The apparatus was designed and installed by the L. H. Prentice Company, of Chicago.

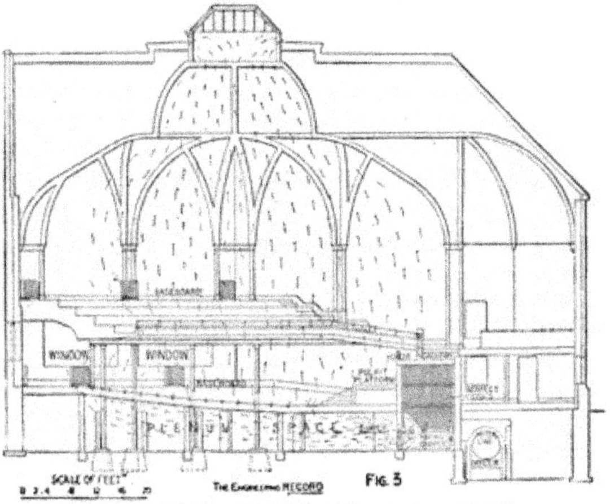

HEATING AND VENTILATION OF A ROCKFORD, ILL., CHURCH.

HOT-WATER HEATING IN A CHURCH AND RECTORY.

PART I.—GENERAL DESCRIPTION, VIEW OF HEATERS AND RECTORY SYSTEM, AND PLAN AND PIPE SYSTEM OF THE CHURCH.

St. Mary's Church at South Amboy, N. J., is a brick building measuring about 60x130 feet, and has a large, high basement used as a Sunday-school room. It was built about 1881, and was furnished with a hot-air heating system, with two furnaces in the basement. In 1890 a large three-story and basement brick rectory was built adjacent to it and a hot-water heating, direct radiator system, was constructed to warm it and replace the old hot-air system in the church. L. J. O'Connor, of New York, is the architect, and Johnson & Co., of Catskill, N. Y., are the contractors.

Two large hot-water boilers are placed in the cellar of the church, and from them the hot water is distributed and returned through two sets of horizontal pipes, from 3 inches to 5 inches in diameter, which are suspended close to the ceiling in the rectory basement and church basement, and are connected by vertical branches to the radiators above, which heat the upper stories. The main pipes are jacketed.

Figure 1 is an isometric view of the boilers and of the main pipes in the rectory basement, including the foot of the rising lines to radiators above. E E are two No. 131 Gurney boilers with a 12-inch smoke flue. These boilers are made in nine sections each, have each about 5½ square feet of grate surface, and a total connected radiating surface, inclusive of pipe mains, of about 4,000 square feet. They are freely connected to the main flow pipe J by the 5-inch branches I I, and to the return main G by 5-inch branches. M is a stop-cock controlling the 1-inch emptying pipe which discharges into the sewer.

The rectory system is all connected to the branches H and I of the church system mains J and K, which, between the two buildings, are laid in a trench about 3 feet below the surface, and protected by the box shown in section in Fig. 2. This box is made of 2-inch tarred pine plank. It is about 7x13 inches inside dimensions, and is packed with mineral wool.

Of the radiators in the rectory, one on the second and one on the third floor are connected to risers N; the risers O connect with one on the second floor; the risers P with one on the first floor; risers Q with two on the second and two on the third floors; risers R with one on the first floor; risers S with one on the second and one on the third floor; risers T and U each with one on the first floor; and riser V with one on the second floor.

Gurney radiators are used throughout, and are decorated in gilt. The radiators and exposed pipes are painted a drab color and are trimmed with gold bronze.

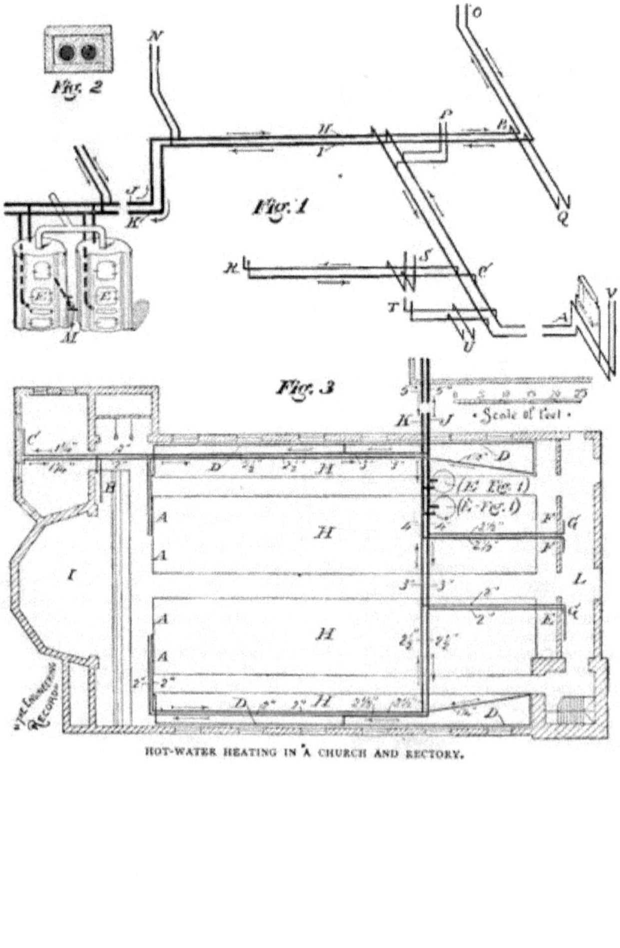

HOT-WATER HEATING IN A CHURCH AND RECTORY.

Figure 3 is a plan of the church, showing in full black lines the flow and return pipes, which are really hidden below the floor, at the extremities of which are the risers to the radiators. J and K are respectively the flow and return mains, shown in Fig. 1, where they are designated by the same letters. H H, etc., are the sittings. The altar is at I and the vestibule at L. A A are Bartlett & Hayward radiators, each with 56 pipes, 6 feet 6 inches long. B is a 67 inch Gurney radiator; C is a 71-inch Gurney radiator; D D, etc., are 4-pipe wall coils, with branches 49 feet and 6 feet long; E E E are Bartlett & Hayward radiators, each with 88 pipes 7 feet long, and G and G are Gurney 60-inch radiators.

PART II.—INTERIOR OF CHURCH, SPECIAL SUPPLY CONNECTIONS FOR WALL COILS AND DOUBLE RADIATORS AND REDUCING FITTINGS, ETC., AT RISERS.

FIGURE 4 is a view of the church interior, showing the position of the radiators, which are indicated by the same letters, A, B, D, E, F, and G, as in Fig. 3. pitched down to O O 49 feet away, thence, at the same grade, to P P, where the bottom pipes almost touch the floor.

Figures 9, 10, and 11 show the connections of the main pipes at points A, B, and C respectively of Fig. 1.

STEAM HEATING IN TRINITY CHURCH NEW YORK.

OLD Trinity Church, Broadway, New York, has been for many years heated chiefly by a hot-air furnace system. This has been supplemented by direct steam radiator coils in the tower and vestibules, and recently a system of indirect steam radiators has been constructed for warming the main part of the building, C. C. Haight, of New York, being the architect, and the Q. N. Evans Construction Company, also of New York, the contractors.

Cold fresh air is drawn in from out-of-doors, passed through filtering screens and up and down and be-

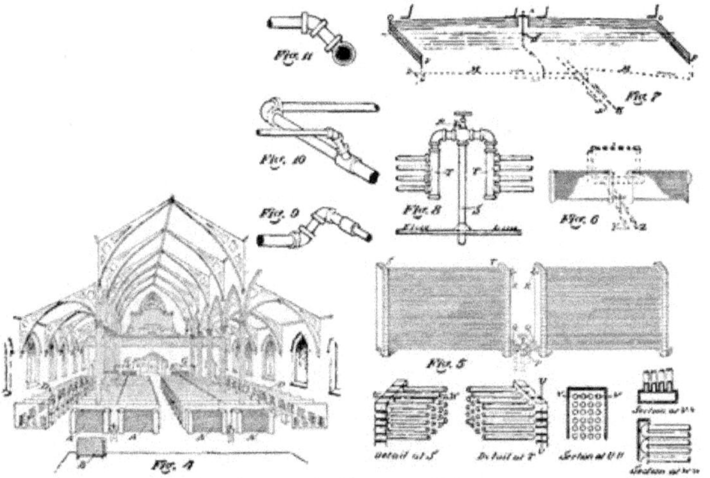

HOT-WATER HEATING IN A CHURCH AND RECTORY.

Figure 5 shows the connections of radiator A A, which are supplied by a single riser P. They have separate return pipes Q Q, and air valves R R. Figure 6 shows the connections of radiators F F and G, Fig. 4, to the flow and return pipes Y and Z. Radiators F F are similar to A A, Fig. 4, and are supplied in the same manner.

Figure 7 shows the connections of wall coils D D, Fig. 4; L is the flow and M M are the return branches. Figure 8 shows how the pipes are connected at N, Fig. 7, to the manifolds T T, and both supplied through one branch S; R is an air valve. The top pipes are raised to the bottoms of the seats at N, and tween the pipes of steam radiators. Thence a blower forces it through a brick passageway under the church floor, where it is carried in smaller brick conduits across the nave and down under each aisle. These conduits are closed at their farther ends and have pipes opening into their walls, through which the air is forced into sheet iron, trough-like conduits, at the floor level along both sides of the aisles, and escapes in six thin, continuous sheets, which extend along each row of sittings. The air is thus distributed all through the room, and, being admitted at the coldest part so as to strike downwards upon the floor, is diffused and uniformly tempered.

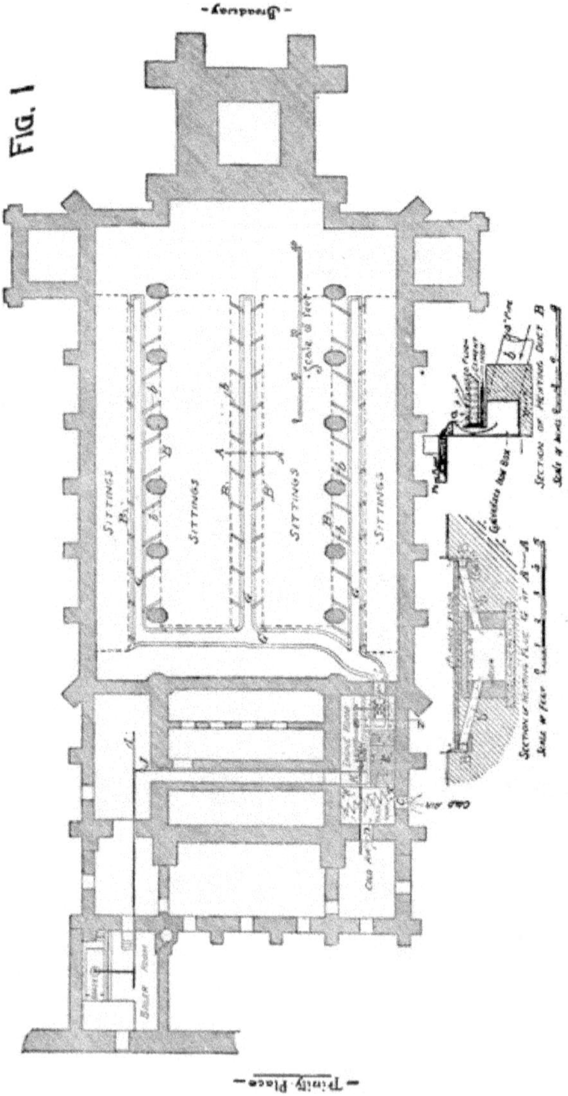

FIG. 1

STEAM HEATING IN TRINITY CHURCH, NEW YORK.

Figure 1 shows the general plan of the building and the location of the heating plant and air ducts beneath the floor.

Figure 2 shows the blower F, Fig. 1, and the masonry in which it is set. Another view of it is given in Fig. 3, which is a sectional diagram of the filter and heating chambers and shows, by arrows, the course of the fresh air as it is admitted, warmed and expelled through conduit G. Q, Q', and I are the steam radiators, of which an enlarged perspective

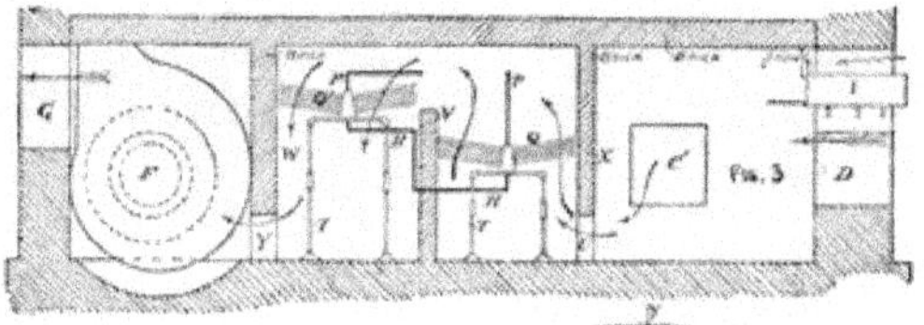

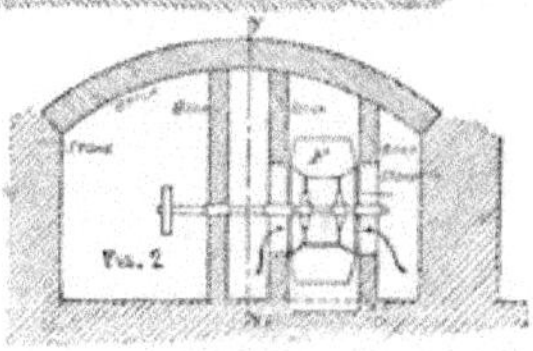

view is given in Fig. 4. This clearly shows their arrangements and connections, and the steam engine H, which drives the blower.

Figure 5 is a perspective of filter-room, showing part of the sieves V, through which the fresh air must pass to enter the heating chambers. These sieves are simply folding screen frames covered with cheese-cloth, and the radiators Q and Q' are essentially cast-iron boxes with numerous pipes, closed at their outer ends, protruding from each of the inclined sides. Steam is admitted to the box and circulates through the tubes.

The temperature is controlled by the Johnson heat regulating system, as constructed by the Metropolitan Electric Service Company, of New York. Two thermostats are used, one set at 63 degrees for the air in the church and the other at 85 degrees for the air in the hot chamber. When the temperature in the chamber rises above 85 degrees the sensitive bar of the thermostat expands in one direction, makes a contact which causes the electric current to open a valve, admitting air under pressure to the steam valve, which is thus closed and the steam shut off from the radiator. Falling temperature causes the sensitive bar to expand in the other direction, reverse the valves, turn on steam, and so on.

Mr. Haight, the architect, reports that with about 120 revolutions of the fan per minute, the normal rate, 12,000 cubic feet of air is forced into the church, and so equally distributed that there is no annoy-

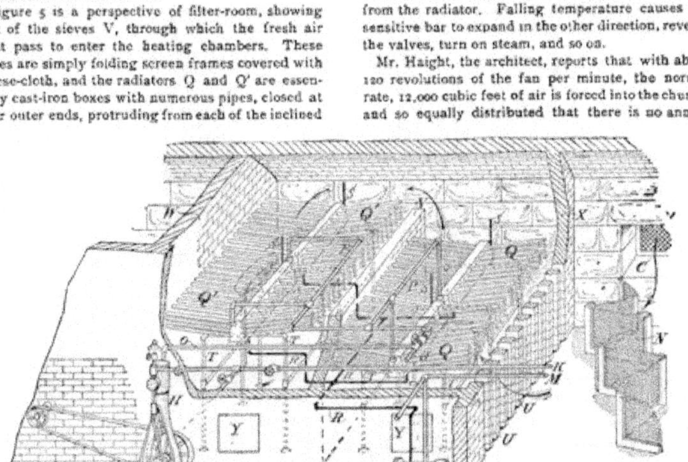

FIG. 4

STEAM HEATING IN TRINITY CHURCH, NEW YORK.

ance from drafts. With crowded congregations the air is fresh and clean, and the temperature varies less than 2 degrees.

Figure 2 is an enlarged section at Z Z of Fig. 1, and Fig. 3 is a section at Y Y of Fig. 2. Fig. 4 is a perspective from W, Fig. 1, and Fig. 5 is a perspective from X, Fig. 1.

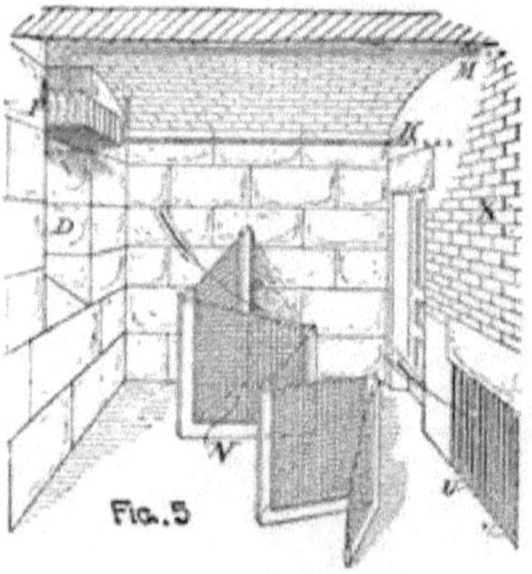

FIG. 5

External air is drawn through gratings at C and D, passes through the screen N (Figs. 4 and 5) in the filter-room, through slits U U, etc., in wall X, and up through radiators Q Q, over partition wall V, down through radiators Q Q', and through the door Y, in wall W, to the fan F. The latter forces it through the 60x60-inch flue G at an actual velocity of about 200 feet per second to the distributing ducts B B, etc., which, at intervals of about 6 feet, through pipes *b b*, etc., diffuse it throughout the aisles.

The duct B has a lip *a* (see detail, Fig. 1), designed to deflect the air against the floor and diffuse it as indicated by arrows. The branches *b b*, etc., have valves, not shown here, which are not generally used, but are provided for controlling the branches separately if necessary.

The fresh-air inlets have a total cross-section of about 12 square feet. The filter chamber and radiator chambers are thoroughly whitewashed, and easily accessible for cleaning, etc.

Steam is generated in a 54-inch steam boiler, 12 feet long, containing 58 11-inch tubes, and is delivered by branch I, Fig. 1, to direct radiators in vestibules, etc., and through J to the branches P P, which supply radiators Q Q', etc., Figs. 3, 4, and 5, and branch O to a 5 horse-power engine H. The latter drives the 72 inch Sturtevant blower F. S S, etc., Fig. 4, are air vents brought together outside the radiator chamber and connected with the sewer. The steam and water of condensation from the radiators Q Q', etc., is returned through pipes R R to the trap S, whence the remaining steam is delivered, by pipe M, to a Gold extended surface radiator I, placed in inlet D, to utilize the remaining heat. Y Y, Fig. 4, are manhole doors, and T T are gas-pipe frames supporting the radiators Q Q', etc.

STEAM HEATING OF A CHURCH.

THE steam-heating plant of the Dutch Reformed Church at Montgomery, Orange County, N. Y., possesses in its details several points of interest. The church is about 80 feet in length by 50 feet in

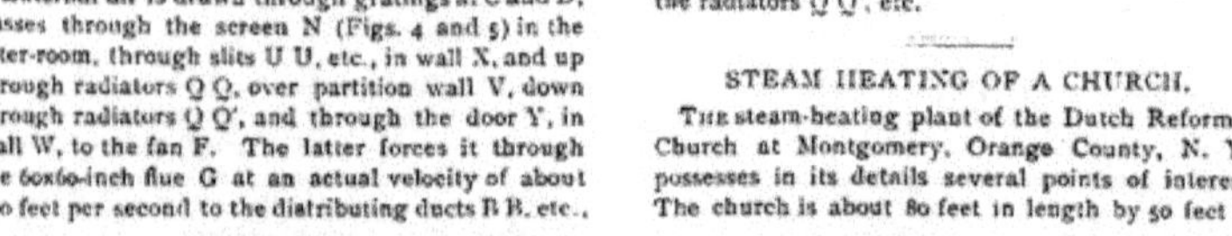

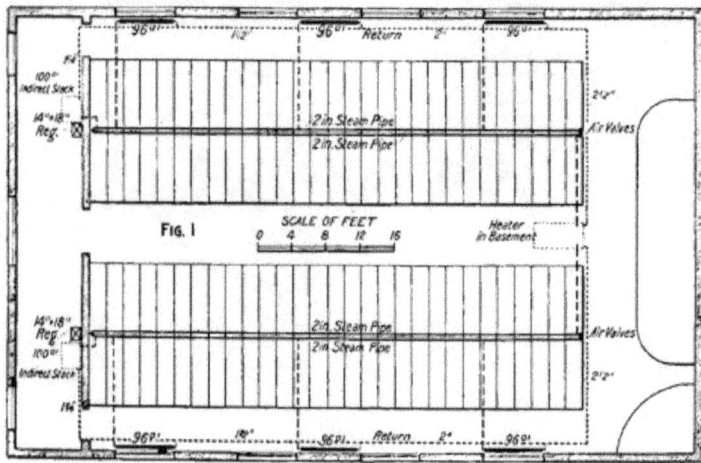

STEAM HEATING OF A CHURCH AT MONTGOMERY, N. Y.

width, the body of the church containing about 96,-000 cubic feet and the vestibule about 4,000 cubic feet.

One of the conditions entering into the design of the heating plant was that the mains supplying the radiators should not be carried along the side wall of the church on account of the room they would occupy, yet it was thought best to run the steam mains above the floor of the church so as to use them as radiating surfaces and also save the cost of pipe covering and extra surface in radiators had these mains been placed in the cellar.

FIG. 2

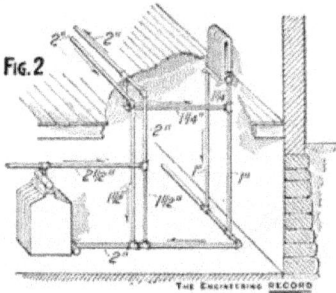

A No. 23 Royal steam heater is placed in the cellar under the rear of the church, and from this heater pipes lead in opposite directions to the two steam mains, each of which is carried down the middle of each block of pews from the rear to the front of the church. Each main starts about 15 inches above the floor and is pitched so as to drop about 6 inches in the length of the building. The steam main then returns, dropping to the floor as it reaches the rear or boiler end of the church. The steam main was carried back in the same manner described so as to give that amount of additional surface in that part of the building. At the front of the church connections lead from the mains to supply steam to two indirect stacks, each containing 100 square feet of heating surface. The air warmed by these stacks is discharged into the vestibule of the church. The body of the church is heated by direct radiation. Three radiators, each containing 96 square feet of surface, are distributed along each side wall. These radiators are connected to the adjacent steam main in the manner shown by Fig. 2. The supply pipe for each radiator drops from the underside of the main into a nearly horizontal pipe running across under the floor to the radiator. This pipe, which is pitched so as to drain toward the radiator, terminates in a tee, one branch of which leads upward to the radiator, while the other, which serves as a drain pipe, drops into the main return pipe which encircles the cellar as shown in Fig. 1.

The plant was designed by Mr. W. M. Mackey, of Hart & Crouse, of New York City. Mr. J. H. Wallace, of Pine Brush, N. Y., was the contractor for the work.

HOT-WATER HEATING OF A CITY CHURCH.

THE Church of the Most Holy Redeemer, in East Third Street, New York City, one of the largest churches in the city, was formerly heated by steam, and while there was ample radiating surface to furnish sufficient heat for the building it was not always noiseless in operation, and required constant attention and a large consumption of fuel to properly heat the building, and when neglected for any length of time, as was often the case, the building cooled down. While the question of improving the apparatus was given considerable attention, the general arrangement of the building prevented any improvements to the steam-heating apparatus.

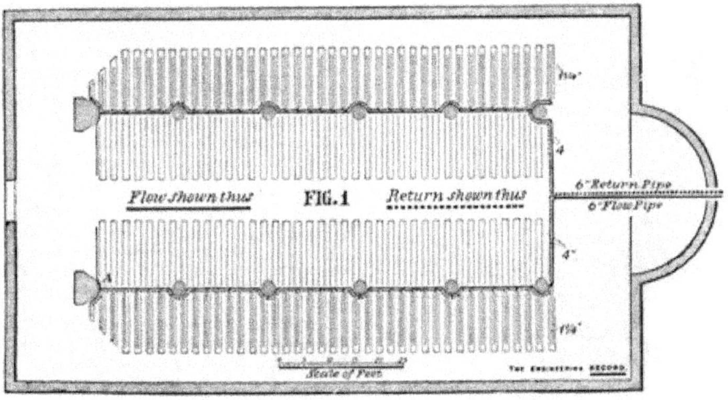

HOT-WATER HEATING OF A NEW YORK CITY CHURCH.

After consulting with Mr. W. M. Mackay, of New York City, it was decided to remove the steam apparatus and replace it with a hot-water heating system. Wishing to keep the aisles free from radiators, radiating surface in the form of 1¼-inch pipe was arranged below each seat, giving a uniform distribution of heat and without discomfort to the occupants of the seats, for with hot water the radiating surface can be kept at a much lower temperature than with steam.

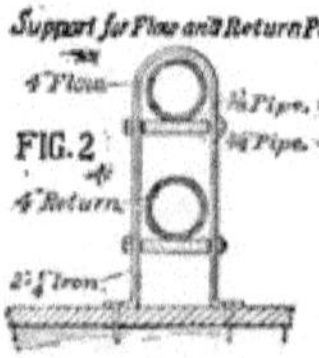

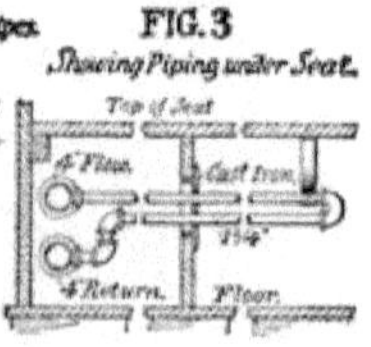

It was decided to utilize the boiler which had been used for heating the building by steam, and a 6-inch flow and return pipe was connected to it and carried about 100 feet, rising into the church at the front of the pews, where it was divided into two 4-inch flow and return mains, each of which was carried along the partitions between the pews below the seats, the flow pipe being carried above the return and connected together through two coils of 2-inch pipe at the entrance of the church. From the side of this flow and return pipe a loop of 1¼-inch pipe 10 feet long is carried along under the seat. There are about 300 feet of 2-inch pipe, 2,000 feet of 1¼-inch pipe, and about 400 feet of 4-inch pipe used for radiation.

Figure 1 shows a plan of the piping, while Fig. 2 shows the method of supporting the 4-inch flow and return pipes. The support consists of a bar of iron 2x¼ inch bent in the form shown, with holes drilled to receive the rollers upon which the pipes rest. Figure 3 shows a view of the piping under the seat. The outer end of the 1¼-inch piping is supported by a cast-iron hook which is screwed into the underside of the top of the seat.

Figure 4 shows the method of connecting the extremity of the 4-inch flow and return pipes (at A, Fig. 1). It also shows the connection with the expansion tank.

Figure 5 shows the special casting employed in running the 4-inch mains around the columns in the church.

The plant is said to have given excellent satisfaction, it being noiseless, economical in operation, and furnishes abundant heat for the needs of the building. Messrs. Albert A. Cryer & Co., of New York City, were the contractors for the job.

In reference to this work "Heating," San Francisco, Cal., writes:

"I have read with interest and profit your description of the hot-water heating system in the Church of the Most Holy Redeemer, New York City, but I cannot see why branch pipes were not put under the pews in the middle aisle. How is the warmed air drawn to the center of the church, especially near the floor? Where is the expansion tank located, if one is used, or do they use only a stand-pipe for a vapor pipe?"

[We have referred the above questions to Mr. Mackay, the designer of the plant, who informs us that provision was originally made for placing pipes under the center as well as the side pews, but these were not found necessary and were never put in. As the radiation is interposed between the exposed walls of the church and the center of the building, the colder currents of air descending close to the walls must pass over the coils to reach the center of the church, and air is thus tempered. If, however, the air was at rest in the center of the church and out of the current produced by the alternate warming and cooling, it would hardly cool, as it is surrounded by a warm body of air, and is not in contact with any cold surface. The described system is said to have heated the church most uniformly. The expansion tank is placed almost over the boiler, and consists of a plumbers' 14-gallon cast-iron tank with an overflow and supply controlled by a ball cock. The vertical pipe shown in the sketch (Fig. 4, p. 194) is only used to relieve the system of air.]

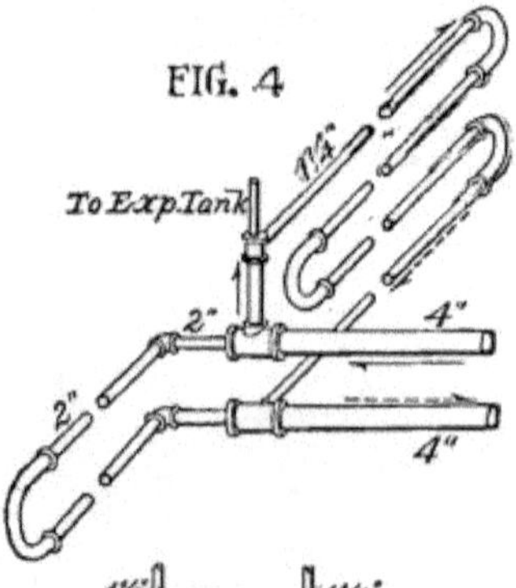

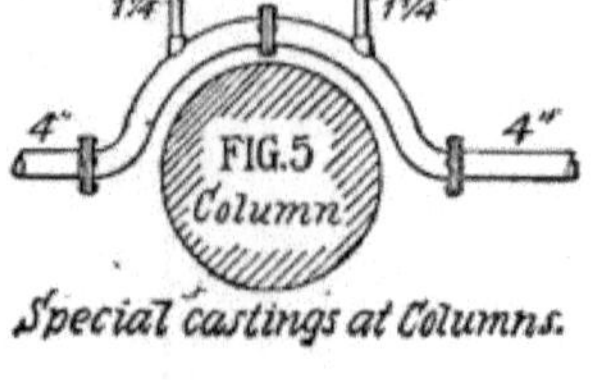

Special castings at Columns.

HEATING OF SCHOOLS.

HOT-WATER HEATING IN THE CONVENT OF THE VISITATION, ST. LOUIS, MO.

PART I.—GENERAL DESCRIPTION AND PLAN OF BASEMENT, PIPE LINES, HEATER-ROOM AND FIRST FLOOR.

The Convent of the Visitation is a large new building of a U-shaped plan, five stories in height above the basement, and is situated on high ground near St. Louis, Mo. The main front is 352 feet with wings, the entire frontage aggregating about 725 feet. The heating system was installed by the Detroit Heating and Lighting Company, from whose original data we have prepared the description and illustration of the work.

The system is of direct hot-water radiators, with double lines of similar pipes for flow and return throughout, and is designed to maintain a temperature of 65 degrees in the interior space which is exposed to the effect of winds and radiation from large window surface on all sides. The water is heated in nine boilers set in batteries of four and five and so connected up that any or all of them may be used at once or disconnected for repairs.

The boilers are of the Bolton pattern, 40"x8', with 100 2-inch tubes each, set in ordinary brick setting. Their draft is governed by automatic damper regulators and expansion and waste of water is provided for by two open-roof tanks, one for each battery, which tanks are supplied by city water through a ball cock. Radiators are used throughout and all their flow and return branches are provided with valves, each of which has a lock and key.

From the 10-inch main header, to which branches from each boiler are connected with valves, there are taken eight main flow pipes which start with diameters of 5, 6, and 8 inches, and run horizontally along the basement ceiling with diminishing sections proportioned to the number of vertical risers to be supplied beyond any given point. The longest horizontal main extends about 210 feet from the boiler, where it is 6 inches, to the extremity, when it becomes 1¼ inches and rises to the third floor. The risers are 1, 1¼, 1½, 2, or 2½ inches in diameter, according to their length and the number of radiators, never more than 10, which are connected to each. The horizontal pipe is supported on rollers to allow expansion and contraction movements, and the vertical lines and their branches are designed to provide for the maximum displacements.

In Fig. 1 the flow mains are shown in full black lines and the return mains, which are in all cases of corresponding size, are indicated by parallel dotted lines alongside. The risers are designated by small circles, marked, for example, thus, 2 R 420 II., III. and IV., indicating that it is a 2-inch riser supplying 420 square feet of radiator surface on the second, third, and fourth floors.

Figure 2 is a plan showing arrangement of rooms on the first floor, their contents in cubic feet and the size and location of riser lines and the position and surface of radiators. The risers are designated as in Fig. 1. The radiators are indicated by small rectangles with cross-hatching; circles at each end show their connecting branches, whose diameter in inches is marked alongside, and in the center of each room is a number designating the volume of the room in cubic feet, as, for example, in the academy refectory the space is 17,664 cubic feet, and there are three radiators, each with 1¼-inch connections, and having 160 square feet of radiating surface, besides a 160-foot coil in the alcove.

PART II.—PLANS OF THE SECOND, THIRD, FOURTH, AND FIFTH STORIES, VOLUME AND ARRANGEMENT OF ROOMS, AND SIZE AND LOCATION OF RADIATORS.

The convent building contains about 1,036,290 cubic feet of space which is heated, and these figures do not include the cell rooms and other rooms not heated. These rooms contain over 85,418 square feet of floor surface. There is 61,462 feet of exposed outside wall surface in the rooms where the heating apparatus is directly placed. The system of heaters has an aggregate of 80 feet of grate surface and a total of 2,325 feet of fire surface.

This building has 168 radiators manufactured by the Standard Radiator Company, of Buffalo, N. Y. These 168 radiators contain a total of 27,421 superficial feet of radiating surface.

To properly connect these boilers and radiators required 8,056 superficial feet of radiating surface, and other mains and risers of the building make a total of 35,477 feet of radiating surface attached to this system of Bolton heaters. The farthest radiator in each wing of the building is located 250 feet distant from the boiler including the main and riser to the third floor.

Figure 3 is a plan of the second floor, the different rooms being designated as follows: A, directress; B, bathroom; C, classroom; D, dormitory, E, mother superior's cabinet; F, assembly room; G, study; H H, etc., cells; I, linen closet; J, museum; K, academy parlor; L L L, special parlors; M, monastery parlor; N, bathroom; O O O, oratories; P, reception-room.

Figure 4 is a plan of the third floor with the rooms as follows: A, pharmacy; B, bathroom; C C, classrooms; D, dormitory; E E, children's infirmary; F, patients' dining room; G G, sacristies; H H, etc., cells; I, chapel; K, sisters' choir; L L, libraries; M M M, sisters' infirmaries; N, ante-choir; O, dining-room.

THE ENGINEERING RECORD'S

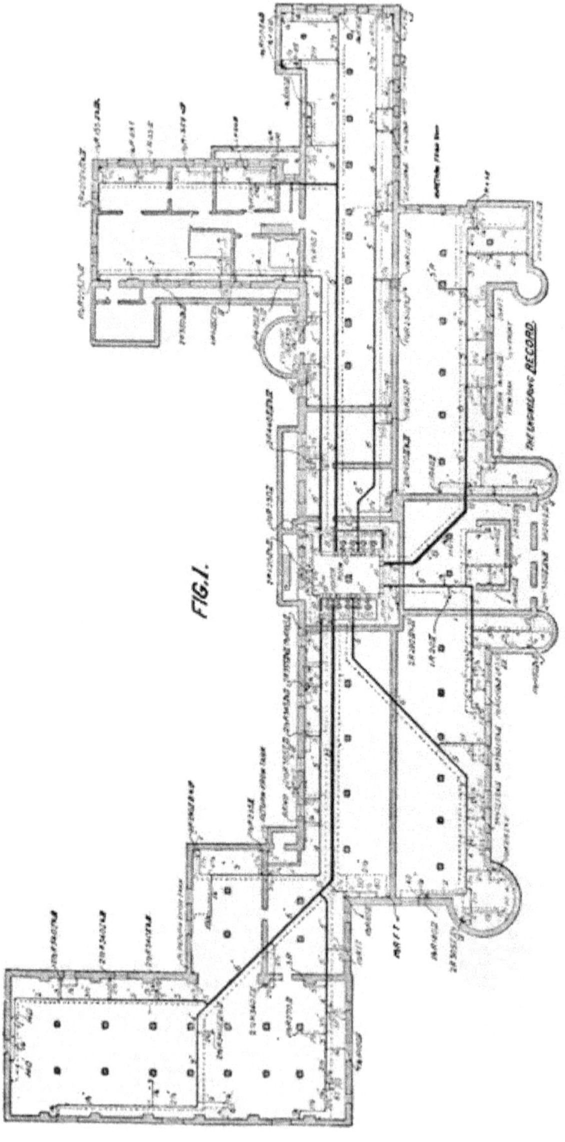

HOT-WATER HEATING OF THE CONVENT OF THE VISITATION, ST. LOUIS, MO.

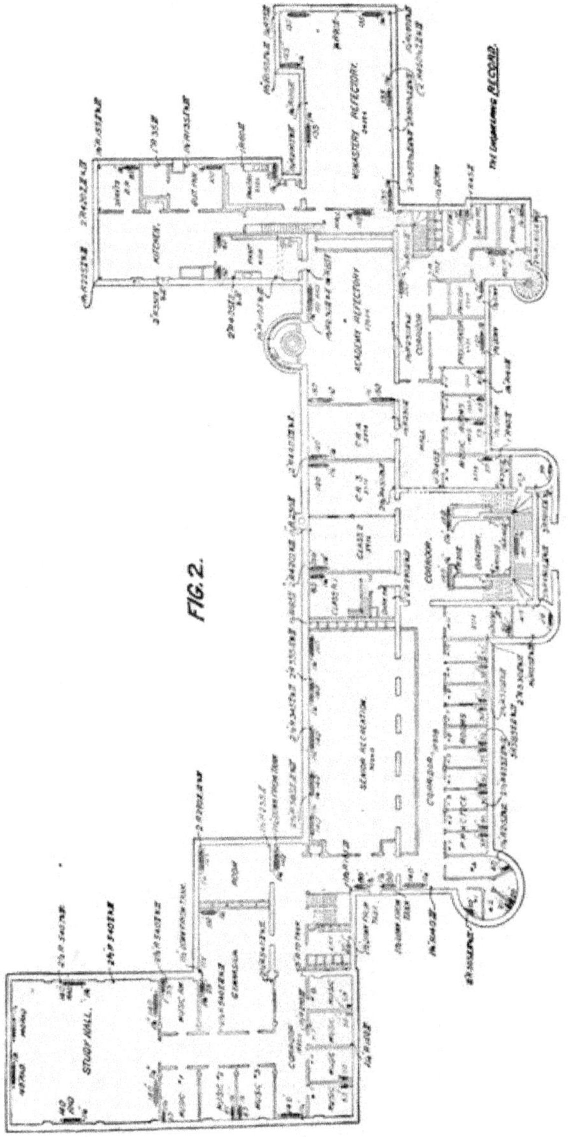
KITCHEN
ACADEMY REFECTORY
MONASTERY REFECTORY
SENIOR RECREATION
STUDY HALL
CORRIDOR
The Engineering RECORD

FIG. 2.

Figure 5 is a plan of the fourth floor. D, dormitory; H H, etc., cells; B, bathroom; Q, chapter-room; T, trunk-room; U, closet; V, locker-room; W, art-rooms; X, students' choir; Y, chapel gallery; Z, novitiates' room.

Figure 6 is a plan of the fifth-story rooms and fourth-story roof, and shows the position of expansion tanks and their branches to the risers from each distribution main line. In Figs. 3, 4, 5, and 6, the radiators are indicated by solid black rectangles, and the adjacent number shows their square feet of radiating surface. They are supplied from riser lines shown in Figs. 1 and 2, but omitted here for clearness. The figures written in the center of the rooms give their respective contents in cubic feet; for example, Room Q, Fig. 5, has a volume of 12,397 cubic feet.

PART III.—ARRANGEMENT OF BOILERS, SMOKE PIPES, FLOW AND RETURN PIPES, AND RADIATOR CONNECTIONS.

FIGURE 7 is a plan of the boiler-room drawn to a large scale. This shows two batteries of boilers set opposite to each other, shows the main 10-inch header and boiler connection, the chimney, which is 24x24 inches, and general arrangement of the piping immediately connected with the boilers in the boiler-room.

Figure 8 is a section of the boiler-room showing the front of the battery of five heaters. Each flow and return header in the boiler-room has a hot-water thermometer giving the temperature of the water in the main. There is also one altitude gauge in the boiler-room to notify of any change in the water level of the apparatus. The heaters are of the Bolton pattern.

The valves used on the main flow and return connections are flanged gate valves; radiator valves are Belknap's quick opening, and the valves are all turned with a key.

The smoke pipes from both batteries are as shown in the plan connected to enter the chimney with a united area of 576 square inches. The draft is governed by a check damper in smoke pipe, and the expansion and waste of water is provided for by two expansion tanks at O O, Fig. 6, at either extremity of

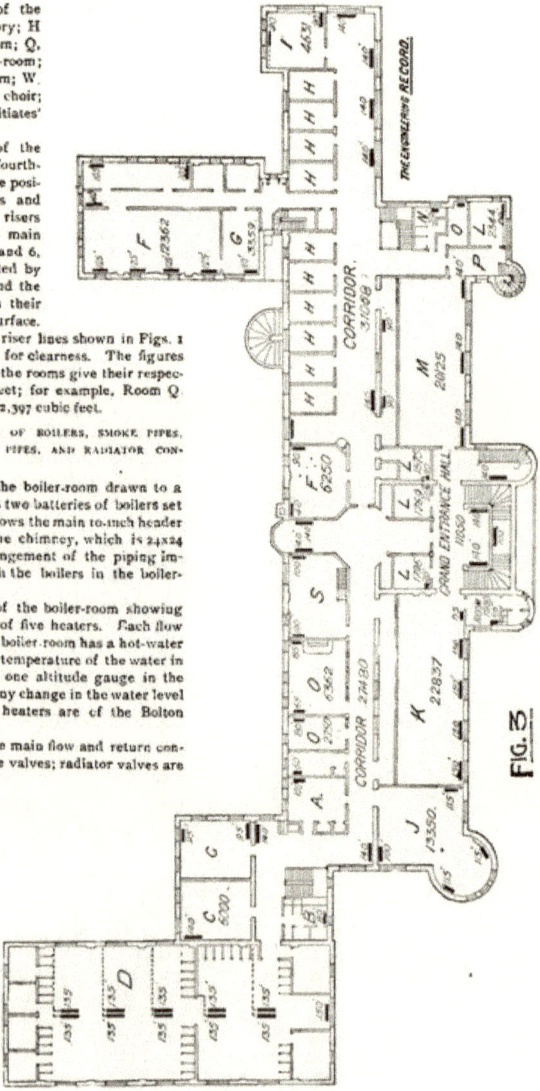

HOT-WATER HEATING OF THE CONVENT OF THE VISITATION, ST. LOUIS, MO.

the main building. These tanks, Fig. 9, are constructed of boiler-iron one-fourth inch thick, are 30 inches diameter by 36 inches high, and each tank has a 3 inch flow pipe from the horizontal main under the building and return pipes of smaller size. Each tank has 3-inch overflow extending through the roof, where the overflow reaches the conductor pipe. Each tank is also furnished with automatic valve for supplying the apparatus with water.

All overflow and return pipes are provided with valves in the boiler-room for shutting off each service without interfering with the main system, and admit of drawing off the water for any possible necessary repairs. It is claimed that this chimney for a boiler of other construction would have failed to have given draft for this amount of grate surface; but that with this peculiar boiler the draft from the chimney is all that is required. Expansion is provided for in the manner in which the branches are taken from the main.

All radiator connections are ground union. The radiators with but few exceptions are standard radi-

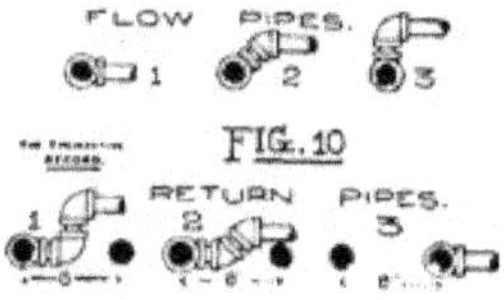

FIG. 10

ators, 37½ inches high. No attempt has been made at ventilation except what is secured by the transoms which are over nearly all the windows.

Figure 10 shows a card with the different methods of connecting branches to mains illustrated. These different connections are designated by a number. Accompanying each set of drawings which have been furnished by the Detroit Heating and Lighting Company, is one of these cards. Each tee connection in the piping system has a number shown on the drawing, which refers to this figure and illustrates to the fitter or others how the connection with the mains should be made in the system of piping. Thus Fig. 1 shows a branch taken off hori-

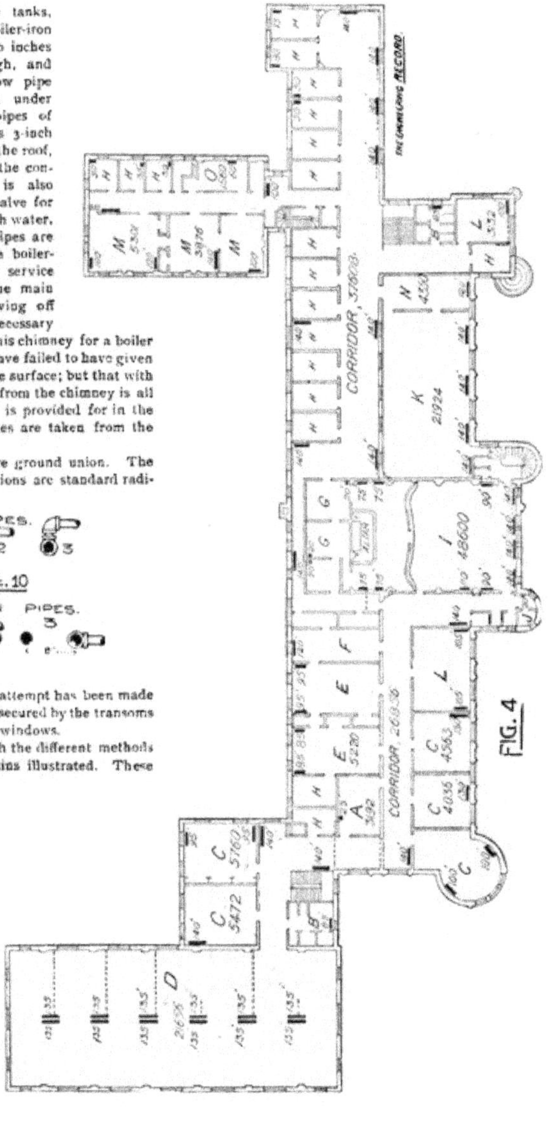

FIG. 4

zontally with the main. This connection would be used when a connection was to be taken off near the boiler or a rapid circulation not desired. Figure 2 shows a connection taken off the main at an angle of 45 degrees; this would increase the flow in the branches and would be used when at a distance from the boiler and a greater circulation was required than where Fig. 1 was used. Figure 3 shows a connection taken vertically from the main. This would give a strong circulation and would be used on large radiators or at a distance from the boiler. The three connections at the bottom of the cut of this figure show that the return connections should enter the main only on the side and the various ways of ac-

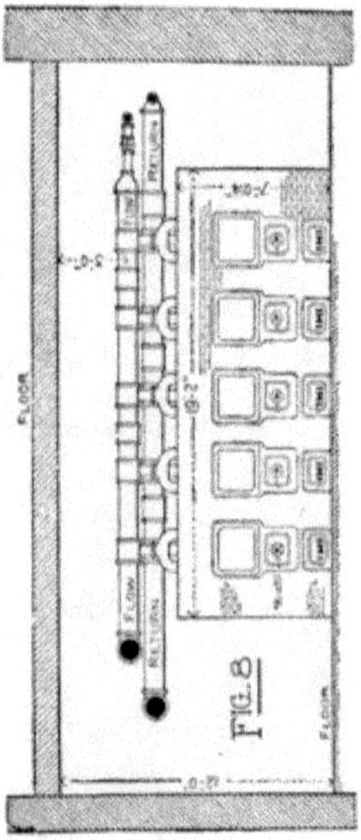

FIG. 8

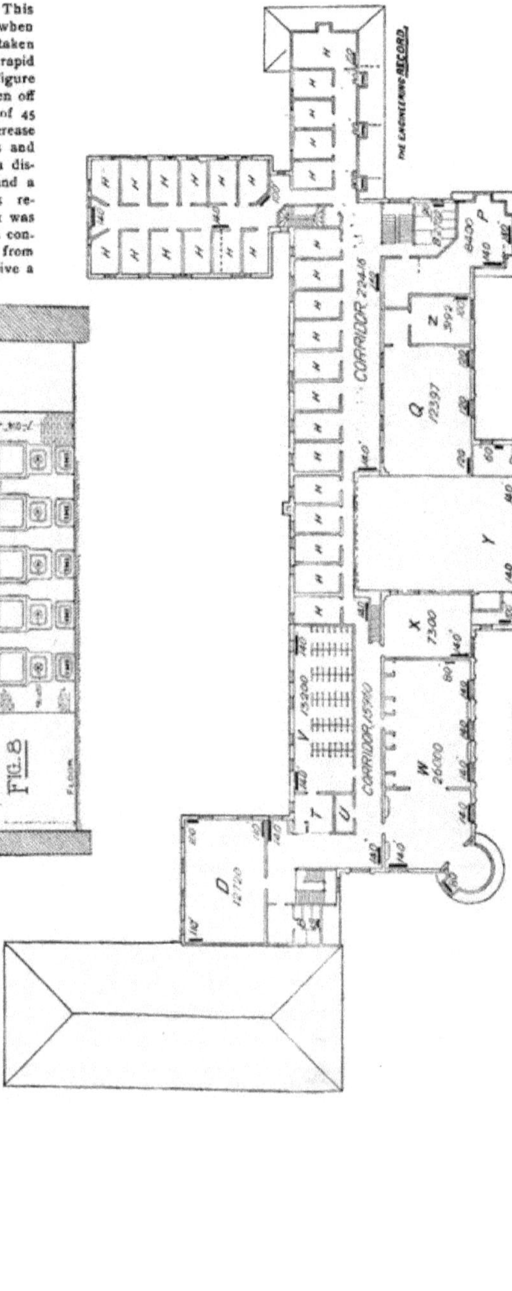

FIG. 5

HOT-WATER HEATING OF THE CONVENT OF THE VISITATION, ST. LOUIS, MO.

complishing that result. This simple card illustrates fully to the fitter how he should make his connection and has been found of great advantage to this company since its adoption.

No special fittings were used in the introduction of this heating apparatus, the ordinary fittings being used with the illustration shown by this figure.

The material used in heating this building aggregated be-

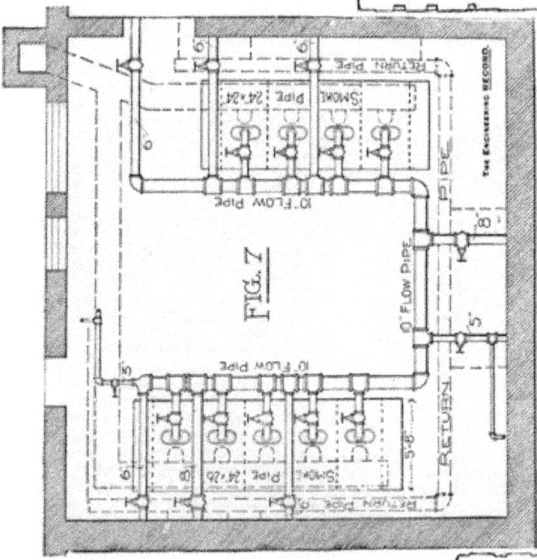

FIG. 7

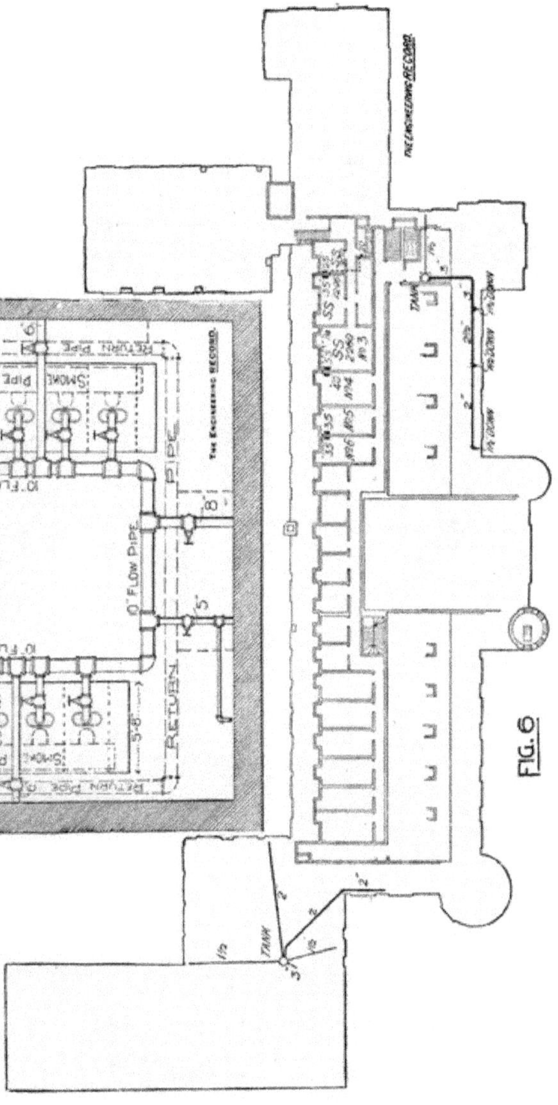

FIG. 6

tween 10 and 12 carloads. The total contract for heating the building amounted to about $25,000. The ventilation is accomplished by the use of transoms at the top of nearly every window in the entire building. The boilers were installed by Messrs. Brislin & Sheble, of St. Louis, from plans furnished by the Detroit Heating and Lighting Company, executed under the direction of their heating

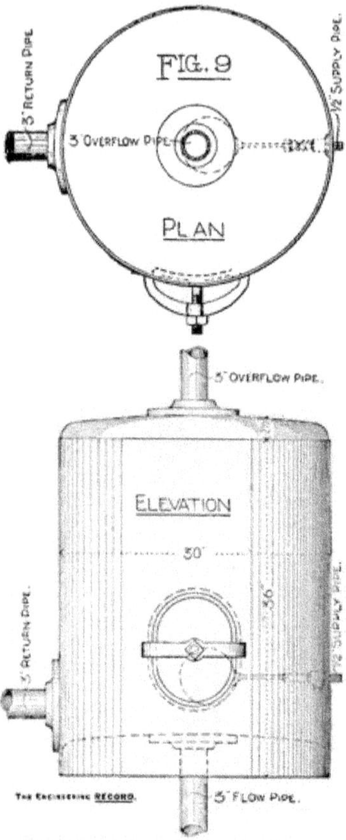

HOT-WATER HEATING IN THE CONVENT OF THE VISITATION, ST. LOUIS, MO.

engineer, Mr. L. S. Daniels. The building was designed by Architects Barnett & Haynes, of St. Louis, and built by General Contractors B. Webber & Son, also of St. Louis.

WARMING AND VENTILATING IN THE COLLEGE OF PHYSICIANS AND SURGEONS, NEW YORK.

Part I.

The steam heating and ventilating apparatus in the building of the College of Physicians and Surgeons, at the corner of Fifty-ninth Street and Tenth Avenue, New York City, embraces features of more than ordinary interest, representing probably the first practical application of what is known as the twin duct system. In this system, as devised by William J. Baldwin, of New York, independent branches of hot and cold air ducts are provided for each room of the building, the hot-air supply being taken from central heating chambers, and the temperature of each room being controllable without affecting the temperature of the air supply entering any other room in the building. In this last respect the system secures the same result realized by the plan of having mixed chambers and valves in combination with independent stacks of radiators for each vertical flue, with this difference, that in the system here described the occupant of a room controls his own heat, no matter how far from the source of supply, without assistance from the engineer or janitor. Where, however, a building covering considerable area is to be heated, the item of cost attached to putting in a large number of such independent radiator stacks becomes considerable, and in such a case the double duct system would seem to commend itself favorably. In the system, as applied in the present instance, there are, moreover, in connection with the hot-air or steam coil chambers, separate chambers containing coils for removing the chill from the entering fresh air. The air supply which passes through the cold-air ducts and through the main heating coils into the hot-air ducts is therefore never cold in the ordinary accepted sense of the word, and hence in the illustrations published the ducts marked "warm" correspond really to cold-air ducts. The illustrations show a general cellar plan and a vertical longitudinal section through the middle building of the group of three, and show the main features of the whole system.

There are, in all, five horizontal tubular boilers which furnish steam for all purposes in this and adjacent buildings. Two 6"x4"x6" duplex pumps, marked P P, serve for pumping water to the top of the house and for feeding the boilers, and are accordingly arranged for pumping hot water which is drawn from the tank into which all water of condensation is discharged. The pumps are arranged so that they can be operated direct by throttle valves, or independently by an automatic pump governor. Four 6x6-inch New York Safety Steam Power Company vertical engines E are used to drive the fans F, each engine being connected with its respective fan in the manner shown. The fans, four in number, supplied by the Nason Manufacturing Company, of New York, are each 6 feet in diameter, and from 1,000,000 to 1,250,000 cubic feet capacity each, according to the speed at which they are run. All the engines and pumps exhaust into a 5-inch main exhaust pipe which extends to the roof of the building. A branch from this pipe leads to a feed-water heater and back again into the pipe. Another branch, 5 inches in diameter, supplies exhaust steam to the coils in the casing A, A', B, C, D, G, and L, after the steam has been passed through a "skimming tank," where the oil from the engines is separated from it. These coils are of the gridiron type, made up of a number of sections each. The sections have an average length of 10 feet, not including the spring pieces which measure about 2 feet. The pipes of the coils are covered with secondary wire surface made of No. 14 square iron wire, and known as Gold's compound coil surface. The coil stands are made of pipes and fittings, and are so arranged that each section can be drawn out for repairs without disturbing those others. Each section, moreover, is fed separately by a 2-inch steam pipe with a valve in the engine-room and the return pipe also come separately into the engine-room. The coils are inclosed in galvanized-iron casings, open at the bottom, and connecting with the air duct at the top, as shown more clearly in the vertical section. Swinging dampers are arranged in the bottoms and near the tops of these casings, so that a mixture of hot and relatively cold air may enter the distributing ducts, the proportions being readily controlled by opening or closing the different dampers.

In front of the four fresh-air inlet windows primary steam coils S, supplied with live steam, are set up. These coils aggregate about 1,600 lineal feet of 1-inch pipe, covered with secondary wire surface, as in the case of the main heating coils. The entering air, which thus receives what may be called an initial heating, reaches the settling chambers, marked on the plan, and from these is drawn by the fans into the four heating chambers, containing the coils and castings A, A', B, C, D, L, and G. It is well to explain here that the coils B and G are not ordinarily supplied with steam, but are designed to be used as substitutes for the coils A and D in case of repairs, in which case the "warm" duct would be used as the "hot" one. It should be remembered also that the air ducts from A and B, and from D and G together lead to twin flues and discharge into the same rooms. The air which passes into the casings B and G, therefore, is not further heated, but is simply at that comparatively low temperature which has been imparted to it in passing through the primary coils. The air which passes through the casings A and D, on the other hand, is further heated to the much higher temperature due to the hot coils within. The hot and warm air supplies from the casings A and B and D and G discharge into the same register boxes, as already intimated, and the proportions of each may be varied by special slide valves at the register face. These valves or dampers are so constructed that they will allow the air from one of the twin flues to escape separately, or admit a part of the air from each flue, one flue opening in the proportion the other is closed. Two streams of air of different temperatures may thus be admitted to the register box, where they mix and thence pass through the register face, and at the same time is beyond the power of the occupants of the rooms to shut off both pipes at the same time. The ducts leading to the dissecting-room, amphitheater, and lecture-room, from the casings C, A', and L, supply only warm air, the temperature of which is regulated by the engineer, the double-duct system not being used for these. The lecture-room duct, as shown in the vertical section, discharges into the space under the seats and from there the warm air issues into the room through openings in front of each row of seats. The warm-air discharge into the amphitheater is effected in the same way. For the dissecting-room a special arrangement of discharge was adopted.

The architect is W. Wheeler Smith, of New York.

PART II.

In heating the dissecting-room the double-duct system is not used, heat and ventilation being se-

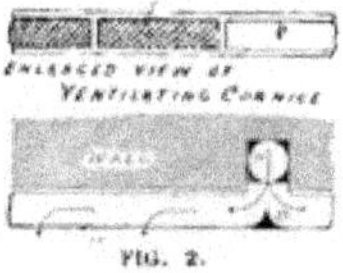

FIG. 2.

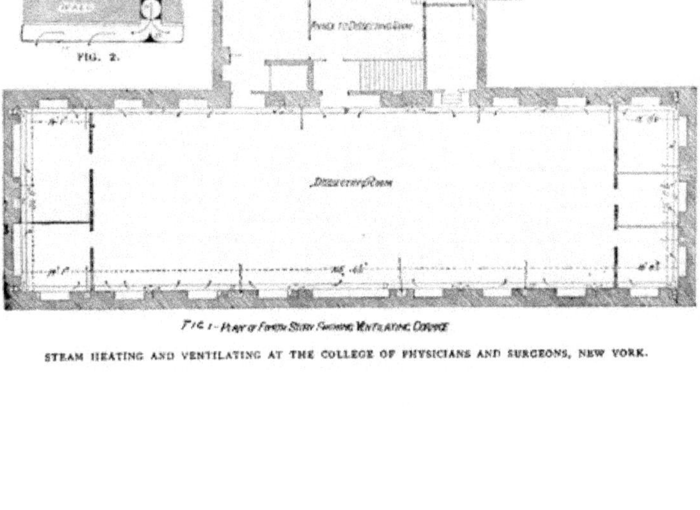

FIG. 1 - PLAN OF FOURTH STORY SHOWING VENTILATING CORNICE

STEAM HEATING AND VENTILATING AT THE COLLEGE OF PHYSICIANS AND SURGEONS, NEW YORK.

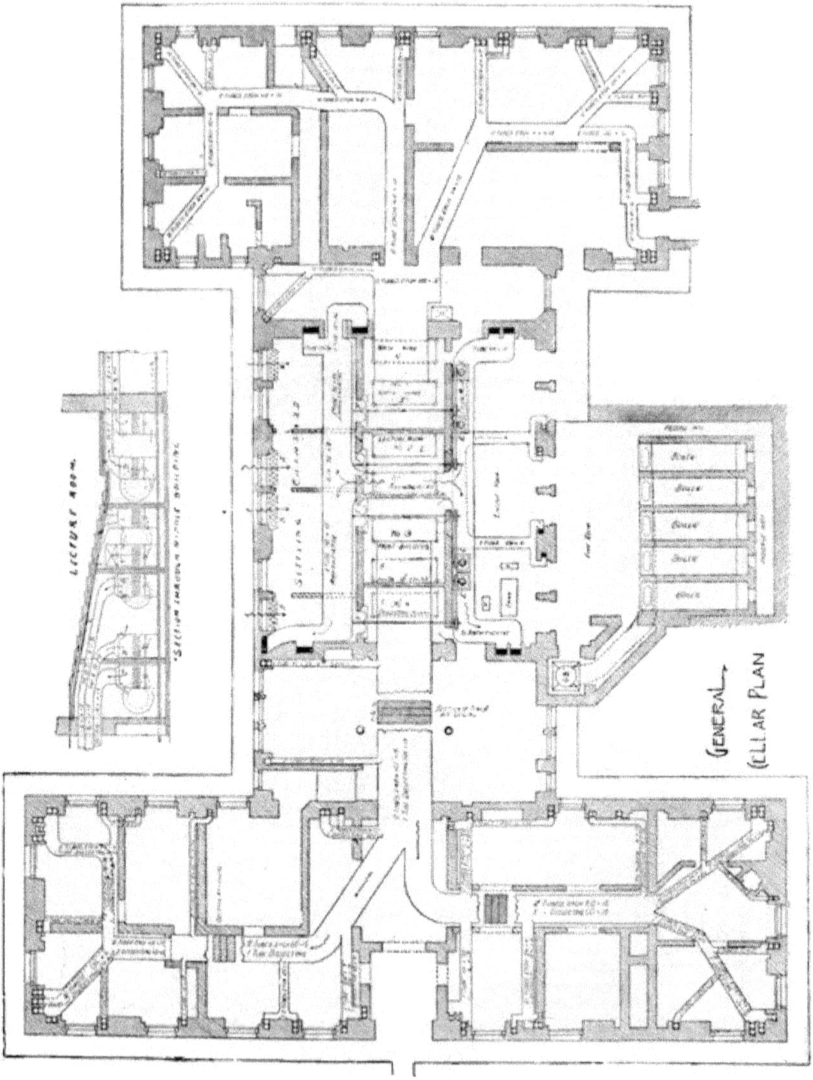
GENERAL CELLAR PLAN
LECTURE ROOM
"SECTION THROUGH MIDDLE BUILDING"

cured by the arrangemen. illustrated in the accompanying illustrations.

The hot-air flues marked H in the plan of the dissecting-room leading from below have no registers, the flue openings connecting with 9x12-inch galvanized iron boxes or cornices which run around the room so as to form a continuous box in appearance. Internal partitions P, Fig. 1, however, divide it up into sections so that the air from one flue cannot pass into the section connected with another flue. The front side of the box or cornice is made with 8-inch ornamental panels, those marked P, immediately opposite the openings of the flues H, being solid, while those at the sides, as P', shown in the enlarged view of the cornice, Fig. 2, being open and provided with wire netting. In the interior of the cornice, moreover, and opposite every flue opening, a deflector D is arranged so that the hot air readily passes to the right and left along the line of the cornice and escapes into the room through the open panels, as indicated by arrows. The office of the blank panel P is simply to prevent the direct flow of hot air from the flue H into the room which would be detrimental to uniform diffusion without drafts. The whole cornice is fastened to the walls and ceiling with plugs and screws.

For the separate ventilation of the dissecting-room, a matter of prime importance, special ventilating flues and ducts were provided. The plan of the space between the ceiling and roof, Fig. 3, explains the nature of the arrangement adopted. The hot air coming from the heating or ventilating cornice, as it is called, ascends, and, coming in contact with the relatively cold glass roof, falls to the floor level, at which the mouths of the various ventilating flues open. These are marked V in the plan of the dissecting-room, and are shown with arrows leading into them. They are shown also in the vertical section through the dissecting-room, together with their connection with the overhead ventilating ducts V. These are arranged in the space between the ceiling and roof, and are made of galvanized iron. The ventilating flues for the dissecting-room are shown black in Fig. 3, the ducts being clearly outlined, and connecting through a main duct with one compartment of the aspirating shaft, shown at the upper right-hand corner. The boiler furnace flue A is carried up inside of this shaft.

It will be noted that besides the black ventilating flues a large number of other flues are shown in the walls, Fig. 3. These flues are for ventilating rooms on the lower floors, and discharge directly into the roof space, from which the foul air passes into one of the other compartments of the aspirating shaft.

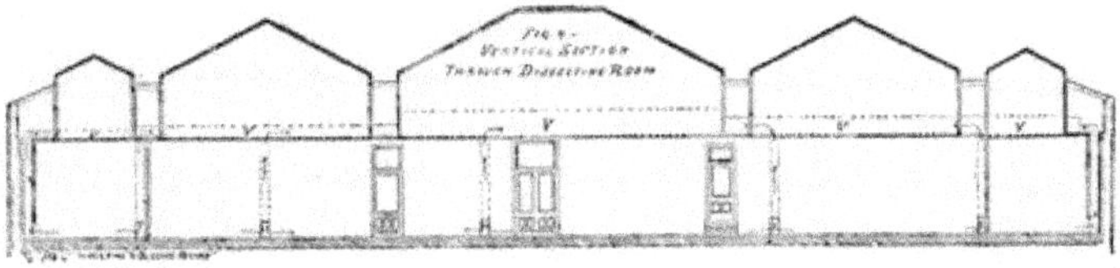

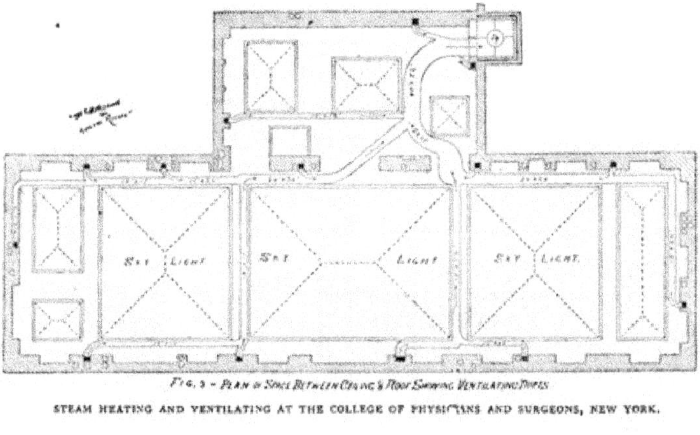

FIG. 3 — PLAN OF SPACE BETWEEN CEILING & ROOF SHOWING VENTILATING PIPES

STEAM HEATING AND VENTILATING AT THE COLLEGE OF PHYSICIANS AND SURGEONS, NEW YORK.

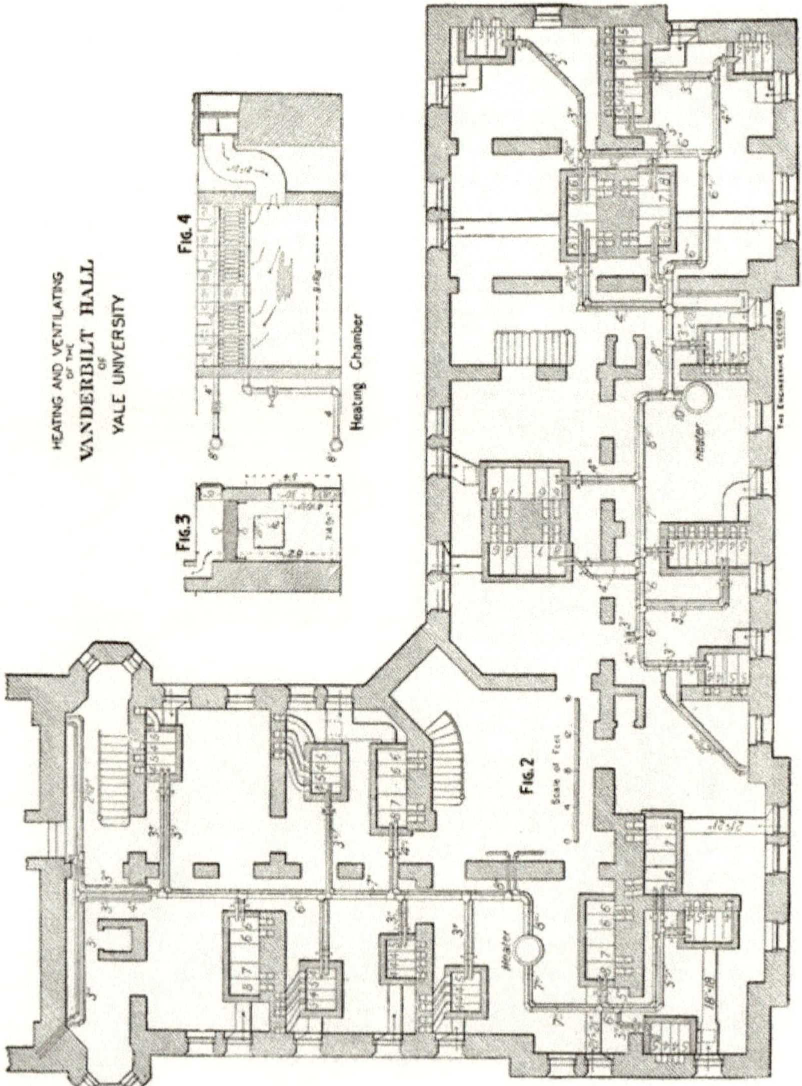
HEATING AND VENTILATING OF THE VANDERBILT HALL OF YALE UNIVERSITY
FIG. 3
FIG. 4
Heating Chamber
FIG. 2
Scale of Feet
Heater
Heater

HEATING AND VENTILATING OF VANDERBILT HALL, YALE COLLEGE.

The new Vanderbilt Hall at Yale College is a large four-story E-shaped building, inclosing a central court or entrance. The building, including the court, occupies an area about 180 feet across the front by 116 feet deep. Its purpose is to afford lodging for students of the college. Charles C. Haight, of New York City, was the architect of the building, and William J. Baldwin, of the same city, the contracting engineer for the heating and ventilating plant.

The general plan of the building is shown by Fig. 1, a plan of the first floor, but in order to present more clearly the heating apparatus in the basement, Fig. 2 shows only one-half of the building, the other half being identical. The building is warmed by the indirect system, hot water being the means of conveying the heat to the indirect stacks placed in various parts of the basement. The water is heated by steam in four ordinary feed-water heaters. The steam is generated in an adjacent central plant, the steam entering the building by a main pipe through the wall of the building to a main stop valve on the inside, and from this stop valve it flows through different pipes to the heaters. Each pipe has its corresponding return pipe, and both are so fitted with valves that each section may be controlled separately and independently from the others. The heaters are the Wainwright water-tube type, and each is rated at 300 horse-power. Each heater supplies the indirect stacks in its immediate vicinity, this subdivision being made to obviate the use of long pipe lines and the consequent chance of a feeble circulation in some of the further radiators.

The building contains 19,136 square feet of indirect radiating surface, distributed as shown on the plans. There are also about 600 square feet of direct radiation in the main halls on the first floor. The location of the heating chambers is shown on the general plan, Fig. 2, and in section by Figs. 3 and 4. They consist, as will be seen, of brick chambers built as near as possible to the base of the clusters of vertical flues leading to the different rooms. The air enters the heating chamber from outdoors by means of a galvanized-iron duct connecting the chamber with the nearest window. The duct contains a sliding damper to cut off the supply of air when necessary.

The indirect radiators are of the Gold pin drum section type, 10 to 12 inches deep, each section containing about 16 square feet of surface. These radiators are in clusters of four or five in brick heating chambers in the basement, each heating chamber containing as many clusters as there are vertical ducts leading from it to the rooms above. Each cluster is separated from the adjoining ones by a galvanized-iron partition suspended from the ceiling. The radiators are supported on I beams resting at each end in the brick wall of the heating chamber. The flow and return pipes run lengthwise in the heating chamber, the former above the radiators and the latter below it, both being connected to the radiators by short nipples with lock nuts. The figures shown on the plan, Fig. 2, show how many radiators are placed in the cluster warming the air for the adjacent duct.

The plan of the different floors is almost the same as the first. Each suite in the dormitory contains a sitting-room and two bedrooms. The former on the first floor contain about 2,800 cubic feet, and on the floors above about 2,400 cubic feet. There is about 1 square foot of indirect radiating surface to every 20 cubic feet of space in the building. The bedrooms on the first floor contain 1,400 cubic feet and on the floors above about 1,280 cubic feet. For the larger rooms the ducts leading to the first floor are 8x16 inches with 8x18-inch registers, and those to the floors

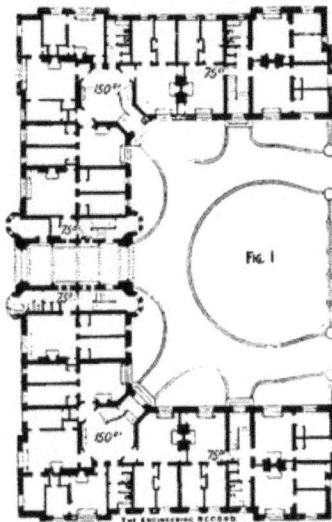

above 8x12 inches in area with 8x15-inch registers. The smaller rooms on the first floor have 8x12-inch ducts with 8x15 inch registers, and for the upper floors 8x8 inches with 8x12-inch registers.

Each room has two registers for drawing off the foul air, one at the ceiling, which can be opened or closed by the occupants of the rooms, and one at the floor, which is always open. The two registers in each room are connected to a common ventilating duct which leads to a large chamber immediately under the roof, which is divided into four sections. The foul air escapes from each of these sections by a ventilating shaft leading to the outer air. A high-pressure steam coil is placed in each shaft to stimulate the draft.

The connection to each indirect radiator or cluster and all branch mains has the full area of 1.4 square inches to each 100 square feet of radiating surface, the return having the same size. All of the warm air flues are lined with No. 26 galvanized iron, and the backs or side of the lining towards the outside walls of the building are doubled with a ⅞-inch pine board between the flue and its back.

Each of the four principal divisions of the heating system is furnished with an expansion tank capable of holding one-twentieth of all the water in the section. Each is provided with an expansion and overflow pipe. Temperature gauges are placed on both the flow and return pipes near the heaters, so that the loss of heat between the flow and return may be ascertained.

STEAM-HEATING PLANT IN THE HILL SEMINARY.

By a recent donation of James J. Hill a Roman Catholic theological seminary has been constructed in the suburbs of St. Paul, Minn., under the designs and specifications of Cass Gilbert, architect, of St. Paul, Minn. From some of his drawings, sketches, and memoranda we have prepared the following account of the low pressure steam-heating system that was installed by Archambo & Morse, Minneapolis, Mr. H. P. Blair, of Chicago, being the consulting engineer for the heating work. The seminary buildings, at present six in number, are mostly five stories in height, substantially built of stone and brick, and each covering on an average about 8,000 square feet of ground area. The buildings are so planned that each room receives direct sunlight. The site is an abrupt bluff 110 feet above the Mississippi River and about 1,200 feet distant from it on a wet clay soil. Stone and brick tunnels connect the buildings as shown on the map, Fig. 1, through which the heating pipes, electric wires, speaking tubes, water, gas, and sewer pipes are carried. The sewer and rainwater connections to the main sewer are all separate from the seepage drains. It was necessary to drain the site, and the soil is accordingly pierced by a complete system of seepage drains through which the spring and surface water is carried to a ravine leading to the river, thus providing a constant flowing brook of clear water, which falls over the bluff to the river in a beautiful little cascade, adding greatly to the beauty of the grounds.

In the basement of the gymnasium building there are two horizontal tubular boilers, each 72 inches in diameter, 18 feet long, having 84 tubes, 4 inches in diameter, 18 feet long, and operating the entire heating system. The system is a low-pressure apparatus, and principally, as far as the individual buildings are concerned, with two-pipe mains and one pipe risers, with drip from each riser to return main. The whole is so arranged as to be under the immediate control of the engineer stationed in the engine-room; this being accomplished by having separate mains for each separate building, both supplies and returns, with the valves in the tank-room readily accessible for controlling the apparatus in each building separately. The aggregate radiation consists of five coils and 397 direct radiators, containing in all 16,750 square feet of radiating surface. Each radiator is of the "Peerless," "Perfection," "Ideal," or the

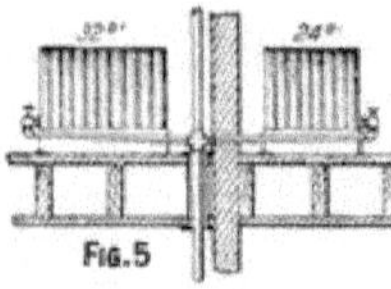

FIG. 5

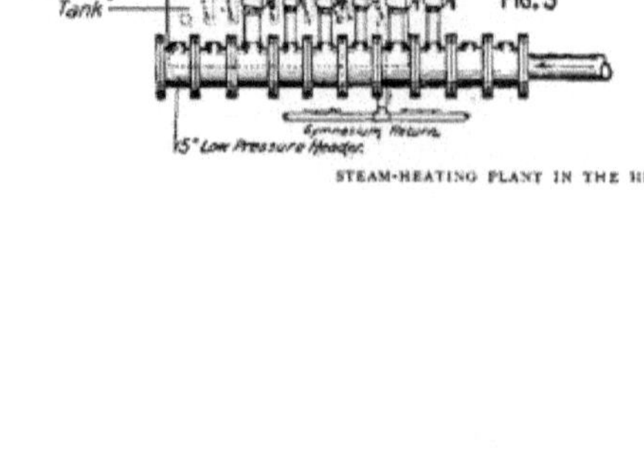

FIG. 3

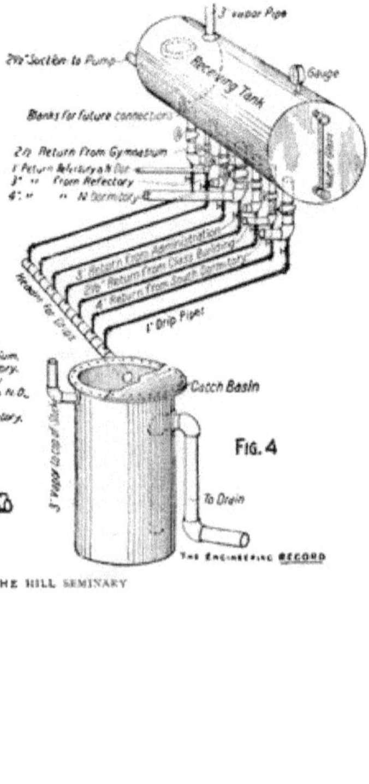

FIG. 4

THE ENGINEERING RECORD

STEAM-HEATING PLANT IN THE HILL SEMINARY

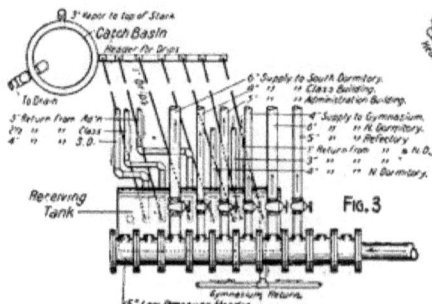

"National" iron vertical loop or tube pattern, and they are connected on what is known as the "single-pipe system," with but one valve to each radiator. Plugged outlets are left in the tops of the risers in the administration building sufficiently large to supply radiators that may hereafter be placed in the attic of that building.

The sizes of the supplies and returns for each section are as follows:

	Supply.	Returns.
Gynasium building	4-inch.	2½ inch.
Dormitory No. 1, or south dormitory	6-inch.	4-inch.
Classroom building	4-inch.	2½-inch.
Administration building	5-inch	3 inch.
Refectory building	5-inch.	3 inch.
Dormitory No. 2, or north dormitory.	6-inch.	4 inch.

Additional provision is also made for heating the chapel, which is to be built at some future time. Connections throughout were uniformly designed with 1-inch pipes to supply radiators of 30 square feet and under, 1¼-inch pipe to supply radiators of 30 to 60 square feet, 1½-inch pipe to supply radiators of 60 to 150 square feet, and 2-inch pipe to supply radiators of 150 to 250 square feet. The return pipes are all led to a receiving tank underneath the supply header, and are valved and dripped into a cesspool that overflows into the main sewer. The contents of the receiving tank, which is open to the atmosphere above the roof, are pumped out periodically into the boiler feed pipes. All heating pipes, both supplies and returns, are covered where they run through the ducts and basements from header to tank, and to the risers and stacks, with H. W. Johns Manufacturing Company's asbestos covering, over which is pasted No. 6 best canvas.

The radiators were disposed as follows:

Name of Room.	Number of Radiators.	Square Feet of Surface.
Gynasium	29 radiators.	4,520
Refectory	26 radiators.	2,440
Dormitory No. 1	148 radiators.	4,220
Dormitory No. 2	145 radiators.	4,230
Class building	7 radiators. 5 coils.	1,670
Chapel	Not yet built.	....
Administration building	51 radiators.	2,690
Total	397 radiators. 5 coils.	16,750 sq. ft.

The contractors guaranteed: First, that steam shall circulate freely through all the pipes and radiators in the whole apparatus, and at the same time with one pound of pressure at the low-pressure header. Second, that the apparatus shall work noiselessly under any pressure which becomes necessary to carry. Third, that the apparatus shall be of ample capacity to heat all the rooms with which it is connected, to the following temperature, when the thermometer outside registers 30 degrees below zero, or warmer, with a pressure of steam not exceeding five pounds at the low-pressure header.

It was required to heat the different rooms as follows: Halls and corridors to 60° Fahr., living and sleeping rooms to 70° Fahr., classrooms to 70° Fahr., office room to 70° Fahr., infirmaries to 75° Fahr., other rooms to 70° Fahr. Hot-water generater tanks for the house service were placed in the several buildings, and heated by high-pressure pipes carried direct to them.

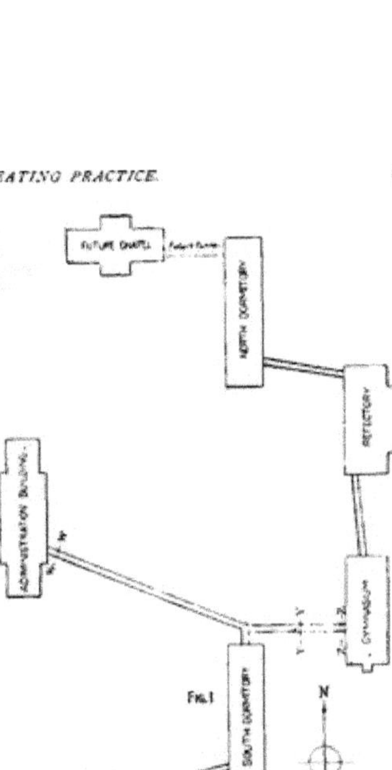

FIG. 1

Figure 2 is a basement plan of the south dormitory showing the arrangement of boilers and general heating plant, and headers, receivers, etc., for the system of steam distribution to the group of buildings. The supply and return mains for the different riser lines serving the groups of radiators in this building are also shown in full and broken lines respectively.

Figure 3 is an enlarged plan at A, Fig. 2, showing connections of distribution mains whose service to the different buildings is under the direct control of the engineman. The header B is made with 10 18-inch special flanged manifolds or tees, four of which are plugged to allow for future connections.

Figure 4 is an isometric sketch showing connections of receiving tank, drips, and cesspool.

Figure 5 shows connections of pairs of radiators to the risers in classrooms, dormitories etc.

Figure 6 is a cross-section of the trench in the basement of the refectory to carry the mains from the north dormitory to the gymnasium. The trench in the south dormitory in which the pipes from the classroom are laid, and the trench in the gymnasium in which the pipes leading from the refectory to the south dormitory are laid, are similar to this figure, except that the south dormitory trench is 18 inches wide and 16 inches deep, and has no cover-plates,

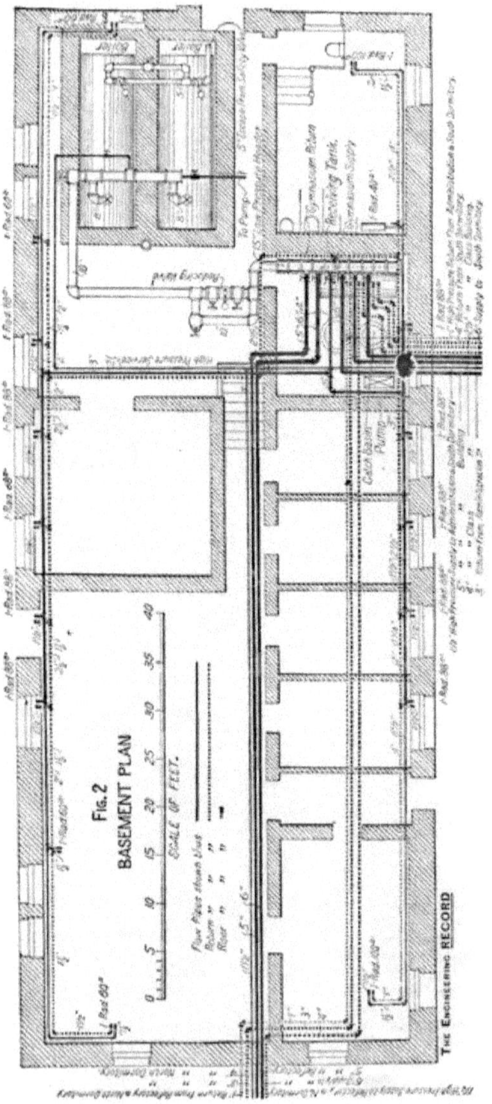

STEAM-HEATING PLANT IN THE HILL SEMINARY, ST. PAUL, MINN.

and the gymnasium trench is 18 inches wide and 21 inches deep. The iron cover is made in 4-foot sections, with studded upper surface and countersunk iron lifting ring.

Figure 7 is a section of the tunnel at Z Z. Fig. 1. Figure 8 is a corresponding section at X X. Figure 9 is a section of the tunnel at Y Y, and the other tunnels are similar to it with slightly varied dimensions.

In construction a concrete footing was first laid in the tunnel trench, and its upper surface was asphalted one-fourth inch thick. Then the stone walls were laid in lime mortar with wrought-iron pipe bars B B, 2 inches in diameter, built in and fitted with cast spools to carry the pipes and facilitate their expansion movements. The roof arches were brick, laid in Louisville cement, and the outside and top surface was entirely covered with a ¼-inch coat of asphalt applied in several coats so as to securely prevent the percolation of surface water through the masonry and to thoroughly water-proof it. The bottom of the tunnel was finished with a 1½-inch Portland cement mortar floor, pitched 1 in 12 to the center line transversely, and graded 1 in 100 longitudinally to trapped strainers discharging to the seepage drain pipes. The trenches were all excavated 2 feet wider than the tunnel itself so as to allow room to asphalt the outside vertical walls.

Figure 10 is a view of the tunnel at W W, Fig. 1, showing the manner in which the longitudinal movement of pipes about 200 feet long is provided for, each joint S S, etc., being intended to act as a swivel.

Figure 11 shows a somewhat simpler connection at the class building, where the main M drops from the low-pressure header, and there was sufficient vertical room to permit the 10-foot section A to swing like a pendulum by twisting on its joints S S.

In the tunnels shown in Figs. 9, 10, and 11 there is a line of 4-inch soft-tile horizontal seepage drain pipe laid outside the foot of each vertical wall, and in the section shown in Fig. 7 there is one such pipe at the offset in the middle of each wall and two at the foot.

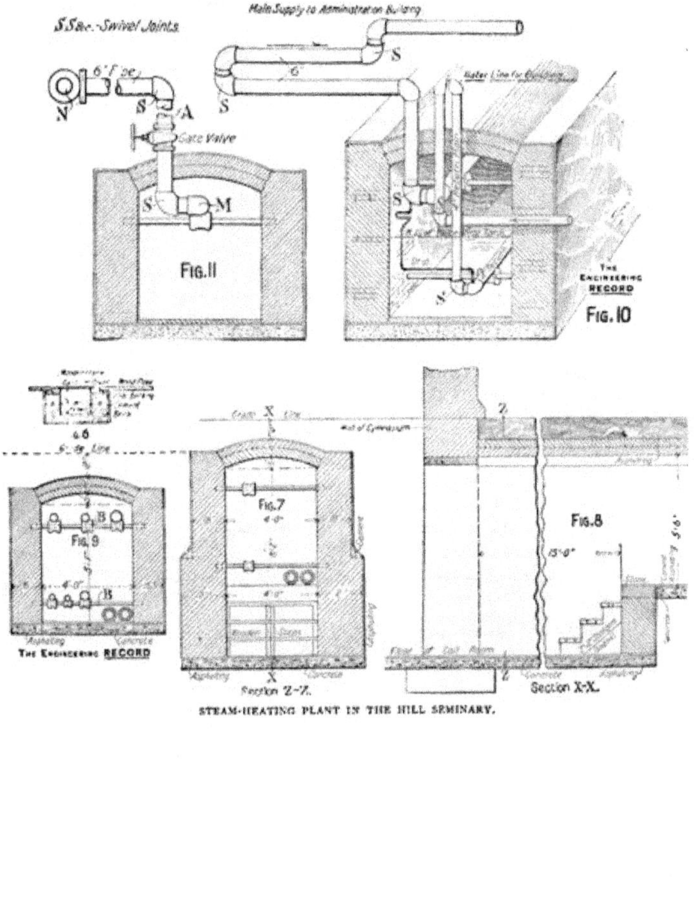

STEAM-HEATING PLANT IN THE HILL SEMINARY.

The total cost of the heating work was about $17,000 and of the tunnels about $4,500. The latter were constructed by James Carlisle & Sons. The contract for the heating work was awarded June 7, 1894, and it was practically completed in about four months.

VENTILATION AND HEATING OF A SCHOOL BUILDING.

The accompanying drawings show the method of heating and ventilating the Medalia High School, from plans made by the architect of the building, Mr. Walter S. Pardee, of Minneapolis, Minn., who, beside laying out the structural details of the building, designed its heating system. The school building covers an area 80x80 feet and is built of Menominee pressed brick, with Kasota stone trimmings.

Figure 1 shows a plan of the basement, which contains rooms for manual training and the toilet-rooms for the pupils. The boiler-room is also on the basement floor and several feet below the same to insure good working of the steam plant. The schoolrooms are all on the first and second floors of the building, and as these are so much alike a plan of but one of them is shown by Fig. 2. This shows the location of the registers and the cubical contents of each room, etc. The cubical contents of each floor is about the same as the one given.

The attic only contains the system of ventilating ducts that are shown in Fig. 3, but is planned for use as a public hall whenever occasion requires.

Turning to the heating and ventilating system we find that the fresh air is taken in at the top of the building through the duct A, shown on the attic plan. The cold air then descends to the base-

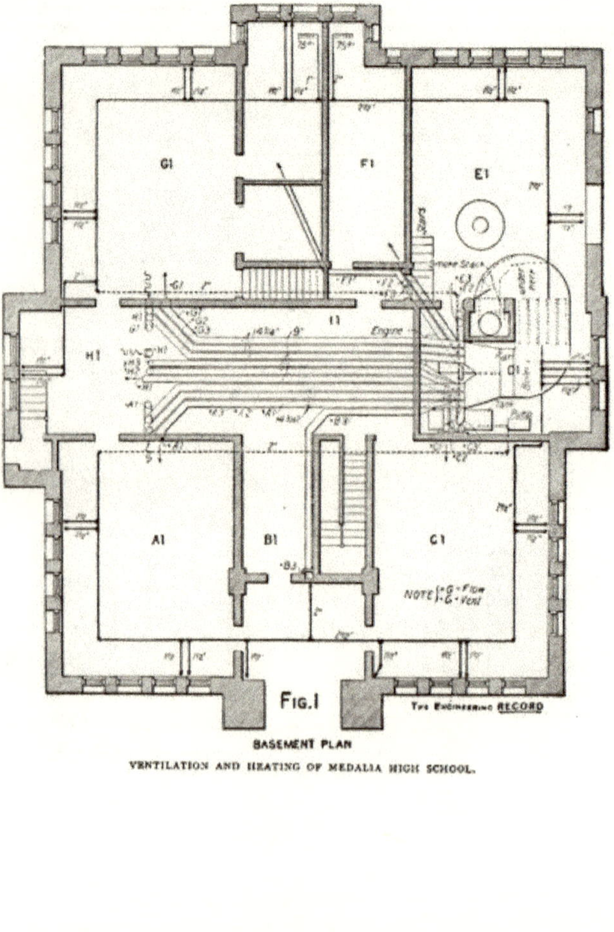

BASEMENT PLAN

VENTILATION AND HEATING OF MEDALIA HIGH SCHOOL.

ment through a 5'3"x3'6" brick shaft, which contains a sheet-iron flue from the boiler 30 inches in diameter. The cold air is slightly tempered by this flue by the time it reaches the basement. There a sheet-iron duct leads the air to heating coils consisting of 1,680 square feet of heating surface divided up among five independent coils, each supplied with steam through its own independent connections.

Beyond the coils is a 6-foot fan driven at any speed from 250 to 350 revolutions per minute, as the case requires, by a 15 horse-power Atlas engine. The fan at 350 turns per minute is capable of furnishing 54,000 cubic feet of air per minute. This same engine drives by a belt a 6-foot exhaust fan located in the attic running at about the same rate of speed as the supply fan. A number of circular sheet-iron ducts lead from the fan chamber both to the rooms in the basement and to the classrooms above.

To aid in tracing the run of the ducts the various rooms in the basement have been marked A_1, B_1, C_1, etc., and the corresponding rooms on the main floor A_2, B_2, C_2, etc., and those on the next story A_3, B_3, C_3. The duct supplying air to these rooms is then marked $+A_1$, $+A_2$, etc., and the ducts venting them $-A_1$, $-A_2$, etc., the plus sign indicating the supply and the minus sign the vent. The attic floor shows the location of the various vent flues, and the circular ducts which, connected to their upper ends, converge to an exhaust fan of the

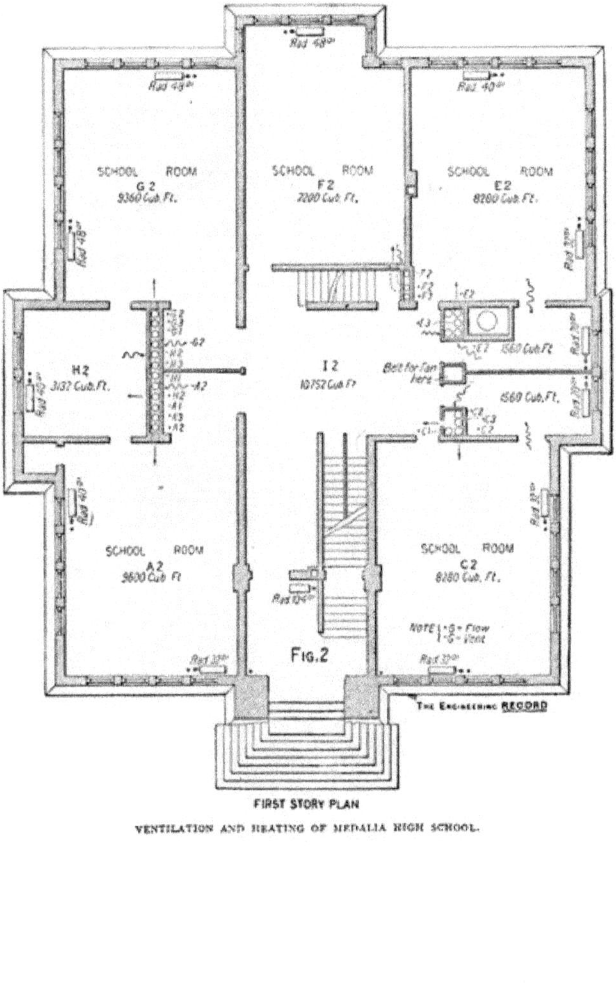

FIRST STORY PLAN

VENTILATION AND HEATING OF MEDALIA HIGH SCHOOL.

same size as the one in the basement. When the building is in use the air is discharged through the duct B, but when not occupied and it is desired to keep the building warm, the switches D and C are so adjusted that the air, instead of being discharged out-of-doors through D goes through the by-pass E down the shaft to the coils in the basement, and thus maintains an internal circulation. The indirect coils are not of course of sufficient capacity to warm the building in cold weather, and so some 896 square feet in direct radiation is provided. One return-tubular boiler, 14 feet in length and 36 inches in diameter, containing 34 3½-inch tubes, supplies steam at 25 to 30 pounds pressure for the fan engines, and when needed, steam at reduced pressure through a reducing valve to the heating system, should the fans be not running, or if the steam exhausted by the fan engine should be insufficient to meet the demands of the heating system. The returns of the direct-heating system are carried back to a return tank 36 inches in diameter and 8 feet long. As the indirect coils are sometimes supplied with steam at full boiler pressure, the condensation from them is led to a steam trap, which then discharges into the return tank. The steam and return mains are inclined 1 inch in every 25 feet, so that the condensation in each instance will tend to flow in the direction of the steam. The base of the risers has a drip into the return main. Jenkins disk valves are used throughout, and the steam pipes are covered where necessary with asbestos covering.

The plant has been successfully operated during the past three seasons, and is economical in operation and simple in management. The contract for installing the apparatus was let to Tunstead & Moore, of Minneapolis, Minn., for $4,400.

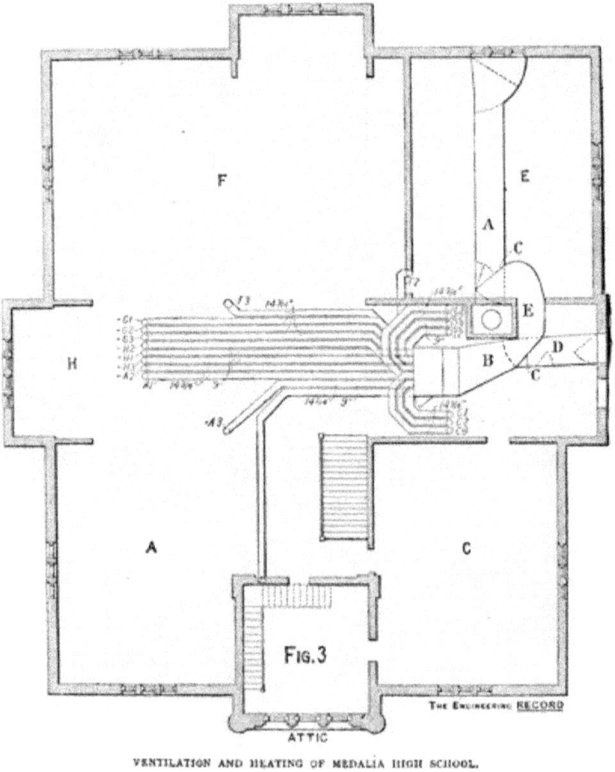

VENTILATION AND HEATING OF MEDALIA HIGH SCHOOL.

VENTILATION AND WARMING OF THE NEW HIGH SCHOOL, MONTCLAIR, N. J.

The High School at Montclair, N. J., recently completed from the plans of Messrs. Loring & Phipps, architects, of Boston, contains many features which attract attention for their novelty as well as utility. Its general shape does not strike one as of the conventional type to which the public is accustomed. The long corridors; the schoolrooms all along one side, with the lavatories, water-closets, etc., not only within the buildings but upon the main corridors on each floor, so placed to avoid running up and down stairs more than is necessary, are features which will strike one as a departure from the common practice.

The building is 212 feet long by an average width of 74 feet, and 2½ stories in height. It is built in the Colonial type of architectural treatment, of buff bricks with terra-cotta trimmings, and cost complete about $100,000. The plans and specifications for the ventilating and heating apparatus were drawn by Fred. P. Smith, C. E., Boston, Mass.

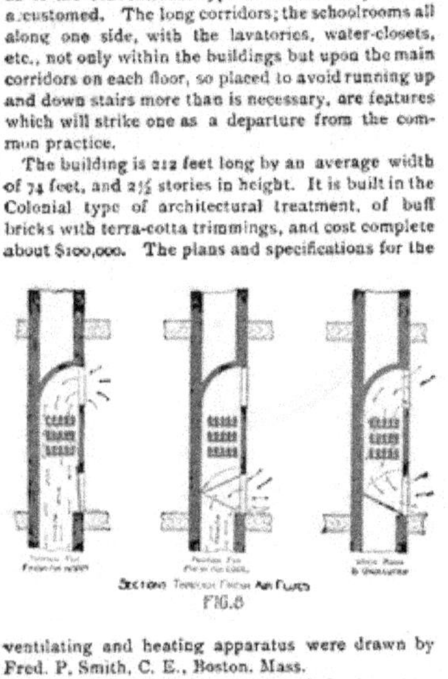

Sections Through Fresh Air Flues
FIG. 8

Figures 1, 2, 3, and 4, the plans of the basement and upper floors of the building, show that the ventilation is by the plenum system, the air being furnished by a fan of the Davidson propeller type, 8 feet in diameter, capable of delivering 2,000,000 cubic feet of air per hour. It is driven at a speed of 120 revolutions per minute by a 10 horse-power Metropolitan engine. The boiler for driving the engine is of the common vertical tubular pattern of 20 horse-power. This not only runs the engine but supplies steam to part of the coils in the heating chamber in the basement. There are two other boilers, each of 60 horse-power, which also supply steam for the heating coils. This primary coil or initial heater at the fresh-air inlet is made up of ordinary cast-iron sections of the "Excelsior" indirect radiators built into stacks and carried on heavy iron beams. The usual practice in plenum systems of making the interstices between the steam pipes in sections of the heater small enough to prevent the too free passage of the air has been abandoned in the system and quite the opposite followed. This was done by the use of special nipples which allowed the surfaces to be placed 1 inch further apart than the usual practice requires.

Figure 5 shows a section through the fresh-air inlet, heating chamber, radiators, and fan. The radiators are divided into five sections.

Figure 5 is a section through the fresh-air inlet, the heating room, radiators, fan and entrance to the main duct. The radiators are divided into five sections, each controllable by a valve. The sections Nos. 1, 2, and 3 are heated by steam from the large boilers, while No. 4 receives live steam from the power boiler and No. 5 condenses the exhaust steam from the engine except when the fresh air requires no heating whatever. Thus from the small power boiler alone the whole building may be made comfortable in damp or chilly weather. When the temperature drops low enough to need it the heating boilers are started up and more surface comes into use. The total surface in the heater is 1,200 square feet, an amount sufficient to heat the entire building if the blower system were employed and the air spaces through radiators were decreased.

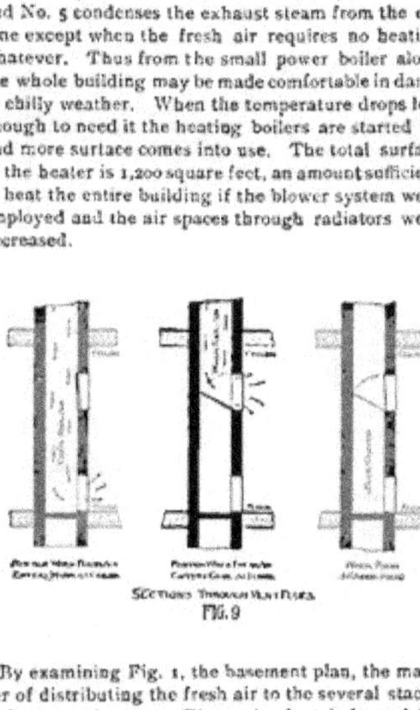

Sections Through Vent Flues
FIG. 9

By examining Fig. 1, the basement plan, the manner of distributing the fresh air to the several stacks of flues may be seen. The main duct is formed by constructing a sub-ceiling over the corridor, in the basement, 24 inches below the ceiling. The air is blown into this space, which is lined with metal, and carried to the several branches shown.

In Fig. 2 the fan is at A, the engine at B, the power boiler at C, the main heating boilers at D D. In the main duct F F F are the deflecting valves H H H for regulating the flow of air to the different rooms. They thus have the means of introducing about 2,000,000 cubic feet of air per hour and the means of warming the air from the temperature of outdoors to that of 120 degrees, and the system is always provided with equalizing valves to distribute the flow of air to the different ducts. It frequently happens, however, that the air for a room in some exposed part of the building may require further heating than that furnished the coils in the heating chamber. To do this without using the ordinary direct radiators, the method shown by Figs. 6 and 7 was followed. Stacks of indirect radiators are placed in each fresh-air flue about 3 feet above the floor. There are openings from the flue to the room both above and beneath the radiators. Valves are so arranged that when

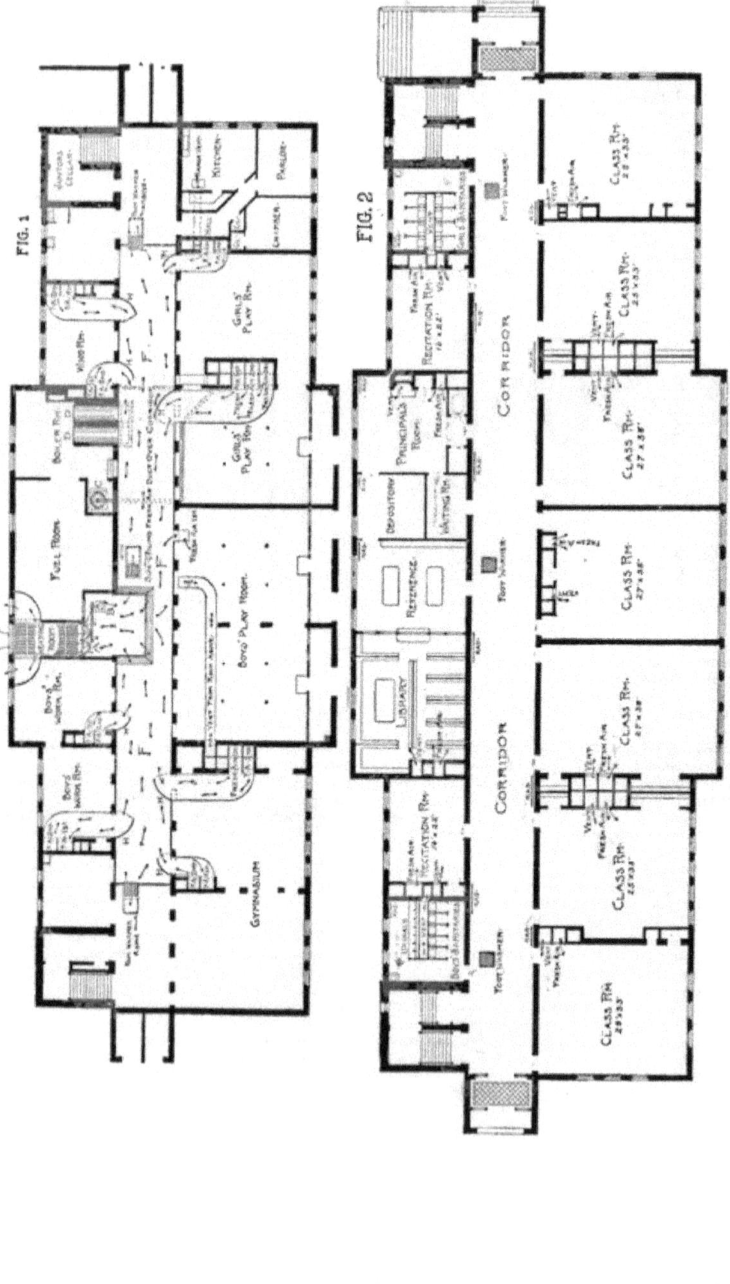

VENTILATION AND WARMING OF THE MONTCLAIR, N. J., HIGH SCHOOL.

FIG. 3.

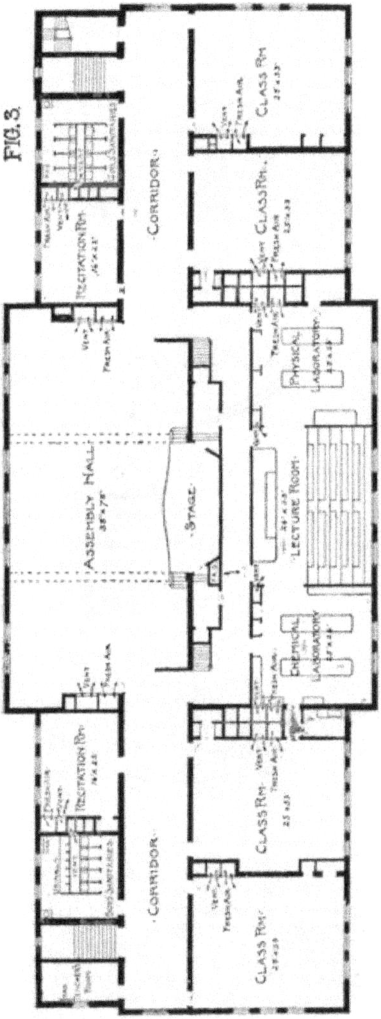
CORRIDOR
CLASS RM.
RECITATION RM.
ASSEMBLY HALL
STAGE
LECTURE ROOM
PHYSICAL LABORATORY
CHEMICAL LABORATORY
VENT
FRESH AIR
CORRIDOR
CLASS RM.
RECITATION RM.

FIG. 4.

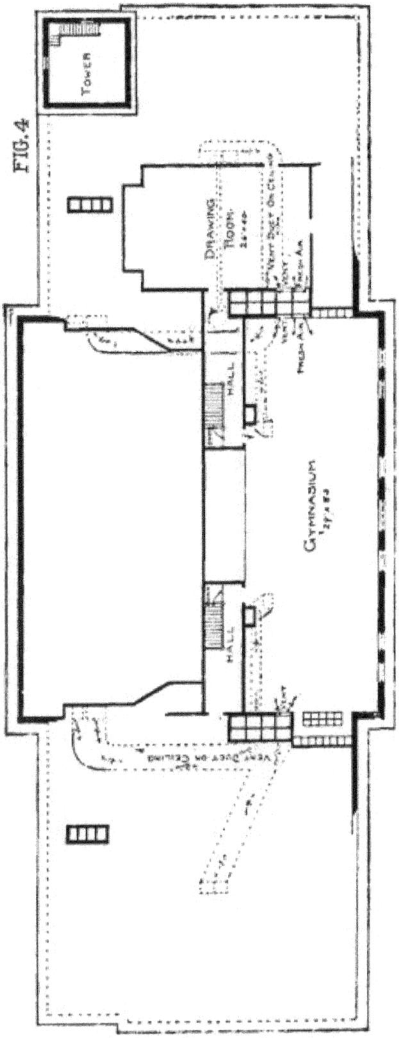
TOWER
DRAWING ROOM
HALL
GYMNASIUM
HALL
VENT DUCT ON CEILING

ventilation is desired the air may enter from the fan, pass up through the radiators, and enter near the ceiling at any desired temperature. If the room becomes too warm, or in warm weather, when cool

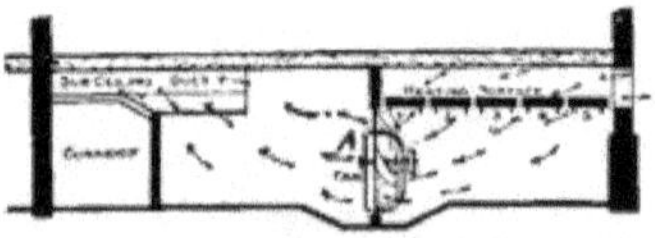

SECTION THROUGH HEATING ROOM & CORRIDOR
FIG. 5.

air is desired, the valve may be so placed that the air enters the room beneath the radiators without being warmed.

In Fig. 6, which is a section through a fresh-air flue to first floor and a vent flue from second floor, the damper valve in the flue to the left at S S is shown so that air from the fan is passing through the radiators and entering, warm, near the ceiling. In the flue to the right the air from the fan is entering, cool, near the floor. For night, or at any time when the rooms are unoccupied, the dampers at S S may be so placed that the air from the fan is cut off and then the air from the room circulates through the radiators, over and over, in this way heating an unoccupied room without wasting fuel by warming a large amount of unnecessary fresh air. The whole building, if necessary, or any part of it, can thus be warmed without running the fan.

Figures 8 and 9 bring out more fully the uses of the radiators within the several fresh-air flues and also the feature of reversing the inlets and outlets as the weather conditions are changed. The idea followed out is that in winter the fresh air enters warm and flows to the ceiling so that the outlet for the

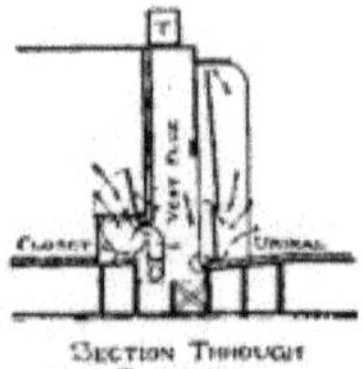

SECTION THROUGH SANITARIES
FIG. 10

vitiated air should be at the floor, but in warm weather when the fresh air enters cooler than the room it falls to the floor, so that in order to secure even distribution and thorough diffusion the summer exhaust should be above the occupants' heads.

The supply of steam is under the control of the system of the Metropolitan Electric Service Company, of New York City. This consists of thermostats which are used by the changing temperature so as to open or close an electric circuit. The electric current operates electric air valves which turn air, compressed to about 10 pounds per square inch, into or out of diaphragms placed on the steam valves. Thus whenever the temperature of any room rises or falls the thermostat automatically turns on or shuts off the steam supply.

In Fig. 2 in the corridors are placed large warm-air registers for warming and drying the pupils' feet. It will be noticed that the water-closets are placed just off main corridors at either end of the building on the first and second floors. There are 26 closets in all ranged as shown by the plans. These are so placed that all plumbing and fixtures are concealed from the pupils but accessible from the rear to janitor and plumber. The urinals are of slate with thorough ventilation, but are not flushed.

Figure 10 shows a section through the urinal and a closet with the flushing tank T. The arrows show how the air from room is drawn towards and through the closets and urinals carrying odors out of the building and thoroughly ventilating the closet rooms.

The Smith Heating and Ventilating Company, Boston, Mass., was contractor for the work.

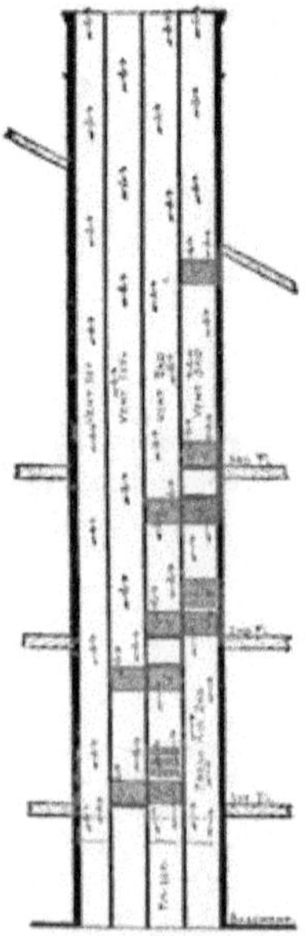

SECTION THROUGH FLUES.
FIG. 7

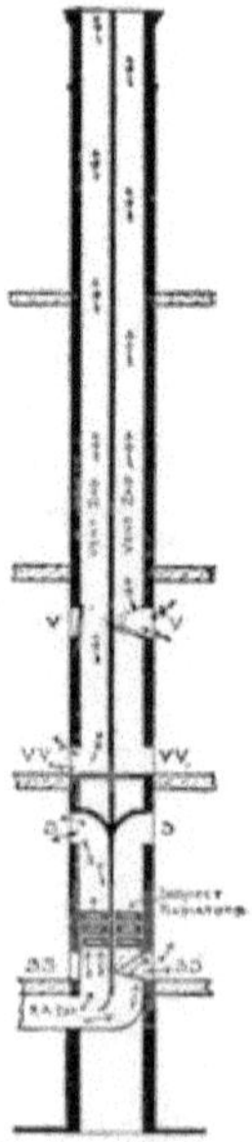

CROSS SECTION.
FIG. 6

STEAM-HEATING AND VENTILATING PLANT IN THE IRVING SCHOOL, WEST DUBUQUE, IOWA.

THE Irving School at West Dubuque, Iowa, is a modern two-story and basement brick edifice about 75x90 feet in extreme dimensions with six small and one large study rooms seating about 450 pupils, besides recitation-rooms, etc. F. Heer & Son are the architects, E. P. Waggoner the mechanical engineer, and W. S. Molo the local contractor who installed the steam-heating and ventilating apparatus, which was designed by the engineering department of the American Boiler Company.

The characteristic features of the plant are the location in the basement of a steam boiler, and pipes and indirect radiators which deliver fresh air either hot or tempered, to the classrooms. These rooms may be additionally warmed by direct radiators which are also supplied in the halls to temper incoming air. Each room is provided with two separate independent ducts, one for delivering fresh air and the other for removing foul air. These ducts are rectangular vertical flues, not in the main walls, but built out like pilasters on the partition walls, and communicate with the rooms served by registers, those for the fresh air being in the upper and those for the foul air being in the lower part of the room. The foul-air ducts all discharge downward into rectangular galvanized-iron ducts on the basement ceiling which open into the brick chimney. This is about 4 feet square inside and contains the 18-inch vertical smokestack from the boilers, radiation from this stack operating to accelerate the circulation of foul air up the chimney. Foul air from the main halls is admitted directly to this vent shaft by registers through its sides in the upper stories near the floors. In the classrooms foul air is drawn in on all sides between the main floor and that of the teacher's platform and passes into the vent duct through an adjustable damper. All the air flues and ducts are made of No. 24 iron and the steam pipes are graded 1 inch in 20 feet, as indicated by the arrows. All the return branches have check valves and there are air valves at all summits. Steam is supplied by a No. 164 heavy-duty double Florida heater, surface burning for soft coal, having two 36-inch firepots and two 3½-inch steam outlets.

Figure 1 is a basement plan showing ducts, indirect radiators, and steam plant. Figure 2 shows some of the details of construction of the indirect radiator stacks. Figure 3 shows the arrangement of rooms, etc., on the first and second floors and the location of registers, flues, and radiators. All hot-air registers are in the wall 8 feet above the floor. Figure 4 shows the arrangement of foul-air exits under the classroom platforms. The perforated metal face plates extend completely around the platform and admit the foul air to chamber C, whence

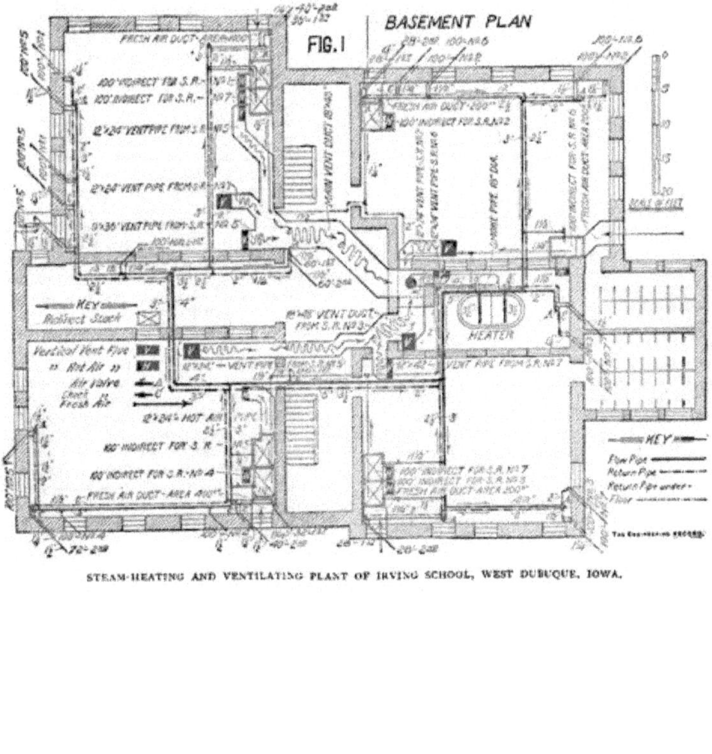

STEAM-HEATING AND VENTILATING PLANT OF IRVING SCHOOL, WEST DUBUQUE, IOWA.

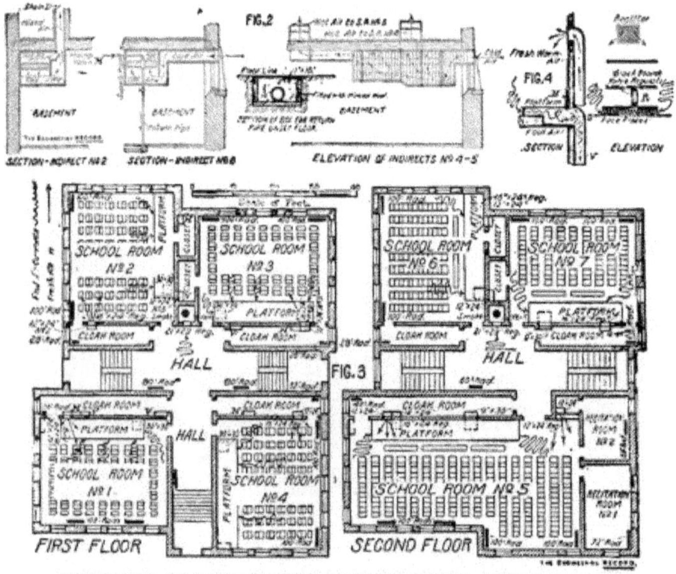

STEAM-HEATING AND VENTILATING PLANT IN THE IRVING SCHOOL, WEST DUBUQUE, IOWA.

it is exhausted through vent duct V. The chain C passes down through the fresh hot-air duct and operates the mixing valve M, Fig. 2. It terminates in a handle H, which may be set at different points on the regulator R and maintain a fixed proportion of hot and cold air.

HOT-WATER HEATING IN THE NEW POLYTECHNIC INSTITUTE, BROOKLYN, N. Y.

PART I.—GENERAL DESCRIPTION OF FLOOR PLAN AND SPECIFICATIONS, BASEMENT PLAN, ELEVATION OF RISERS, PERSPECTIVE AND ELEVATIONS OF BOILERS AND CONNECTIONS.

THIS building, erected for use in connection with the adjacent building, is of brick, about 80x100 feet square, and has five stories above the basement. All the ceilings are very high, that in the upper story being pointed, with its apex 40 feet above its floor and about 120 feet above the basement floor.

W. B. Tubby, of New York, is the architect, and G. C. Blackmore, of New York, is the contractor for the heating. Four of the Richardson & Boynton Company's largest size "Perfect" hot-water boilers are used, the plans for the work having been made by their engineer, W. M. Mackay, who also superintended the erection of the work. The boilers used were made especially for this building. Each contains 315 square feet of fire surface, has 9½ square feet of grate area, and a 12-inch smoke connection.

For the information of those who are interested in the proportioning of radiating surface for this building, we give here the cubical contents and the amounts of radiating surface placed upon the several floors, the surface being larger than if the system were direct radiation, on account of the cold air taken into the building. The basement contains 136,925 cubic feet of air space, and is heated by the mains; the first floor contains 175,286 cubic feet of air space, and has 3,187 cubic feet of radiating surface; the second floor contains 96,406 cubic feet of air space, and has 2,085 cubic feet of radiating surface; the third floor contains 119,308 cubic feet of air space, and has 2,270 cubic feet of radiating surface; the fourth floor contains 98,480 cubic feet of air space, and has 2,114 cubic feet of radiating surface; the fifth floor contains 223,300 cubic feet of air space, and has 2,675 cubic feet of radiating surface; total cubic contents, 849,735 cubic feet; total radiating surface, 12,331 square feet, exclusive of mains.

The low pressure direct-indirect system is employed, and there are throughout the building vertical tube hot-water radiators of the Detroit ornamental pattern. These radiators are placed under windows, and provided with fresh-air inlets arranged in such a manner that the air will pass between the tubes of radiators just above the base. Each fresh-air inlet

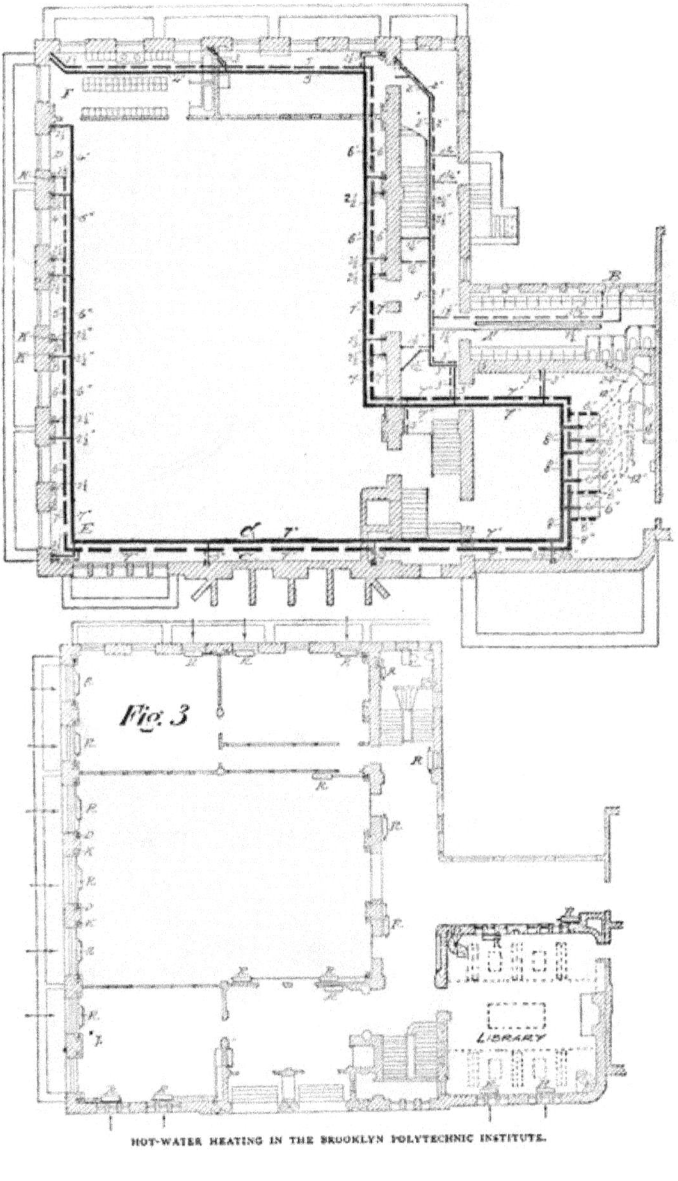

HOT-WATER HEATING IN THE BROOKLYN POLYTECHNIC INSTITUTE.

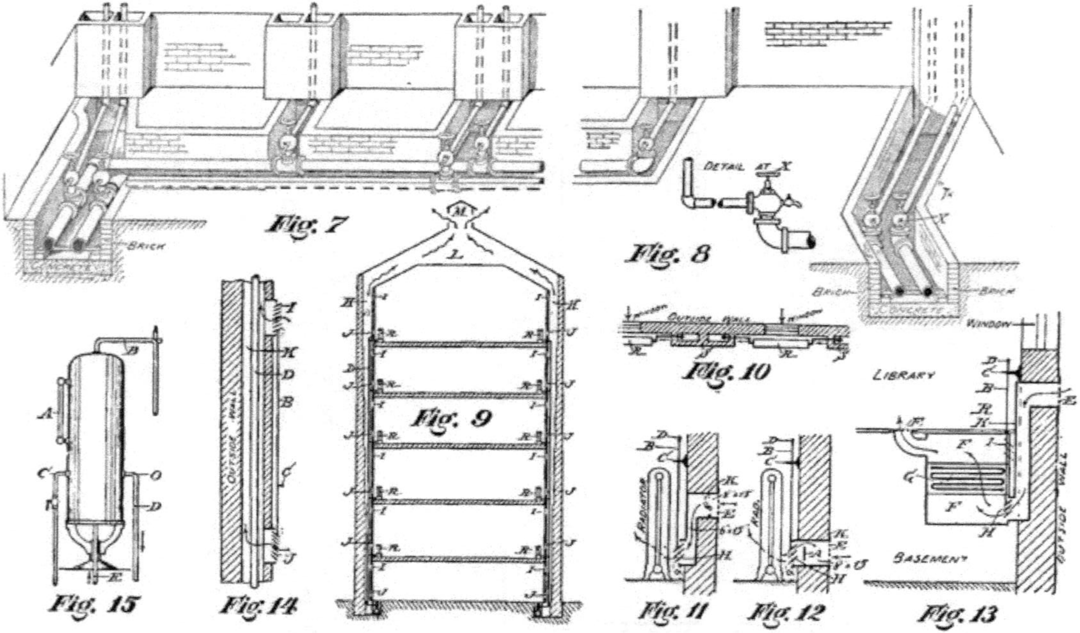

HOT-WATER HEATING IN THE BROOKLYN POLYTECHNIC INSTITUTE.

has an 8x15-inch black japanned register face, with reverse beveled frame on the outside of the wall. In the front wall registers are placed directly under windows, and have a galvanized-iron pipe connection to a damper placed on the inside of the wall, which is operated by a rod and adjustable handle to graduate the supply of air.

All branch connections from risers to radiators are made under floors where practicable, each set of risers having valves at the bottom and draw-off cocks, so that they can be shut off and the water drawn out without interfering with the rest of the apparatus. The risers are run in the ventilating flues to insure a positive circulation of air at all times.

The ventilating registers are of white japanned finish and are securely fastened to cast-iron, reversed beveled frames, built into brick walls. The upper registers are fitted with cords and indicating handles to open and close.

In the fifth floor there is an expansion tank with plumbing connection, feed-cock, self-closing cock and funnel, gauge glass and overflow pipe. The bottom opening of the tank is connected to return opening of heaters, with 1-inch branches, from a 1½-inch main expansion pipe. Each branch has a keyed valve at the heater.

Where pipes pass through walls, floors, or ceilings, the openings are covered with cast-iron floor and ceiling plates, properly secured in place, and are protected with tin tubes according to the rules of New York Board of Fire Underwriters.

Figure 1 is a plan of the basement, and shows the flow mains in full heavy lines; the return mains by broken lines, and the risers by black circles. In the toilet-room the flow pipe A has its length increased so as to serve as an overhead radiator, and obviate the necessity for one on the floor, which could not be set there without too much obstructing the limited area. An ordinary radiator on the floor above, at B, gives a circulation head for the pipe A.

The gymnasium ceiling is about 15 feet high, and the room about 80x100 feet square. It is warmed by radiation from the mains C C, etc., which, for this purpose, are made of a larger diameter than the radiator system alone requires. These mains are laid on rollers in a brick-lined trench and are covered by iron gratings.

Figure 2 is a diagram of some of the riser lines. The rest are similar to them. Figure 3 is a plan of the first floor, showing arrangement of rooms and location of radiators.

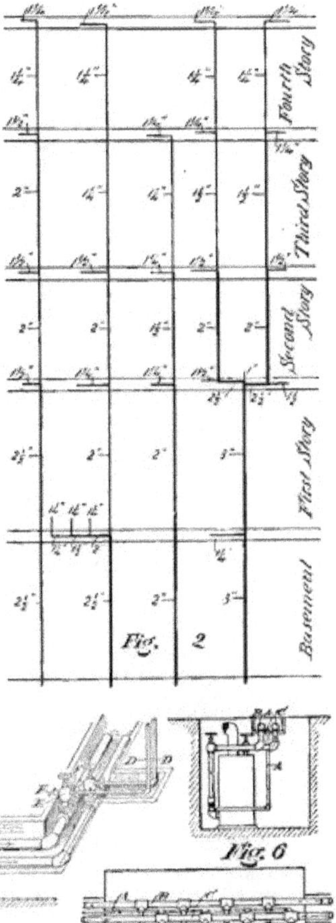

Fig. 2

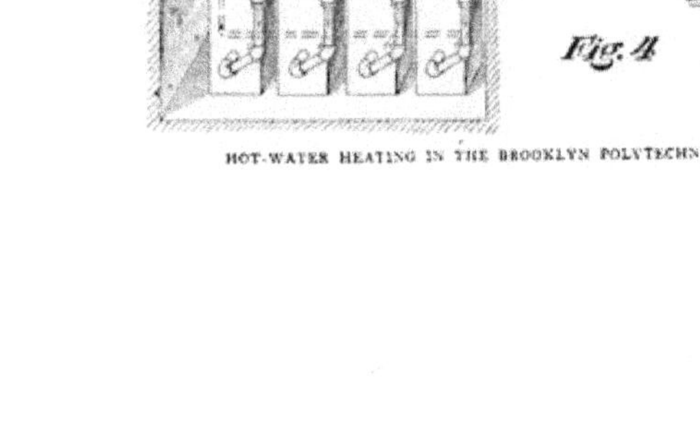

Fig. 4

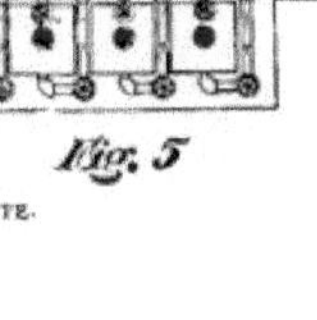

Fig. 6

Fig. 5

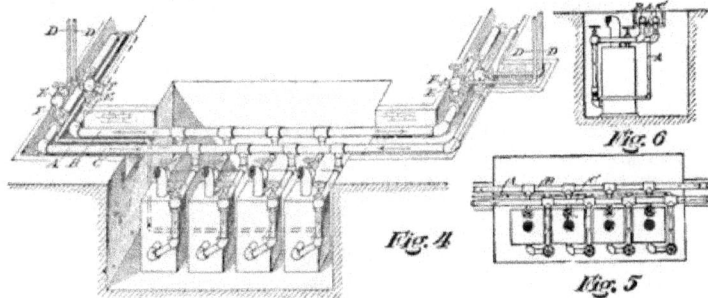

HOT-WATER HEATING IN THE BROOKLYN POLYTECHNIC INSTITUTE.

The other floor plans are similar. All the upper floors are warmed entirely by direct-indirect radiators. The top story has excessive exposure through very large skylights in the sloping ceilings.

Figures 4, 5, and 6 show the arrangement of heaters, which are set in a pit below the basement floor. Figure 4 is a section and perspective, Fig. 5 is a plan, and Fig. 6 is a cross-section. B is the 8-inch flow and C the 8-inch return main; A is the 1½-inch expansion pipe connected separately to each boiler, D D, etc., are risers connected to the mains by special angle valves E E, etc., and flange joints, and with cocks F for emptying the lines when the valves are closed.

PART II.—DETAILS, FLOW AND RETURN MAINS AND RISER CONNECTIONS, VENTILATION DUCTS AND REGISTERS, EXPANSION TANK, COLD-AIR INLETS AND CONDUITS AND INDIRECT RADIATORS.

FIGURE 7 shows the section of the pipe trench in the gymnasium and riser connections at E, Fig. 1. Figure 8 shows the termination of mains at F, Fig. 1.

Figure 9 is a sectional diagram of the building, showing, on an exaggerated scale, the vent flues K K, which, maintaining a vertical face inside, follow the offsets of the wall and increase in size upwards proportionately with the increased duties of the upper parts. They discharge into the chamber L and thence above the roof through the ventilating stacks M, which are 5 feet in diameter and about 2 feet high above the ridge.

Figure 10 shows the arrangement of radiators R R, etc., between shafts S S, so that each of the riser connections is made to the nearest line. Figure 11 shows the general arrangement of fresh-air inlets in the outside walls. Figure 12 shows another arrangement where the wall hole is opposite the bottom of the register and a galvanized-iron diaphragm A, with 40 square inches area, is put perpendicularly in the center of galvanized-iron flue, to prevent undue blasts of cold air, while allowing full passage-way area in the annular space of 80 square inches around the obstruction.

In both figures the inside dampers H are operated by rods B and handles D, which can be secured by set screws C; E E are ordinary scrolled register frames protecting the wall openings.

Figure 13 shows the arrangement of the two indirect radiators for the library, Fig. 1. E is same as in Figs. 11 and 12. K is a gavanized air duct, partly built in the wall; G is a Perfection pin radiator, 12 inches high, with air chambers F F, above and below it.

Hot air is delivered to the library through two registers E E (only one being shown here), and the cold-air damper H is controlled by the rod B. Another damper, I, has been suggested, to be so attached to B as to open when H is closed, and *vice versa*, thus permitting cold or tempered air to be used; but this damper has not yet been adopted.

It was originally intended to warm the library by radiators set at R, as in Fig. 12, but as it was afterwards determined to preserve the floor unobstructed by radiators, the inlet E was retained, and the fresh air conducted through special duct K to the chamber F.

Figure 14 shows the arrangement, in pairs, of all ventilation registers I and J, Fig. 9, which are so connected by a rod B that when either one is open the other must be closed, and *vice versa*.

The figure shows both partly open, so that moving rod B down by means of its handle C will set I wide open and J tightly closed. Moving B up will open J and close I. Both registers cannot be closed at once, but will always jointly maintain a uniform inlet opening equal to the full area of one of them. K is the ventilation duct, and D the 2½-inch hot-water riser, shown in Figs. 1, 2, and 9, the ventilation circulation being natural, except as promoted by radiation from pipes D.

Figure 15 shows the 82-gallon riveted galvanized steel expansion tank, placed just beneath the roof. A is the water glass; B the 1½-inch overflow emptying into the house storage tank; C is the 1-inch supply from city pressure; D is a 1½-inch circulation pipe to prevent danger of freezing, and E is the 1½-inch expansion pipe that is separately connected with each of the heaters. H is a vacuum valve to prevent syphoning.

HEATING AND VENTILATION OF THE JEFFERSON SCHOOL, DULUTH, MINN.

PART I.—GENERAL DESCRIPTION, FLOOR PLANS, STEAM PLANT, AND PIPING.

THE accompanying drawings show the means employed in the heating and ventilating of the Jefferson School building, of Duluth, Minn. Messrs. McMillen & Radcliffe, of Duluth, were the architects of the building, while the heating and ventilating apparatus was put in according to the plans and specifications of Edgar G. Barrett, C. E., Chicago, Ill. The Pond & Hasey Company, of Minneapolis, Minn., were the contractors for the work.

The building is about 175 feet in length by 100 feet wide, and contains a basement, first, second, and attic floors. Plans of each of these are shown, with the exception of the second floor, which is very much like the first-floor plan. As this building extends lengthwise down a hill, four rooms are used as schoolrooms in what is really the basement, but they are about 4 feet above the ground, while at the opposite end of the building the first story is only a few feet above the ground, there being a very pronounced difference in level between one end of the building and the other.

The building is heated both by direct and indirect radiation, the air being circulated by mechanical means. The fresh air is drawn down the air shaft at the side of the building and warmed by tempering radiators in the fresh-air chamber to a temperature not exceeding 70 degrees. It is then forced by a 7-foot Blackman propeller fan through warm-air ducts extending to the various vertical flues, at the base of which are located additional indirect heating coils. The speed of the fans is such as to give 2,000 cubic feet of air per hour to each pupil, figuring 60 pupils to each classroom.

In these coil boxes the air is forced either through the steam coil or around it, as desired by the occupants of the rooms above, and is then forced through the vertical flues to the rooms to which they lead, the air being admitted near the ceiling. After circulating through the room, the air is drawn out through a register located at or near the floor up through flues to a separate system of ducts in the attic into an exhaust chamber at the center of the building, in which is located a 7-foot Blackman exhaust fan discharging the foul air out through the roof.

The heating is so arranged and proportioned that in mild weather when very little heat is required, only the tempering coil for warming the air is used. Then as it becomes colder the indirect radiators at the base of the flues are put into service, raising the temperature of the air to a higher degree. In extreme cold weather direct radiation in the building is turned on, so that for the few days of extreme temperature, say 20 to 30 degrees below zero, the entire heating apparatus is in use, but under ordinary circumstances only the indirect system is necessary.

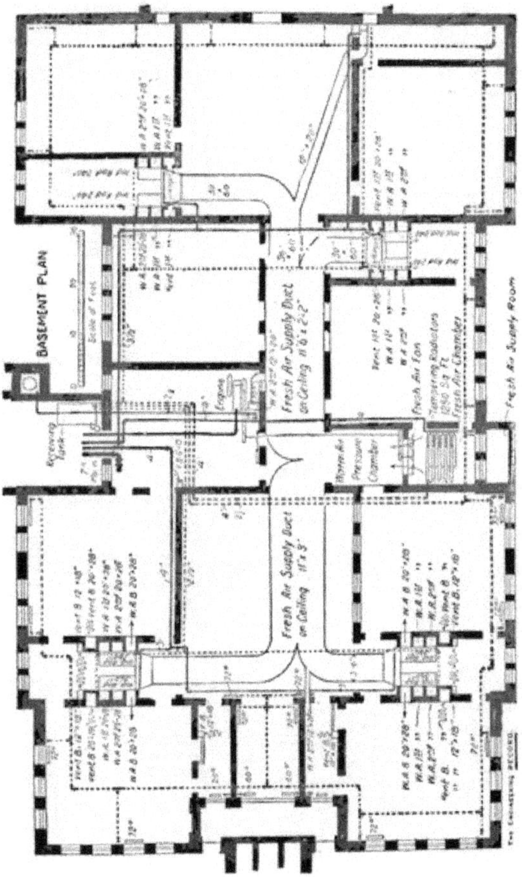

HEATING AND VENTILATION OF THE JEFFERSON SCHOOL, DULUTH, MINN.

Steam for the entire plant is generated in two Babcock & Wilcox boilers of a total of 130 horse-power. Owing to the construction of the building the overhead system of piping is used. A 7-inch main pipe from the boiler-house (not shown) is carried through the wall of the school building and led up to the attic, where it branches into two main supply pipes, making a circuit of the building as shown on the attic plan. The farther ends of these pipes are carried down to the basement, where they connect below the water line and thence extend back to the receiving tank.

The 7-inch riser is supported at the base by a yoke supported by a piece of 4-inch pipe, the lower end of which rests upon a stone foundation. The steam mains for the indirect radiation and the tempering coils are hung from the basement ceiling, and are extended as shown in the basement plan. The returns from the former are carried below the basement floor and are finally collected at the receiving tank. As the tempering coils will be run at a pressure different from the other heating systems, the return from these coils is connected into a float and lever trap, discharging into the receiving tank. The tempering coil in the heating chamber contains 10 radiators of about 125 square feet of surface each, the details of which will be described in the following part.

PART II.—DETAILS OF STEAM AND EXHAUST PIPING AND RADIATORS.

The steam drums of the two boilers are connected by a 1½-inch pipe and the mud drums by a 2-inch equalizing pipe, but both pipes are provided with a valve so that each boiler may be made independent of the other.

The 6-inch supply pipe, shown in Fig. 1, from each boiler runs into an 8-inch steam header, from which various branches lead to the different heating systems, engine, etc. One of these branches starts as a 5-inch from the header and is provided with a stop valve and pressure-reducing valve, and then increased to a 7-inch pipe, from which the 7-inch and 5-inch branches are taken for the direct and indirect radiation respectively. A second branch pipe, 3 inches in size, also provided with stop valve and pressure-reducing valve, increasing at the reducing valve to 4 inches, supplies the tempering coils with steam. A third branch 2½ inches in size supplies the engine. The exhaust from the engine is carried up the shaft surrounding the smoke pipe from the boiler. A 2½-inch connection was made between the exhaust pipe and the tempering coils so that they could be heated by exhaust steam. The exhaust pipe contains a Davis back-pressure valve and the 2½-inch pipe a grease extractor. All of the engine drips, blow-off connections from the boiler; and grease extractor are carried to a catch-basin in the boiler-room. This is 24 inches in diameter and 48 inches deep and is connected to the sewer. A 2-inch vapor pipe is led to the main exhaust pipe. The return water from the heating system is discharged into a No. 3 Worthington automatic feed pump and receiver.

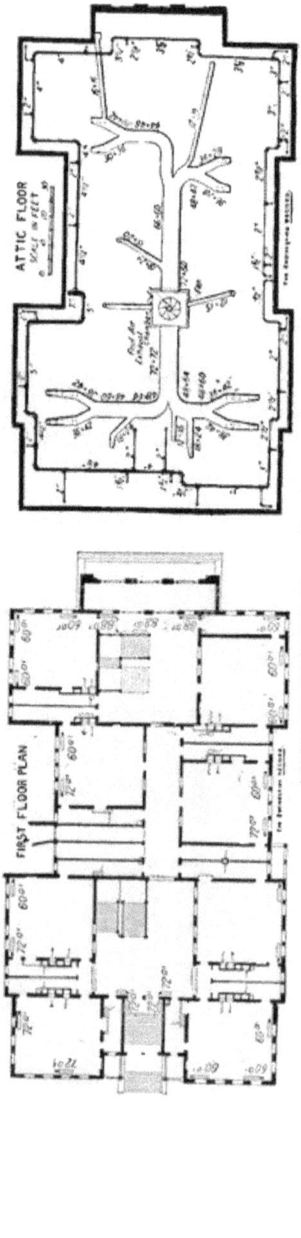

HEATING AND VENTILATION OF THE JEFFERSON SCHOOL, DULUTH, MINN.

The tempering radiators are 10 in number, each containing 125 square feet of heating surface. They are made of 1¼-inch pipes screwed into heavy cast-iron bases, tapped for steam supply and return at opposite ends. The pipes are set staggered in the bases. The radiators are connected into two batteries of two radiators each and two batteries of three radiators each. Each battery is provided with a valve on the flow and return pipes so that any combination can be used as desired. The indirect radiators consist of 1-inch pipes screwed into a cast-iron base. The indirect radiators are supported by iron hangers from the ceiling joists.

There are eight large indirect coils in the basement, four of them containing 240 square feet and four of them 360 square feet of heating surface each. There are three smaller coil boxes, each containing 72 square feet of surface. There is a by-pass around each indirect stack, controlled by a switch operated from the classroom to which the duct leads, so that the temperature of the air can be regulated without diminishing the supply. They are so constructed that the entire volume of air can be heated or not, as desired.

Figure 3 shows a sectional view of one of the ducts with the indirect stacks and the switch. Figures 4 and 5 are elevations showing the location of the supply and exhaust fans, fan engine, etc. It will be noticed that the exhaust fan in the roof lies in a horizontal plane and is driven by means of rope transmission by the engine in the basement.

The bottom of each radiator box is put on with stove bolts so that it can be removed when necessary without destroying the rest of the casing. The bottom is also provided with a hinged door so that the radiators may be easily reached from below.

The entire cost of the heating plant was $5,800. We are indebted to Mr. E. G. Barrett, the designer of the plant, for the data and drawings from which this description was made.

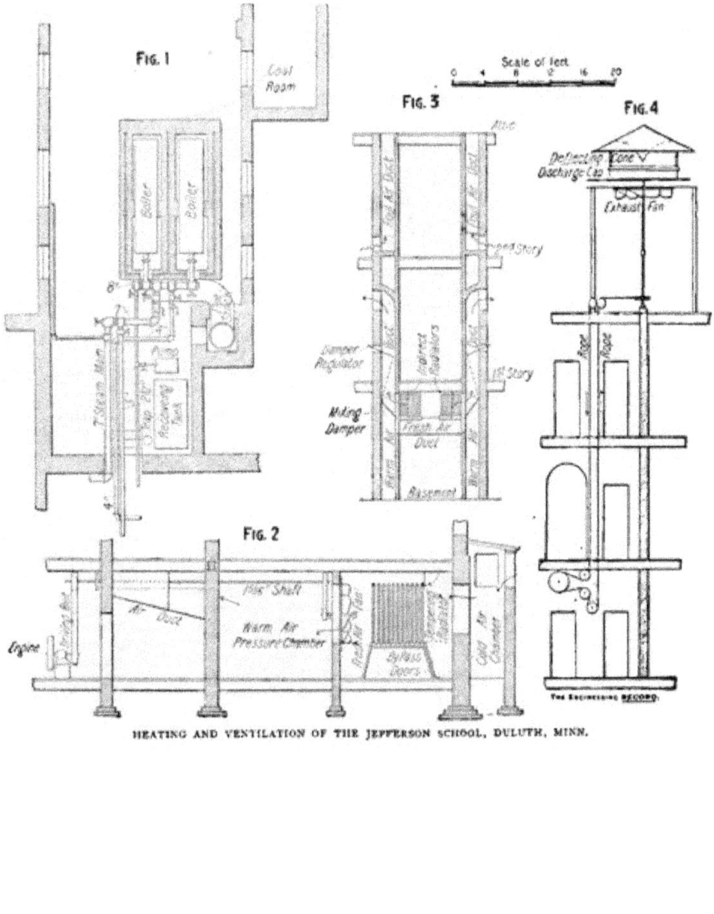

HEATING AND VENTILATION OF THE JEFFERSON SCHOOL, DULUTH, MINN.

HEATING AND VENTILATING A MILWAUKEE SCHOOL.

St. Mary's School, Milwaukee, Wis., is a four-story brick building about 100x64 feet, which was recently completed according to the plans and specifications of T. Schultz, architect. The building is equipped with a direct and indirect system of hot-water heating and accelerated ventilation of foul air, that was installed by Cordes & Treis, of Milwaukee, at a contract price of about $6,000. The conditions were those of a large graded school, with classrooms, lunch-rooms, and a large auditorium hall to be warmed and ventilated during the severest winter weather in a climate where an external temperature of —20° Fahr. or less and sharp lake winds are to be anticipated. It was required to maintain the classrooms at 70 degrees in the most severe weather, and to keep the hall at 68 degrees and the corridors at 65 degrees, besides providing for the continual withdrawal of foul air. The natural and unavoidable inflow through opening doors and windows, and the pressure through all cracks and openings to restore equilibrium was relied upon to supply sufficient fresh air continually.

All the rooms were heated by wall coils or cast-iron radiators, and the principal ones above the basement had also a supply of fresh-tempered air delivered through floor radiators from wall ducts centrally located to connect with indirect stacks that draw air from the wide basement corridor. Other wall ducts had ventilating registers near the ceilings of the rooms, and through them the foul air is drawn through a system of galvanized iron conduits in the attic to a shaft discharging above the roof. This contains a hot-water radiator coil intended to heat the air sufficiently to produce a positive circulation. Four No. 22 Bolton heaters were set in the basement and connected by two pipes each with the flow main. The two principal branches were taken out of each end of the 8-inch header by angle gate valves and carried directly to 5-inch risers that each ran around one side wall and part of two end walls just above the attic floor, diminishing to a final size of 2½ inches as drop pipes were successively taken off to supply the radiators below. The lower ends of these drop pipes were similarly connected to belt pipes parallel with each main wall of the building. These were laid under the basement floor and increased in size as they gathered successive branches. They entered the ends of the 8-inch return header through 5-inch angle valves. Each drop pipe was commanded by top and bottom valves (not here shown), near the supply and return mains so that it could be cut off without interfering with the rest of the system, and all of the boiler connections to the headers were valved so as to be independently operated and used separately or together as required, thus enabling one or more to be thrown out of service in moderate weather. There is a special 4-inch supply and return run to the eight indirect stacks on the basement ceiling, and the boiler-room floor is sunk to such a level as to bring the return pipes below the basement floor. Each boiler has a valved connection to a 1½-inch supply of city pressure water, that is also valved directly into the return header. The return header and the return pipe from the indirect system are both connected to the 40-gallon expansion tank in the attic vent shaft, and there is also a vent pipe from the return header to an open pipe above the roof. The boilers have thermometer, pressure gauge, and key valve for emptying into the sewer. The indirect radiators are of the "Perfection" pin type cased in galvanized-iron

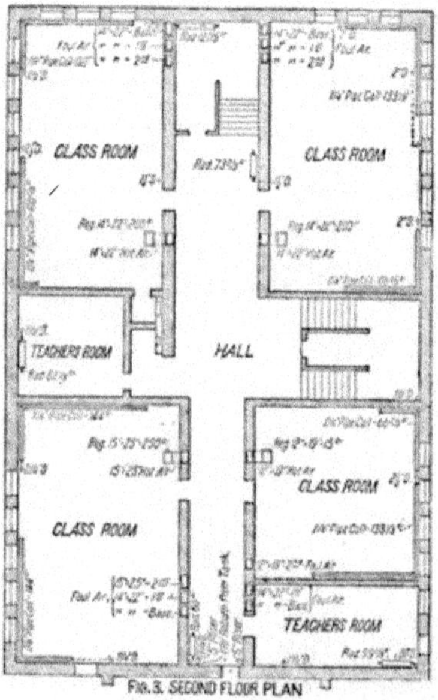

FIG. 3. SECOND FLOOR PLAN

boxes. The direct radiators are either coils of straight and right-angled pipes on the walls with special cast headers, or are cast radiators. All radiators are provided with one valve (not shown in the general plan) on the flow pipe.

Figure 1 is a general plan of the basement showing the position of the flow main and risers, drop pipes from lines of radiators, and location of the return mains, besides the special service for the indirect system and the arrangement of boilers, and the wall coils for basement rooms.

Figures 2, 3, and 4 show the location of risers, registers, and hot-air and ventilation flues and arrangement of radiators for the first, second, and third floors respectively. The dotted lines at A A, Fig. 4, show where the ventilation ducts are carried under the floor in the auditorium hall from the inlet registers to the vertical ducts that are run up along the partitions to the collecting ducts above.

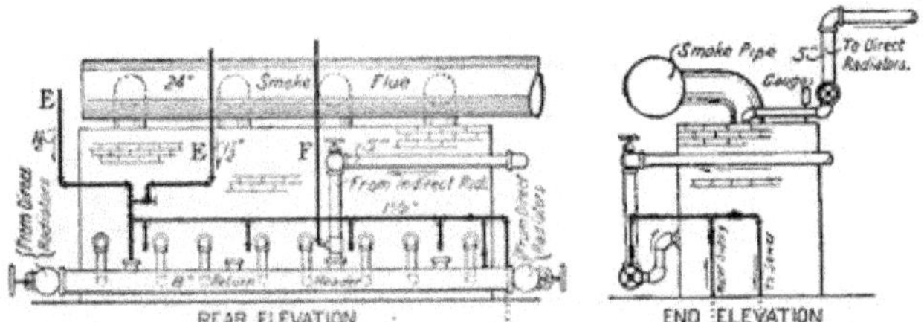

FIG. 7

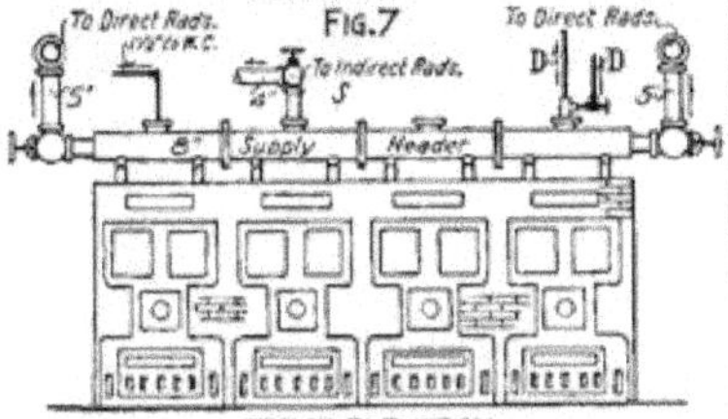

FRONT ELEVATION

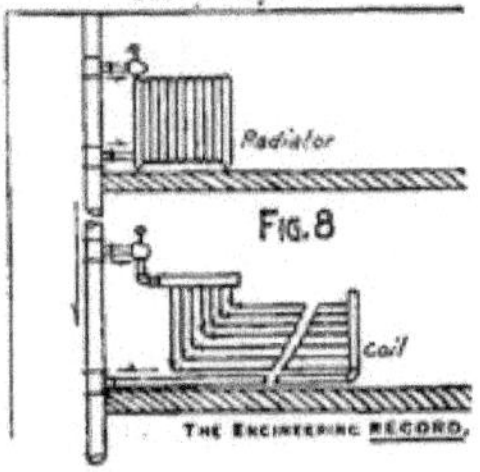

FIG. 8

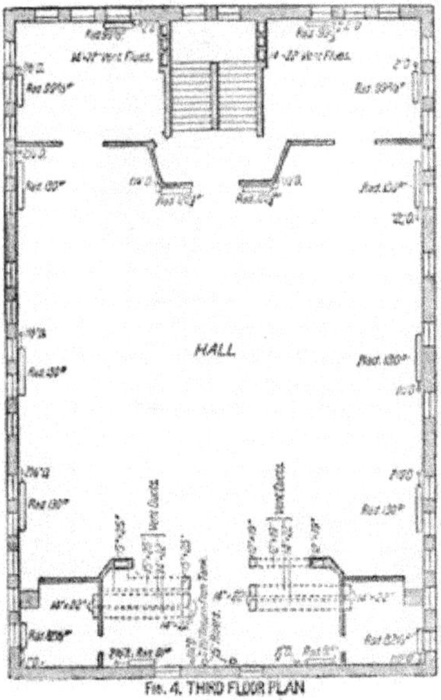

FIG. 4. THIRD FLOOR PLAN

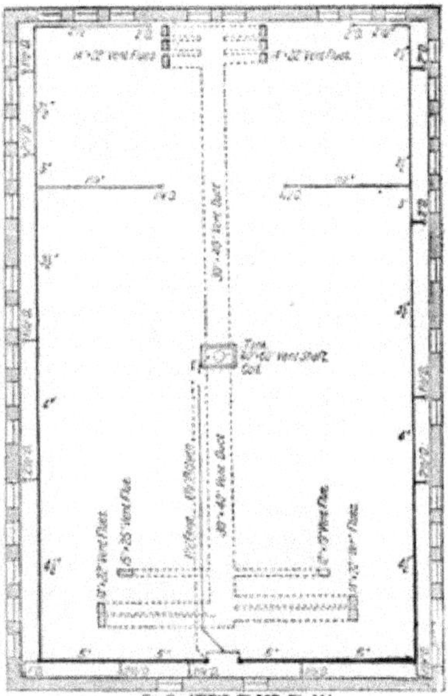

FIG. 5. ATTIC FLOOR PLAN

HEATING AND VENTILATING A MILWAUKEE SCHOOL.

Figure 5 is a floor plan of the attic showing the run of the feed pipes, the drop pipes supplying radiators, and the main risers at B from the supply header. All the pipes are graded to summits C C, whence vent pipes extend to free inverted openings above the roof and have small return branches to the boiler to secure constant circulation, as shown in Fig. 6. The coil in the ventilating shaft consists of six circuits (about 100 feet) of 1½-inch pipe that is connected top and bottom with the expansion tank which has an open vent and overflow above the roof. The horizontal rectangular galvanized-iron ventilation ducts are made with locked and riveted joints, shown by the broken lines, rest directly on the floor

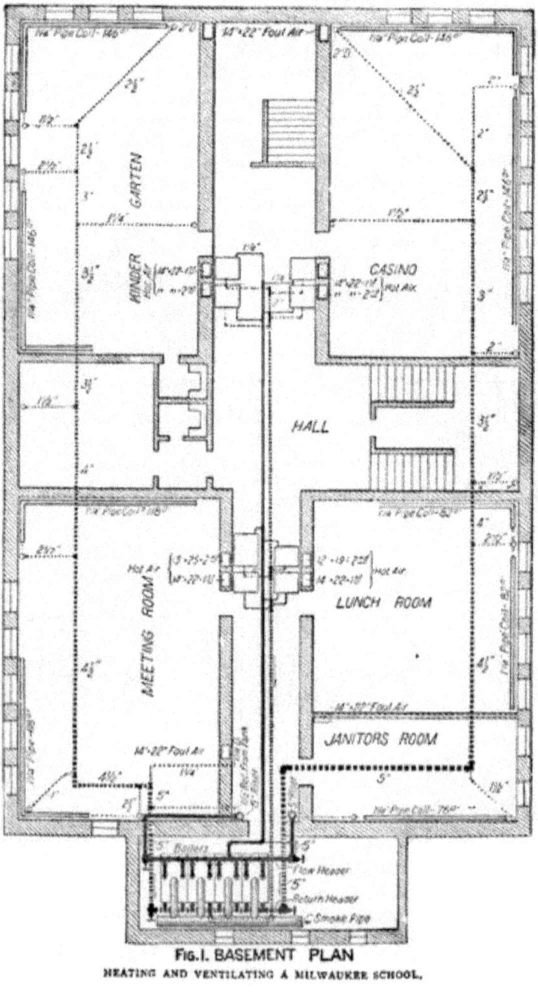

FIG. I. BASEMENT PLAN

HEATING AND VENTILATING A MILWAUKEE SCHOOL.

and serve the vertical metal flues that extend to the lower-story registers, and are indicated here in full-line rectangles. The cross sectional area of the vertical galvanized-iron outlet stack is equal to the sum of the sections of its branches, and it terminates in an overhanging umbrella head 4 feet above the roof, and elevated a foot above the stack so as to allow free egress on all sides for the foul air. It was intended to promote the discharge of this air by an electrically-driven exhaust fan, but it has not yet been considered necessary to provide this.

Figure 7 is a set of elevations showing the arrangement and principal features of the connections of the battery of boilers shown in plan in Fig. 1. F is a

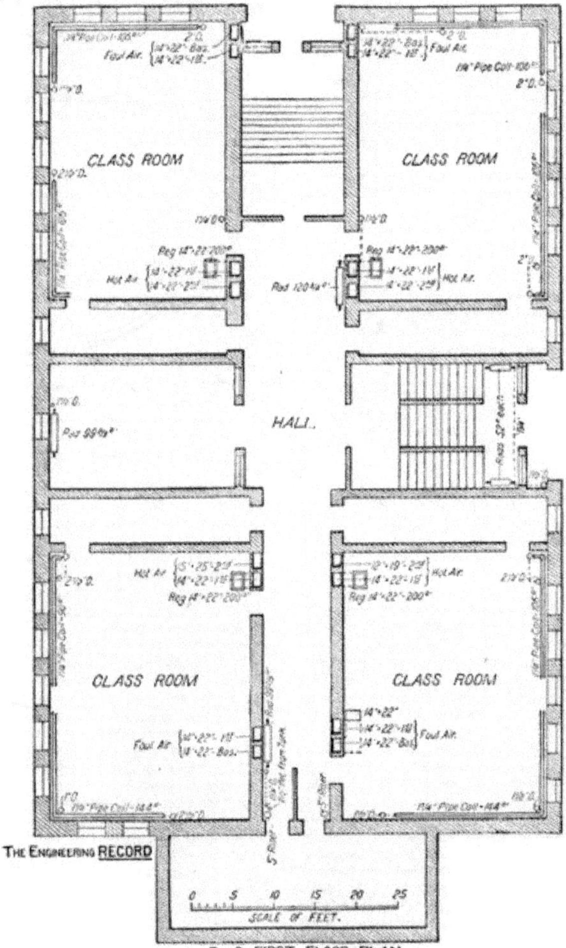

FIG. 2. FIRST FLOOR PLAN

HEATING AND VENTILATING A MILWAUKEE SCHOOL.

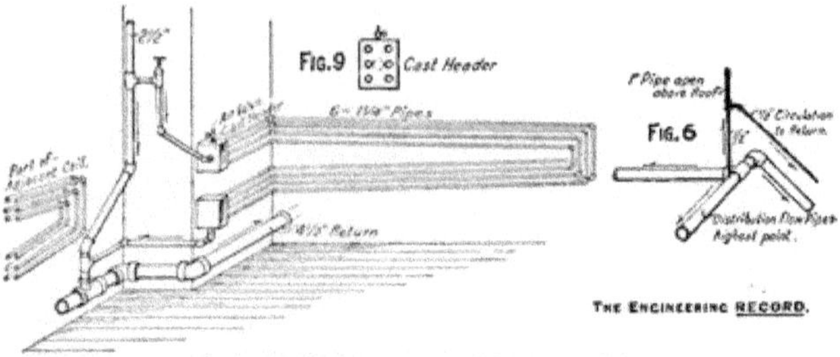

HEATING AND VENTILATING A MILWAUKEE SCHOOL.

circulation pipe connected with a ½-inch branch at each of the two vent pipes, one of which is shown in Fig. 6.

Figure 8 shows the method of connection of cast and coil radiators to the drop pipes that serve them.

Figure 9 shows the special wall coils in the basement classrooms. The headers are simply cast-iron boxes tapped to receive the pipes, and in this case the drop pipe is shown in an out-of-the-way corner of a wall offset just beyond the loop end of a similar adjacent wall coil.

HEATING AND VENTILATING IN THE ENGINEERING BUILDING OF THE MASSACHUSETTS INSTITUTE OF TECHNOLOGY.

The system of heating and ventilating in the engineering building of the Massachusetts Institute of Technology, at Boston, was designed by Prof. S. H. Woodbridge, and to him we are indebted for the following particulars and annexed illustrations relating to it.

The building is without interior walls other than light partitions, and all available external wall space is demanded for piers and windows. Locations were allowed for eight vertical flues, varying from 9 to 12 square feet in cross-section, for the supply and discharge ventilation of the 30 and more rooms. The rooms were arranged by their intended users, with only partial reference to the fixed location of flues and connecting air ducts were disapproved as unsightly. The great value of the basement floor space imposed a limit of 10x12 feet on the area to be surrendered to ventilating purposes. The use of the concrete floor of the sub-basement for apparatus, above and about which the basement floor is removed, precluded the use of this space as a distributing air chamber, and compelled the building of a continuous duct about the perimeter of the sub-basement, Fig. 1, with one cross-duct beneath the fan at A A, and into which it discharges. The main duct, except where engine beds B encroach upon it, is 15 square feet in cross-section. The cross-duct has nearly twice that area. The control of the air quantities to be moved in one direction or another within these ducts is effected by movable deflectors, one under the fan, and one at each end of the cross-duct.

The perimeter ducts have for three of their sides the foundation wall of the building, the sub-basement concrete floor, and the wooden floor of the basement. The fourth side is of galvanized iron, secured by nailing to wooden strips set in the concrete and nailed to the wooden floor and beams. A free use of elastic cement was made in all joints between metal and wood, and of paint in all locked or other joints of the sheet metal, and provision made for a possible settling away of the concrete from the wooden floor.

To clear the laboratory ceiling and the floor space of all possible obstructions and the unsightly appearance of piping, the steam mains and branches, traps, etc., are placed within the air conduits. All such steam pipes are carefully pitched and drained and covered.

In the plan of the sub-basement floor, Fig. 1, these pipes have not been shown, since the small scale of the drawing might have tended to create some confusion.

The eight vertical ducts F (see also Fig. 3, which is an elevation of the ducts), are of necessity made to serve the dual purpose of supply and discharge. To adapt them to such purpose, a diaphragm is fixed in such way as to provide two channels having areas proportioned to the quantities of fresh and spent air to be moved through them to and from the successive stories. These diaphragms are made of sheet iron, which is secured by methods effectually preventing the leakage of air from the plenum into the exhaust conduits. Wherever practicable, the diaphragm is so placed as to remove the supply conduit from the outer walls, and to bring the discharge conduit against them.

Because of the small space occupied by the entire system, velocities of the air moved must be high. To secure to each register of the lower stories its proportion of air, and to prevent its going by such register under the momentum of its movement, deflectors are used, the area of each and the angle at which it is set controlling the air volume issuing from each register. Similar deflectors, set in a reverse position, are used for the outlets from the upper stories. To thoroughly break up and diffuse the swift flow of cool air in solid current from the register, diffusers, such as are shown in Fig. 1, are used.

The building accommodates some 30 students, and the air supply is nearly 2,000,000 cubic feet per hour, the fan running at 250 revolutions.

The warming is effected by three systems. Because of the great amount of steam work done in the basement, air must be supplied in large quantities, and at a temperature ranging from 45 to 55 degrees, according to laboratory work and outside conditions of weather. The eight distributing flues cannot supply air to the several floors or rooms at different temperatures. They must supply it at the temperature required by that room above the basement most easily warmed to the point desired. Therefore it becomes necessary to provide means for supplying air through one system of conduits to the basement at, say 50 degrees, and to all rooms above the basement at 70 degrees, and to further warm the air by direct means, in such rooms as require supplementary heat.

The air is heated before it reaches the fan to 50 degrees, a "Standard" metallic thermometer mounted in the fan case indicating the temperature, which is controlled by regulating the steam pressure in the coil. In moving under pressure through the sub-basement conduits the air leaks into the basement, as was anticipated, through innumerable small vents, the current being nowhere sensible, and yet the aggregate volume amounting to some 750,000 cubic feet per hour. Reaching the base of the flues the air passes through steam coils so made and placed that the flue area is not obstructed (Fig. 4). The control of steam to these coils is by means of the Johnson electric regulating apparatus, the thermostat being hung before the supply register on the third floor. Whatever the temperature in the sub-basement conduits, the air supply to the rooms may be maintained at 70 degrees or 72 degrees, the range being confined within these limits by the automatic action of the regulator.

Within the rooms are placed wall steam pipes, the steam supply to which is regulated by the Johnson automatic apparatus, the thermostat being exposed within the rooms. For the quick warming of the building the sub-basement conduit temperature may be run up to 100 degrees, and the flue thermostat may be swung away from the register front. Air at such times may be circulated through the building instead of being taken from the outside.

The construction and arrangement of the auxiliary heater at the base of the flues is a matter of interest, because well suited to a successful working of the automatic method of steam supply. The steam enters at the top and through a valve so throttled that when the main conduit air is at its coldest the steam flow will be nearly continuous. The coil drains through a check valve. Without such arrangement, Professor Woodbridge explains, the temperature within the flue would fluctuate through a considerable range, for on the wide opening of the supply and return valves steam would enter freely at both ends and suddenly heat up the coil and the flue. It is desirable that the steam flow should be as nearly continuous as possible, and sufficient in quantity to warm the air passing through the coil. If the supply valve is throttled the drip valve must be closed

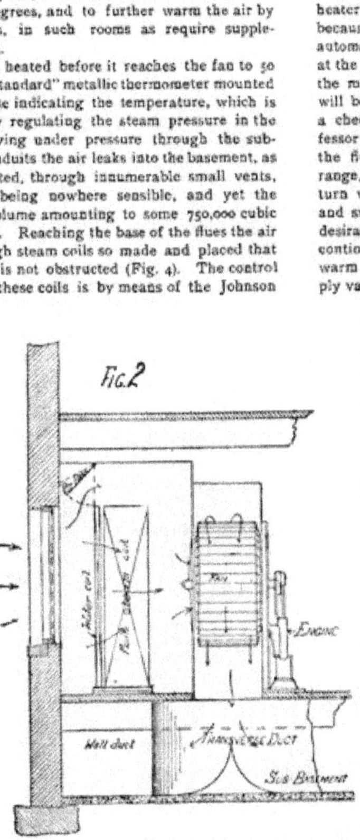

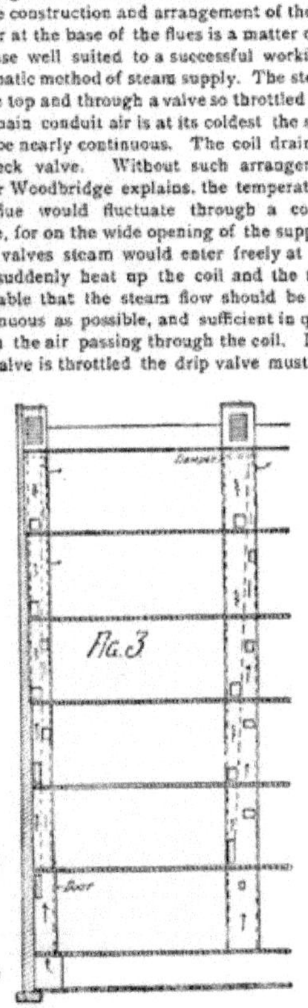

STEAM HEATING AT THE MASSACHUSETTS INSTITUTE OF TECHNOLOGY.

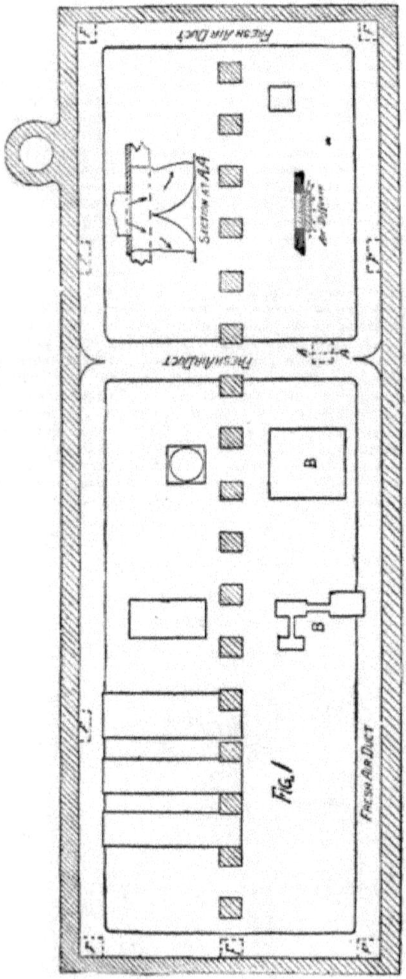

STEAM HEATING AT THE MASSACHUSETTS INSTITUTE OF TECHNOLOGY.

until the pressure within the coil is sufficient to force the accumulated water outward against the steam pressure. A throttled drip valve would allow steam to back into the coil and cause pounding. But the check valve holds back the steam and allows the condensation to collect until its weight and the steam pressure combined force the valve open and the water out. The filling of the pipes with condensed water serves also the useful purpose of automatically regulating the length of their heated parts, and aids in maintaining the even temperature sought in the flues.

The heating is for the most part done by the exhaust steam of engines and pumps used in the building, and to avoid the possibility of returning oily water to the boiler the condensation is passed into the sewer. For the purpose of cooling this water, and of utilizing its heat, it is passed through 800 feet of continuous 1¼-inch pipe, made into a trombone coil 38 pipes high, 7 feet long, and 3 pipes deep, placed before the inlet window. In mild weather the condensation is so small that it goes to the sewer cold. When the outside temperature is low the temperature of the chilled water is higher, the increase in the rate of condensation slightly exceeding that of the chilling. The maximum rate of flow is about 1 cubic foot per minute.

The fan and combined heater, Fig. 2, with directly attached engine, is of the Sturtevant pattern and make, with a large by-pass over the heater. The fan is 6 feet in diameter, and at 250 revolutions per minute supplies 33,000 cubic feet of air.

The water of condensation is taken care of by a syphon trap made of a 4-inch pipe, 18 feet long, driven vertically into the ground, bushed at the top and tapped at the side. Through the bushing runs a 2½-inch pipe to within 1 foot of the bottom of the large pipe. This pipe is bushed at the top, tapped at the side and open at the bottom. The tap receives the water from the returns. The bushing receives a 1-inch pipe which drains the supply main at a higher pressure than the return, and runs inside the 2½-inch to within 1 foot of the bottom of the large pipe. Within the trap there may, therefore, be two pressures and two heights of water columns on the steam side, one vent discharging the water of both. The only resistance or friction is that due to the flow of water through the large pipes.

All steam for the building is brought from the Rogers building through an underground 6-inch pipe, about 1,000 feet long. The water condensed in the heating apparatus is metered, and the record preserved for the purpose of record and investigation.

The cost of the complete installment was nearly as follows:

Fan, engine, main coil (1,000 square feet), coolers, etc.	$1,445
6,380 square feet of direct steam surface, flue coils, mains, fittings, and placing	4,490
Construction of ducts and sheet-iron work	900
Johnson's electric service	1,355
Pump, Locke's regulators, sunken syphon trap, covering mains, etc.	1,775
Total	$[illegible]

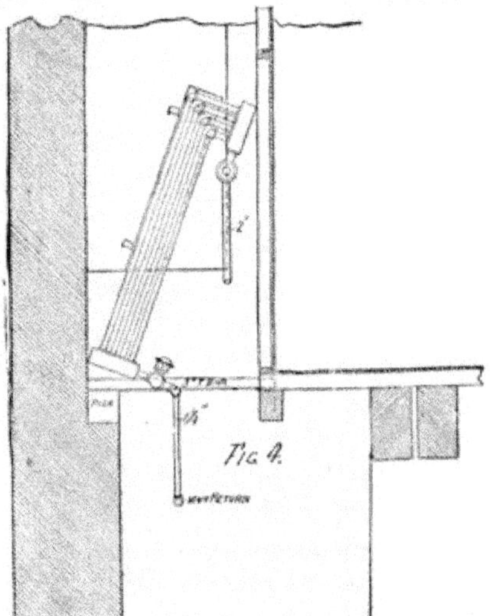

STEAM HEATING AT THE MASSACHUSETTS INSTITUTE OF TECHNOLOGY.

The direct heating surface is as great as though the heating of the building depended solely upon it, as insufficient boiler power threatened to make the use of the ventilating system impracticable in severe weather. Furthermore, if air is passed into the rooms at the temperature at which it is desired to keep such rooms, to maintain that temperature the direct surface must be as large as would be required for heating by direct radiation.

The system is practically a dual one, the capacity of either part being enough for the heating of the building. The ventilating system includes the main heater, cooler, fan, engine, duct, supplementary heater in flues, etc. Its cost may be put at $3,500, and the balance may be charged to the heating plant. The total heating surface is about 1 square foot to 110 cubic feet of space.

Handicapped by the conditions imposed, the work is of interest not so much as an illustration of a perfect system, as of what may be accomplished under difficulties.

The following summary will prove of interest:

Area of inlet windows	33	square feet.
Area through steam coil	20	"
Area of fan mouth	13.3	"
Area of fan discharge	12.2	"
Area of floor occupied by fan-room and heating chamber	120	"
Area of heating coil	1,200	"
Area of flues for supply and discharge of air	56	"
Number of flues for supply and discharge of air	8	"
Air volume supplied, cubic feet per hour	1,950,000	

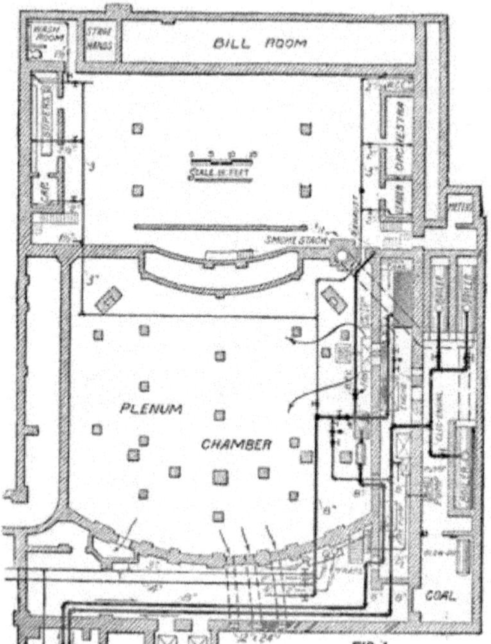

FIG. 1
BASEMENT PLAN

VENTILATION AND HEATING OF THE AMERICAN THEATER, NEW YORK CITY. (See next page.)

VENTILATION AND HEATING OF THE AMERICAN THEATER, NEW YORK.

PART I.—GENERAL DESCRIPTION OF THE HEATING AND VENTILATING PLANT.

The American Theater, at the corner of Forty-second Street and Eighth Avenue, New York City, was designed by Mr. Charles C. Haight, architect, of New York, while Mr. William J. Baldwin, of the same city, was the engineering contractor for the heating and ventilating plant. The theater proper occupies an area about 175x100 feet, and it is one of the largest in New York, there being seating capacity for about 2,700 persons. In designing the ventilating plant no expense was spared to make the system a most perfect one, and though the principle involved is not original it is said to be carried out in a more thorough manner than in any theater in this or any other country.

The theater is heated mainly by the indirect system, while a few direct-heating radiators are placed in the dressing-rooms, lobby, the rear of the stage, and other places where the heated air that is blown into the body of the theater would not be liable to penetrate. There are about 1,400 square feet of heating surface of direct radiators in the building, and about 2,500 square feet of heating surface in specially designed coils for the heating chamber in the basement. About 2,000,000 cubic feet of air per minute is drawn from the heating chamber by the fan and forced into the theater, thus giving about 660 cubic feet per person per minute, assuming the theater to hold 3,000 people.

The fresh air for the indirect system enters the building by a loggia or open gallery near the roof and descends to the heating chamber in the basement by means of an 8½x3-foot duct. An iron damper in this, controlled from the heating chamber, prevents an upward current when the fan is at rest. The air enters at one end of the chamber and near the floor, and rising, passes between the inclined coils to the fan. There is, however, an unobstructed passage at one side of the coils which allows the greater part of the air to pass directly to the fan. This passage can be closed by a switch valve or door swinging on a vertical axis, and by the partial opening or closing of this door the temperature of the air entering the theater can be regulated. This, however, is not the only way in which this can be done, as the coils are in separate sections, each controlled by a valve, so that any number may be in use at the

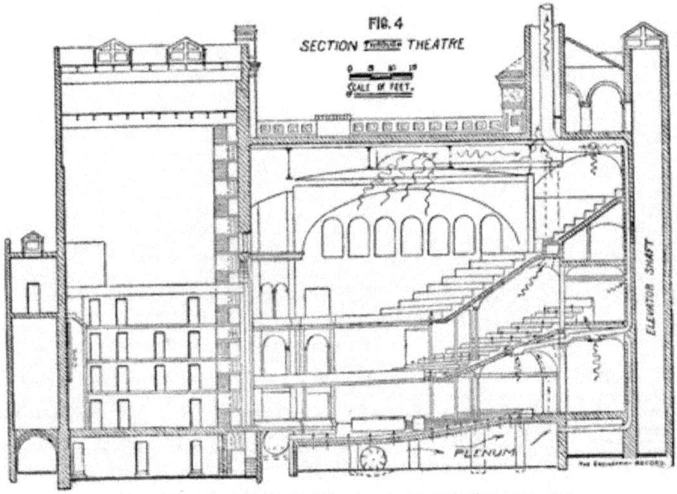

VENTILATION AND HEATING OF THE AMERICAN THEATER, NEW YORK CITY.

will of the operator. An opening through the wall of the coil chamber allows the passage of air to the plenum chamber. An 8-foot Sturtevant cone-wheel fan is placed opposite the opening, the shaft carrying the fan being supported by a pillow and spider bearing. The fan is driven by a belt from a 9x10-inch Ames engine.

The plenum chamber occupies all the space in the basement under the main floor of the theater, as will be seen by Fig. 1, the basement plan. The air is delivered to the lower floor by means of 341 openings that pierce the main floor in the locations shown by the small circles among the seats. Figures 2 and 3 show the locations of these openings, Fig. 3 being the plan of the balcony, which is ventilated by the same method. These openings are approximately under every seat on the ground floor and under every third seat in the balcony. A hood is placed over each opening to diffuse the air so that it will not interfere with the comfort of the occupants of the seats by causing a draft about their feet. Each opening has a sectional area of 7 square inches. The hoods will be described in detail later on.

delivered to the hall above that discharged into the theater proper. The latter is supposed to be at 70 degrees, or a little less, while the air for the hall will be heated by this radiator, or secondary coil, to about 140 degrees. The fresh air in all cases is shown by straight and the foul air by wavy arrows. By following the latter in the sectional view, Fig. 4, the foul-air ducts will readily be distinguished. The foul air underneath the balcony is carried off by ducts that run horizontally to the front wall, where they rise to the roof and there finally combine to form one circular flue 30 inches in diameter. The foul air underneath the gallery is carried off by two vertical flues, one of which is shown by the vertical dotted lines. The main exit for the air in the theater, however, is by a bell in the ceiling. A horizontal duct leads from the bell to a vertical masonry shaft which finally discharges the foul air into the atmosphere. The horizontal duct is provided with a damper controlled by the engineer from the plenum chamber, so that ventilation through the bell may be stopped at any time. The vertical shaft is 49 square feet in section.

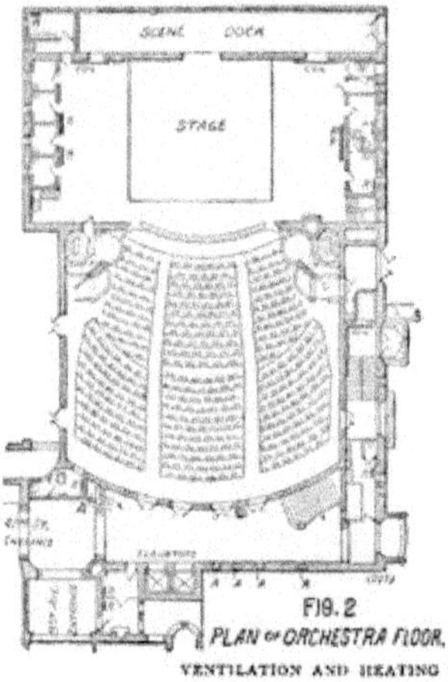

FIG. 2
PLAN OF ORCHESTRA FLOOR.

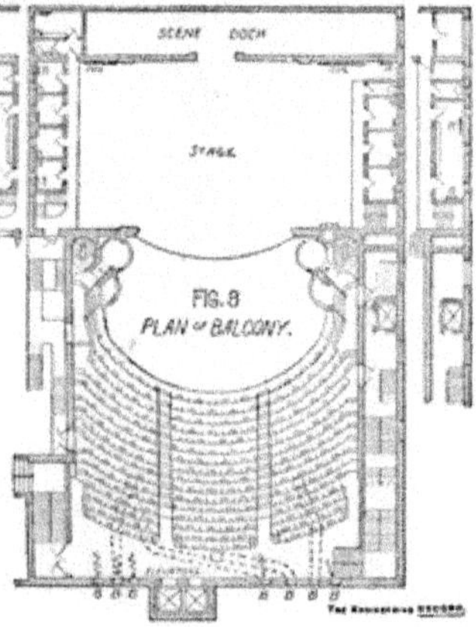

FIG. 3
PLAN OF BALCONY.

VENTILATION AND HEATING OF THE AMERICAN THEATER, NEW YORK CITY.

Several horizontal ducts connect the plenum chamber with the vertical ducts built in the front wall of the building. The latter leads the air into the space under the floor of the balcony, from which it is finally discharged by the hoods under the seats.

The locations marked A on the orchestra floor show the location of the vertical ducts furnishing fresh air to the balcony. The largest of these ducts has a sectional area of 16 square feet. This duct (marked C) beside supplying air to the space under the seats in the balcony also furnishes air to the main hall. A radiator is placed in the branch serving for this purpose to increase the temperature of the air

PART II.—DETAILS OF BOILER PLANT, HEATING COILS, AND AIR INLET HOODS.

The boiler plant consists of three horizontal return-tubular boilers 5 feet in diameter and 18 feet in length. The tubes are 64 in number, 18 feet in length and 4 inches in diameter. The shell of the boiler is made of ⅜-inch flange steel with a tensile strength of 60,000 pounds to the square inch. The reduction in area of the test specimen was not to have been less than 45 per cent. and the elongation less than 22 per cent. in 8 inches. The heads are of ½-inch steel, and each is braced by five gusset braces made of ⅜-inch steel plate. Owing to the narrowness of the boiler-room the boilers were supported by four 10½-inch I beams, each of sufficient length to have a foot on each end rest in the walls of the building. The I beams support each boiler by means of four wrought-iron hangers, the lower ends of which are riveted to the shell of the boiler by six ¾-inch rivets. Figure 1A shows an enlarged view of the method of connecting the hangers to the I beams.

Figure 1 shows the longitudinal section through one of the boilers and the cross-section through the two that are set in a battery. The setting of each is lined with firebrick. The steam leaves the boiler through an 8-inch main (see Fig. 2) leading to the electric light engines in another part of the building. Various branches supply the elevator, boiler feed, and house pumps. The exhaust returns through an 8-inch pipe to a 200 horse-power Wainwright feed-water heater and then passes through a grease extractor, designed by Mr. Baldwin. After passing through the grease extractor the exhaust line has several branches, one of which leads the exhaust steam to the direct radiators located in different parts of the building. A second branch leads to the heating coils in the fan or coil chamber, while a third pipe, 8 inches in diameter, serves as a free exhaust to the atmosphere. Each of these lines is provided with a Jenkins back-pressure valve. A live-steam connection is run from the main steam pipe to the branch leading to the direct radiators and also one to the pipe supplying the heating coils. The connection to these pipes is in each instance between the back-pressure valves before mentioned and the system which is to be heated. Each of the pipes supplying live steam has a Davis reducing valve set to open at three pounds pressure. A third back-pressure valve A will also be noticed placed in the line leading to the direct radiators. A glance at the sketch (Fig. 2) will show that each heating system, that supplying the heating coils and that supplying the direct radiators, is protected by two back-pressure valves, so that if one becomes deranged a second one is in readiness to prevent live steam from blowing through the reducing valve and out of the building through the free exhaust.

Figure 3 shows a plan and elevation of the heating coils that are placed in the coil chamber to heat the air forced into the theater. These coils contain about 2,500 square feet of radiating surface composed of 1-inch pipes placed 2½ inches between centers horizontally. The headers are of cast-iron specially designed for this plant. They are square in section, the inlet or supply pipe coming in one end of the header and near the top while the return leaves the header from the other end nearer the bottom. The air passes in a downward direction around them. Figure 4 shows the construction of the standard for supporting the coils, while Fig. 5 shows the miter coils placed against the rear wall of the stage.

Figure 6 shows several drawings of the specially designed cast-iron hoods that admit the air from the plenum chamber into the theater, these being placed under the seats. The hood or cover is held in position by the guides. A projection in the outer edge of a pair of these seats itself into slots cut out of the

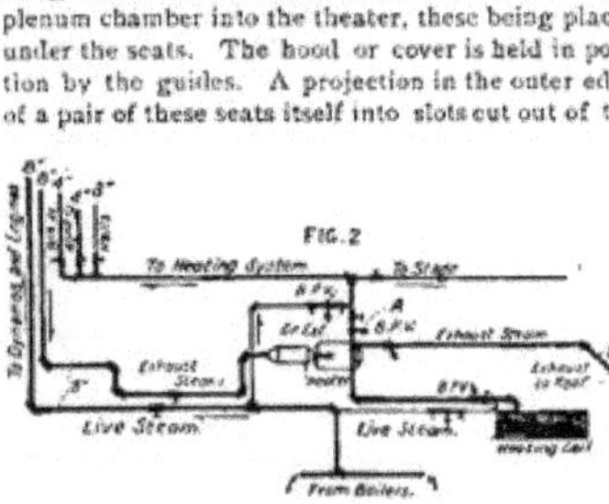

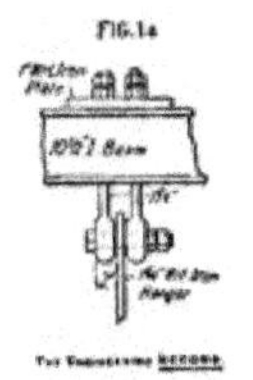

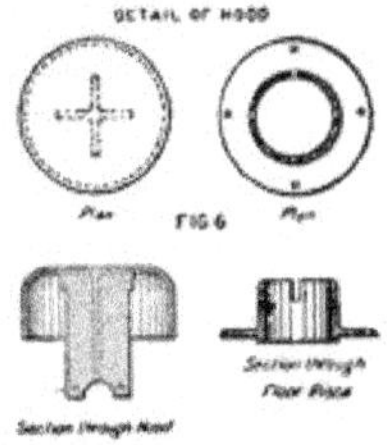

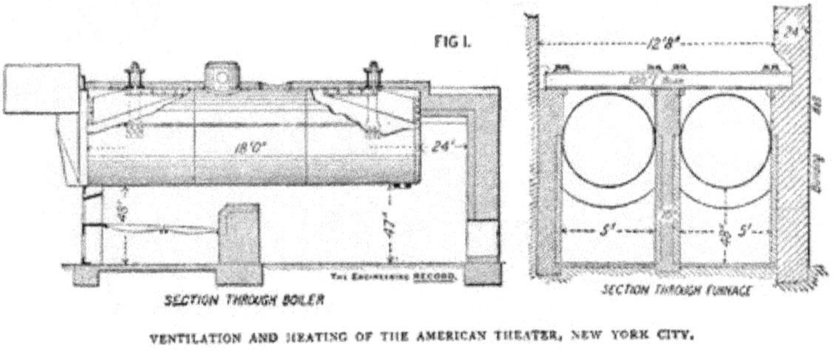

VENTILATION AND HEATING OF THE AMERICAN THEATER, NEW YORK CITY.

upper edge of the floor piece. These slots are cut to varying depths, thus allowing the hood to be adjusted so as to admit the desired amount of air.

HEATING AND VENTILATING THE NEW YORK MUSIC HALL.

The New York Music Hall, founded by Andrew Carnegie, and controlled by the Music Hall Company of New York, Limited, is located at the corner of Seventh Avenue and Fifty-seventh Street, New York City, and was opened to the public for the first time on the evening of May 5, 1891. The building, as its name indicates, is to serve, in the main, the purpose of musical entertainment, and has been designed to accommodate large audiences. One of the several important problems connected with its construction has, therefore, been, as might be readily anticipated, that of adequately heating and ventilating its several portions, and the exposition of the chief characteristics of the manner in which this has been accomplished is the purpose of this article.

Before taking up this matter, however, it may not be amiss to direct attention to some of the structural features of the building. Brick has been employed for the exterior, the decorations being of terra-cotta. The principal doorways are approached by a series of steps, 80 feet broad. The chief feature of the interior is the main concert hall, shown in Figs. 1 and 2, with a seating capacity of 3,000, and standing room for about 1,000 more. The entrance to this is on Fifty-seventh Street through a vestibule 70 feet long, with a vaulted ceiling 25 feet high. This hall was designed purely as a concert hall, and is, therefore, not equipped in any way with theatrical devices, and has neither drop curtain nor footlights. The parquet, capable of seating 1,000 persons, has nine exits upon the corridors surrounding it, the corridors, as shown, continuing entirely around the building. Above the parquet are two tiers of boxes, the dress circle, and the balcony. The arrangement of these several tiers is different from the usual method, in that they do not extend entirely around the three sides of the house, stopping at the line of the proscenium, but terminate on the side walls at points further and further back from the front of the auditorium, gradually expanding the hall and advantageously displaying the magnificent ceiling which spans the great space. The prevailing colors in the decoration of the main hall are in ivory-white, gold, and rose.

The second large room in the building is known as the "Recital Hall," and is located under the main hall.

Its general disposition is shown in Fig. 3. Its seating capacity is 1,200, and the decorations are similar to those of the main hall. Above the main hall, and extending also laterally, are a series of lodge-rooms, smoking and committee rooms, banquet hall, etc., arranged in the manner shown in Fig. 4.

The heating and ventilating, in the main, is accomplished by a plenum-vacuum system, both blowers and exhausters being used to handle the immense quantities of air requisite for the purpose.

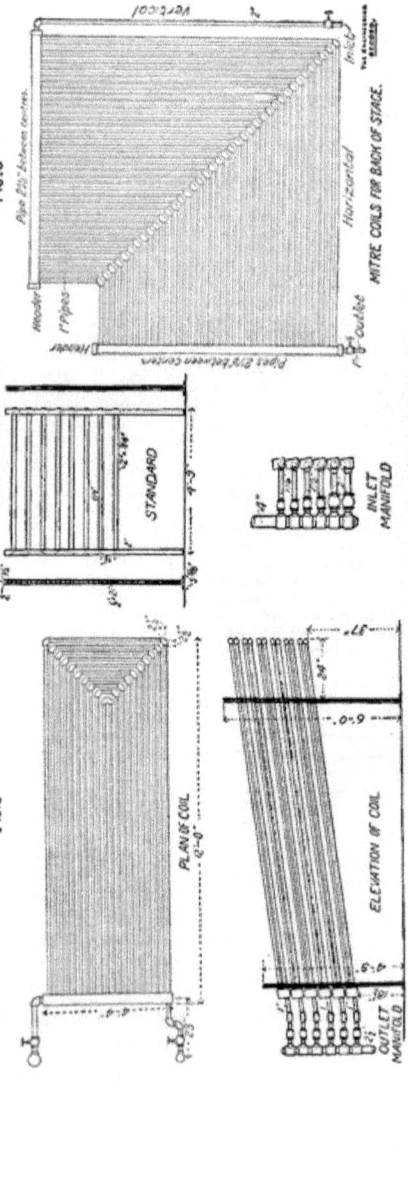

VENTILATION AND HEATING OF THE AMERICAN THEATER, NEW YORK CITY.

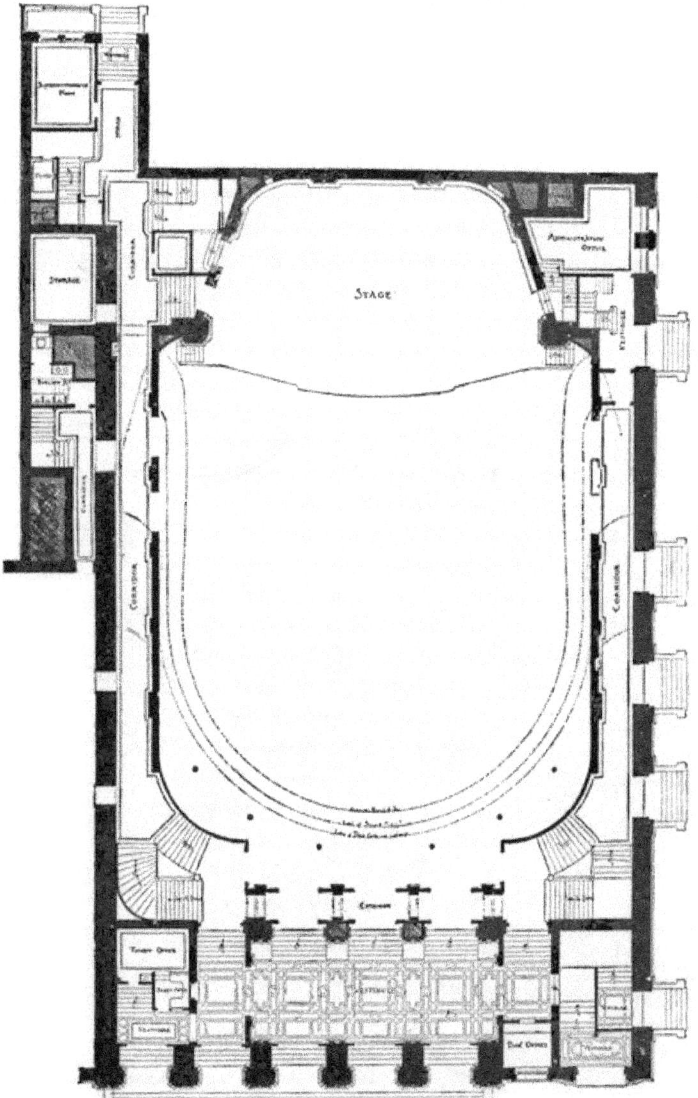

FIG. 2.—HEATING AND VENTILATING THE NEW YORK MUSIC HALL.

A noteworthy feature of the manner of air supply is found in the fact that the fresh air, at any temperature desired, is made to enter through perforations in or near the ceilings, passes downward through the halls and different rooms, and passes out through exhaust registers near the floor or through perforated risers in the latter. Ten million cubic feet of air per minute are supplied to the building, representing a complete change of air six times per hour, and the aggregate area of inlet and outlet openings amounts to about 2,000 square feet. This large area brings the velocities of influx and efflux down to about 1 foot per second, avoiding objectionable drafts.

The general arrangement of blowers, fresh-air ducts, engines, etc., is shown in Fig. 5, which represents a partial plan of the cellar portion of the building. The air supplied to the building is taken through a vertical inlet shaft, marked "fresh-air shaft" in Figs. 4 and 5, and simply "air shaft" in Fig. 1. This extends from the top to the bottom of the building, and the air is led from it to four Sturtevant pressure blowers B B and B' B'. That part of the building known as the music hall proper, and comprising the main portion outlined in Fig. 1, is heated and ventilated by a duplex system complete in itself. The parts of the structure known as the lateral building, and also the Fifty-sixth Street building, shown in outline by the upper and lateral projections in Figs. 4 and 5, also have jointly an independent duplex system.

The Music Hall system comprises the two 7-foot blowers B B, Fig. 5, driven by the 10x12-inch horizontal engines E E, making from 150 to 175 revolutions per minute, and placed in front of each fan under the air duct, one being right hand and the other left hand. The fans are 4 feet wide at the inlet, and have a top horizontal discharge measuring 4x4 feet. Both outlets discharge into one main duct F suspended from the ceiling, dampers D D being fitted at the intersection to equalize the blast from each fan. In case only one fan is working, the pressure caused by it will automatically close the duct from the idle fan. The main duct, it will be observed, is fitted with branch ducts F^1 and F^2, each having a damper D, and supplying the recital and main halls, and also the air-supply duct G to the stage. The larger portion of the air supply through the duct F is however delivered into the shaft C, which supplies the boxes in the music hall proper, and the space above the music hall main ceiling, from which, as already explained, the air passes through perforations in the ceiling. The main duct F is made of No. 14 iron, and has a free area of 32 square feet. The openings J J permit the escape of some fresh air into the engine-room.

Before entering the blowers B B the air, on its way from the fresh-air shaft, passes through Sturtevant heaters H H, consisting of four groups containing 13,664 lineal feet of 1¼-inch pipe, equivalent to 6,620 square feet of heating surface, the bases and fittings included.

The lateral and Fifty-sixth Street building system has also two Sturtevant blowers, B' B', each 6 feet in diameter, with 42x42-inch inlets. Each has a top horizontal discharge measuring 42 inches square

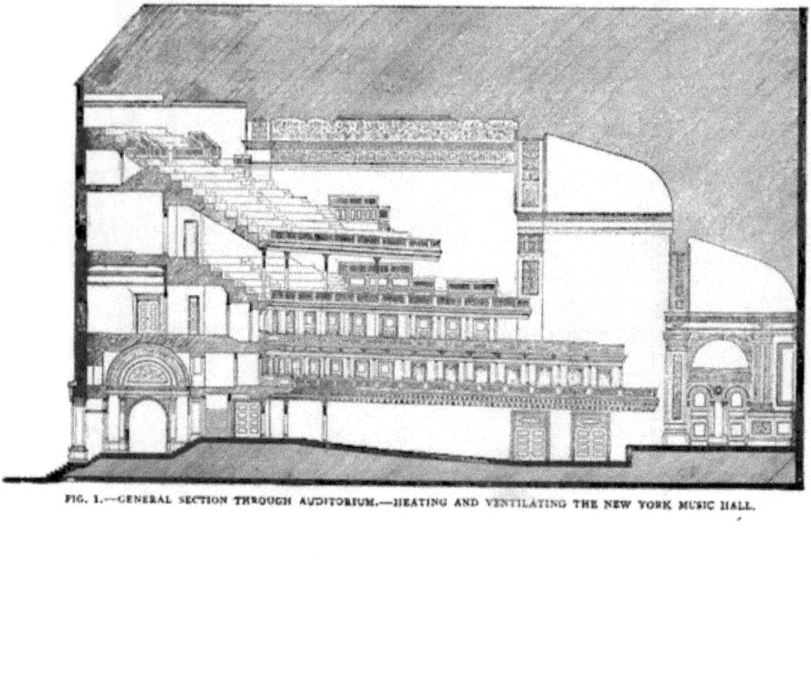

FIG. 1.—GENERAL SECTION THROUGH AUDITORIUM.—HEATING AND VENTILATING THE NEW YORK MUSIC HALL.

in the clear and giving, thus, an area of opening of 12¼ square feet. Each fan is driven by a vertical 9x12-inch engine E', making from 125 to 150 revolutions per minute. The outlets from these two blowers also discharge into one main duct, F^2, having a sectional area of 30 square feet, and suspended from the ceiling. After leaving the blowers this duct branches into two ducts, in opposite directions, one of 12¼ square feet area, going to the Fifty-sixth Street building, and the other of 24 square feet area, leading to the lateral building. Suitable dampers D are here again provided to properly regulate the flow of air. Besides these two main branches there is a third branch leading to the air-supply duct C for the music hall stage. The duct for the Fifty-sixth Street building discharges into the shaft U, from which the air is further distributed to the several points to be supplied.

The fresh-air heater H' for this second system is also of the Sturtevant type, and consists of two groups containing 6,832 lineal feet of 1¼-inch pipe, equivalent to 3,310 square feet of heating surface, including, as before, bases and fittings. The heaters are incased in steel plate of No. 12 gauge, and are so connected to the respective inlets of the several blowers that all the air drawn in by them will have passed through the heaters. Each heater is under control by individual steam and return connections, valves, etc., and the temperature of the warm air supplied to the buildings is regulated by turning on or shutting off the steam supply to the heater sections, of which there are 42. One square foot

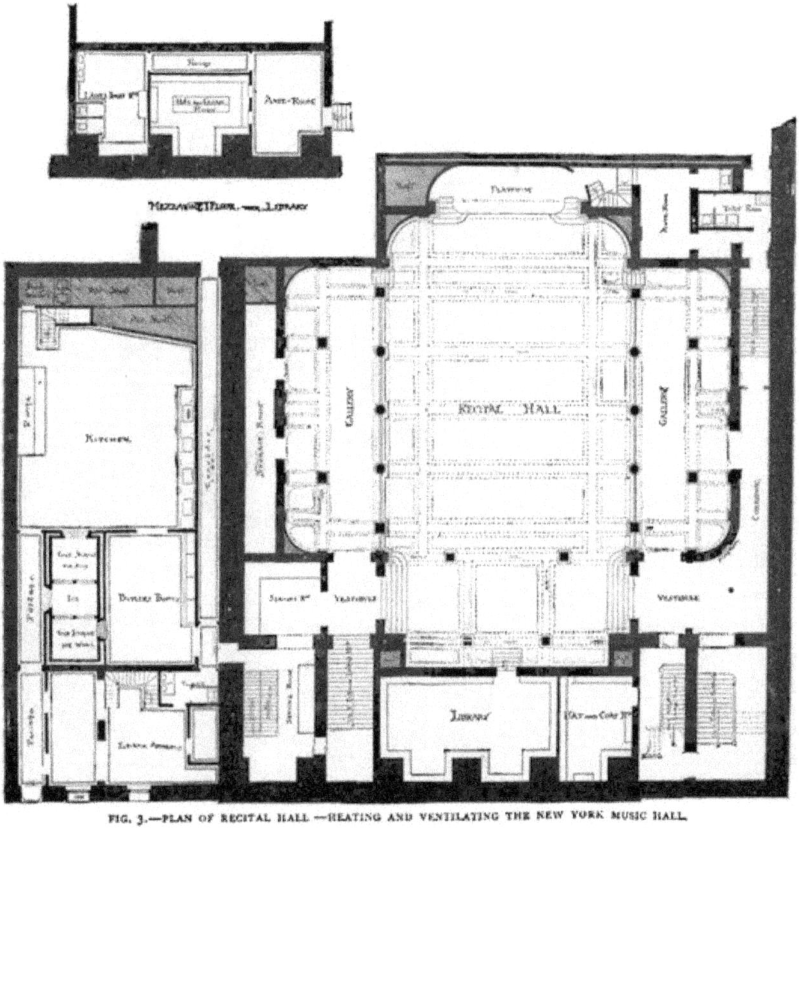

FIG. 3.—PLAN OF RECITAL HALL.—HEATING AND VENTILATING THE NEW YORK MUSIC HALL.

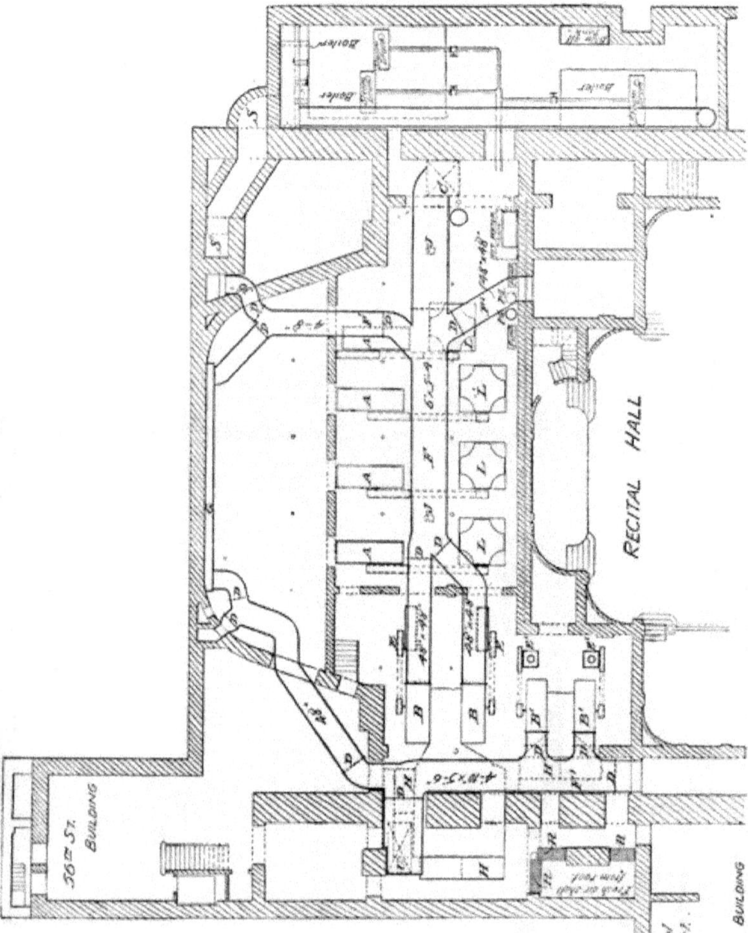

HEATING AND VENTILATING THE NEW YORK MUSIC HALL.

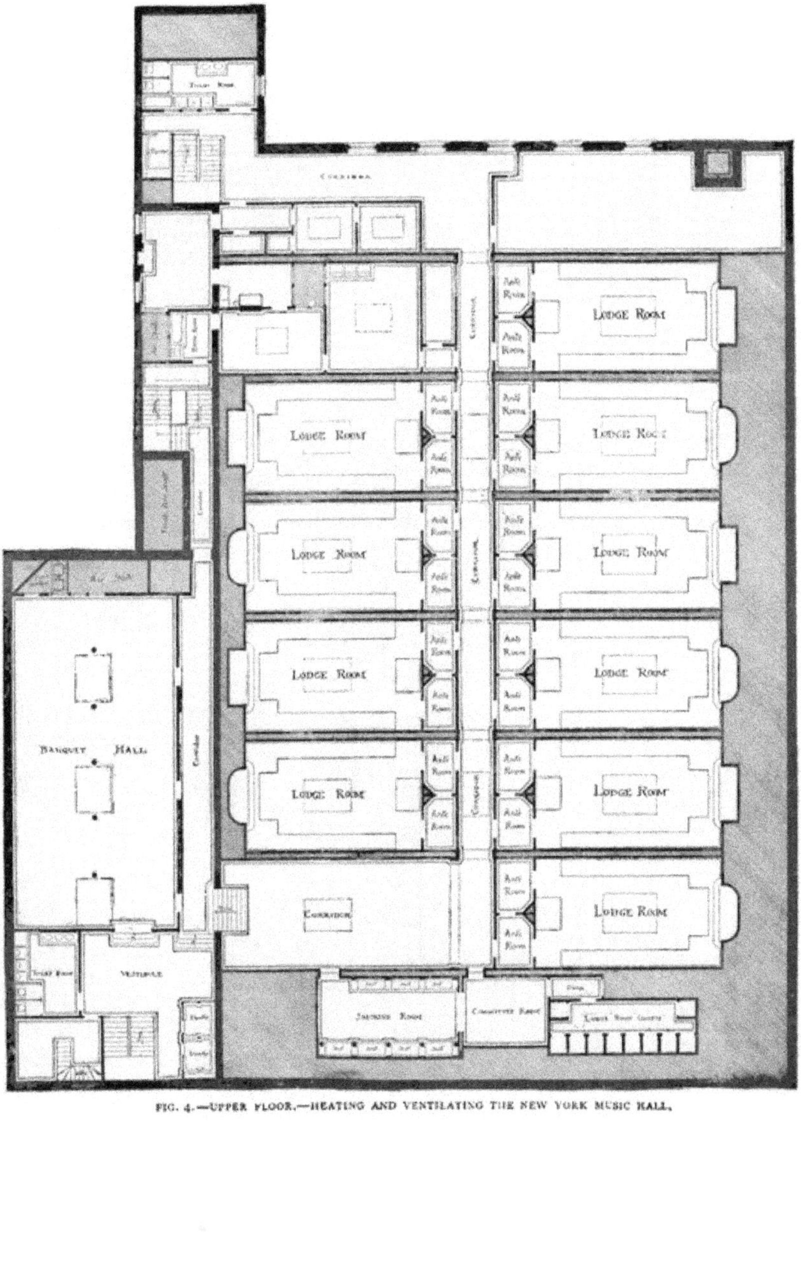

FIG. 4.—UPPER FLOOR.—HEATING AND VENTILATING THE NEW YORK MUSIC HALL.

of heating surface is provided for every 1,000 cubic feet of air entering the buildings. For the purpose of cooling the air supplied to the building in warm weather, ice racks R R R are provided, capable of holding six tons.

The horizontal blower engines E E were built by the Porter Manufacturing Company, of Syracuse, N. Y., and the vertical engines E' E', by the New York Safety Steam Power Company, of New York.

As previously stated, all the air supplied to the different parts of the building is again drawn out by a separate fan system, the exhaust taking place at or near the floor levels. This vitiated air is led through a duct system to an exhaust shaft, Fig. 5, situated in the lateral building, and shown also in Figs. 3 and 4. This shaft leads to the roof of the building, where the exhaust plant is located. A plan of this roof portion, showing the location of engines and fans, is given in Fig. 6. The whole engine and fan compartment is, as will be readily understood, simply an enlargement of the upper exhaust duct which leads from the upper end of the exhaust shaft at the rear end of the lateral building to the front end. There are, as shown, three 6-foot Sturtevant fans F, driven by three 9x12-inch horizontal Porter engines E. The fan outlets O are connected each with a separate No. 18 galvanized-iron duct, extended about 4 feet above the roof and furnished with a protecting cap. The small shaft D, of triangular section, discharging directly above the

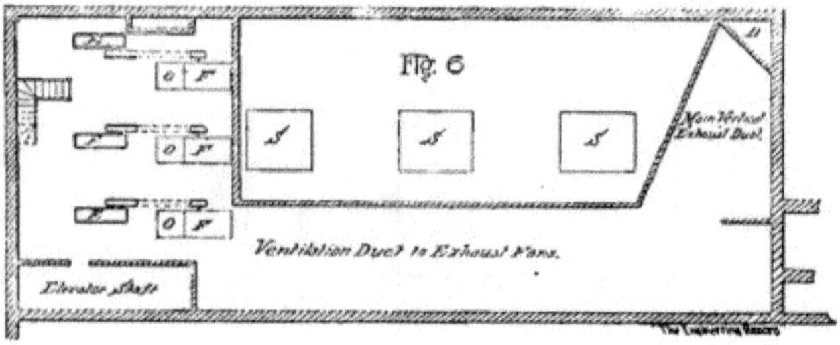

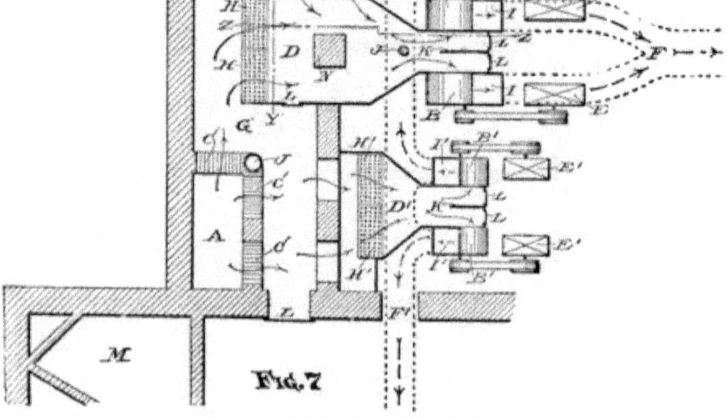

HEATING AND VENTILATING THE NEW YORK MUSIC HALL.

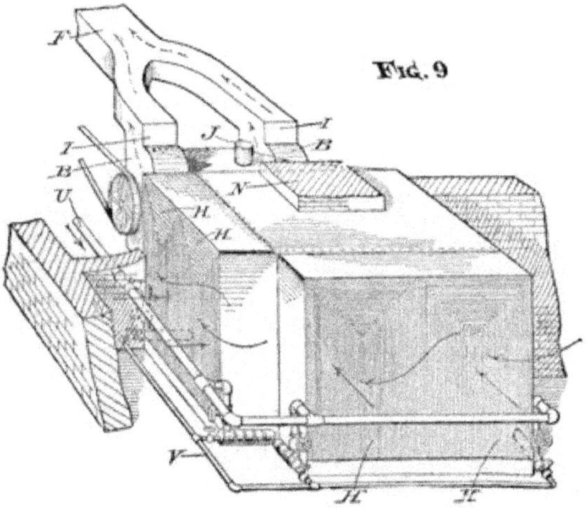

FIG. 9

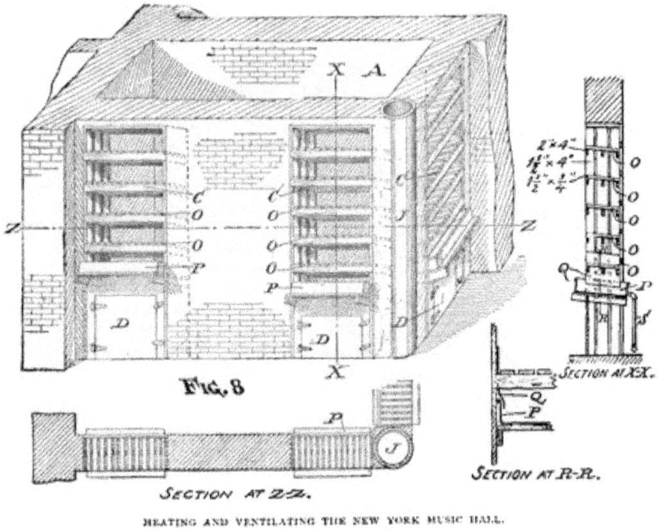

FIG. 8

HEATING AND VENTILATING THE NEW YORK MUSIC HALL.

building, is an exhaust shaft leading up from the kitchen, where a Blackman fan is located for ventilating purposes. S S S are skylights.

The heating of the building throughout is designed principally for the use of exhaust steam, but, in addition, live-steam heating connections are supplied. A number of direct radiators are also placed in different parts of the building. The details of this part of the work, however, we have not attempted to enter into in this article.

The location of the three boilers, which are of the sectional type, supplied by the Abendroth & Root Mfg. Co., of New York, is shown in Fig. 5. S is the smoke flue. The boilers are arranged in two batteries, the single boiler being of 175 H. P., and each of the other two is rated at 150 H. P., making 475 H. P. boiler capacity in all. The boilers supply steam to four Straight-Line electric light engines A, driving Thomson-Houston dynamos L, and to the seven engines connected with the heating and ventilating system.

The architect of the Music Hall is Mr. William Burnet Tuthill, of New York. Messrs. Adler & Sullivan, of Chicago, are the associate architects. The steam power, heating and ventilating plant was designed, and its installation supervised, by Mr. Alfred R. Wolff, of New York; and Messrs. Johnson & Morris, of New York, were the contractors for the work.

PART II.—DETAILS OF HEATING AND COOLING APPARATUS FOR FRESH-AIR SUPPLY.

All the fresh air, a full delivery of about ten million cubic feet per minute, is taken above the roof and brought down the shaft A, Fig. 7, which is 6x12 feet in size, to the basement distributing chamber G, which supplies the fan suctions. Figure 7 is an enlarged reproduction of a part of Fig. 5, and shows the heating, cooling, and blowing plant. In warm weather ice is placed in racks C C C, over which the air must pass to enter the chamber G. From the chamber G the air is drawn by the blast fan into

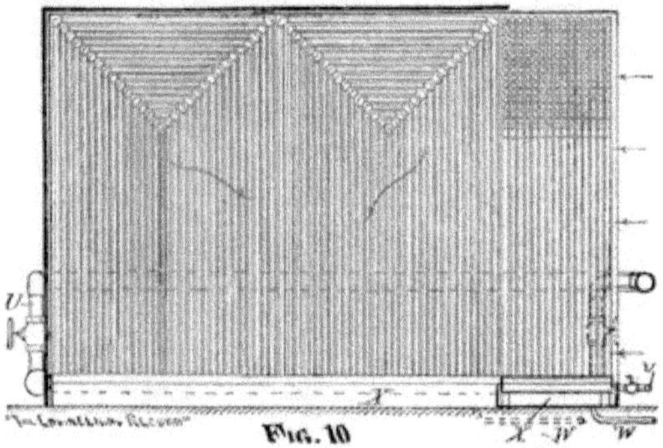

Fig. 10

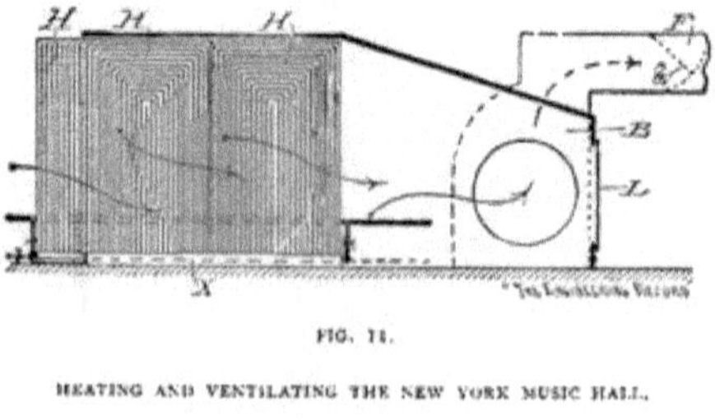

FIG. 11.

HEATING AND VENTILATING THE NEW YORK MUSIC HALL.

chambers D and D_1. Two engines E E with 10x 12-inch cylinders drive the 7-foot blowers B B, which deliver the air at a pressure of about three ounces to the main duct F, which has a net area of 32 square feet, and supplies the music hall and the main building. The fresh air can, at all times, only enter the chambers D and D_1 by passing through the steam radiators H H and H_1 H_1, the former having a total radiating surface of about 6,600 square feet, and the latter of about 3,300 square feet. Of course, in warm weather the steam is partly or entirely cut off from the radiators, so that the air passes through them with little or no rise of temperature. The two 6-foot fans are driven by engines E_1 E_1, with 9x12-inch cylinders, and deliver into the blast main F_1, which has a net cross-section of 30 square feet, and supplies the rest of the building. I I and I_1 I_1 are the outlets of the fans and communicate with the blast mains F and F_1, which are shown dotted, because they are close to the basement ceiling and really above the plane of section for this figure. K K are diaphragms to distribute the air between the two fans of each pair. L L, etc., are doors, J is a column and N a foundation pier. M is the ventilation shaft which receives foul air from a basement gallery and ducts on the upper floors and discharges above the roof through an exhaust chamber and fans.

Figure 8 shows the bottom of the fresh-air shaft A and its outlets C C C, through which the air must pass into the distribution chamber G. The ice is placed on racks O O, etc., and drips into galvanized-iron pans P P, etc., which slide in galvanized-iron guides

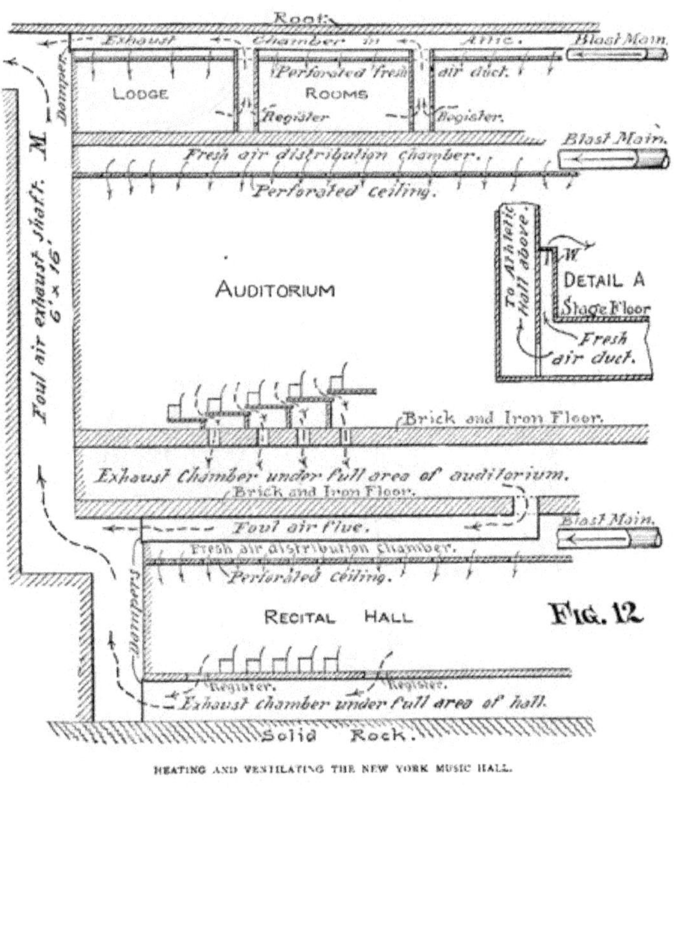

FIG. 12

HEATING AND VENTILATING THE NEW YORK MUSIC HALL.

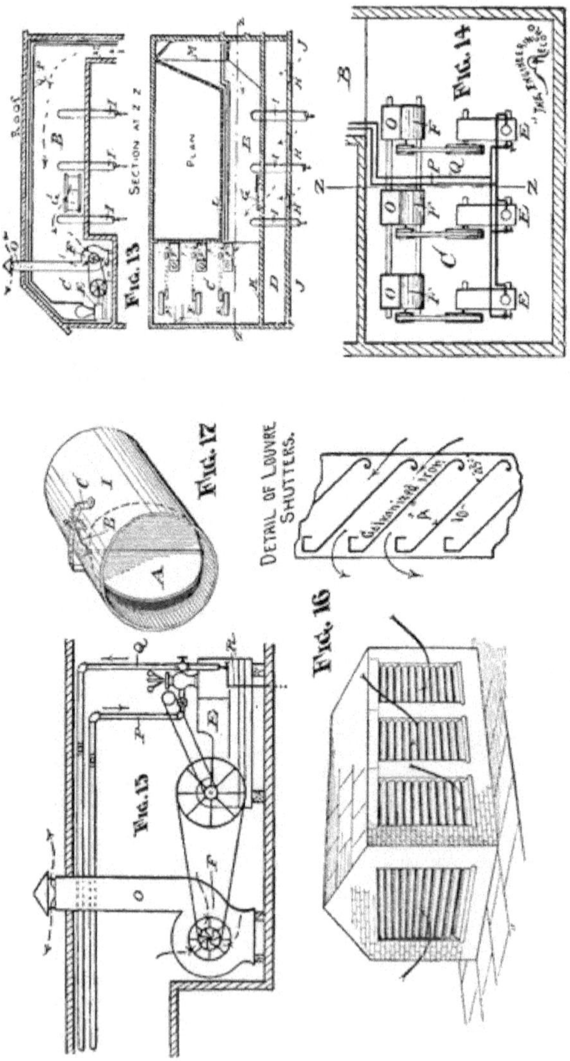

HEATING AND VENTILATING THE NEW YORK MUSIC HALL.

Q Q, etc. S S, etc., are waste pipes, and D D D are doors. Figure 9 is a perspective view from T, Fig. 7, of the chamber D, two sides of which are composed of radiators H H, etc., and the top and remaining sides of No. 12 steel plates. U is the steam supply and V the drip pipe with independent connections to each of the four radiators. Figure 10 is a section and elevation at Y Y, Fig. 7; X X are iron safes, and W W their wastes. Figure 11 is a section at Z Z Fig. 7, showing the inlet into fan B, and a check-valve damper & in the branch to main F, which opens with the blast, but closes against any back pressure so as to prevent the possibility of suction through it if its fan was stopped and the companion was running.

PART III.—DISTRIBUTION OF FRESH AIR, VENTILATION SHAFT AND EXHAUST CHAMBERS, EXHAUST ENGINES, DETAILS OF LOUVER AND DAMPER.

FIGURE 12 is a general vertical section of the main building, not to scale or accurate position, but intended as a diagram to show the distribution of fresh air and the withdrawal of foul air in the principal rooms. Detail A shows the method of supplying extra heat and air to the stage through perforations in the horizontal top of the 6-foot wainscoting W, around the walls. Figure 13 shows the top of ventilation shaft M, Figs. 7 and 12, and the gallery B connecting with the attic exhaust chamber C. The foul air is withdrawn as indicated by the dotted arrows, by the 6-foot fans F F F, driven by the 9x12-inch engines E E E, which are served by branches (here omitted, to avoid confusion) from steam pipes P Q. The air withdrawn is discharged through the stacks O O O above the roof. J J are lodge-rooms which receive fresh air through ducts I I I, and discharge foul air through registers H H H into the exhaust chamber D, whence it is drawn through sliding damper G into the gallery B. Pipes I I I do not really appear on the section at Z Z, but are here shown as if on the wall L, instead of on wall K, so as to indicate their position. Figure 14 is an enlarged plan of exhaust chamber C, Fig. 13, showing connections of the steam pipe P and exhaust pipe Q. Figure 15 is an enlarged elevation and section at Z Z, Fig. 14. Figure 16 shows the house on the roof, which covers the top of the fresh-air shaft A, Fig. 7. The house is about 10x18x10 feet high, with brick walls and a tin roof. Figure 17 shows the butterfly damper A, in the fresh-air main I, Fig. 13. Its axis is fastened to a crank B which moves on a horizontal radial guide C, at any point of which it may be secured by a pin.

HEATING OF PUBLIC BUILDINGS.

REMODELING THE VENTILATING PLANT IN A NEW YORK COURT-HOUSE.

The New York County Court-House, which was erected before 1870, contains a ventilating plant which was entirely remodeled in 1892. The building was formerly warmed and ventilated by a fan system divided into four sections, so that each quarter of the building was provided with its own independent fan driven by a belt from a small slow-speed Wright engine. Steam was then supplied to the engines and heating coils by four locomotive boilers located under the sidewalk and in the same location as the boilers shown on the plan, Fig. 1.

Each fan drew its supply of air from a cellar window and discharged it into a brick distributing duct that was built under the cellar floor and which terminated in coil chambers containing large coils of 1-inch pipe. Flues led from these coil chambers to the various rooms to be heated. Nearly all of these flues were 18"x2'6" in size.

Owing to poor circulation of steam in the coils the entering air in excessively cold weather froze the water in the coils and burst the pipes. To prevent this it was necessary to stop the fan and only allow such a quantity of air to flow through the coils as would be moved by natural ventilation. The quantity of air being too small for a proper ventilation, numerous complaints were made, which ended in the renovating of the entire plant. Tempering coils were placed in the cold-air ducts leading to the fans, sufficient in surface to warm the entering air to 70° Fahr. before passing to the newly put in pin coils at the base of the flues. The flues which take their air supply from these coil chambers can either take the air after it has passed through these secondary coils or before it has been in contact with them, this being permitted by the extension of the old vertical flues downward to the lower part of the coil chamber, as shown by the sectional view, Fig. 2. The direction of the current of air through the coils is controlled by a damper A actuated by a Johnson thermostat placed in the room to be heated. When the room becomes too warm the damper is so moved that the air enters the flue through the bottom opening B. As each flue has its independent damper controlled by the thermostat in the room to which the flue leads, it will be seen that when a room becomes too warm the air entering it is at a temperature of 70 degrees, while the heating power of the secondary coils is entirely used in heating some other room which is at a lower temperature. To avoid any possibility of a back-flow of heated air from the coils down to the lower flue openings C, a thermostat is placed in the lower part of the heating chamber so that if the air at that point becomes too warm it will shut off the supply of steam from the coils. Impure air is removed from the rooms through registers at the floor and ceiling, both entering into a common flue. A damper D is placed in the upper one, and it is moved in conjunction with the damper A in the hot-air flue. When the room gets too warm the damper in the vent register opens as the damper A closes so that the upper layers of air in the room which are the warmest will pass out as quickly as possible. When, on the other hand, the room is too cool the damper in the vent register closes, thus compelling the warm air as it enters the room to gradually sink and force the cold air at the bottom out through the vent shaft.

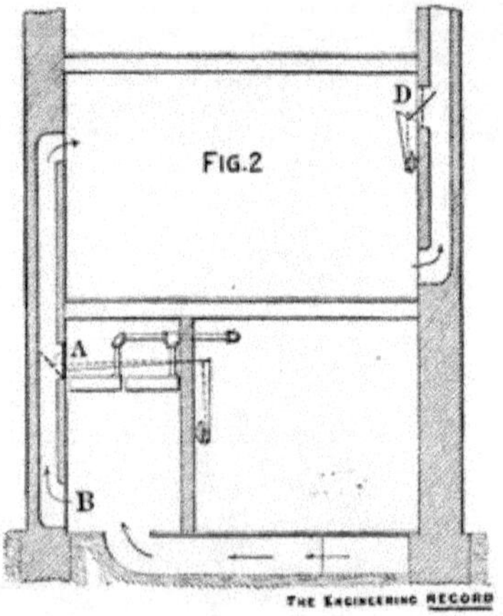

Other improvements made in the apparatus were the removal of the old locomotive boilers and putting in four horizontal tubular boilers. The vertical engines were replaced by two Skinner engines, and so connected by countershafts and belts that either engine may run the four fans.

The work of improving this plant was done by the Q. N. Evans Construction Company, of New York City, and to them we are indebted for the data from which this description was prepared.

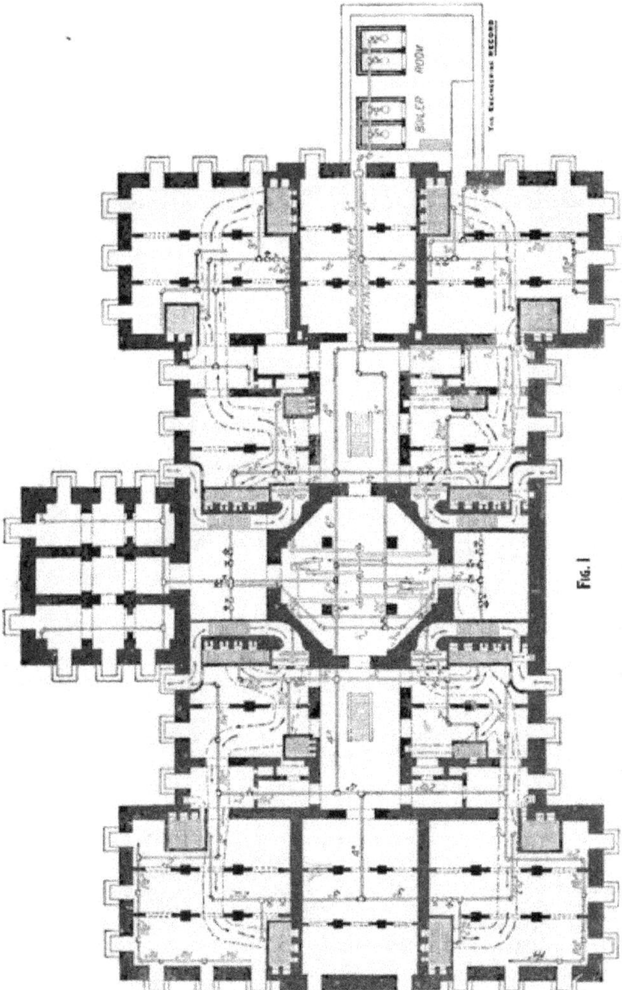

FIG. 1

REMODELING THE VENTILATING PLANT IN THE COUNTY COURT-HOUSE, NEW YORK CITY.

HEATING AND VENTILATION OF THE SUFFOLK COUNTY COURT-HOUSE.

PART I.—GENERAL DESCRIPTION, PLAN AND VERTICAL SECTION OF BOILER-ROOM.

AMONG the new public buildings of Boston, Mass., the structure completed in 1892 for the accommodation of the county officials and courts of Suffolk County is prominent. It is a brick and iron building about 450x180 in extreme dimensions and is located on the eastern acclivity of the hill which is crowned by the gilded dome of the State House. The building has many public and legal halls, offices, single and in suites, beside the city jail, which forms a part of the same construction and is included in the heating and ventilating system, which was installed by Samuel I. Pope & Co., of Chicago, Ill., in accordance with the plans and specifications of Bartlett, Hayward & Co., of Baltimore, Md. George A. Clough was architect of the building. The above-named contractors executed their work for a little over $122,000, and believed it to be the largest hot-water job that had at that time been let in one contract, although they had received $128,879 for the Cincinnati Post-Office building, which was done when materials were more expensive. Among other jobs of a nature and magnitude to invite comparison are the Chicago and St. Louis Post-Offices, which were done by Bartlett, Hayward & Co., of Baltimore, Md., and the Army and Navy building in Washington, D. C., which was done by sections at various times.

The Suffolk County Court-House has a volume of about 4,500,000 cubic feet, and by reason of its situation is somewhat exposed to sea winds, and is subject to low temperatures in the colder months. There are many external walls and numerous large windows present great areas of radiating surface for the loss of internal heat, so that it was decided beside furnishing warm fresh air throughout the corridors, large rooms and halls, to provide local heaters in the offices and in the exposed positions under windows, etc., to temper the incoming currents of air.

The heating plant is a low-temperature hot-water system, operated by boilers situated in the sub-basement, furnishing the heated water to the radiator coils of cast and wrought-iron pipe, distributed throughout the building, the indirect radiators being located in the basement, and the direct and direct-indirect radiators being mostly set in the recesses of the windows. These radiator coils warm the fresh air, which is introduced beneath the cast-iron sub-sills on the various stories, and the exterior windows and opening of basement.

The boilers for the heating apparatus are 12 in number, 48 inches in diameter and 16 feet long, containing 72 3½-inch flues. These 12 heating boilers are set in one battery, with 14-inch pipes for flow and return pipes, connecting into a 30-inch main flow pipe. This main flow pipe starts at 30 inches from the boilers, and is gradually reduced as branches are taken off. It is of cast iron from 30 inches to 6 inches inclusive. There are also two boilers for power purposes, 54 inches in diameter and 15 feet long, containing 47 3½-inch flues. These boilers are for operating elevators and for pumping water for supplying the building.

The air supply to the direct radiators being on a level with the tops of the radiators fresh-air ducts are formed in the window recesses behind the radiators to deflect the air to the bottom of the radiators. These ducts are formed and protected from the radiant heat of the radiators by non-conducting aprons; the amount of fresh air introduced being controlled by a damper situated beneath the bottom of the apron. The indirect radiators are situated in the basement. They are set upon brick piers, and encased by brick walls, the air supply being conducted to the chamber beneath the radiators by brick and galvanized-iron ducts, and the supply of air being controlled by

Fig. 2

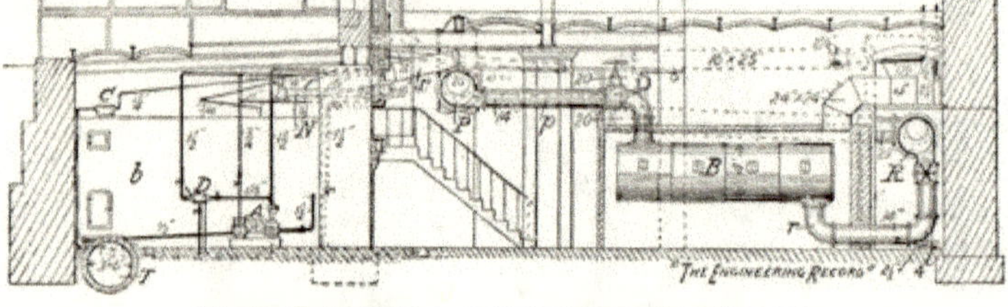

HEATING AND VENTILATION OF THE SUFFOLK COUNTY COURT-HOUSE.

Fig. 1

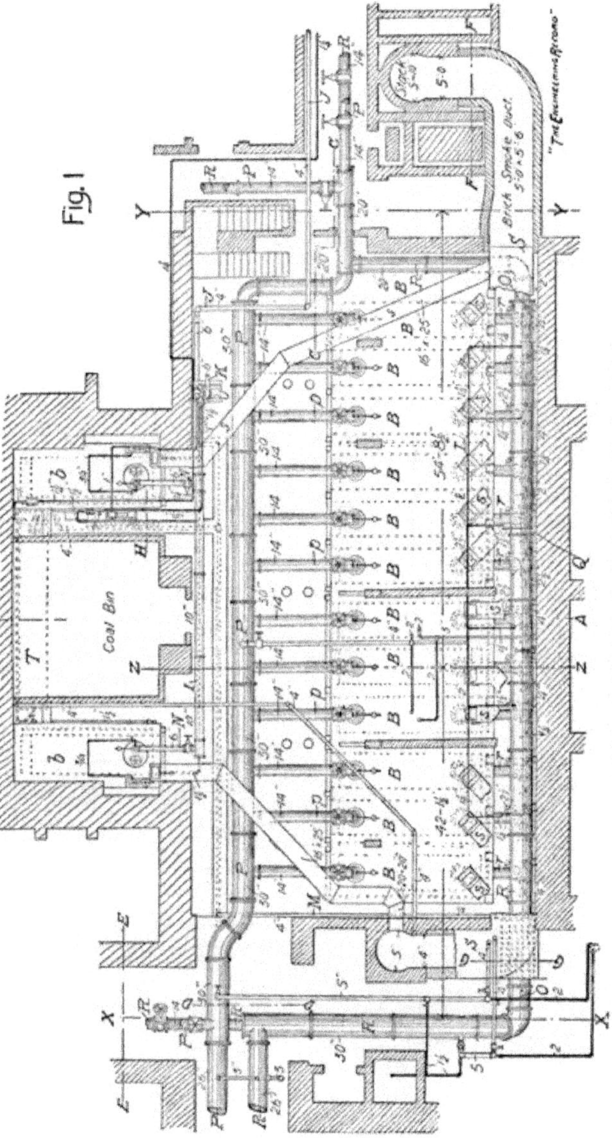

HEATING AND VENTILATION OF THE SUFFOLK COUNTY COURT-HOUSE.

dampers. There are about 80,000 square feet of indirect radiation placed in brick chambers, and about 80,000 lineal feet of direct and direct-indirect radiation. The indirect radiation is 3-inch cast-iron coils; the indirect and indirect-direct radiation is 1-inch horizontal radiators.

All of the vitiated air in the building is expelled through ventilating flues, which have a forced circulation and discharge above the roof. The schedule of material submitted by the commissioners for the contractors to estimate from covered 60 pages, and the contract provided for 80 per cent. payments monthly, as the work progressed, for material delivered at the building; 10 per cent. of the retained amount to be paid when the job was completed and 10 per cent. after the apparatus was run one heating season.

Figure 1 is a plan of the boilers and connections in the boiler-room in the sub-basement, and Fig. 2 is

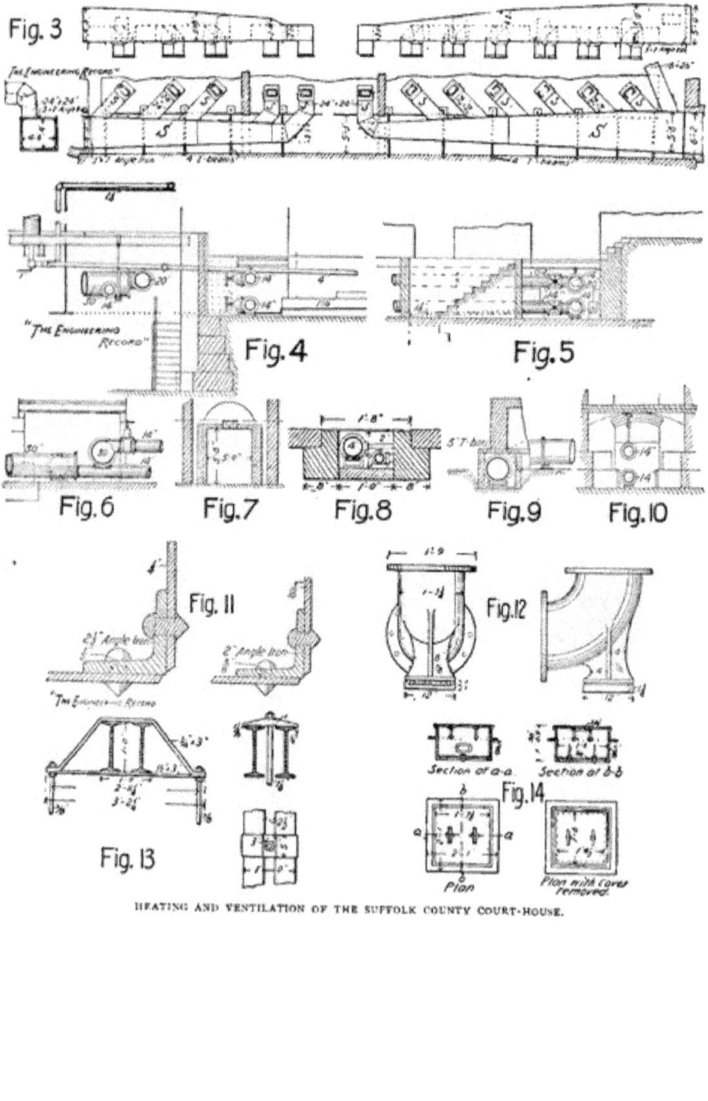

HEATING AND VENTILATION OF THE SUFFOLK COUNTY COURT-HOUSE.

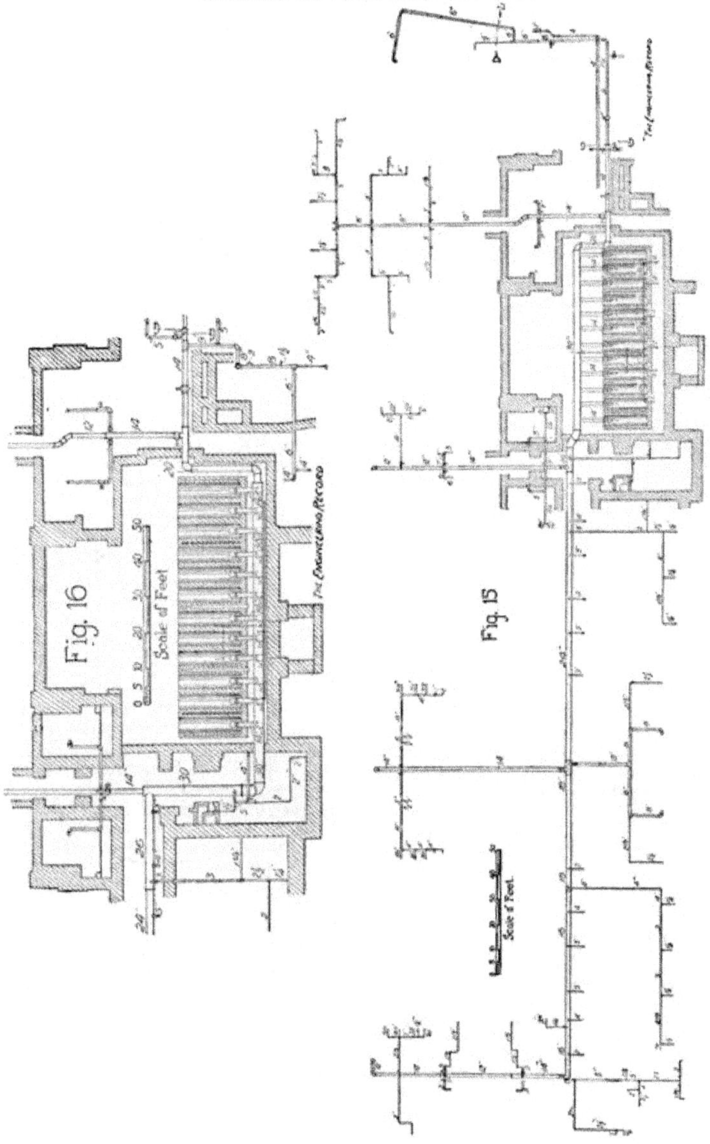

HEATING AND VENTILATION OF THE SUFFOLK COUNTY COURT-HOUSE.

a vertical section at A A, Fig. 1. B B, etc., are the hot-water boilers with 14-inch connections *f* and *r* to the 20-inch flow and return mains F and R respectively, so arranged that each boiler can be cut out at will without interfering with any others in the battery. In these figures *b b* are the steam boilers, with branches E E to a main supply pipe I. T is a flow-off tank, and P is one of two Knowles No. 4 pumps for boiler feed and house and fire service, which are so arranged and connected as to be interchangeable for these different duties. In these figures *s s* are branches to the large flue S which delivers the smoke to the brick stacks 5 feet and 5 feet 10 inches in size, and 100 feet high. F is a heater for supplying hot water to the toilet-rooms, etc., in the building. C is a Jones steam trap, D is a No. 7 Korting injector, double tube, front lever style, and H is a damper regulator for the steam boilers.

PART II.—DETAILS IN THE BOILER-ROOM, AND DIAGRAMS OF FLOW AND RETURN MAINS.

THE smoke breeching in the boiler-room is shown in plan and elevation in Fig. 3, *s s* representing the branches from the several hot-water boilers to the large flue S, which delivers the smoke to the brick duct leading to the stacks shown in Fig. 1. Figure 4 is a section at C C, Fig. 1. Figure 5 is a section at X X. Figure 6 is a section at D D. Figure 7 is a section at F F, showing the manhole entrance to the smoke tunnel. Figure 8 is a section at E. Figure 9 is a section at G G, and Fig. 10 is a section at E E. Figure 11 shows the details of construction of the smoke breeching. Figure 12 shows the expansion pedestal Y Y of the return drum. Figure 13 shows the different methods of suspending the pipes from the iron floor beams above. Figure 14 shows the details of the iron manhole frame and cones shown in Fig. 7.

Figure 15 is a plan of the flow mains, and Fig. 16 is a partial plan of the return mains, the sizes and location of branches, risers, valves, etc., corresponding very closely to the flow system shown in Fig. 15. Here the connections to the boilers are really from the under side and would be hidden, but are shown in full lines, as if above, to avoid confusion.

PART III.—PLANS OF BASEMENT AND FIRST FLOORS.

THE locations of the indirect radiator stacks, the direct radiators and the register risers, together with some of the radiator connections, are shown in Fig. 17, which is a plan of the basement. The "second transverse center south" is represented by H H, and the "first transverse center south" by I I. Here A A, B B, C C, D D, E E, F F, K K, and G G are section lines from which detailed elevations will be given in succeeding articles. J J is the "first transverse center north," L L is the "second transverse center north," M M is the "center of north entrance," N N is the "third transverse center north," and O O is the main longitudinal axis of the building.

The arrangement of the first floor and location and sizes of radiators, connecting pipes, registers and flues are shown on the plan, Fig. 18. The upper floors are arranged in a manner substantially similar to this.

PART IV.—DETAILS OF RISER LINES, ACCELERATING COIL AND ARRANGEMENT OF AIR DUCTS IN ROOM 24.

AN enlarged plan of the connections and radiators in the basement at riser 88, Fig. 17, is shown in Fig. 23. Figure 22 is a section at E E, Fig. 23; Fig. 20 is a section at Y Y, Fig. 22, and Fig. 21 is a section at F F, Figs. 20 and 23. In these *r r* are radiators, and D² and D³ are fresh-air ducts to second and third floors respectively.

Figures 24 and 25 are respectively elevations at the foot of risers 27 and 51, Fig. 17. Figure 26 shows the plans and elevations of the air ducts in room 24, Municipal building, Fig. 17. Figure 27 shows plan, elevation and support of the accelerating coil in the foot of the vent shaft of the women's prison. Figure 28 is an elevation, looking south, and Fig. 30 is one looking north from B B, Fig. 17.

PART V.—SECTIONS AND ELEVATIONS OF MAIN PIPES, BRANCHES AND CONNECTIONS IN BASEMENT.

A PARTIAL elevation looking north from A A, in Fig. 17, is given in Fig. 29; it shows the arrangement and relative positions of the flow and return mains, where they rise from the tunnel. In this figure *r* is a radiator stack. An elevation looking north from C C, Fig. 17, is shown in Fig. 31. In this figure *r r r* are radiators tacks. W is an inlet to the fresh-air shaft S, the dampers of which, D D, command the ducts to the radiator stacks. An elevation section looking west at D D, Fig. 17, under the cells in the women's prison, is shown in Fig. 32, and one looking south from G G in the Municipal building, Fig. 17, is shown in Fig. 33. This shows the eccentric joints of the mains and the manner of running branches from their upper sides, which are also shown in side view in Fig. 34. This figure is a section and elevation through the lower part of the first-floor cells and of the flow and return mains for the men's prison, looking south from K K, Fig. 17.

PART VI.—PLANS AND ELEVATIONS OF RADIATOR STACKS AND DETAILS OF REGISTERS, LOCK MECHANISM, DAMPER AND VALVE.

AN enlarged plan of the radiators, air chambers, ducts and pipe mains in the basement at riser 37, Fig. 17, is shown in Fig. 35. A vertical section at E E, Fig. 35, given in Fig. 36, makes clear the arrangement of the fresh-air duct to the three upper stories. Figure 37 is an elevation at H H, Fig. 35, corresponding to Fig. 36. Figure 38 is a vertical cross-section at C C, Fig. 35, through the radiators and hot and cold-air chambers. Figure 39 is a vertical section and elevation at G G, Fig. 35, showing the details of connecting the branches with individual valves to the flow and return mains in the tunnel below the basement floor. The main flow and return branches B B' have secondary branches *b b'* to the risers No. 37 which serve direct radiators on the upper floors. D D are dampers controlling the fresh cold-air supply; C is the cold and H is the hot-air chamber; A is the basement corridor; F a galvan-

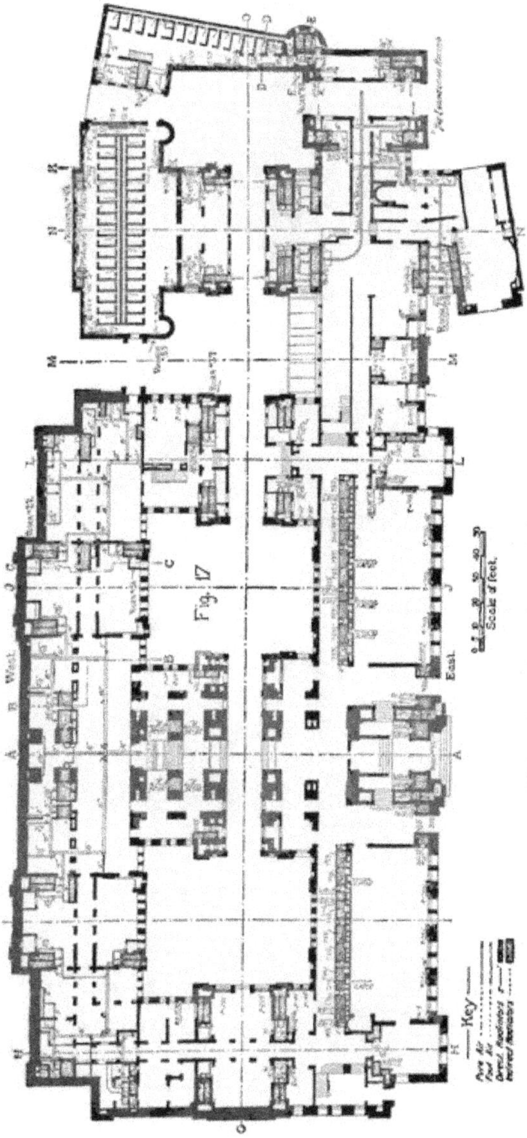

HEATING AND VENTILATION OF THE SUFFOLK COUNTY COURT-HOUSE.

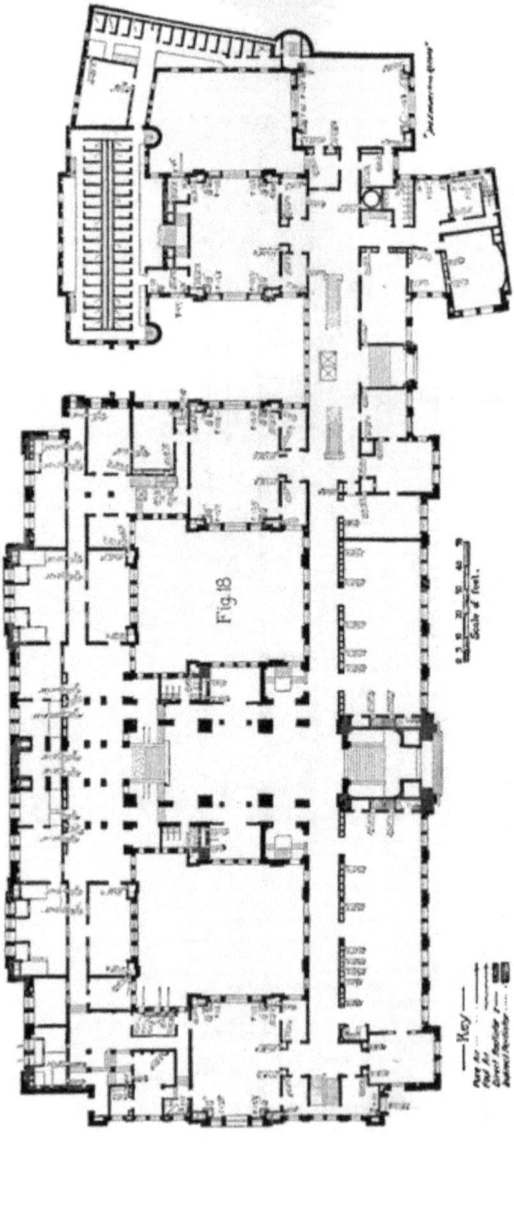
Fig 18
Scale of feet.
Key

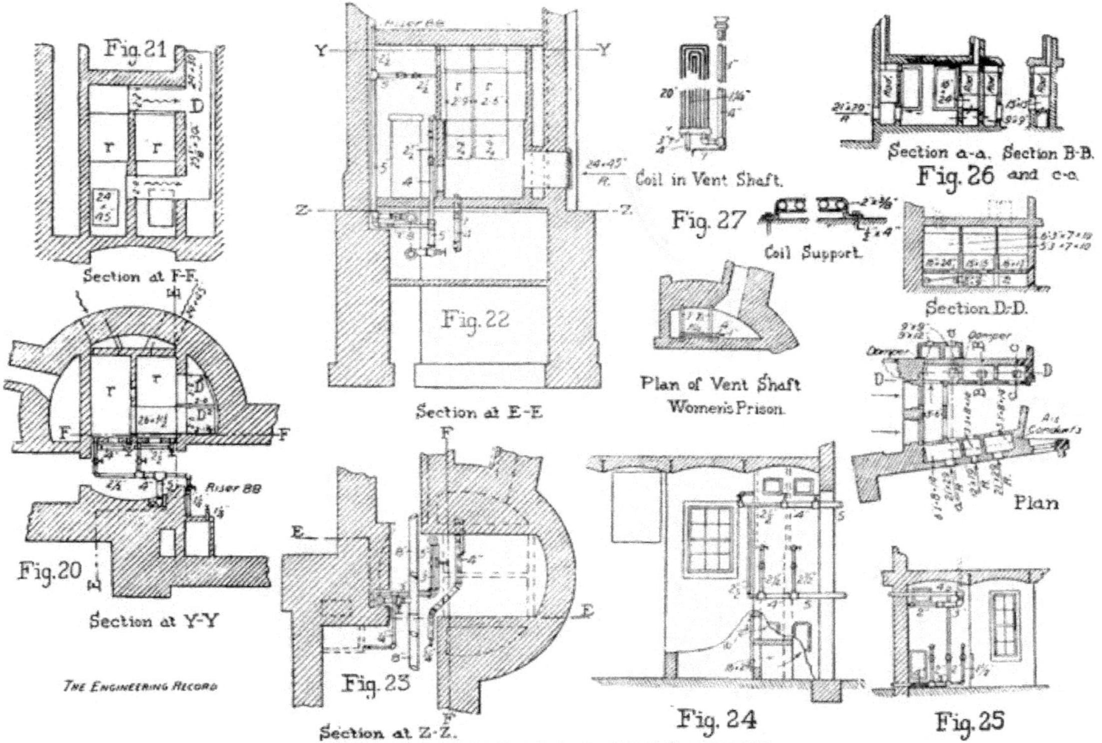

HEATING AND VENTILATION OF THE SUFFOLK COUNTY COURT-HOUSE.

ized-iron rectangular fresh-air duct. R R, etc. are registers, in the face of which a handle is set operating crank M, which through rod L commands regulating valve V, which is shown in position to exclude cold and admit hot air through port F. When, however, it is revolved to dotted position V' it conversely excludes the hot air and admits cold air only.

Figure 40 is an enlarged elevation of a part of the front of a register face R, Fig. 37, and the mechanism which commands through valve rod L the mixing valve V. The rod M, Fig. 41, revolves in bearings B B, and is keyed to the crank A and lever N. The former can be locked at any position on arc C, and the latter is connected by pivot P with valve rod I.

Figure 42 shows the mechanism by which crank A is automatically locked to arc C. Bolt B is set in a chamber H in crank A, and is extended in the position shown by spiral spring Q, which incloses its spindle E, and bears on crosshead F. There is a slot J to allow the bolt to slip by pin G, which carries a cam P, playing in recess I of bolt B, and always opposed to its crosshead F. When pin G is turned by key N, the revolution of cam P, compressing spring Q, pushes back the bolt until its spindle E comes to E', and B is disengaged from rack C, around which the crank A may be revolved to any position, still loosely grasping it with guides D D. When key N is removed, the spring Q is released, and engages the bolt at the first slot in the rack.

Figure 43 shows the details of mixing valve V, Fig. 36, and Fig. 44 shows the construction of damper D, Fig. 36.

PART VII.—ASH CAR, OVERHEAD ASH TRACK, CAR TROLLEY AND CAR TURNTABLE.

The ashes from the boiler-room are moved tnrough the basement corridor aud delivered to the scavenger in wrought-iron dump cars which are suspended by trolleys from an overhead track, shown in Fig. 17. This car, Fig. 45, has two wheels W W, upon which it can be rolled along the boiler-room floor, and its bail B, is pivoted upon an axis A, below its center of gravity, so that by throwing the lock D to the position D', it will automatically revolve and empty itself into the receiving cart, remaining suspended the while

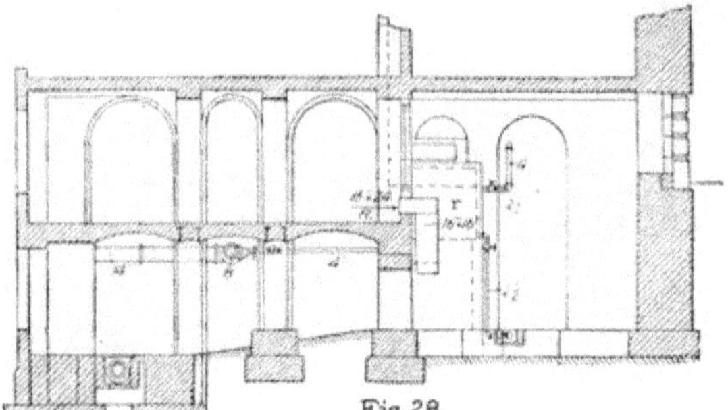

Fig. 28

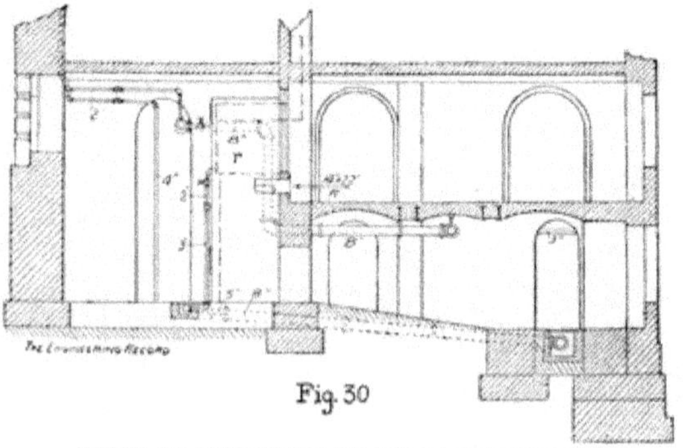

Fig. 30

HEATING AND VENTILATION OF THE SUFFOLK COUNTY COURT-HOUSE.

by link C from hook H of the trolley T, Fig. 46, which travels upon the rails R R, of the overhead track. The trolley has four single-flanged wheels, F F F F, and a forged boss B, which is bored to give free clearance to the stem S of the hook H. This stem is tapped to fit the female thread in the hub D of handwheel C, and hub D has a faced bearing at Z Z on the carrier bar. When, therefore, the car K, Fig. 45, is rolled underneath the track and its ring C is engaged with hook H, the latter is screwed up by handwheel C until it lifts the car to a sufficient height above the floor to insure clearance and permit it to be run off. Stem S is kept from turning by the spurs G G engaging in the keyway O

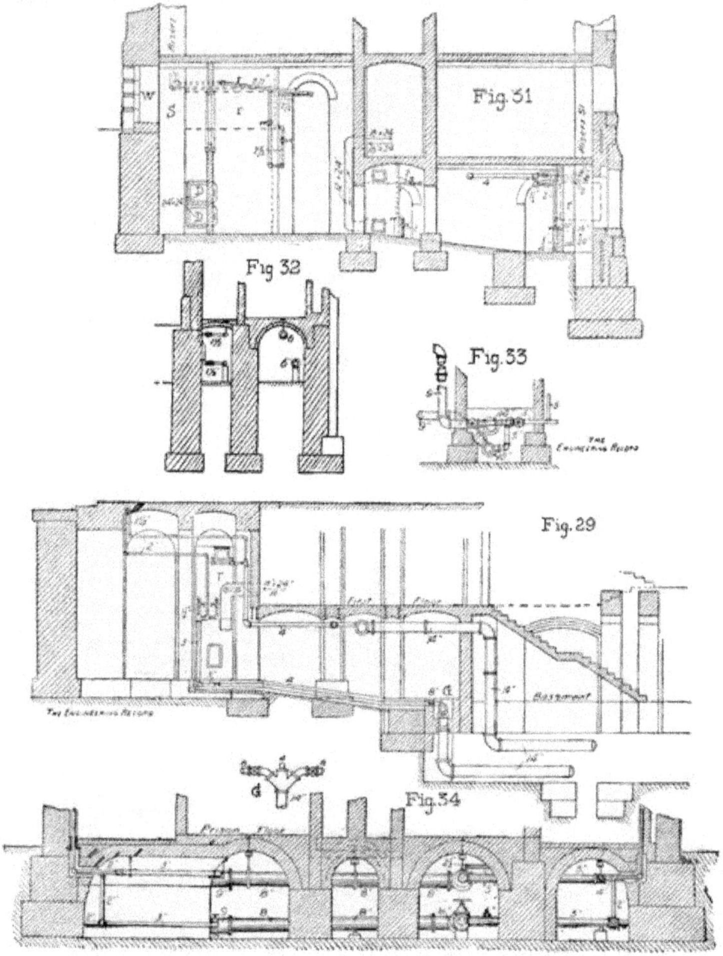

HEATING AND VENTILATION OF THE SUFFOLK COUNTY COURT-HOUSE.

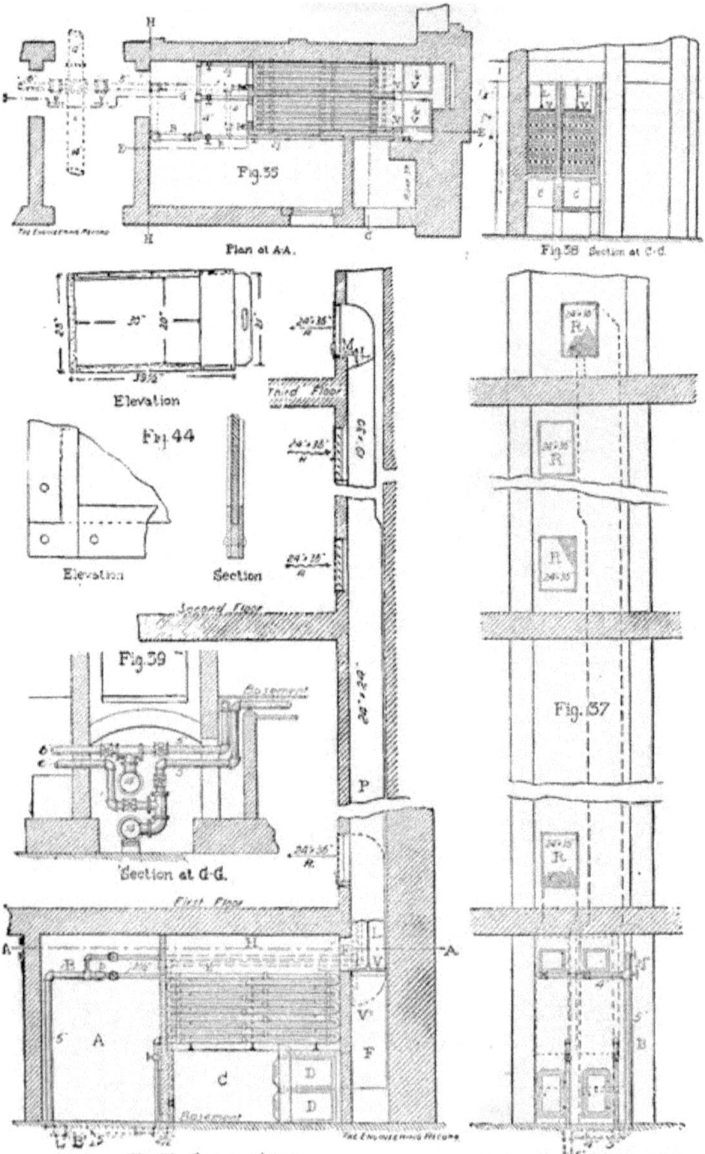

Fig. 36 Section at E-E Sectional Elevation H-H

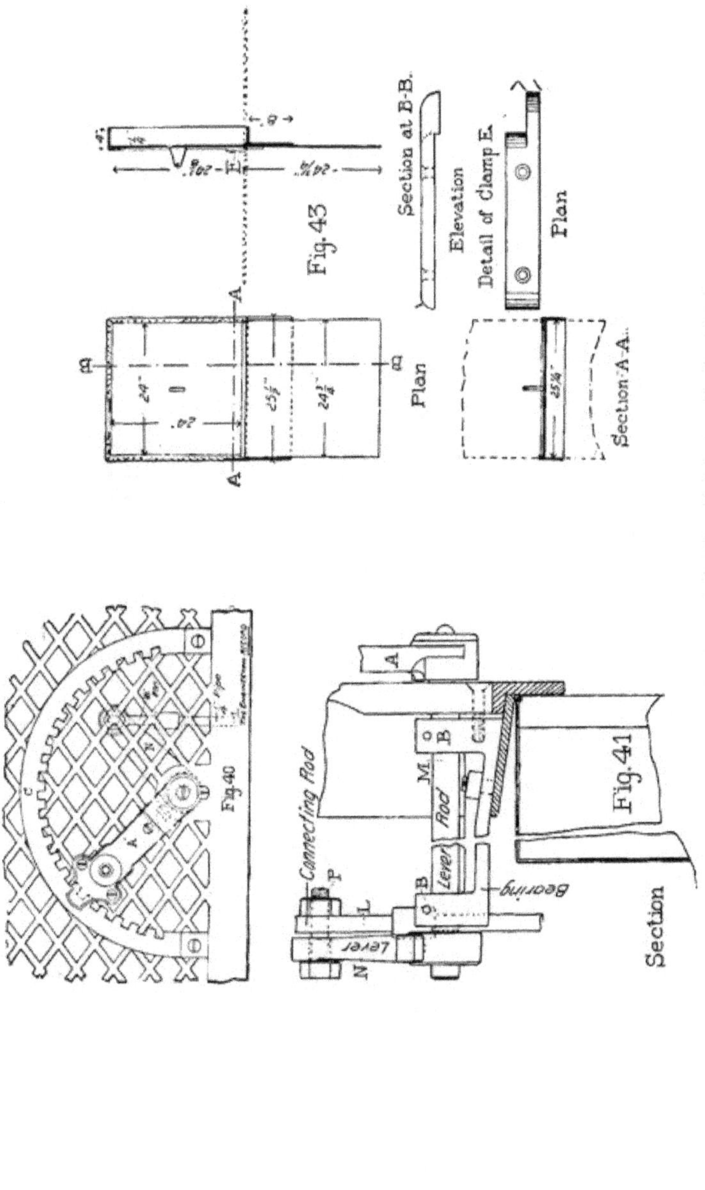

HEATING AND VENTILATION OF THE SUFFOLK COUNTY COURT-HOUSE.

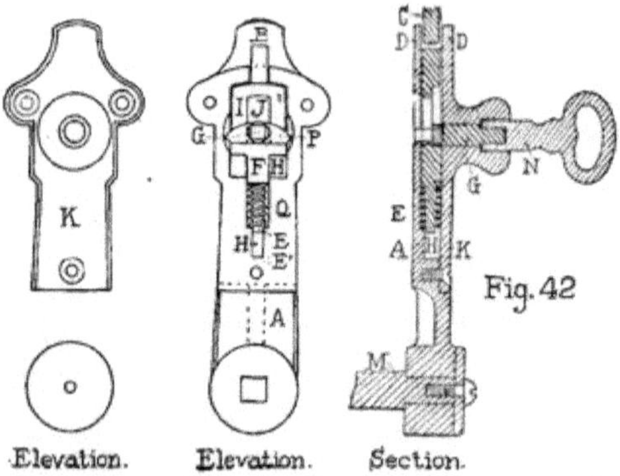

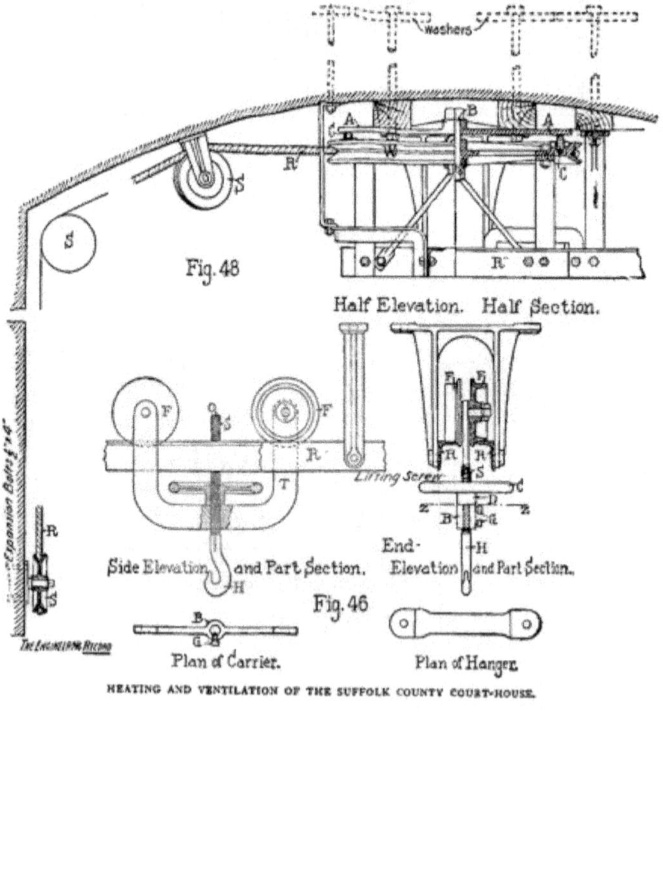

HEATING AND VENTILATION OF THE SUFFOLK COUNTY COURT-HOUSE.

Horizontal and vertical views of the turntable, shown in Fig. 17, are indicated by Figs. 47 and 48. By this turntable a car may be transferred from the rails R R of one track to those R' R', of another at right angles to it. The horizontal wheel W is suspended from a vertical axis fixed in the cast plate A so as to turn freely and engage the rails R' R', which are suspended from it with those of either of the tracks, R or R'. A is a bearing plate to prevent the wheel from tipping, and C C are friction rollers. The turntable may be operated from the floor by means of the rope R passing over guide sheaves S S, etc.

PART VIII.—DETAILS OF DIFFERENT METHODS OF SUPPORTING PIPES, EXHAUST HEAD, EXPANSION TANK, THERMOMETER, WATER GLASS AND HEATER.

LARGE pipes are suspended from the ceiling in the manner shown in Fig. 49, the object sought being opportunity for them to expand freely in the direction of their length. Large mains are suspended from the ceiling in the manner shown in Fig. 50, so that they are free to move in two directions at right angles to each others. Rigid overhead suspension of pipes from 3 to 5 inches in diameter is accomplished in the manner shown in Fig. 51, and Fig. 52 shows the manner of rigid overhead suspension of pipes 4 inches in diameter and less. The details of the clamp which supports the pipe suspender from the lower flange of a rolled I beam is shown in Fig. 53. Figure 54 shows the loop and bearing plate used in suspending pipes from brick arches, and Fig. 55 shows the method of supporting large pipes on the floor, so as to enable them to move freely in the direction of their lengths and at right angles thereto. Figure 56 shows the method of floor support for all pipes, enabling them to move

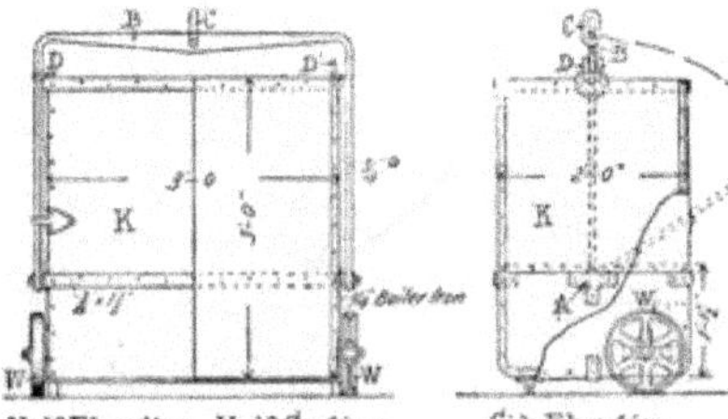

Half Elevation. Half Section. Side Elevation.

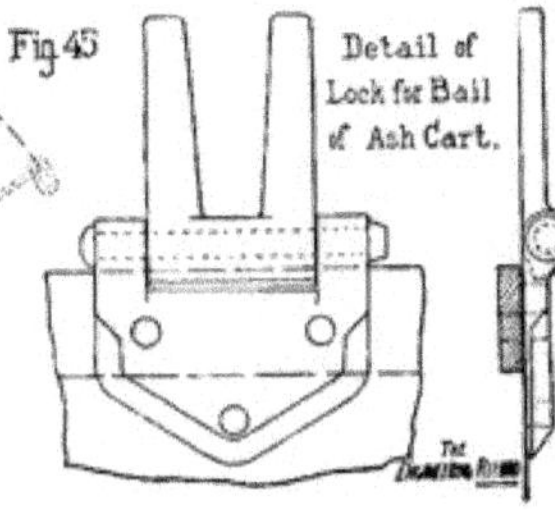

Fig. 45

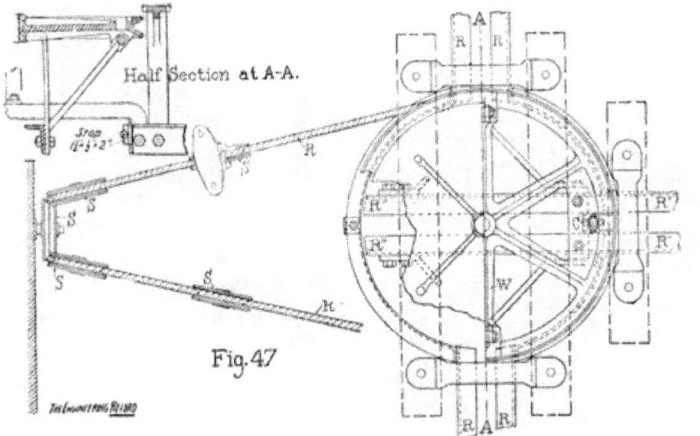

Fig. 47

Half Plan of Top Plate with portion removed. Half Plan showing Pulley.

HEATING AND VENTILATION OF THE SUFFOLK COUNTY COURT-HOUSE.

freely in the direction of their length only. Figure 57 shows the details of the cast-iron return bend for 3-inch radiator coils. Figure 58 shows a part of the checkered cast-iron floor plate covering the basement pipe trench. Figure 59 shows the details of the copper exhaust trap; Fig. 60 the wrought-iron expansion tank; Fig. 61 one of the thermometers used; Fig. 62 the water glass on one of the main boilers, and Fig. 63 the steam heater.

PART IX.—ARRANGEMENT OF DIRECT RADIATORS AND DETAILS OF SCREENS, DAMPERS, ETC.

FIGURE 64 is a vertical section at Z Z, Fig. 65, and shows the method of setting radiators R at the basement windows. Figure 65 is a plan, half section and half elevation, corresponding to Fig. 64. Fresh air is admitted from outdoors, underneath the window sill at S, and being deflected downwards by the iron wall C, passes through damper D, and up through the body of the radiator, escaping into the room through a grating G in the upper part of an otherwise solid cast-iron panel P.

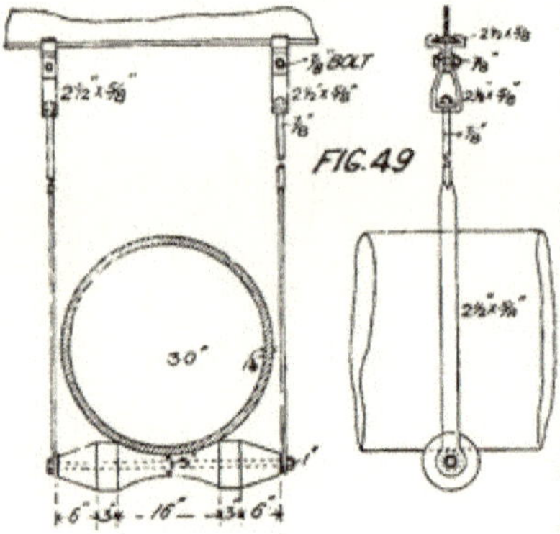

Figures 66 and 67 are vertical sections through window radiators in the first mezzanine story. The arrangement is similar to that shown in Figs. 64 and 65, and the reference letters have the same significance as in those figures. Figure 68 is a vertical section at Z Z, Fig. 69, and shows the arrangement of a long radiator set close to the cast-iron front of the second mezzanine story. Outside air enters through perforations in the cast-iron window sill T, passes down behind screen C, through damper D and after being warmed in radiator R escapes through grating G. Figure 69 is a plan in section and elevation corresponding to Fig. 68.

Figure 70 is a vertical section at Z Z, Fig. 71, and Fig. 71 is a horizontal section at X X, Fig. 70. Cold external air enters freely through cast-iron grating A, passes through wall duct F and register B to radiator R and is delivered to the room through open grating G in the cast-iron panel P. The flat radiator is set in a shallow alcove or wall recess W so that its front casing is within the inside wall line.

Figures 72 and 73 are respectively vertical and horizontal sections, showing an arrangement which is similar except in admitting the fresh air underneath the radiator. The reference letters here have the same significance as before. Figuro 74 is a vertical section at Z Z, Fig. 75, and shows the arrangement of a slightly projecting window radiator on the second mezzanine floor. Figure 76 is an elevation at Y Y, Fig. 74, and Fig. 75 is a corresponding plan in

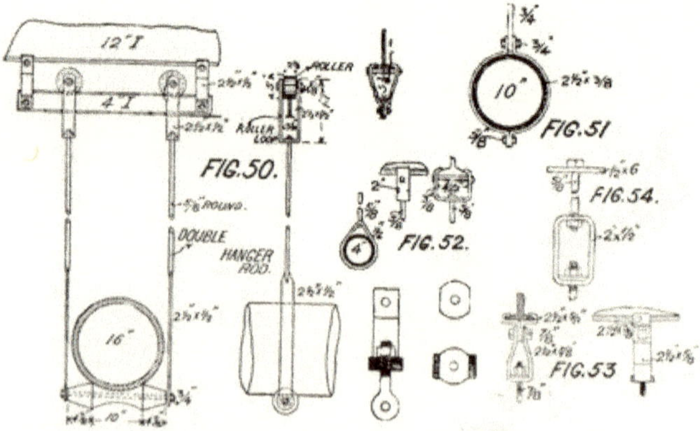

HEATING AND VENTILATION OF THE SUFFOLK COUNTY COURT-HOUSE.

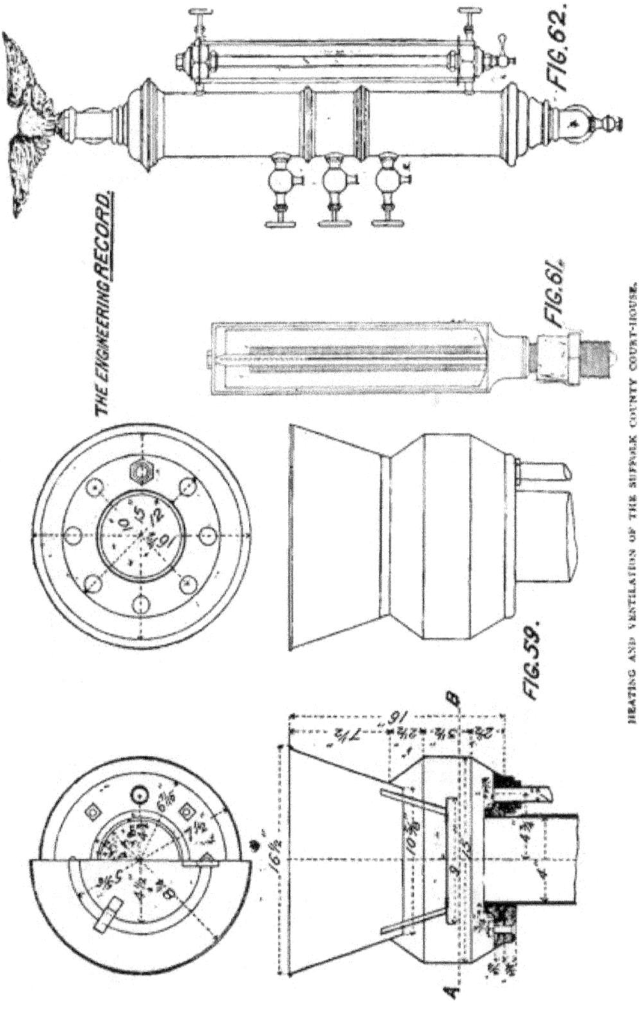

HEATING AND VENTILATION OF THE SUFFOLK COUNTY COURT-HOUSE.

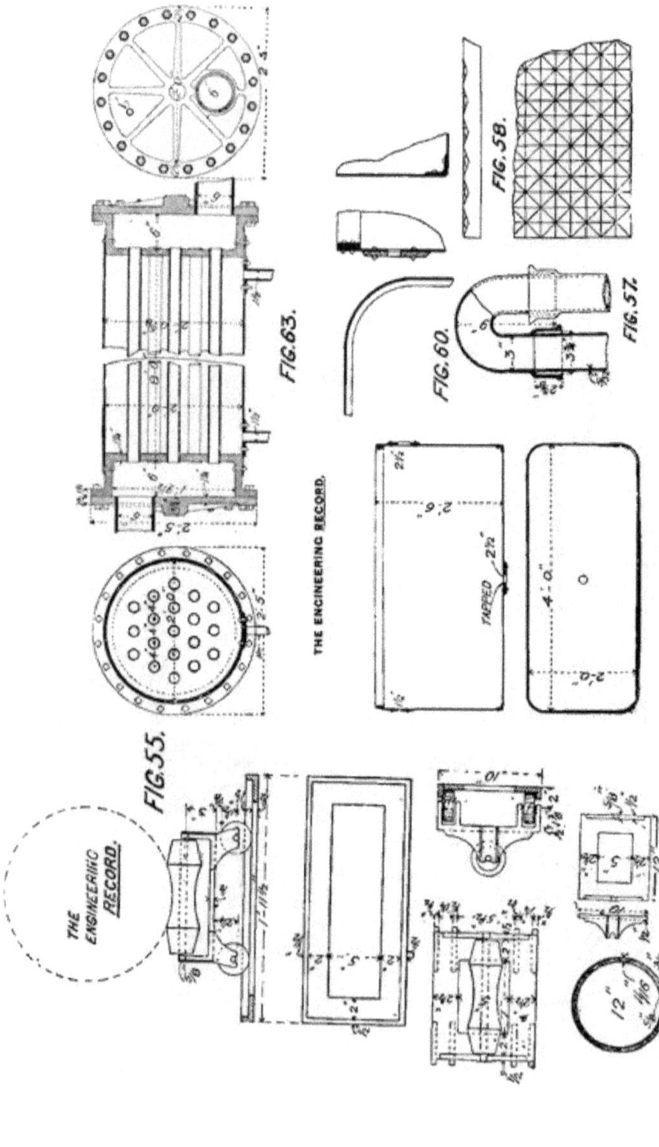

HEATING AND VENTILATION OF THE SUFFOLK COUNTY COURT-HOUSE.

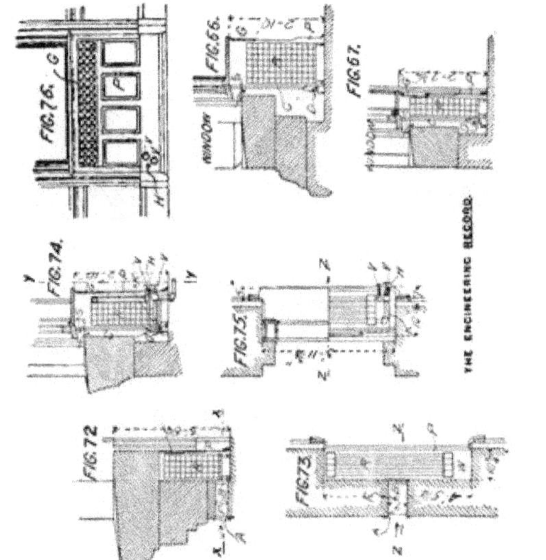

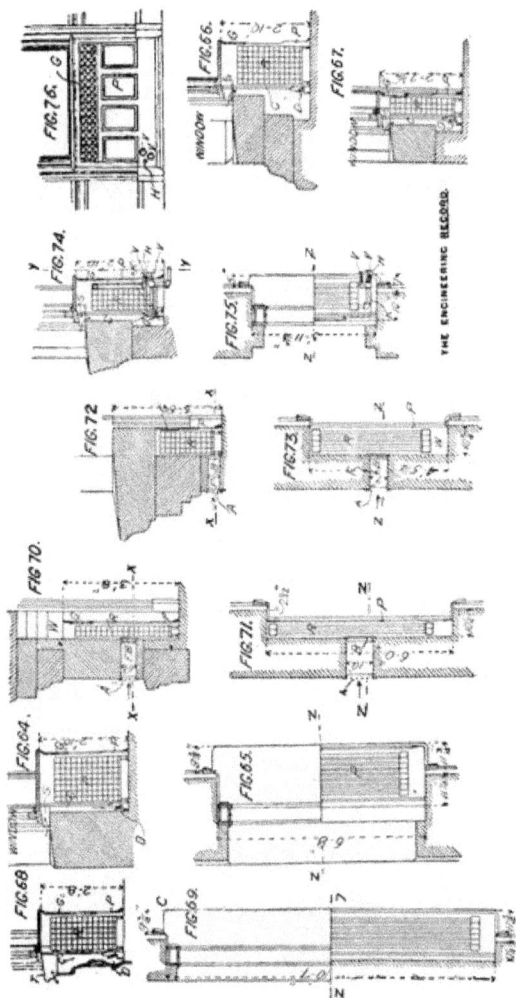

THE ENGINEERING RECORD.

HEATING AND VENTILATION OF THE SUFFOLK COUNTY COURT-HOUSE.

elevation and section. V V are the handles of the hot-water flow and return valves, and H is the handle of the damper rod K. The other reference letters have the same significance as in the preceding figures. Some of these radiators do not receive a direct supply of cold external air. In such cases inlet S and damper D are omitted, and the lower parts of panels P have gratings similar to G, thus admitting the coolest interior air and keeping up a continued circulation.

Figure 77 is a section at Z Z, Fig. 78, and shows the setting of recessed window radiators between court-room pavilions. Figure 78 is a corresponding horizontal view, and the reference letters have the same significance as in preceding figures. Figure 79 is a detail of cast screen G, Fig. 77, for window breast. Figure 80 is a detail of the non-conducting apron L, Fig. 77. Figure 81 shows the detail of damper rod K and handle H, Fig. 77, and its escutcheon. Figure 82 shows the construction and operation of cast-iron damper D, Fig. 77.

PART X.—RADIATOR CASES AND SMOKESTACKS.

FIGURE 83 shows the construction of one of the galvanized-iron cases in which the third-floor radiators are set. The case D is made of sheet iron riveted to

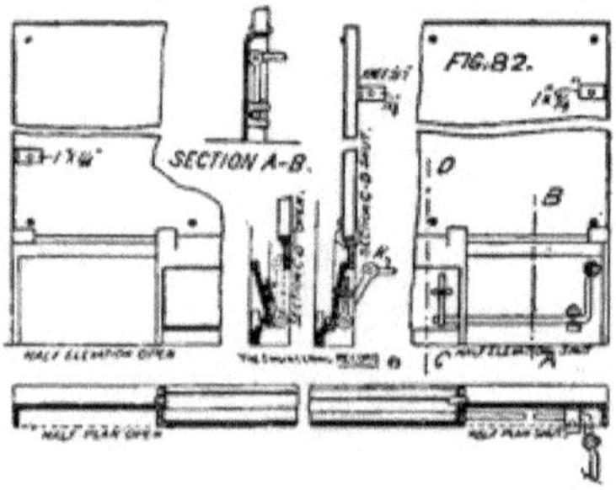

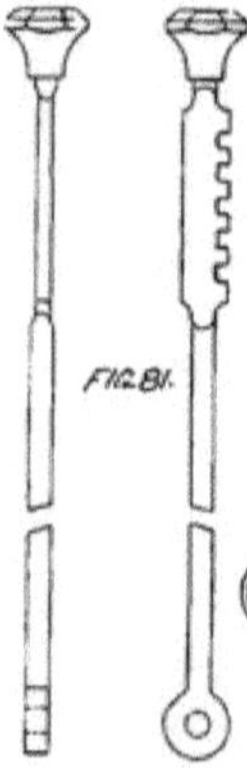

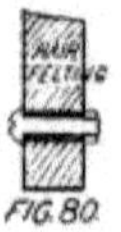

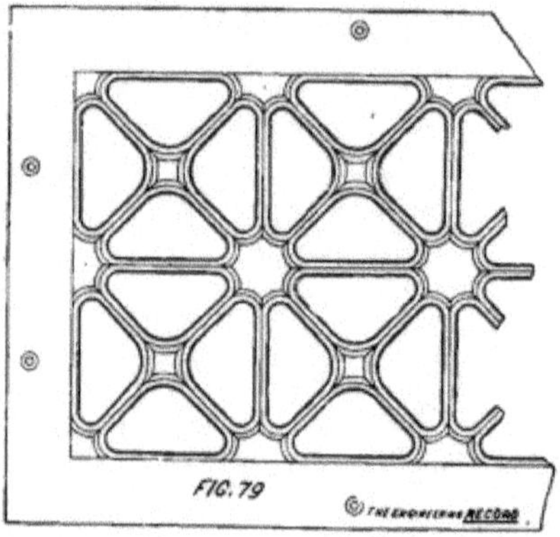

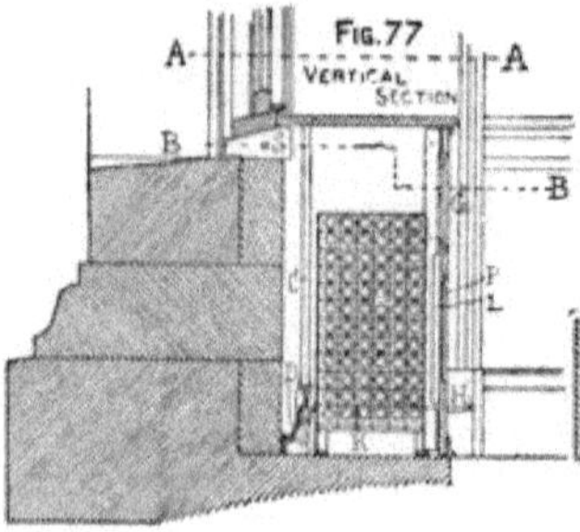

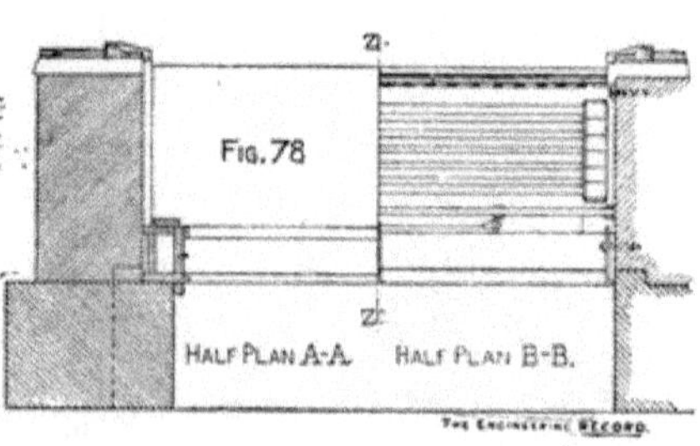

HEATING AND VENTILATION OF THE SUFFOLK COUNTY COURT-HOUSE.

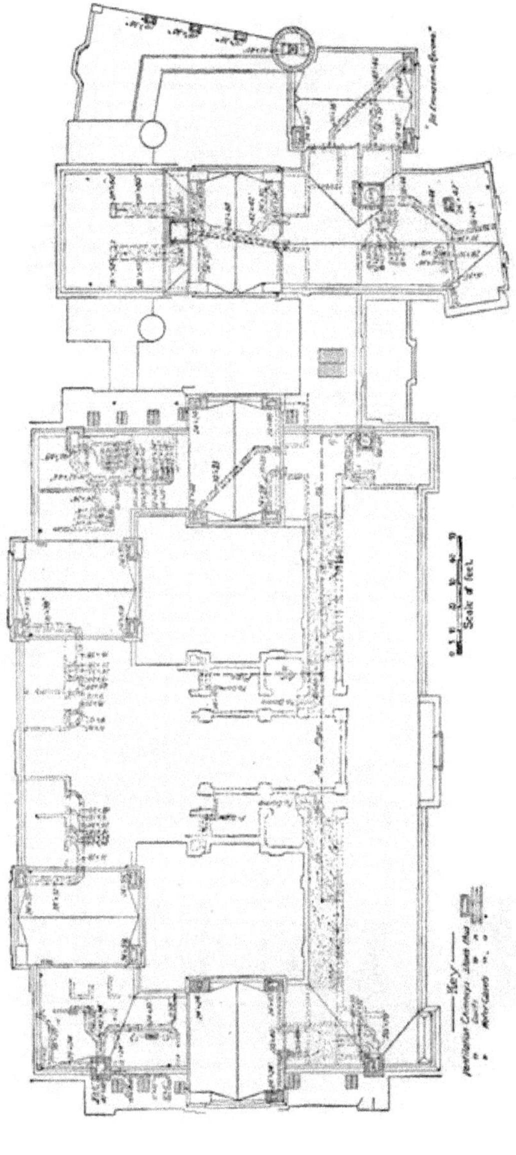

FIG. 92.—HEATING AND VENTILATION OF THE SUFFOLK COUNTY COURT-HOUSE.

angle iron corner pieces, and serves as a tight hood inclosing the radiator upon all sides. Cold air received through the outside grating A and wall flue F is admitted through the register B, which is operated by the rod R and lock handle H (illustrated in Fig. 81). After passing through the radiator the air is delivered into the room through the cast-iron grating C.

Figure 84 is a vertical section of one of the main smokestacks which has brick walls faced with stone and contains a wrought-iron cylindrical smokestack around which is a space into which ventilation ducts open and have their discharge promoted by the accelerating effect of the radiation of heat from the smokestack.

Figure 85 is a section at the foot of the stack, showing the arrangement of main steam pipe P beneath it. Figure 89 is a vertical section at right angles to Fig. 85, showing the horizontal main smoke flue. Figures 86, 87, and 88 are respectively horizontal sections at B B, A A, and E E; F is a ventilation flue and G is a soot door.

Figure 90 is a vertical section of another smokestack of similar construction and Fig. 91 is a corresponding horizontal section. In this case the double walls are both of brick up to near the roof line, beyond which a 30-foot iron cylinder forms the continuation of the smoke flue and leaves a larger space between it and the outside walls. Below the upper ceiling this space is not utilized, but above it it serves as a discharge for vent flues.

PART XI.—VENTILATION DUCTS IN THE ATTIC AND VENTILATING CHIMNEYS IN THE MEN'S AND WOMEN'S PRISONS.

FIGURE 92 is a roof plan showing the location of ventilating ducts in the attic and their delivery to

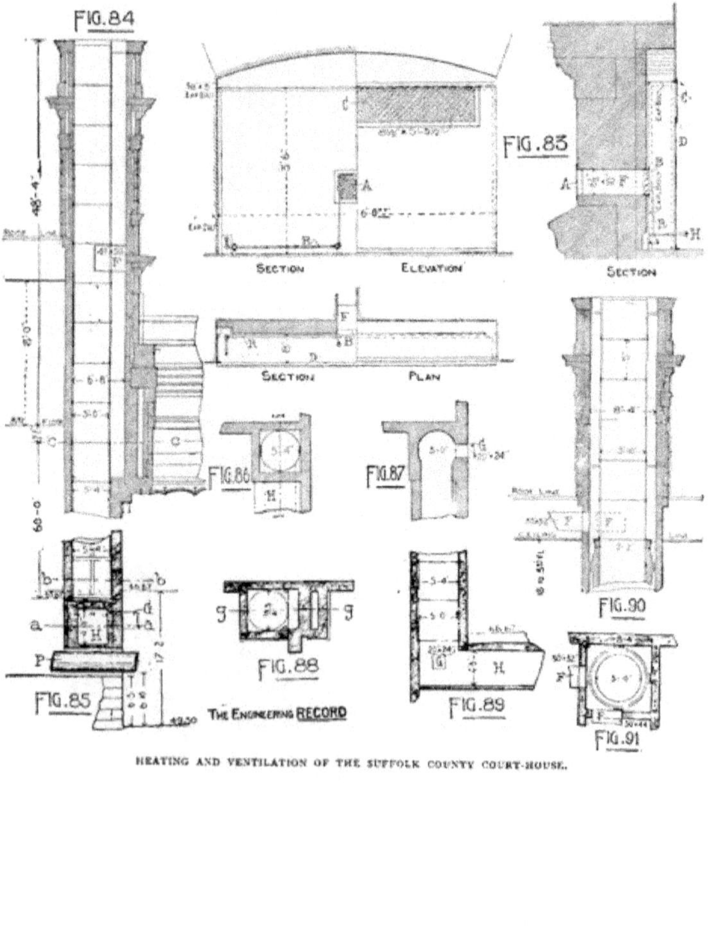

HEATING AND VENTILATION OF THE SUFFOLK COUNTY COURT-HOUSE.

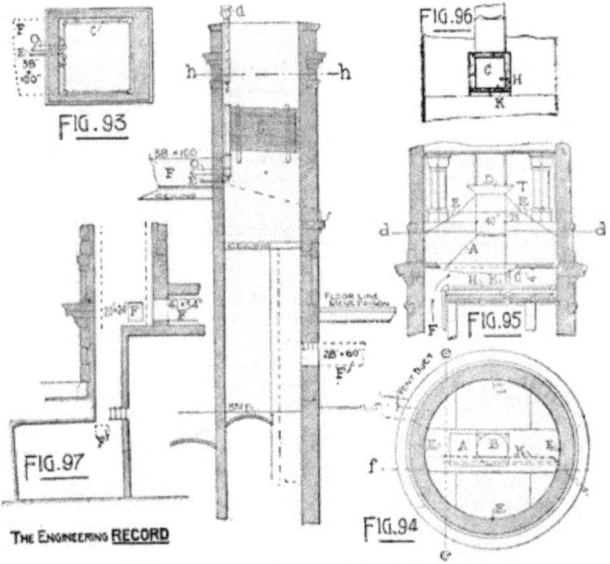

HEATING AND VENTILATION OF THE SUFFOLK COUNTY COURT-HOUSE.

small chimneys and through large main ducts in the center to the dome, which is not here shown. At the right of the figure are shown the large square and circular ventilating chimneys which serve the men's and women's prisons respectively.

Figure 93 is a section of the base of one of the regular small ventilation chimneys. Figure 94 is a horizontal section at *a a*, Fig. 95, through the ventilating chimney of the women's prison. Figures 95 and 96 are vertical sections at *f f* and *e e* respectively, Fig. 94. The vent flue F terminates in a lateral branch C from which the air escapes through a copper duct A and passes out of the cylindrical head into the tower T and thence through louvers into the open air. D is a protecting cap and E E, etc., are ¾-inch braces. H is a 4-inch steam exhaust pipe running to a roof trap and intended to promote circulation in the duct F; K is a 1¼-inch drip pipe.

Figure 97 is a vertical section of the ventilating chimney for the men's prison, and Fig. 98 is a corresponding section at right angles to Fig. 97. F F are foul-air flues, C is an accelerating coil of 14 lines of 1¼-inch pipes, F is a 5-inch and E is a 1¼-inch pipe and G is an exhaust head.

HEATING OF HOSPITALS.

HOT-BLAST HEATING OF ST. LUKE'S HOSPITAL, ST. PAUL, MINN.

The building known as St. Luke's Hospital, at St. Paul, Minn., was designed by Mr. Clarence H. Johnston, architect, of St. Paul, while the Huyett & Smith Manufacturing Company, of Detroit, Mich., were the designers of the heating and ventilating plant, and to them we are indebted for data from which this description was prepared.

The building is built of brick and stone and stands in a very exposed quarter. It is 150x50 feet on the ground and three stories high above the basement. Each story is 13 feet and the basement is 11 feet high. The entire heating and ventilating apparatus, which consists of a blower system, is located in a sub-basement, which is 10 feet high.

Manufacturing Company's hot-blast apparatus, consisting of a bank of steam coils, encased by a jacket of sheet steel and having attached to one end a Smith steel disk fan 72 inches in diameter, which forces cold air against the surfaces of the steam coils. In addition to this bank of steam coils is a tempering coil containing 690 square feet of surface in the fresh-air inlet on the second floor, Fig. 4. This coil is designed to raise the temperature of the air to about 60 degrees in severe weather, and is sufficient to heat the entire building in mild spring weather.

Between the fan and coils, in the apparatus, is an opening in the base of the same so that the tempered air, as it comes from the fan, can be forced under and past the coils and into the lower, or cold-air, conduit. The condensation from the coils is returned to the

FIG. 1.

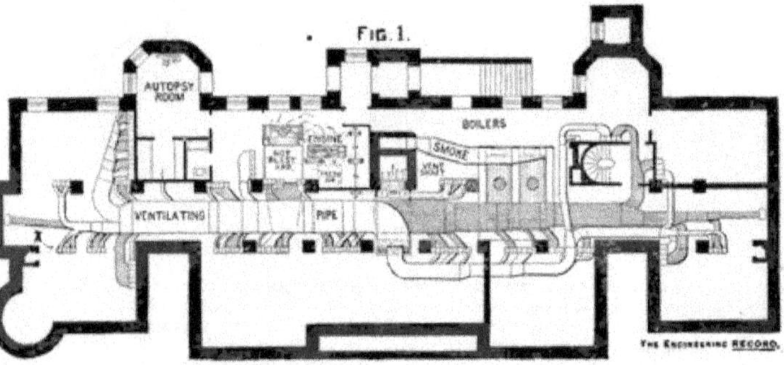

HEATING OF ST. LUKE'S HOSPITAL, ST. PAUL, MINN.

Figure 1 is a plan of the sub-basement, Fig. 2 of the basement, Fig. 3 of the first story, Fig. 4 of the second story, Fig. 5 of the third story, and Fig. 6 a side elevation (part in section) of the heater, showing a portion of the fresh-air shaft, how cold air passes the coils or through them, and how the conduits are attached to convey hot and cold air.

Figure 7 is a section through foul-air shaft and chamber showing the position of the ventilating fan and relative positions of the hot and cold-air conduits and foul-air exhaust conduit. Figure 8 is an elevation of the fresh-air and ventilating risers, showing how fresh air is introduced and foul air exhausted from the various rooms or apartments.

The plant is designed to heat the building to 70° Fahr. when outside temperature is 40 degrees below zero. This is accomplished by a Huyett & Smith

boiler with one Morehead return steam trap and the condensation from the radiators with another. There is also a Worthington boiler feed pump 4½"x 2¾"x4" and a receiving tank so the water can be returned by this means if found desirable or necessary.

All the halls, corridors, small wards which have an extraordinary exposure, autopsy-room, kitchens, laundry, and nurses' room are provided with direct radiation; the total surface in radiation being 1,585 square feet. In some instances the direct radiation is in addition to the other system, especially in the wards. The fresh air is obtained through an opening in the back wall of the second story, and is conveyed to the fan by a shaft ending at the ceiling of the sub-basement alongside the heater and fan. After passing through the coils the air is conveyed to the various parts of the building by galvanized-

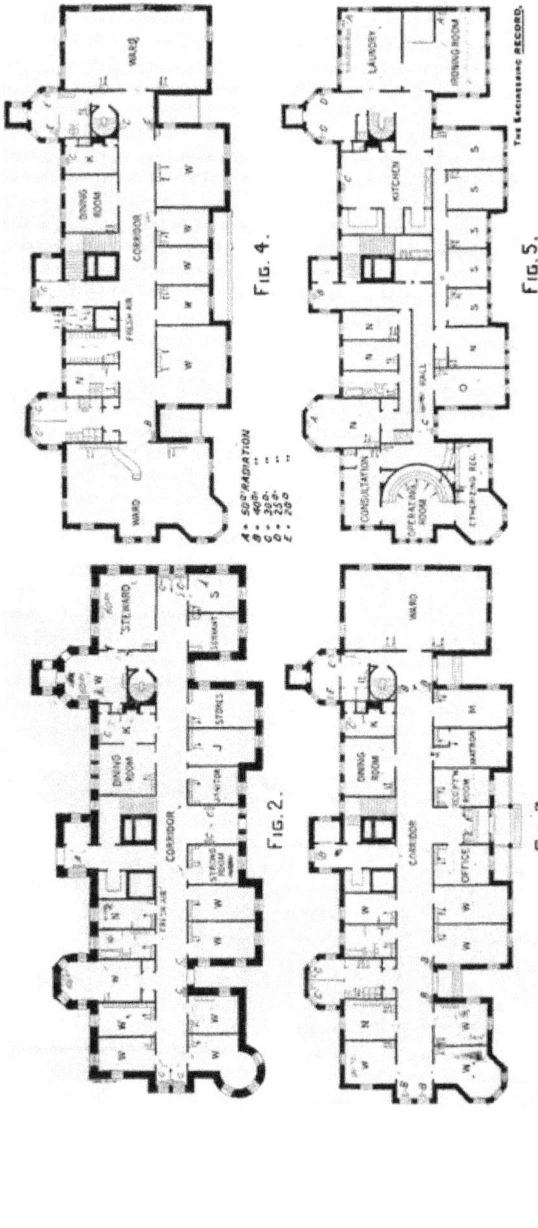

HEATING OF ST. LUKE'S HOSPITAL, ST. PAUL, MINN.

iron conduits. All the risers or flues throughout the building are of galvanized iron.

The hot-air and vent flues are of the same size for each room. Those leading to the large ward-rooms are 16x24 inches, while the ducts to the medium-sized wards (Fig. 4) and the operating and consulting rooms (Fig. 5) are 12x16 inches.

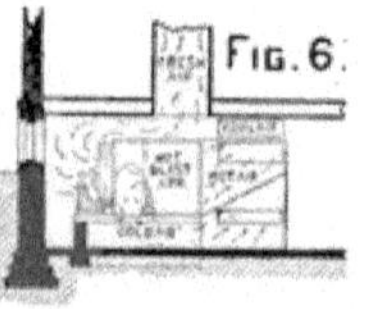

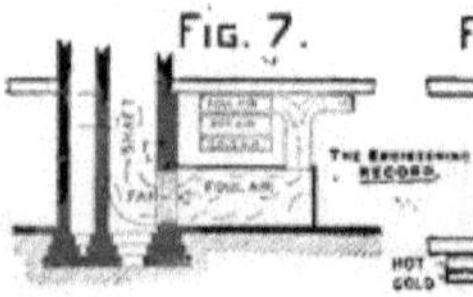

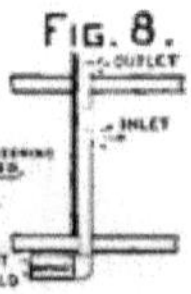

To all other rooms, including small wards, nurses' rooms, servants' rooms, and matrons' office, and toilet-rooms, the ducts are each 10x10 inches. At the base of the fresh-air flues, the hot and cold-air conduits connect as shown by Fig. 8; at the point of connection are dampers controlled by the Johnson Electric Service Company's regulator, which is not to allow the temperature of the air to rise above 70 degrees, but does not cut off the supply of fresh air to any part of the building. Each patient in all the wards is provided with a constant supply of 45 cubic feet of fresh air per minute. In all other parts of the building the air is designed to be changed once in every 15 minutes.

The fresh-air inlets are 8 feet above the floor, while the vents are at the floor, and in most instances directly under the inlets. The air is blown towards the coldest walls and traverses back across the floor to the vent flue.

The building is ventilated by a steel disk fan 60 inches in diameter. This fan is located in the sub-basement at the base of a large vent stack, which extends several feet above the highest point of the roof. All the vents throughout the building are carried down to the sub-basement and are attached to the foul-air conduit. The ventilating fan draws the foul air out of this conduit and discharges it into the vent stack.

Steam for power and heat is supplied by two 60 horse-power, horizontal, multitubular boilers. The plenum fan has a capacity to deliver 56,000 cubic feet of air per minute and the exhaust fan a capacity of 40,000 cubic feet. The exhaust fan is smaller than the plenum fan, so that a slight outward pressure can be maintained by the latter, to prevent drafts and to counteract wind pressures from the outside. The building contains about 364,000 cubic feet of space to be supplied with fresh air by the plenum fan; thus a complete change of air can be effected every 6½ minutes, though this is rarely done, as the fan is not usually run at more than half its specified speed.

These fans are driven by belts from a countershaft, which in turn is driven by a 7x10-inch horizontal automatic engine. On the accompanying plans N designates nurses' room, S servants' room, and W ward.

The past winter was the second winter of service for this plant, and it is said to have given excellent satisfaction.

The scale was omitted in the drawing, but Fig. 1 is on a scale of 26 feet to the inch, while Figs. 2, 3, 4, and 5 are on a scale of 38 feet to the inch.

HEATING AND VENTILATING OF A RECEPTION HOSPITAL.

The accompanying plans show the building erected by the city of New York at the foot of Sixteenth Street, East River, to serve as a reception hospital for patients with contagious diseases, where they are confined until they can be taken to the Wards Island Hospital. The building was designed by Messrs. Jackson & Warner, engineers and architects, of New York, while Baker, Smith & Co. were the contractors for the heating plant.

The hospital is in two parts, as is shown by Fig. 1, one containing 12 wards, each 14x24 feet on the ground by 16 feet in height. The other building contains two wards, each about 20x24 feet.

The hospital is entirely fireproof and the wards are separated by a fireproof wall running up for a

PLAN OF BOILER HOUSE

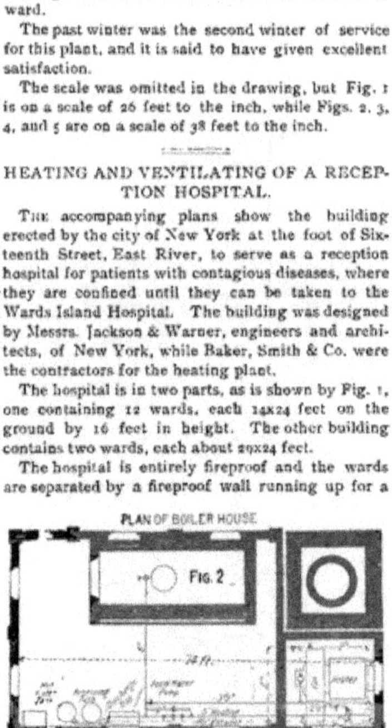

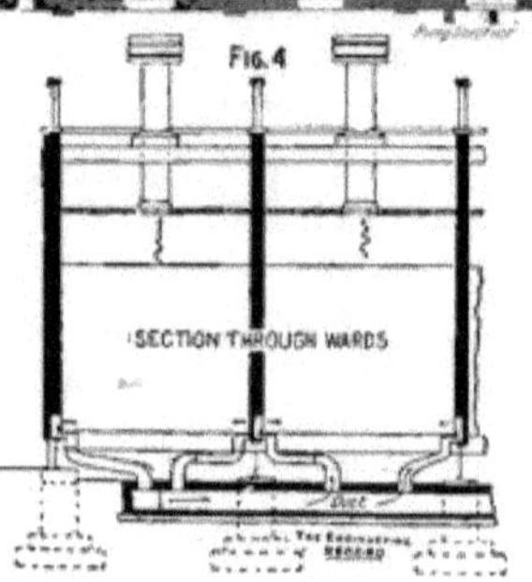

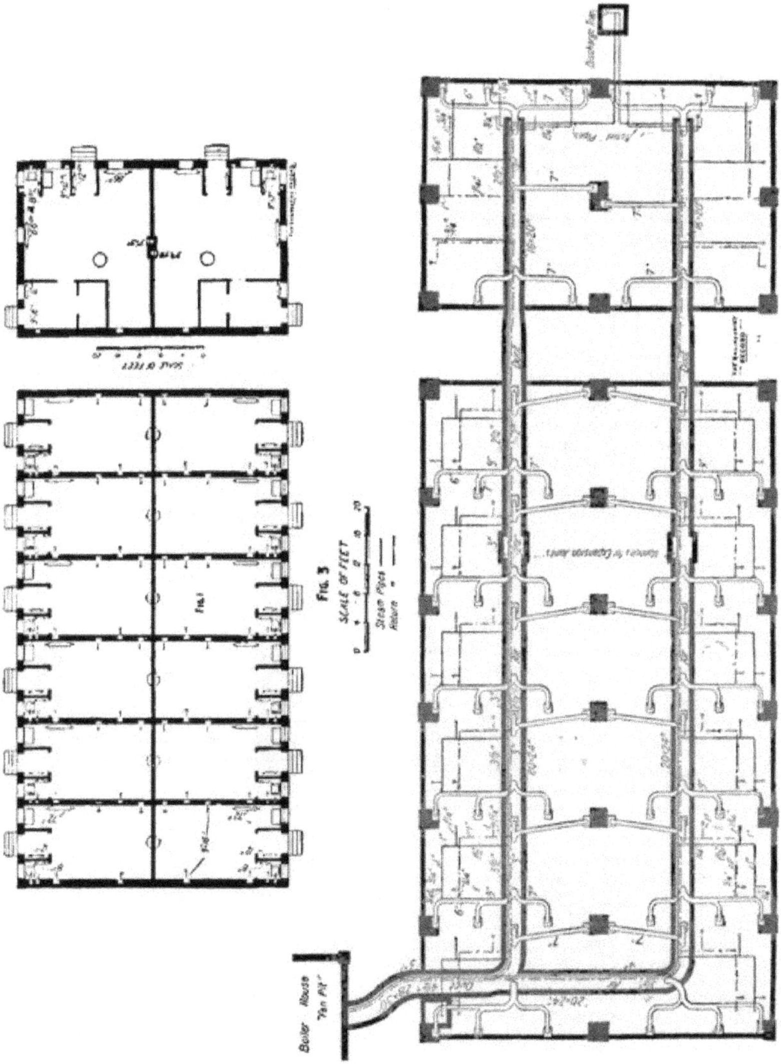

HEATING AND VENTILATING OF A RECEPTION HOSPITAL FOR THE CITY OF NEW YORK.

distance of 12 feet. The side walls are lined on the inside with enameled brick so that they can be washed down with a hose.

The building is warmed and ventilated partly by a blower system and partly by direct radiation. The boilers and fan engine, etc., are located in a separate building a few feet distant from the hospital proper. Steam for warming the building and driving the fan engine and pumps is generated in a 100-horse-power return-tubular boiler. The boiler is 6 feet in diameter, 16 feet long, and contains 132 3-inch tubes. The air is warmed and forced into the hospital buildings by a Sturtevant hot-blast apparatus containing 1,000 square feet of heating surface and a 4½x6¼-foot fan driven by a Sturtevant vertical engine. It is estimated that the fan is capable of delivering 15,000 cubic feet of air per minute when making 200 revolutions in the same length of time. The warmed air is distributed by means of brick ducts, shown in Fig. 3, resting upon concrete foundations. The duct is 28x30 inches as it leaves the boiler-house. Its walls are 8 inches in thickness, and the foundations are 6 inches thick and of such a width as to extend for 6 inches beyond the outer edge of the duct. The duct is lined with cement three-fourths of an inch thick, while the outside was given a coat of Trinidad asphaltum applied hot. After entering the hospital the duct divides into two branches, as shown by Fig. 3. Circular galvanized-iron flues lead from the arched top of the brick ducts to specially designed cast-iron wall frames (see Fig. 4), which terminate in registers of two sizes, either 9x16 inches or 9x9 inches in area. Each of the smaller wards is provided with a closet and sink, while each is supplied with air by four 9x16-inch registers and one 9x9 inches in size. A 9x9-inch vent draws off the impure air from the closet, while each pair of adjacent wards is ventilated by a circular vent in the ceiling 30 inches in diameter. Each ward is also supplied with three direct radiators containing 8, 12, and 72 square feet of radiating surface. Each of the larger wards, of which there are two, contains two 9x16-inch, one 9x12-inch, and three 9x9-inch supply ducts, with five radiators containing 176 square feet of surface in all. The vent shaft for each of these wards is 30 inches in diameter.

The steam for the direct radiators leaves the boiler-room through a 5-inch main, and the condensed water is carried back to a Kieley automatic pump governor in the boiler-room. This is connected to a 4½"x2¾"x4" Worthington duplex pump, which returns the water to the boiler. A feed pump supplies the boiler with water of an amount equal to that used by the fan engine. A tank containing a steam coil is provided to supply the hospital with hot water. The condensation from the coil is discharged into the blow-off pipe of the boilers outside of the blow-off cock by a Kieley steam trap, the blow-off pipe being extended to the river.

The steam main supplying the direct radiators that run in the brick ducts are pitched so that the condensation runs to the further end from the boilers, where it is also taken off by a trap. The return is pitched so that it drops as it nears the boiler.

HOT-WATER HEATING AT SANFORD HALL.

PART I.—DESCRIPTION, GENERAL PLANS, AND SECTION.

Sanford Hall, Flushing, L. I., is a large brick building, mainly two stories in height, the central part of which was originally a private residence, but has for many years been used as an insane asylum, and has been added to from time to time, chiefly, as occasion required additional separate rooms for patients. At first individual stoves and fireplaces were used throughout for heating and ventilating, then hot-air furnaces were provided to warm most of the building, and within a few months the necessity of heating a large new wing has occasioned the adoption of a complete uniform system of hot-water heating, which has been designed and executed by the Blackmore Heating Company, of New York. Two large Richardson & Boynton Company's "Perfect" heaters were set at convenient positions (about 49 feet apart) in the basement, and their main flow and return pipes, carried along the ceiling, were connected together so as to permit free circulation between them. Auxiliary return pipes, which were necessarily carried below the level of the boilers, were not connected together.

Figure 1 is a basement plan showing the sizes and positions of nearly all the horizontal flow and return pipes, the former being indicated by full and the latter by lines broken with one dot, thus —— · —— In the men's wing A, and in wing B, the horizontal flow pipe is specially arranged to be in the upper story, and in the uncompleted women's part C, the work is not yet executed, so the pipes do not appear on this plan. H H are the heaters, and D and E the circulation pipes which connect them. F G and G are coils for indirect stacks which heat the main entrance hall and a reception-room, private office, etc. Stack F is placed in the brick chamber of an old hot-air furnace. The rest of the heating system is composed entirely of direct radiators, chiefly of the Bartlett-Hayward and Detroit patterns. Figure 2 is a plan of the second floor, and Fig. 3 of the third floor. Figure 4 is a partial section through the building (not exact in scale or positions, but intended as a sketch diagram to illustrate the system and arrangement), showing the method of running horizontal mains, risers, and branches to the direct radiators in different parts of the building. H is one of the boilers connected with the other boiler by circulation pipes D E (see Fig. 1), and having a flow main I, from which risers L L supply the radiators of the main part of the building. Main I rises by vertical branch J to the top of the wing, along which it extends horizontally at K as far as possible above the upper floor radiators, so as to carry the pipe out of the way on the ceiling, and because it was thought that the radiation from the small supply branches M M, etc., being much greater than from the single large main J, would cause sufficient difference in the densities of their respective water columns to materially promote the circulation. The returns are taken, in the wing through pipes N N, etc., to main O, which drops about 5 feet to connect with main P, which receives returns from the basement radiators

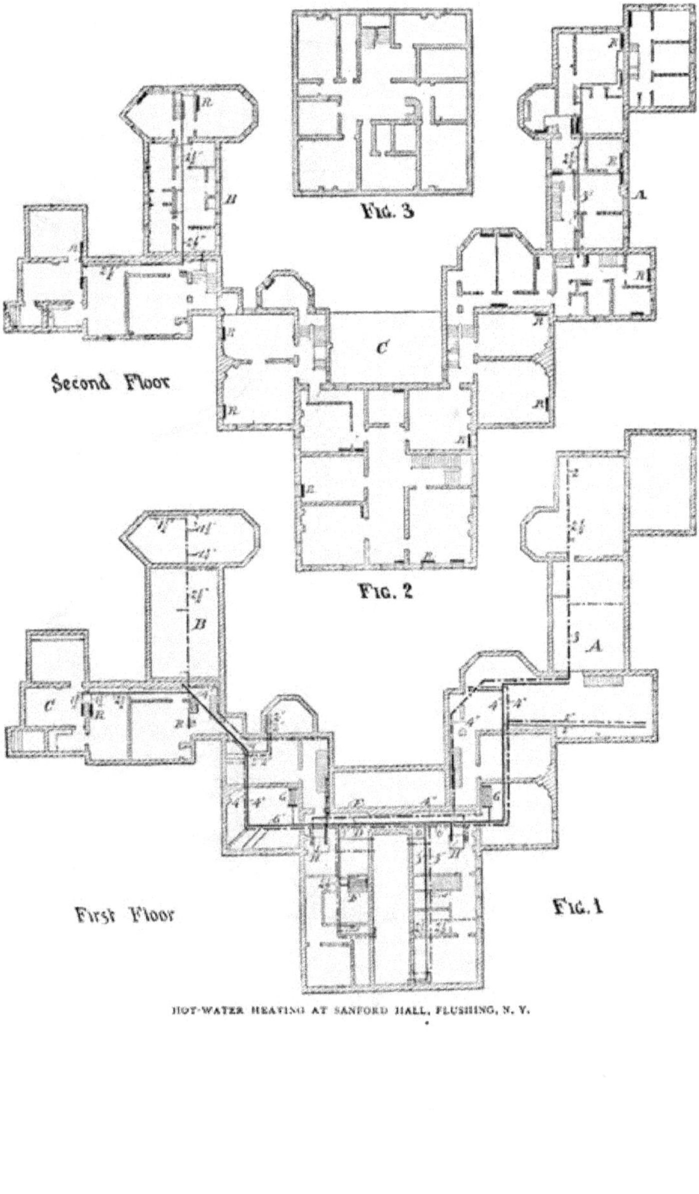

HOT-WATER HEATING AT SANFORD HALL, FLUSHING, N. Y.

Fig. 6

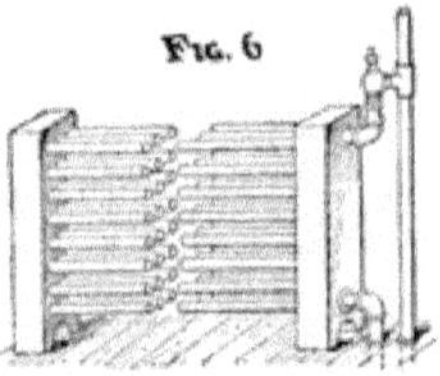

Fig. 7

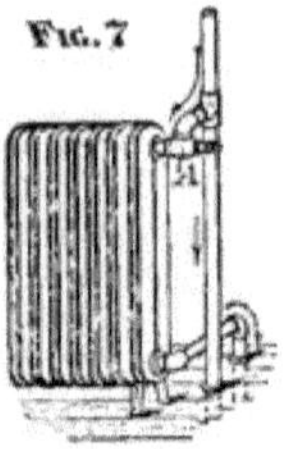

Fig. 8

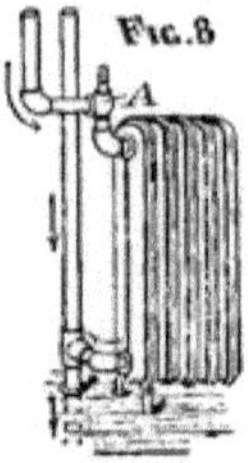

Fig. 4

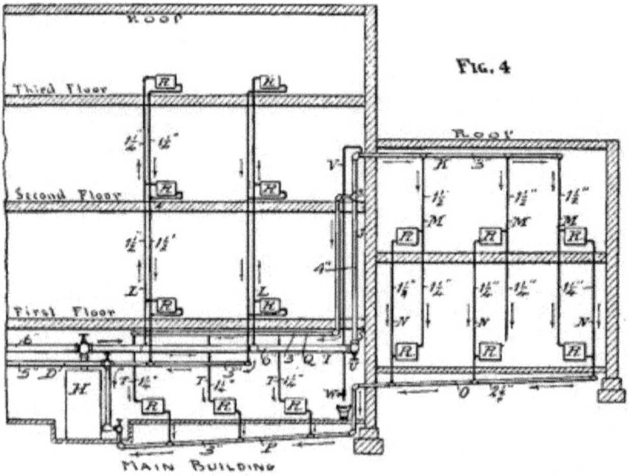

Fig. 5

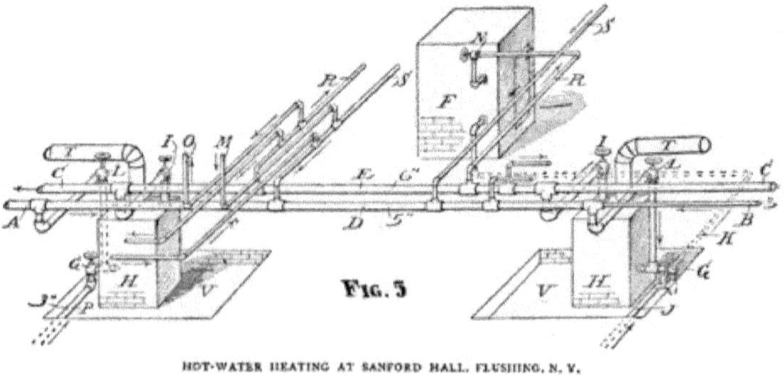

HOT-WATER HEATING AT SANFORD HALL, FLUSHING, N. Y.

only. These radiators are supplied on the same principle as those in the wing, from a branch Q taken at R from riser J. In this case the circulation would have been essentially the same if Q had been connected at S, or would have been accelerated if Q had been placed at the level of R, where advantage would have been taken of the less radiation from the single large pipe J than from the small pipes T T, etc. In this case the small pipes would also have been longer, but it was determined not to allow them to appear, except in the basement. It will be noticed that all of the supply branches, L, M, etc., are continuous straight lines from the main to the last radiator, being connected to the intermediate radiators by side branches, thus giving an independent supply to each one, and the returns are similarly arranged. An air pipe V is connected to the highest point of main K, and if it collects water it may be drawn off at W at intervals. U is a valve controlling the system in the wing.

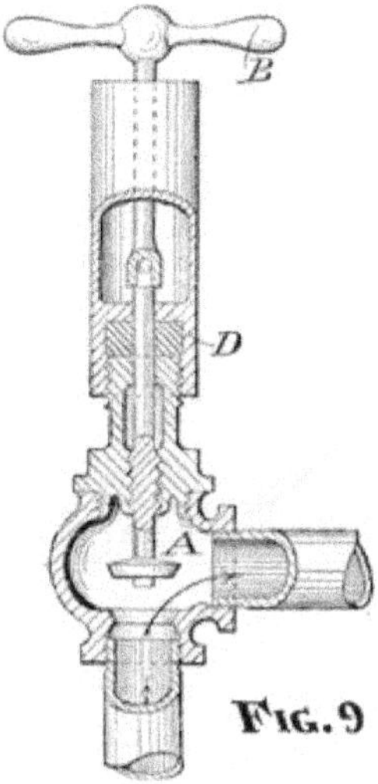

FIG. 9

FIG. 12

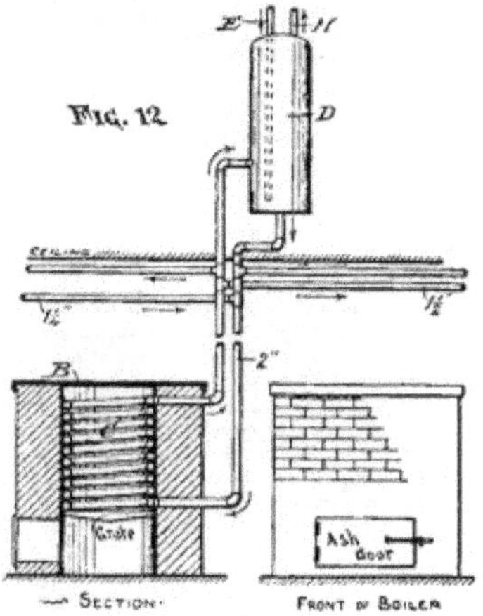

FIG. 10

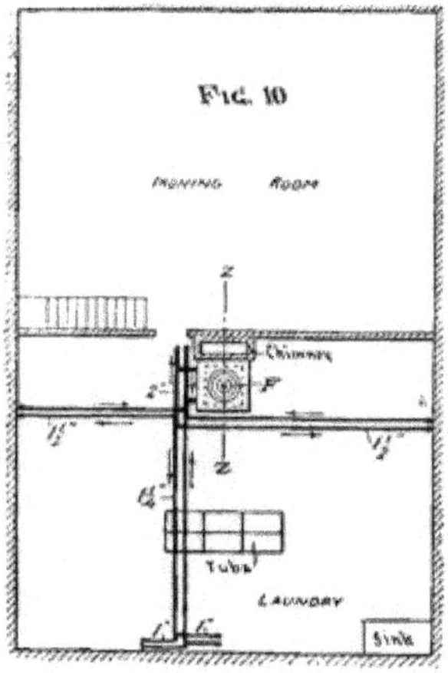

FIG. 11

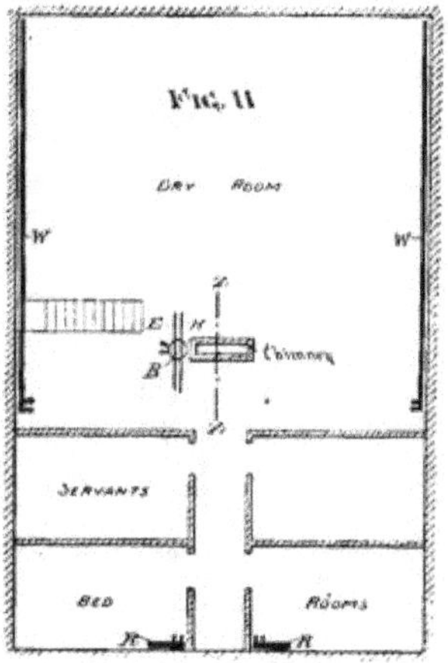

HOT-WATER HEATING AT SANFORD HALL.

PART II.—ARRANGEMENT OF HEATERS, CONNECTIONS OF RADIATORS, AND GENERAL DATA.

Figure 5 shows the connections of the No. 31 Richardson & Boynton "Perfect" heaters, each having a fire-chamber area of about 1,300 square inches. To secure necessary head room, the heaters were placed in pits V V. Ordinarily both heaters are used together, free circulation between them being maintained through pipes D and E (see Fig. 1), but either one can be cut out by simply closing its main flow and return valves I and L, and the other will operate the whole system, though it is not intended to be adequate for very cold weather. P, J, and K are separate returns from the men's wing, and from the indirect stacks F and G, Fig. 1. If valves G G, N N, and U, Fig. 4, are closed only the main part of the building will be heated, and by proper manipulation of the valves any section of the building may be cut off, and the rest heated, or *vice versa*. S S and R R, etc., are return and flow pipes for different parts of the house, from which vertical branches O and M, etc., serve the groups of radiators, which are in general placed in the same vertical lines. T T are smoke flues.

Figures 6 and 7 show slightly different connections of the upper floor radiators in the men's wing A, Fig. 1. Figure 8 shows the connection of the lower floor radiators in the same wing. Figure 9 shows the valves A, Figs. 6, 7, and 8, which are operated by a key B, the stem being protected by a long sleeve D, specially designed to prevent the patients from meddling with the heat. There are altogether 54 direct radiators with a total surface of 6,000 square feet, which are intended to heat about 250,000 cubic feet of air to an average temperature of 70 degrees, with a possible outside temperature of 0 degrees. Most of the radiators were placed in private rooms 12 or 14 feet square. The radiators in the two indirect stacks each contained 900 square feet, and are designed to heat about 30,000 cubic feet of air.

The laundry, dry-room, and servants' quarters are in a detached building, which is provided with an independent hot-water supply and heating system. Figure 10 is a plan of the lower floor of the laundry house, about 60x40 feet in size, and shows also the pipe lines on its ceiling. Figure 11 is a plan of the second floor, and Fig. 12 is a partial section at Z Z, Figs. 10 and 11, showing connection of furnace and boiler. In this F is a brick furnace, 4 feet 8 inches square, with a cylindrical fire-box containing a 2-inch iron extra heavy spiral coil C, in which the water is heated and rises to bedroom radiators R R and wall coils W W, the latter each 35 feet long and containing 10 1¼-inch pipes. The branches are also connected with an 80-gallon laundry boiler D, which receives cold-water supply for the system through pipe E, and delivers hot water to the fixtures through pipe H.

The work was adapted to the existing features and conditions of the old construction and arrangement, and was consequently done at some disadvantage, but is said to be satisfactory and efficient. The contract price was about $8,000.

HEATING AND VENTILATION IN THE JOHNS HOPKINS HOSPITAL, BALTIMORE, MD.

PART I.—ESTABLISHMENT OF THE HOSPITAL, MAP, ARRANGEMENT AND CONSTRUCTION OF BUILDINGS, GENERAL DESCRIPTION OF HEATING AND VENTILATING SYSTEMS, THEIR OPERATION AND RESULTS.

The Johns Hopkins Hospital has been established and maintained from the income of an endowment of about three million dollars given by Johns Hopkins, in 1873, and since controlled by a Board of Trustees. The hospital is situated on a 14 acre plot of ground on top of a hill, about 100 feet above mean tide in Chesapeake Bay, in the heart of the city of Baltimore, Md. The institution is intended to ultimately provide for 400 patients, and to afford a training school for female nurses and for the medical college of the Johns Hopkins University.

The Board of Trustees advised concerning the design, construction, management and operation of the hospital with the following physicians: Dr. Norton Folsom, of Boston; Dr. Stephen Smith, of New York; Dr. Caspar Morris, of Philadelphia; Dr. Joseph Jones, of New Orleans; and Dr. John S. Billings, U. S. A.

The external designs of the buildings were furnished by the architects Cabot & Chandler, of Boston, and the heating, ventilating and plumbing systems were designed and installed by Bartlett, Hayward & Co., of Baltimore, Md.

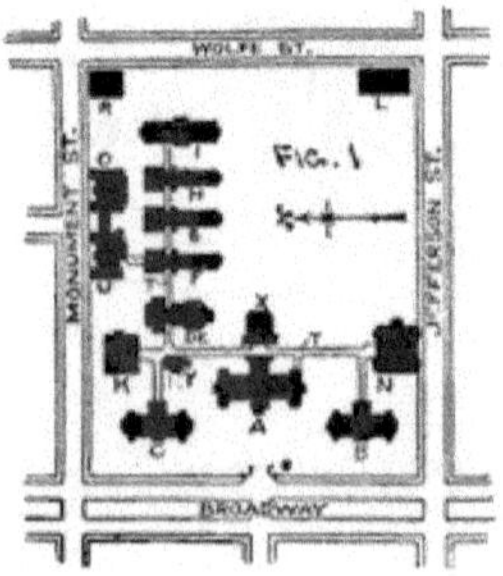

At the formal opening of the hospital, May 7, 1889, its history, construction and conditions were briefly presented in an address by Dr. John S. Billings, Medical Adviser to President Francis T. King, of the Board of Trustees, and to the Building Committee, and this was subsequently followed by an elaborate report of the construction systems and appliances, with general descriptions and illustrations. From this address and report we shall quote most of the general description, and prepare general illustrations, accompanied by special features and details, described and illustrated from notes and sketches recently made at the hospital by us.

Figure 1 is a general map of the hospital grounds and buildings. A is the administration building; X, apothecaries' building; C, male pay-ward; B, female pay-ward; N, nurses' home; K, kitchen; Y, bathhouse; D E, octagon ward; F, common wards; I, isolating ward; U, amphitheater; O, dispensary; R, pathological building; and L, laundry.

In addition to these buildings now erected and in use, the original complete plan provides for a row of five wards on the south side, opposite to and corresponding with the octagon, common, and isolating wards now constructed on the north, thus partially inclosing the large central lawn or garden. The original plan also provides for the erection of a large greenhouse on the south front, midway between the laundry and nurses' home. The open space on the east, fronting on Wolfe Street, is reserved for tents or temporary wooden buildings, in case of the outbreak of an epidemic. The buildings intended for

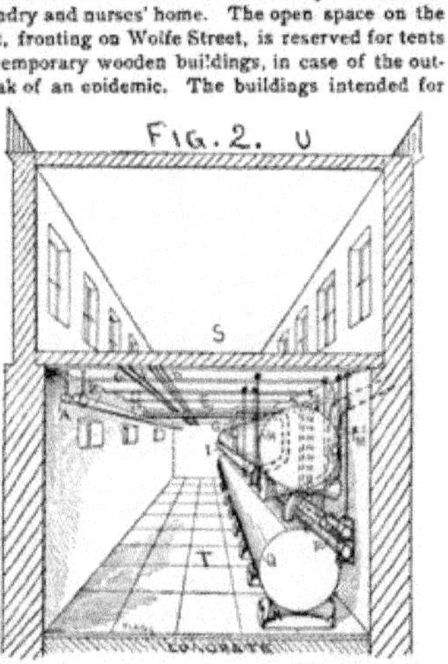

administrative purposes are of a size suited to the original complete plan, and will be ample when all the wards are erected.

All the buildings except the gate lodge, the pathological laboratory, the laundry, and the stable, are connected by a covered corridor T, as shown in Figs. 1 and 2. The floor of this corridor is at the uniform level of 114 feet above mean tide, being the level of the main floors of the administration and apothecaries' building, the kitchen, nurses' home and bathhouse. The top of this corridor is nearly flat, forming an open terrace walk at the uniform level of 124 feet above mean tide, being the level of the ward floors. This arrangement permits a perfectly free circulation of air between and around the buildings above the level of the ward floors, and secures the best influence of the prevailing southerly winds. As will be seen in the description of the ward buildings, it is not possible to pass to or from the octagon or either of the common wards without going into the free external air, so that there can be no communication between the air of different wards. Beneath the corridor is a passage-way containing the pipes for heating, lighting, water supply, sewage, etc., which is called the pipe tunnel, although it is above the level of the ground for more than half its height.

A sectional and perspective view of this tunnel is shown in Fig. 2. U is the terrace walk; S is the covered passage, about 10 feet wide, with an iron and brick floor, and T is the pipe tunnel, about 8 feet high. All main pipes are arranged as shown, suspended or on rollers. A and B are, respectively, 2-inch hot and 3-inch cold water supplies; C is ½-inch gas pipe; V is for pneumatic service; E, waste from bathrooms; G, steam distribution; H, 26-inch hot-water, main flow; I, 3-inch main steam flow; L, soil-pipe branch; K, bathroom waste branch; J, hot-water return branch; M, steam return branch; N, 4-inch main bathroom waste; O, 2-inch steam return main; P, 5-inch soil pipe; Q, 26-inch hot-water return main; R R, etc., are roller saddles; and D, electric wires.

All foundation and interior walls are of hard brick, laid in Cumberland cement below the ground level, at which point they are covered by a layer of heavy slate. Lines of drain tile are laid around the foundations, and for all the buildings having cellars or half basements, the outer surface of the walls beneath the ground is sheathed with overlapping slates. Above the the horizontal layers of slate at grade the walls are hollow, with a 2-inch air space 9 inches from the inner surface. This air space is closed in for two or three courses of brick at the top.

The floors of the principal buildings, and of the corridors, are formed of molded hollow blocks of hydraulic lime of Teil, laid between iron beams of suitable size and covered with wood, concrete or asphalt. Such floors are fireproof and are much lighter than those constructed with solid brick arches. The floors of the basement are of artificial stone laid in large blocks, and underneath all heat coils is placed a heavy coat of asphalt to prevent the passage of ground air up through the coil.

HEATING AND VENTILATING.

All the wards, the administration building, the nurses' home, the apothecaries' building, and the kitchen, are heated mainly by a system of circulation, through iron pipes, of hot-water of comparatively low temperature and pressure, the heat being furnished by boilers at the kitchen and nurses' home. In many of the rooms in these buildings, including all the private, or isolating, rooms for patients, and all living rooms in the administration building, open fireplaces are also provided, but these will probably be rarely used. The amphitheater, dispensary and bathhouses are heated by steam furnished from boilers at the kitchen building. The pathological laboratory and the laundry are heated by steam, each having its own boiler. The general distribution of the hot-water and steam systems is shown in Fig. 3.

The hot-water boilers for heating are six in number, four being in the vaults at the kitchen building and two in the cellar of the nurses' home, and all are on precisely the same level—viz., 85 feet above mean tide. Each of these boilers is 5 feet in diameter and 16 feet long, and contains 106 3½-inch tubes or flues. From the boilers the heated water passes into the great outflow main, which is a cast-iron pipe 26 inches inside diameter, hung on rollers from the ceiling of the pipe tunnel, as shown in Fig. 2, and provided with expansion joints just outside the kitchen building. From this main flow-pipes are given off at each building, and from these smaller mains the pipes in the heating coils are supplied. From these heating coils the cooled water returns by a similar system of pipes and mains to the boiler. This circuit is practically a closed one; none of the water being drawn off, or used, at any point, so that there is very little loss. The force which produces this circulation is a small one, being the difference in weight of a column of heated water from that of a similar column of water of from 8° to 15° Fahr. lower temperature, each column being about 29 feet high, which is the difference between the level of the water in the boilers and that of the top of the heating coils. By means of valves on all the mains, and on the supply and the discharge pipe to each coil, the rapidity of the circulation can be controlled for each building and for each coil, thus giving a corresponding control over the temperature of the coils themselves, since this is dependent on the amount of water of a given temperature which passes through the coil in a given time. The entire system of hot-water heating contains about 175,000 gallons

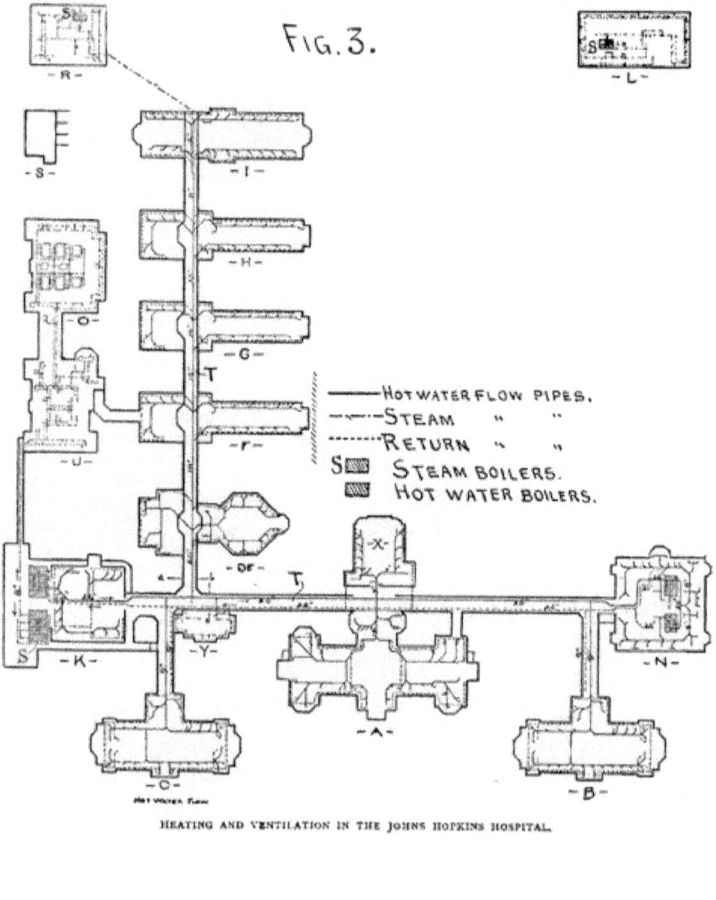

HEATING AND VENTILATION IN THE JOHNS HOPKINS HOSPITAL.

of water, and practical trial has shown that it produces an equable, agreeable temperature in all the buildings to which it is supplied, in all conditions of cold weather, and with the fullest ventilation desired. To prevent loss and waste of heat from the mains in the pipe tunnel, and in the basements of the several buildings, these pipes are covered with felt, enveloped in asbestos paper, and the whole is inclosed with stout canvas thoroughly painted. The effect of this protection is marked and satisfactory—very little heat is lost, as is shown by the temperatures in the pipe tunnel, and a great saving of fuel is thus effected. The heating coil most distant from the kitchen boilers is that in the southeast end of the isolating ward, being 763 feet away, as measured along the tunnel and basement of the ward.

The great advantages of this system of heating for rooms constantly occupied by the sick, in the climate of Baltimore, are its uniformity of action, the comparatively low temperature of the heating surfaces over which the air is passed, the ease with which different temperatures may be secured in different rooms, or even for different beds in the same room, and, above all, that it insures the delivery of a large supply of air heated to the temperature required for comfort, without the risk of overheating or of sudden changes.

Closely connected with the heating apparatus are many of the arrangements for ventilation. The external temperatures at the hospital have a range of from 102° Fahr. in summer to 6 degrees below zero Fahr. in winter, these extremes occurring about once in 10 years. To provide for these requires buildings and apparatus which would be satisfactory in either Calcutta or St. Petersburg. Let us first consider the arrangements for ventilation in cold weather. In the wards and rooms occupied by the sick, the sizes of flues and registers and the amount of heating surface have been arranged for a supply of 1 cubic foot of fresh air per second for each person in the ward, with the possibility of doubling this supply for a short time in flushing out the ward, as will be presently explained. In the pay wards, where each patient has a separate room, making it more difficult to secure thorough distribution, the supply of air is to be 1½ cubic feet per second per head. In the isolating ward, designed for cases giving rise to offensive odors or in which a large amount of organic matter is thrown off, or in which, for other reasons, a large amount of air is desirable, the air supply is fixed at 2 cubic feet per second per head. Finally, three rooms in the isolating ward are arranged with perforated floors for an air supply of 4 cubic feet per second per head, with capacity for doubling this if desired. For all the wards the air is warmed in cold weather before it is admitted to the room, forming the so-called method of heating by indirect radiation or by air convection. All registers and flues for fresh air are of such sizes as to permit the passage of the requisite amount of air with a velocity not exceeding 1½ feet per second under ordinary circumstances. Air currents of this velocity, having a temperature of from 70° to 75° Fahr., are barely perceptible by the hand, and create little or no discomfort. The fresh-air registers are, as a rule, placed in the piers in the outer walls, at a height of 9 inches from the floor, one register being allowed to each pair of beds. Besides these there are registers beneath the windows in the wards, which are only used in very cold weather to check the down drafts produced by the chilling of the air through the glass of the window. The chief register being that in the pier between each pair of beds, is so arranged that the nurse, by turning an iron arm upon its face, can reduce the temperature of the incoming air nearly to that of the external air, or can increase it to the maximum which the heating coil affords, but without changing the quantity of the air admitted. Ordinarily, as is well known, when a room heated by indirect radiation becomes too warm, the only way to shut off the heat supply is to close the register and thus shut off the air supply also, but in these wards the temperature can be regulated at the different registers, in different parts of the room to suit the needs of the different patients, without interfering with the air supply.

Table Showing the Radiating Surface in the Coils for Buildings and Rooms Heated by Steam.

Building or Room.	Cubic Feet of Space Heated.	Square Feet of Radiating Surface.
Amphitheater, main room	55,614	1,216
Dispensary, main room	96,013	4,724
Pathological building, amphitheater	26,019	498
Laundry building, total	50,544	2,071
Ironing-room	15,094	160
Drying-room for patients' clothing	1,664	340

Table Showing the Number of Square Feet of Radiating Surface in the Hot-Water Coils for the Principal Buildings and Rooms Supplied.

Building or Room.	Cubic Feet of Space Heated.	Square Feet of Radiating Surface.
Administration building, total	490,441	17,881
Superintendent's office	14,073	630
Pay ward, total	147,554	7,947
Single room for patients	2,836	162
Nurses' home, total	229,104	9,304½
Each room	1,360	65
Octagon ward, ward	144,197	2,400
Bay window in ward	2,400	60
Nurses' closet and bathroom	3,124	150
Dining-room	6,313	200
Common ward, total	78,880	5,107
Main ward	39,766	3,150
Private ward	3,570	262½
Dining-room	3,570	262½
Water-closet and lavatory	2,244	103
Footplate in sun-room	1,886	132
Isolating ward, total	92,195	4,545½
Single room	2,145	162
Apothecaries' building, total	79,463	3,419

The accompanying memoranda were furnished by Dr. A. C. Abbott as the results of observations made of the workings of this apparatus in one of the common wards during the month of December, 1889, the average number of patients in the ward being 24.

In the main ventilating shaft of the ward, the accelerating coil being heated, the velocity of the ascending current was 3.8 feet per second, giving a

Table Showing the Average Temperature, the Mean Relative Humidity and the Mean Dew Point of the Outside Air as Compared with the Corresponding Figures for the Air in the Wards.

Month.	Temperature of Outside Air.			Temperature of Air in Ward.			Mean Relative Humidity.		Mean Dew Point.		Mean Temperature in Coils.		Velocity of Incoming Air.
	Max.	Min.	Mean.	Max.	Min.	Mean.	Outside.	Inside.	Outside	Inside.	Flow.	Return.	Average.
November ..	67°	27°	44.9°	75.5°	60.4°	70.4°	70.7%	31.9%	34.7°	38.5°	119.7°	110°	3 feet.
December ..	50.1°	33.1°	43.6°	74.5°	67.3°	70.5°	73%	34.9%	34.8°	37.8°	134.8°	129.7°	3.3 feet.

total flow of 95 cubic feet of air per second, being nearly 4 cubic feet per second per head. When the accelerating coil was not heated, the velocity was 2.8 feet per second. In the water-closet shaft of the common ward, measuring 24x36 inches, the velocity of the ascending current was 183 feet per minute. The velocity of the air currents entering through the registers varied from 1.6 to 3.3 per second, depending upon the adjustment of the valve for admitting the outer air freely to the coils.

The carbonic acid determinations made by Dr. Abbott to determine the distribution of the fresh air within the ward are not yet sufficient in number to give positive results, chiefly owing to the fact that very great variations are found in the proportion of the carbonic acid in the external air due to the direction of the wind and to other circumstances. In general, however, it may be said that the proportion of carbonic impurity due to the respiration of patients in the ward is about 2 parts in 10,000, and the above table, showing the comparison of dew points and relative humidity of the outer air and the air in the ward, indicates that the vapor and other impurities added to the air by the respiration of the patients in the ward is removed almost as rapidly as it is formed. As regards temperature, it is certain that the wards can be kept at a temperature of 70° Fahr. in the coldest weather, while at the same time such ventilation is being secured in the ward that a person with a normal sense of smell coming from the fresh external air shall at no time perceive any trace of musty or animal odor. A systematic series of observations of temperature, dew point, and relative humidity of the external air in the central garden and in the wards are now being carried on, the observation being made at 8 A. M. and 8 P. M., and from these taken in connection with the corresponding series of carbonic acid determinations in the wards and in the external air to be made by Dr. Abbott throughout the coming year, all of which will be duly published, the operation of the heating and ventilation apparatus can be determined with scientific precision. At present it is sufficient to say that it amply fulfills the purposes for which it is designed, and furnishes and properly distributes within the wards an amount of fresh air heated to an agreeable temperature, which is, if anything, in excess of the requirements laid down by the best authorities on hygiene.

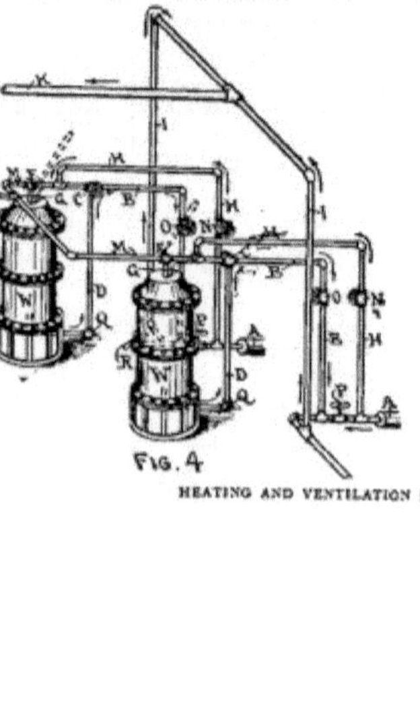

FIG. 4

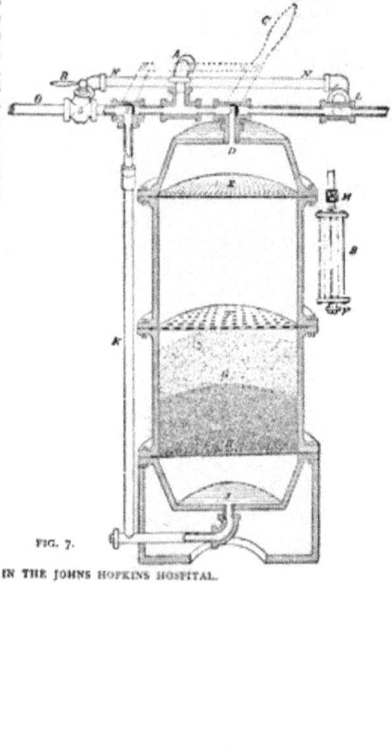

FIG. 7.

HEATING AND VENTILATION IN THE JOHNS HOPKINS HOSPITAL.

PART II.—FILTERING PLANT, PLAN AND PERSPECTIVE OF MAIN BOILER-ROOM, SPECIAL HOT WATER AND STEAM MAIN EXPANSION JOINTS.

THE water supply of the hospital is taken from a general supply of the city of Baltimore, mainly through a 6-inch pipe, which enters the center of the street front of the boiler vaults at the kitchen building. In these vaults it passes through two large filters of the Loomis patent. These are iron cylinders, nearly filled with clean sand. At intervals of about 24 hours the direction of the current through these filters is reversed, and the accumulated impurities are washed out into the street. The straining effect of these filters is shown by the bacteriological examinations of the water which have been made in the hospital. Before filtration the water contains an average of 39 micro-organisms per cubic centimeter; after filtration the average number found is six per cubic centimeter.

Figure 4 is a diagram showing the position of the Loomis filters W W, Fig. 8. The water from meters A A passes through the 4-inch pipes H H, valves E E, and branches G G into the filters W W; thence through 4-inch pipes D D and valves C C into branches B B, to the 4-inch branches I I, which connect with the 6-inch branch K which supplies boilers, house tank, etc. For cleaning the filter, valves E E are reversed and water enters through pipes D D, and escaping through branches M M is

The water enters from the meter A, Fig. 4, through three-way valve No. 1, passes down into filter at D, through perforated plate E, where the coarser impurities floating in the water are strained out, thence into the main body of the cylinder and down through the filtering material G and perforated plate H. The filtered water passes out of the filter at I, up pipe K, through valve No. 2, and into the delivery pipe O.

Where it is desirable to use alum or other chemical for the purpose of bringing the impurities in solution into a state of suspension, a trace of it is introduced into the water from air chamber B through valve M as the water passes it.

When valve 1 is reversed by lever C the water is made to enter at J, and is discharged through D to waste pipe L, and the filtering material G is forced through the cutting plate F and thoroughly broken up and washed in the upper part of the filter. The perforated plates E and H prevent the escape of the filtering material with the effluent water. The arrangement of pipes in Fig. 4 was necessitated in order to have the filters set in an alcove and carry the pipes along the walls out of the way.

Figure 10 is a side elevation of the vertical section of hot-water main Z, which is arranged to provide for expansion movements at the entrance to

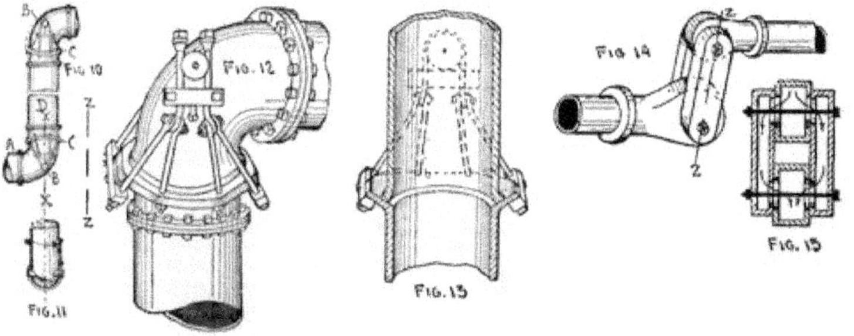

HEATING AND VENTILATION IN THE JOHNS HOPKINS HOSPITAL.

discharged into the sewer through the waste pipes L L. If an unobstructed flow of water is required, as in case of fire, or for any other purpose that demands a great delivery, valves O O and N N are closed and P B are opened, thus reversing their usual position and providing a by-pass around the filters so that water is delivered directly from meters A A to the pumps, etc. C C are check valves, closing toward the filters; Q Q are key valves, and R is a draw cock for the boiler-room supply.

Figure 7 is a vertical section of one of the filters, which is made of cast iron with brass cocks and is 96 inches high and 36 inches in diameter.

pipe tunnel T, Fig. 8; each joint, A A, permitting rotary motion in the cylindrical surface C about center B, so that the normally vertical section D may assume an inclination either way to conform to either expansion or contraction of the horizontal mains.

Figure 11 is part section and part elevation at X X and Z Z, Fig. 10. Figure 12 is an enlarged view of top joint A, and Fig. 13 is a vertical section through same. Figure 14 shows an expansion joint of one of the steam mains, and Fig. 15 is a section at Z Z, Fig. 14, the arrows indicating the course of the steam.

PART III.—SYSTEM IN THE ADMINISTRATION BUILDING, SYSTEMS IN THE COMMON WARDS, GENERAL DESCRIPTION, OPERATION, PLAN OF WARD, CROSS-SECTION AND LONGITUDINAL SECTION.

THE administration building, A, Fig. 1, is three stories high, besides basement, finished attic and dome, and has extreme dimensions of about 184x171 feet. It contains general offices, library, board rooms, apartments for the superintendent and resident physicians, and students' bedrooms. It is heated throughout by the hot-water system, mainly by indirect coils in the cellar, each room having a separate fresh-air flue by which, as well as the open fireplace and chimney flue, it is ventilated.

The common wards, H, G, F, Fig. 1, are essentially alike, the main rooms being each 99'6"x27'6"x16' clear height, giving to each of the 24 beds 7 feet 6 inches of wall space, 107 square feet of floor area, and 1,769 cubic feet of air space.

At the south end of the ward is a large bay window, the sash of which comes nearly to the floor, forming a sort of sun-room, and in the floor, near the walls and windows of this bay, are laid iron plates, covering an iron box, in which run hot-water pipes, thus giving a warm floor to prevent the downflow of air chilled by the window surface, and thus to insure warm feet and comfort to the convalescents sitting in this space in cold weather. Right angles in the ward are avoided as far as possible; all corners are rounded, the junction of the ceiling with the walls forms a quarter-circle, and the same occurs at the junction of the walls and floors, this last being effected by the use of a curved strip of hard wood instead of the ordinary washboard.

The most important feature of the ward is the method of heating and ventilating. The heating is effected by hot water coming from the mains in the pipe tunnel and passing through coils of 3-inch cast-iron pipe, arranged in stacks in the basement against the other outer walls. Under ordinary circumstances, in cold weather, the average temperature in these coils is 150° Fahr., but the temperature in any coil, or set of coils, can be lowered to any degree above that of the external air by lessening the velocity of the current of hot water passing through it, which is readily effected by valves placed on the flow and return pipe of each coil.

The fresh-air supply is admitted through openings in the exterior walls of the basement of the ward, coming from over the green lawn which surrounds the wards. These openings in the wall are protected by wire nettings and communicate with galvanized-iron flues which pass downward to open in the chambers beneath the heating coils, and also upward directly to the fresh-air registers in the ward. In each flue, opposite the external opening, is a cast-iron valve or damper, operated from the ward above, by means of which the incoming fresh air can be either directed wholly downward, so that it must all pass through the heating coil, or wholly upward, so that it passes directly to the ward without being heated, or partly upward and partly downward, so as to produce a mixture of any desired temperature. The lower end of the galvanized iron fresh-air flue, which opens beneath the radiator, has a valve which can be closed to regulate the amount of incoming air, when this is necessary in cold and windy weather. The heating coils are inclosed in brick chambers, which have at the top, in front of the coils, a large plate composed of two sheets of galvanized iron with felt between. These plates or doors fit tightly, but can be readily removed, thus giving the freest access to the pipes for the purposes of cleansing or of repair.

FIG. 17.

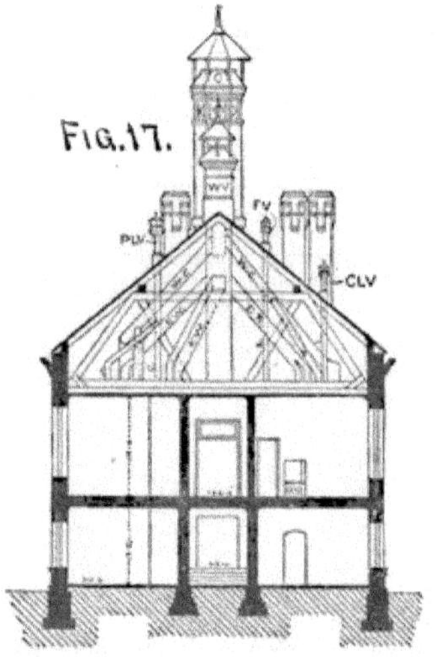

Two systems of exit flues are provided to remove foul air from the ward. The first has a series of circular openings in the floor of the ward, one beneath the foot of each bed. These openings are 12 inches in diameter, and are each covered with a nearly hemispherical dome of wire netting to prevent the dropping of small objects into the flues beneath. Each of these openings communicates with a galvanized-iron tube, 12 inches in diameter, which passes obliquely on the ceiling of the basement to enter the lower foul-air duct, which runs longitudinally beneath the ward floor to enter the ventilating chimney. The main longitudinal foul-air duct is constructed of wood, lined with galvanized iron. At the end most remote from the chimney it measures internally 1'10"x1'3", and from this point it gradually enlarges to provide for the additional flues entering it, until, at the point where it enters the ventilating chimney, it measures internally 4'4"x2'10". The ventilating chimney is 4 feet 2 inches in diameter and 75 feet high.

The upper system for the escape of foul air has six openings in the center of the ceiling of the ward, each measuring 2x2 feet and placed 13 feet apart. These open into the upper foul-air duct, which runs longitudinally in the attic above the ceiling of the ward and enters the ventilating chimney, corresponding to the lower duct described. The ceiling of the ward is 1 foot higher in the center than at the sides. The openings in the ceiling into the upper foul-air duct are controlled by shutters, which are raised or lowered at pleasure by the movement of an iron lever in the ventilating chimney. In this main ventilating chimney or aspirating shaft there is placed, just above the entrance of the upper longitudinal air duct, a coil of pipe, heated by high-pressure steam, which serves to increase the velocity of the upward current of air in the chimney, and is therefore called the accelerating coil. Under ordinary circumstances, in cold weather, only the downward ventilation is used, as this tends to save heat; but whenever the ward becomes overheated, or it is desired for any reason to pass a large quantity of air through it, the ceiling registers are also opened. In moderate and warm weather both sets of registers are open. The velocity of the upward current in the ventilating chimney, and therefore its aspiratory power, may be increased, as above stated, by means of the accelerating coil. It may also be regulated by a pair of valves near the top of the chimney, which can be closed or opened by an iron lever in the chimney, accessible through a small door just opposite the door of the ward.

In addition to these means for producing and regulating air currents, the common ward, nearest the octagon, is provided with a propelling fan, situated in the basement, at the south end. From this fan a duct is led with a branch which enters each coil chamber at the floor and turns upward for a short distance. The diameter of the air duct as it leaves the fan case is 4 feet. It divides into two main

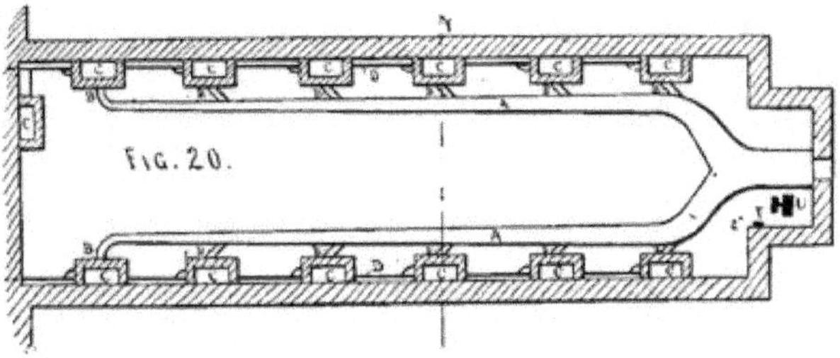

Fig. 20.

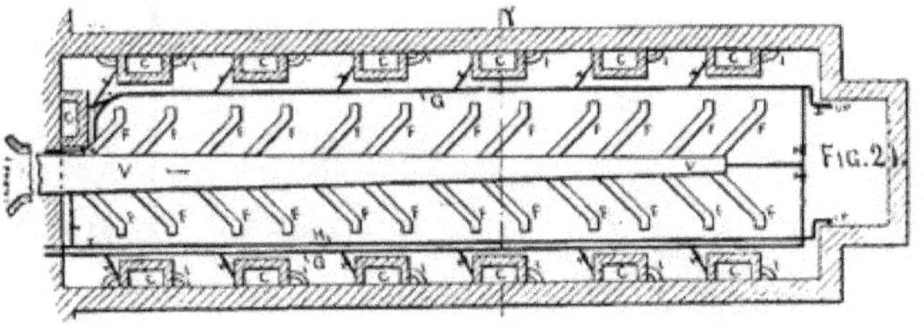

Fig. 21.

Fig. 22.

HEATING AND VENTILATION IN THE JOHNS HOPKINS HOSPITAL.

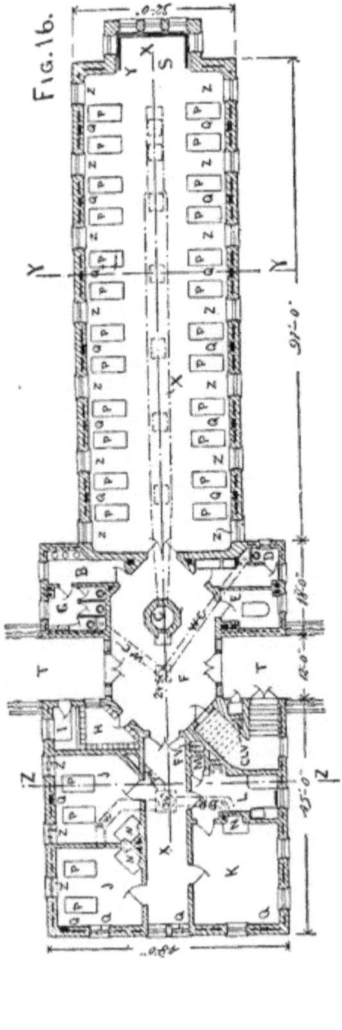

FIG. 16.

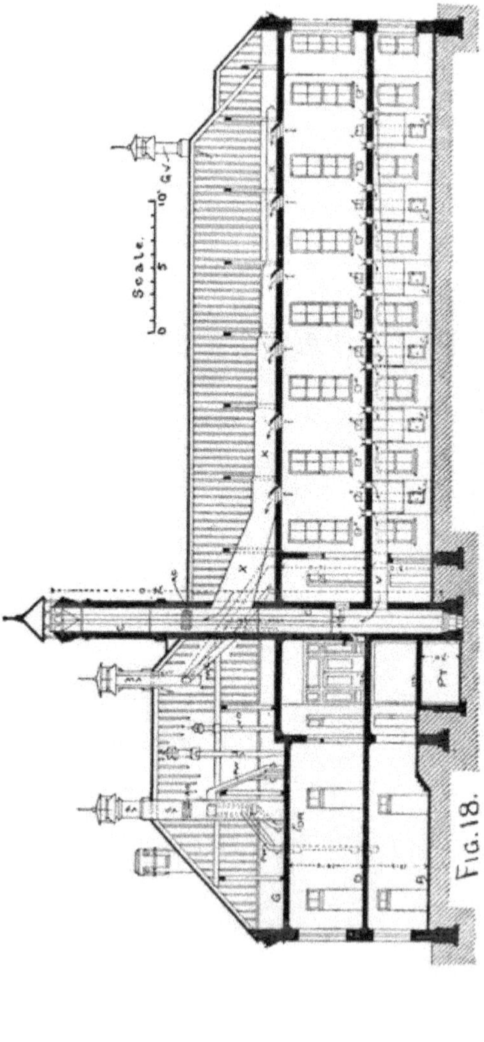

FIG. 18.

HEATING AND VENTILATION IN THE JOHNS HOPKINS HOSPITAL.

branches, each 2 feet square, and from these are given off the branch ducts to each coil, each being 10 inches in diameter. By running this fan for a few moments a very large amount of air can be forced into the ward, securing a thorough air flushing and the prompt removal of unpleasant odors. In hot, still weather the currents of air produced by the fan in the ward are very grateful to the sick. The other wards are fitted to receive similar fans and ducts, but are not yet supplied with them.

The whole system of ventilation, sizes of ducts, flues and registers, and provision of power to insure movement of air is intended to secure 1 cubic foot of fresh air per second for each of the 24 beds in the room, and to provide a reserve capacity of doubling this supply if it be desired to do so. The capacity of the apparatus is in excess of this. In this connection it should be borne in mind that these wards are not intended for cases of contagious disease, nor for cases such as uterine cancer, etc., which give rise to very offensive odors.

Figure 16 is a first-floor plan of a common ward; Fig. 17 is a section at Z Z, Fig. 16; Fig. 18 is a section at X X, Fig. 16. In Figs. 16, 17, and 18 the following reference letters are used: C, brick ventilating chimney, 4 feet diameter inside. The different ventilating flues are: W C and W C, 16 inches and 20 inches in diameter, from the water-closets; P W and P W, 18x22 inches, from the private wards; D R, 17x24 inches, from the dining-room; C L V, 10 inches diameter, from elevator shaft; E V, 10 inches diameter, from dumb-waiter shaft; P L V, from linen and

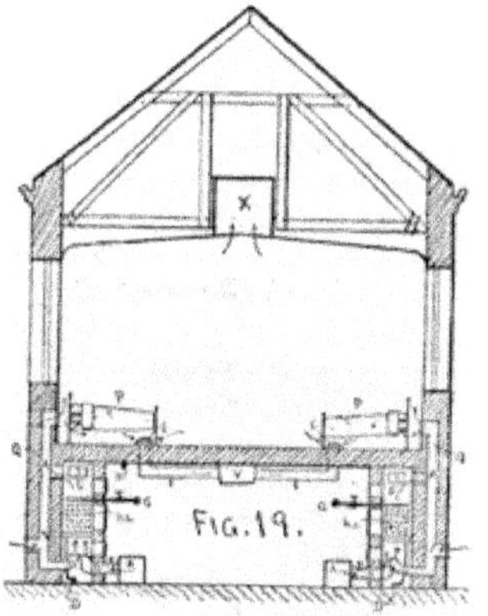

FIG. 19.

clothes closets; A C, A C, etc., accelerating steam coils; *h c*, *h c*, etc., hot-water coils; G V is a 24-inch diameter attic ventilation duct; V is a ventilating duct on basement ceiling; X is the attic main ventilating duct; and V W is a 24-inch ventilating shaft for water-closets.

In Fig. 16, B is the lavatory; G, patients' water-closet; D, nurses' water-closet; E, bathroom; F F, halls; H, patients' clothing closet; I, linen closet; J

FIG. 23.

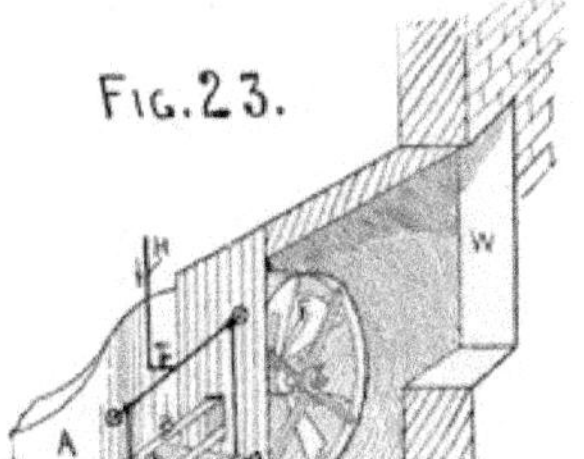

FIG. 24.

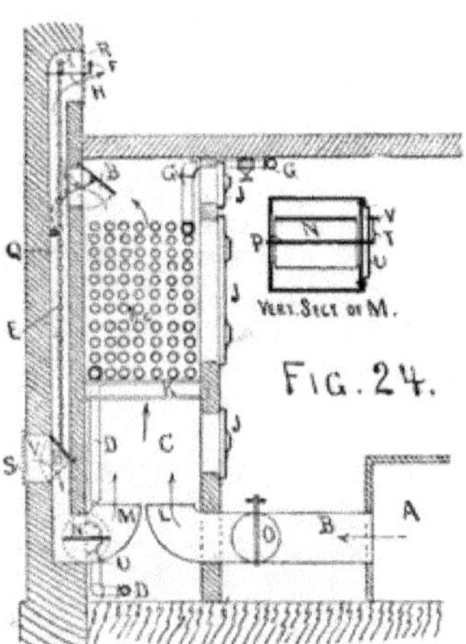

FIG. 25.

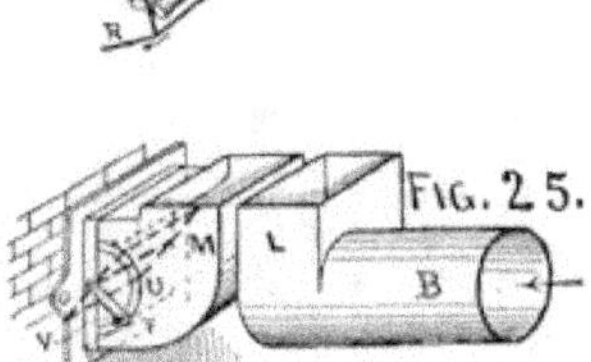

HEATING AND VENTILATION IN THE JOHNS HOPKINS HOSPITAL.

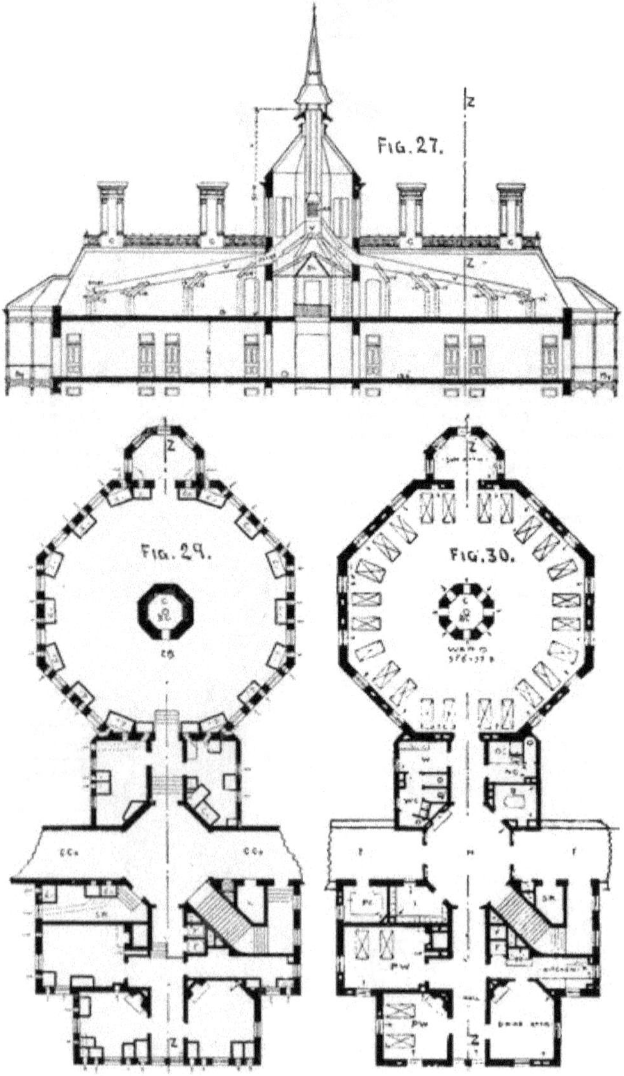

HEATING AND VENTILATION IN THE JOHNS HOPKINS HOSPITAL.

J are private wards; K is the dining-room; L, tea kitchen; M, elevator; N N, etc., are fireplaces; P P, etc., beds; Q Q, etc., 9x22 inches, vertical fresh-air ducts, with 14x22 registers; Z Z, etc., 9x13 inches, vertical fresh-air ducts, with 11x17-inch registers; W is a direct radiator; Y Y are steam pipes; S is the sun-room; T T is a covered terrace (see Figs. 1 and 2).

PART IV.—THE COMMON WARDS CONTINUED, DETAILS OF SYSTEMS, BASEMENT FLOOR AND CEILING PLANS AND PERSPECTIVE, TRANSVERSE SECTION THROUGH WARD, SECTION OF RADIATOR CHAMBER, VIEW OF FAN, AND DETAIL OF DUCTS.

FIGURE 19 is a general transverse section of the common ward building at Y Y, Fig. 16. Figure 20 is a floor plan of the basement under the ward. Figure 21 is a ceiling plan of the same. Figure 22 is a perspective view from Z, Fig. 20. In Figs 19, 20, 21, and 22 the references are the same—viz., C, ventilating chimney; *h c*, hot-water coils; V and X, ventilating ducts.

and, by belt B, drives the 60-inch fan F, which draws in fresh outside air through window W, and forces it through duct A.

Figure 24 is an enlarged section, not to scale, through chamber C, Fig. 19. Cold air is admitted through the outer wall, at L, and through duct M to chamber C; thence, rising through heating coils *h c*, it passes through duct Q and register R into the ward room. The shaft H, commanded by handle F, carries an arm I, to which is attached the chain E, which controls dampers B B. Turning handle F in one direction lowers the chain and the upper damper is gradually closed, and the lower one simultaneously cuts off the air from duct M, and admits it directly to duct Q, so that the dampers may easily be set to admit any desired proportion of the cold supply to the heater without diminishing the total volume, as the inlet V can never be closed; and when one damper B is shut the other must be open. H is a register, and S a screen; *h c* is the hot-water radiator, containing about 50 3 inch cast-iron pipes, 4 feet 6 inches long, supported on iron beams K, and con-

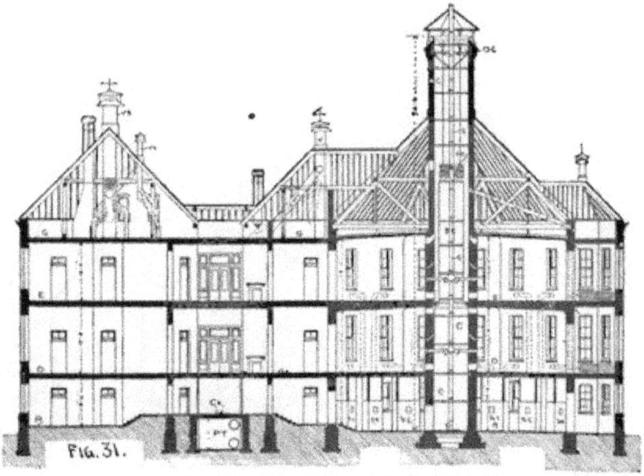

HEATING AND VENTILATION IN THE JOHNS HOPKINS HOSPITAL.

P P are ward beds; E E, shields over the ventilation duct branches F F; C C, brick radiator chambers; G G, hot-water flow; D D, return pipes; M M, outdoor fresh-air inlet; Q Q, fresh-air flues; J J, etc., iron doors; A A are the special fresh-air forcing ducts which are not yet placed in any ward, except F (Fig. 1); B B, are the branches, 10 inches in diameter; H is a live-steam pipe.

In Fig. 20 U is the fan engine forcing air through ducts A A, etc., and T is a steam trap on the exhaust pipe.

Figure 23 shows the engine at U, Fig. 20. It receives steam from pipe H, exhausts through pipe R,

nected to the flow and return branches G and D. J J J are cast-iron doors.

In ward F only a special supply of fresh air can be forced in through duct A and branches B B, etc., governed by damper O. The fresh-air damper N is made to be set and fastened at any required position, or it may be entirely closed, and tempered air from basement be admitted through the upper door J. Damper N is mounted on a horizontal axis P (see section), which has an arm T which moves on a semicircular guide bar U. To this it may be clamped in any position by a screw V. Figure 25 is an enlarged view of the inlets L and M, Fig. 24.

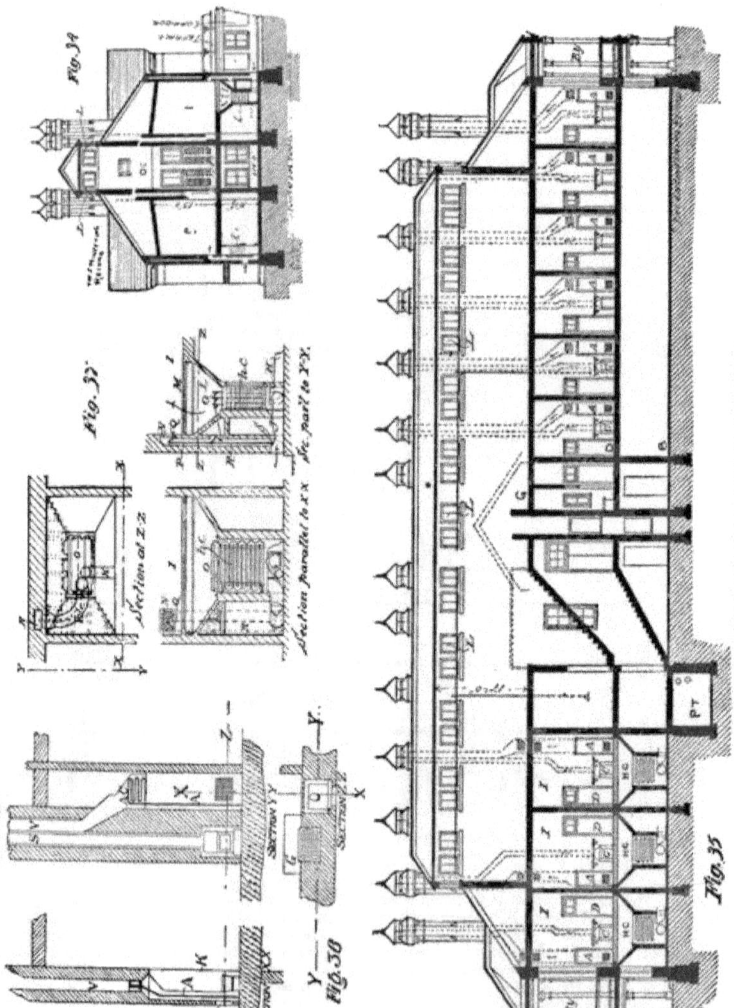

HEATING AND VENTILATION IN THE JOHNS HOPKINS HOSPITAL.

PART V.—DESCRIPTION, TRANSVERSE AND LONGITUDINAL SECTIONS OF THE PAY WARDS, FRESH-AIR CONDUITS, OCTAGON WARD, DESCRIPTION OF SYSTEMS, FIRST-FLOOR AND BASEMENT PLAN AND LONGITUDINAL SECTION.

THE pay wards B and C, Fig. 1, are two-story buildings, each 49½x130 feet, with one-story wings, containing bathrooms, etc. The patients' rooms measure about 12x15 feet, and open from each side of a central corridor. Each room has an open fireplace and smoke flue, and in the corridor wall, next the fireplace, are exit flues connected with an attic galvanized-iron flue, discharging into a central, perpendicular, ventilating shaft, that contains an accelerating steam coil. The heating system is the same as described in Part III. for the common wards, except that fresh air is admitted from the basement instead of directly through external openings, as shown in Fig. 26, which is a section of the radiator chamber C, corresponding to Fig. 24.

Tempered air from the basemant A enters the brick chamber C through inlet V, and passes up through flue Q, to enter the patients' room P through a register R. The valves B B are simultaneously controlled by chain E, operated by crank I of shaft H, which is turned by handle F. When both valves are fully raised, as shown, only heated air through S is admitted at R. If both are fully lowered, as indicated by dotted lines, only cold air through T is admitted at R; but if the dampers are only partly raised, the hot and cold air is mixed, and the full volume of fresh air is always admitted at any desired temperature between the limits. The cast-iron radiator pipes *h c* are supported on iron beams K. There are doors, not here shown, from chamber C to the basement.

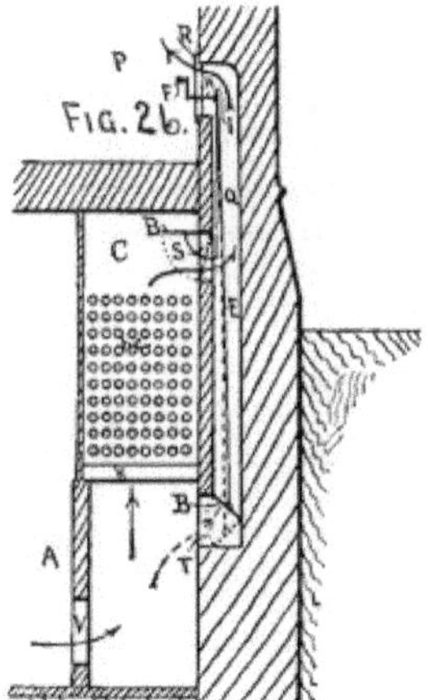

FIG. 26.

Figure 27 is a longitudinal section through the corridor of the upper part of the pay ward.

Figure 28 is a section at Z Z, Fig. 27. A, B, D, and G are respectively the cellar, basement and main and attic floors; V V are ventilating flues; C C, etc., smoke chimneys; H C, H C radiator chambers; S L is a skylight; B Y a balcony; V V are corridors.

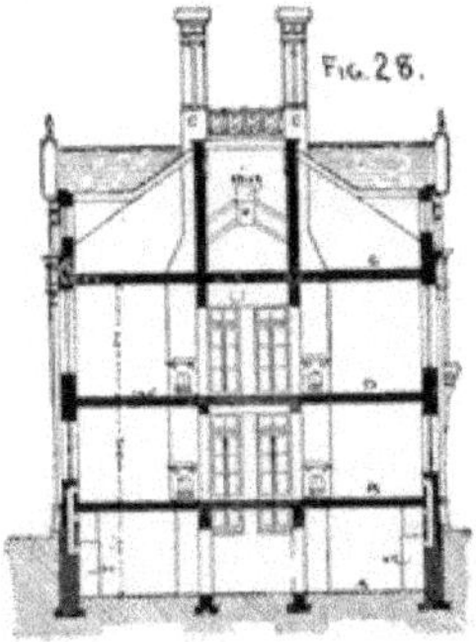

FIG. 28.

The octagon ward D E, Fig. 1, is 57 feet 8 inches in diameter, has two 16-foot stories above the basement, and provides 1,760 cubic feet of space for each bed. The heating of the ward is arranged substantially as in the common ward before described, but the system of ventilation is altogether different. Rising through the center of the ward is an octagonal brick chimney 8 feet in diameter internally, and with walls 2 feet 6 inches thick, making a total external diameter of 13 feet. Upon each face of this chimney are two openings from the ward, one near the floor, the other near the ceiling, each measuring 20x26 inches. Those in the lower ward open directly into the central shaft.

Within this brick chimney is set a boiler-iron tube, 5 feet 9 inches in diameter, resting on a projecting cast-iron base built into the walls, which tube extends from the floor of the lower ward to above the ceiling of the upper one. Into the space between this boiler-iron flue and the outer chimney the openings from the upper ward enter. Just above the top of the boiler-iron flue is placed a ring of steam pipe to act as an accelerating coil. Through the center of the chimney rises a cast-iron pipe 12 inches in diameter, which is intended to serve as a smoke flue for the open fireplaces to be placed in the wards against the central chimney, if these are found to be desirable. This smoke flue extends to the basement floor, where a large opening into it is provided to permit of the removal of soot swept down. Above the smoke flue extends through and projects a little above the fixed cowl which caps the top of the chimney.

In these wards the general direction of the air currents is from the circumference towards the central shaft. In cold weather the air passes either entirely or in part through the heating coils, the tem-

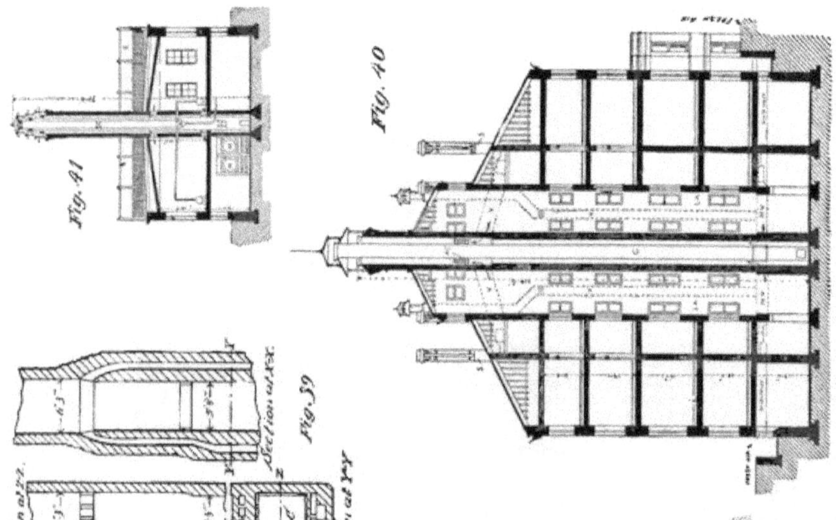

HEATING AND VENTILATION IN THE JOHNS HOPKINS HOSPITAL.

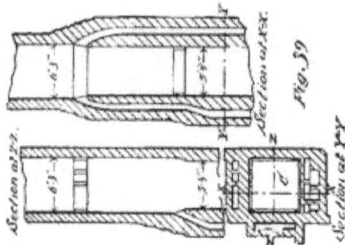

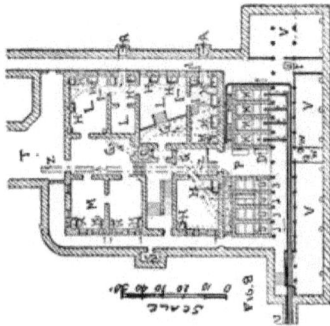

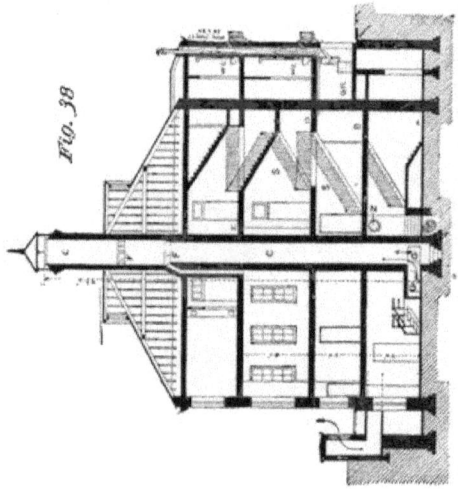

perature being regulated by a valve, as described for the common wards, and is allowed to escape during the greater part of the time through the openings near the floor in the central shaft, in order to secure uniform diffusion of the fresh air and to prevent undue loss of heat. During warm weather, or when it is desired to rapidly change the air of the ward, the upper registers in the central shaft near the ceiling are opened, in addition to the lower ones. The general arrangement of the central shaft is shown in the longitudinal section of the ward, Fig. 31, which also indicates by dotted lines the position and course of the hot-air flues in the piers between the windows in the external walls. The area of clear opening at the top of the central shaft can be diminished by valves.

Figure 29 is a plan of the basement. Figure 30 is a plan of the first main floor, and Fig. 31 is a section at Z Z, Figs. 29 and 30. C is the ventilating chimney, 8 feet diameter; B C is a 6-foot iron cylinder; C C is a covered corridor; H C are heat coils; wing that contains nurses' rooms, bathrooms, etc. Figure 32 is a plan of one end of the main floor, and Fig. 33 is a plan of the basement underneath. Figure 34 is a section at Z Z, Fig. 32; Fig. 35 is a full longitudinal section at X X, Fig. 32. O O is a corridor extending the full height of the building, opening into the open air at B Y, and having movable glass louvers L L, etc.

There are 17 patients' rooms P P, etc., each measuring 11x13 feet, and three rooms I I I, 13'x13'x10'. Each has an open chimney and fireplace C, and is entered by double doors D. A A, etc., are small closets, each with one door into the patient's room and one door into the corridor through which a commode may be removed without entering the patient's room. This closet is lined with galvanized iron and may be easily purified by flame.

Figure 35 shows the arrangement of closet A and chimney C, which is over 40 feet high. G is the grate; A C, an accelerating coil in the 28x28-inch closet flue V; S, the smoke flue; H, the door into the

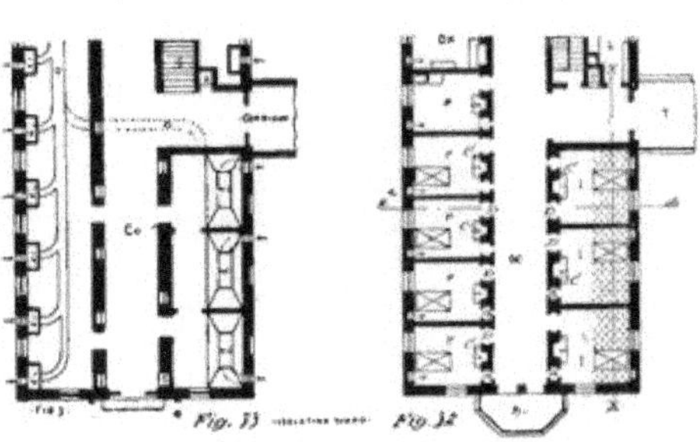

Fig. 33 ISOLATING WARD. Fig. 32

F F, food lifts; V is a ventilating shaft for same; C L is a coal and soiled clothes lift; V is a 10-inch ventilation duct for same; H is a hall, P W, P W, etc., are private wards; W is a lavatory; B, bathroom; W C, water-closets; P C are patients' clothing closets; L is a clean linen room; N C is a nurses' closet; R, range; K, sink; D C, drying closet; S R, storeroom; T, open terrace, see Fig. 1; A C are accelerating steam coils; V W C is a 24-inch water-closet vent; V, 32-inch vent for water-closet, bathroom and lavatory; V S, 42-inch ventilator for special wards; V L, 14-inch vent for linen closet and clothes room; P T, pipe tunnel; D C, chimney damper; S, smoke pipe; B, basement floor; D, main floor; E, second floor; G, attic floor.

PART VI.—DESCRIPTION, PLANS, TRANSVERSE AND LONGITUDINAL SECTIONS OF THE ISOLATED WARD, AND DETAILS OF ITS VENTILATING CHIMNEYS AND HEATING CHAMBERS.

THE isolating ward I, Fig. 1, is one story high, about 45 feet wide by 160 feet long, and contains 20 rooms for patients, arranged on both sides of a central corridor; and K, the door into the patient's room. As there are no common rooms it is intended to isolate each patient from every other patient.

Fresh air enters the patient's room through registers in the corners, in the outer wall; the arrangements for heating and regulating the temperature of the incoming air being substantially the same as those described for the common wards, but the amount of heating surface is greater, being calculated for a constant supply of air amounting to 2 cubic feet per second for each room. The excreta removed from closets A A, etc., can be thoroughly disinfected before it is taken to a sink that is inclosed by glass doors, and has special ventilation and air supply.

In the three rooms marked *I*, Figs. 32, 34, and 35, the fresh incoming air, instead of entering through a register in the side wall, enters through the floor, which, for a distance of 7 feet from the outer wall, is perforated with ¼-inch holes, giving over 94 square feet of floor, having 50 holes to the square foot. These holes are slightly funnel-shaped and 20 are estimated as equal to 1 square inch of clear inlet. There are 5,000 such holes in each room.

The arrangement for heating purposes in these rooms is shown in Fig. 37, which gives enlarged views of heating chamber H C, Fig. 35. The object is to supply a large amount of air, about 4 cubic feet per second, to each inmate, and to have this air pass constantly upward, so that no portion of it shall be rebreathed or come a second time in contact with the patient, thus placing him in the condition of being out-of-doors in a very gentle current of air. Fresh air is admitted either directly from out-of-doors through conduit J, or under pressure from the fan conduit through branch K. The figures show the cold air passing directly through the heating coil N C, register O, and perforations M, into the patient's room I; but, by turning handle Q, rod P will be depressed and close register O and turn damper S, so that the cold air, instead of entering duct J, will pass directly up flue R and through regulating register N into room I, or, by partly depressing rod P, the hot and cold air may be mixed in any desired proportions.

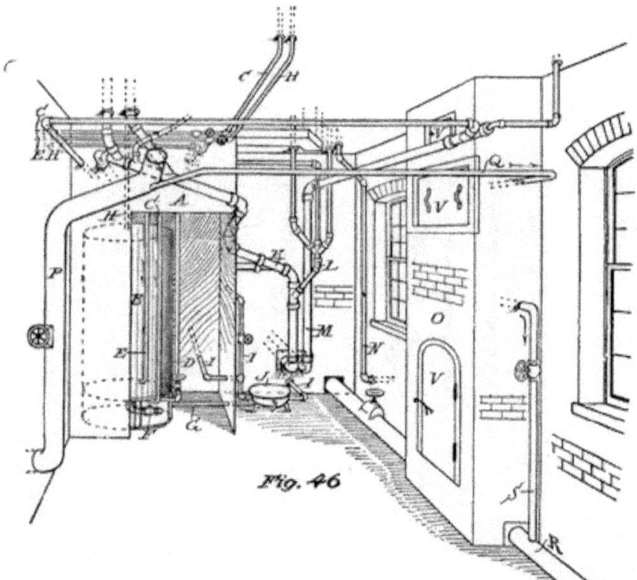

Fig. 46

HEATING AND VENTILATION IN THE JOHNS HOPKINS HOSPITAL.

PART VII. — DESCRIPTION OF THE KITCHEN BUILDING SYSTEM, GENERAL PLAN AND SECTION, AND DETAILS OF CHIMNEY AND FLUES, DESCRIPTION OF SYSTEM IN THE NURSES' HOME, TRANSVERSE AND LONGITUDINAL SECTIONS OF BUILDING.

THE kitchen building K, Fig. 1, is 75 feet square and three stories high above the cellar, and contains the main steam and hot-water boiler batteries, machine shop, room for a future electric light plant, bakery, kitchen, scullery, refrigerator, and garbage rooms, dining-rooms for housekeeper, cooks, assistants, etc., and vaults with a storage capacity of about 50 tons of coal.

Referring again to Fig. 8, C is a smoke and ventilating chimney, 5'8"x5'8"; L L, etc., are electric light rooms; E is a fuel elevator, 37x37 inches; S is a pipe shaft for riser lines in this building; H H, etc., are heating coils; A A are fresh-air inlets, 42x42 inches; V V are coal vaults; U is a steam-pipe tunnel, 3'6"x 5'9" high, to the amphitheater and dispensary; W W are two water filters; T is the main pipe tunnel (see Figs. 1, 2, and 3); P is the pump for house tank; D is the pump for steam boilers; G G, etc., are smoke flues; I I, etc., are fresh-air conduits; J J, etc., are hot-water boilers; K K, etc., are steam boilers; Z Z are the 26-inch hot water mains.

Figure 38 is a partial section through the building F F are ventilating flues, and the other reference letters are the same as in Fig. 8.

Figure 39 shows the construction of chimney C at the third floor. C is main smoke flue from boilers.

In the section Y Y the left-hand flue measures 9x62 inches and is for kitchen ventilation; the upper left-hand flue, 9x34 inches, is for the kitchen ranges. The upper middle flue, 9x13 inches, is for the pastry oven; the upper right-hand flue, 9x13 inches, is for the hot-water boiler. The lower flues, 9x30 inches, are for the bake oven; the 9x24-inch flue is for the bakery ventilation, and besides there are several other ventilation flues. The entrance of the small flues into the main chimney so far above the boiler flue is intended to prevent any irregularities of draft in the kitchen and bakery fires by reason of variations of temperature in the large central chimney flue, according to whether all the boilers, or only a portion of them, are being fired.

The nurses' home N, Fig. 1, is 90 feet square and four stories high above the cellar, and contains hot-water boilers, fuel vaults, storerooms, dining-room, training kitchen, lecture-room, sewing-rooms, parlor, library, and apartments for superintendent, nurses, etc. In the center of the building is a ventilating chimney, within which is the brick smokestack of the boilers in the cellar. This chimney is square, measuring 6 feet 3 inches on each side internally, and having walls 2 feet thick. The thickness of the walls of the internal circular brick stack is 8 inches. The inside diameter of the circular smokestack is 4 feet; the outside diameter 5 feet 6 inches, thus giving 15.3 square feet of clear area for ventilating purposes between the chimney and the stack.

The building is heated by hot water, all radiators being in the cellar. The outlet air flows from the rooms on the basement floor from openings on the inner walls into flues which run downward and then horizontally, being suspended from the ceiling of the cellar, and enters the space in the central ventilating chimney between its inner surface and the outer surface of the central smokestack. The ventilating flues from the airing-rooms on the upper floors pass upward in the inner walls to the attic, where they unite in large galvanized-iron flues which enter the central chimney. The central corridor is ventilated by openings direct into the central chimney.

Figure 40 is a vertical section of the nurses' home. C is the ventilating chimney; H V, H V are horizontal ventilating flues, 30x36 inches and 30x30 inches; V V, etc., ventilating flues, 24x32 inches and 48x48 inches; S S, smoke flues; L A, L A light and air shafts, 7x11 feet; A C accelerating steam coils.

The laundry, Fig. 1, measures 56x115 feet, and contains boiler and engine rooms, washing and drying rooms, ironing-room, hair-cutting and bed-making room, airing-room, dressing-room, etc., and disinfecting-room.

Figure 41 is a cross-section; S C is one of two smoke and ventilating chimneys 5 feet square; H H are boilers, and E is a drying room.

PART VIII.—SYSTEMS IN AMPHITHEATER, DISPENSARY, PATHOLOGICAL BUILDING AND BATHHOUSE, SECTION OF DISPENSARY BUILDING AND DETAILS OF HEATING CHAMBERS AND VALVES AND DESCRIPTION OF STEAM DISINFECTING CHAMBER.

The heating of the amphitheater and dispensary is effected by low-pressure steam, furnished by boilers at the kitchen building and conveyed through pipes carried direct from the boiler vaults

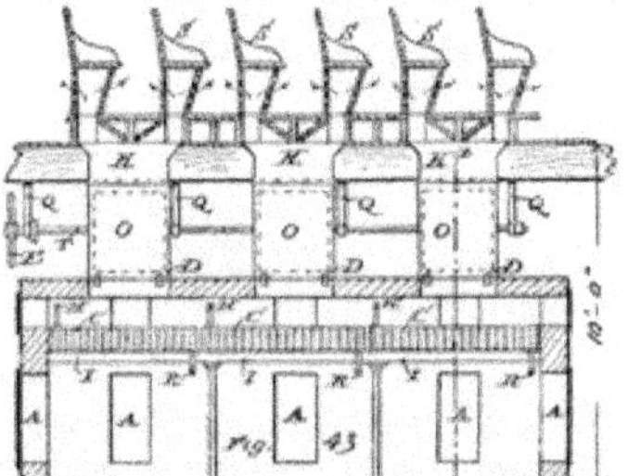

Fig. 43

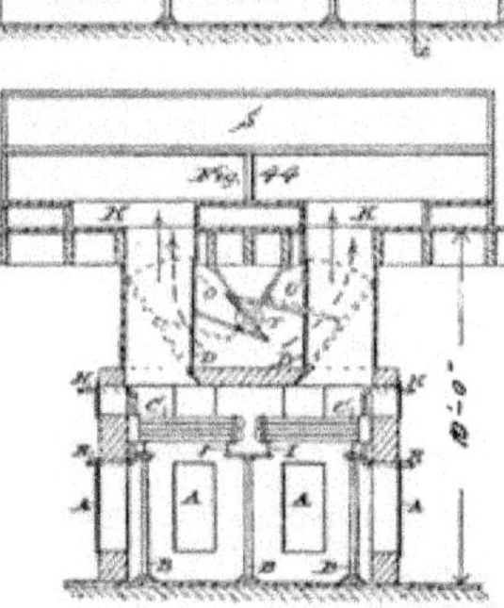

Fig. 44

to the amphitheater, through an underground tunnel 8 feet high, specially constructed for that purpose. Steam heating was selected for these buildings partly because they are not constantly occupied, and it is desirable to have the means of raising the temperature in them more rapidly than can be done by the circulation of hot water, and partly because it was desired to have the means of careful comparison of the two systems of heating for experimental and teaching purposes.

The amphitheater building U, Fig. 1, is 91x75 feet square, with one story and basement, and contains the amphitheater room, 52x47 feet, seating 280 persons; an operating-room, 18x26 feet; an etherizing-room, recovering-room, surgeon's room, accident reception-room a three-bed ward, and a photographer's room. The heating of the amphitheater is effected by steam coils placed in the space below the seats, the fresh, warm air entering through the risers. The foul air is drawn off into the ventilating chimney, 6 feet square, which is in the center of the south side of the building. The air may be taken into this chimney either below, near the floor, or above, near the ceiling.

The dispensary O, Fig. 1, is 91x75 feet square, and contains a large central waiting-room, 52 feet square, rooms for the reception of general and special diseases, bath and toilet rooms, a pharmacy, and a janitor's room. It is heated by steam coils in the cellar, the fresh, warm air being delivered through the risers and backs of the benches, as shown in Figs. 43 and 44. The temperature of this air can be regulated without diminishing the quantity by the use of the valves, as shown in Fig. 44. The extraction of the foul air from the dispensary is effected by a large ventilating shaft on the south of the general waiting-room, this shaft being 6 feet square internally. The air may enter this shaft either through a large opening near the floor level or through a large duct which communicates with the skylight, and its flow is made constant by means of an accelerating steam coil placed just above the upper opening.

Figure 43 is a section through a brick heating chamber II at X X, Fig. 42. Figure 44 is a section at Z Z, Fig. 43. The hot-water coils C C, etc., are supported on iron beams I I, etc., on columns B B, etc., and are connected to the supply and return pipes H and R, etc. As shown in the figures, cold air from the cellar is admitted to the heating chambers through doors A A, etc., and, being heated by coils C C, etc., passes up through flues K K, etc., and is discharged under seats S S, etc., as indicated by the full black arrows. There are openings O O in the insides of the flues K K, that are closed only by the dampers D D in the positions shown in Figs. 43 and 44. These dampers are operated by shaf. T which turns in journals Q Q, etc., and is controlled by hand-wheel E; if they be turned to the position shown by broken lines, they will close conduits K K to the heating chamber, and, uncovering the openings O O, etc., admit only cold air, as shown by the dotted arrows.

The pathological building R, Fig. 1, is 58x78 feet square and two stories high, besides attic and cellar. It contains a morgue, waiting-room, autopsy-room, private research-rooms, bacteriological workrooms, laboratories, museums, photograph-rooms, and rooms for keeping animals. The building is heated by steam coils in the cellar, and is ventilated by two chimneys, each 3'x3'x6' square and provided with steam accelerating coils. The autopsy table is ventilated downward by a tube. A cremation furnace for small animals, etc., is provided with a special ventilating flue, as is the animal room.

The bathhouse Y, Fig. 1, is about 65x31 feet square, and one story high above the basement. It is heated by direct steam radiators, and ventilated by a central chimney, 4 feet square, with an accelerating steam coil. The disinfecting chambers are situated in the basement of the laundry building, and comprise two rooms so arranged that the clothing, or other articles to be disinfected, when taken into one room A, Fig. 45, pass thence into the disinfecting oven C, or boiler, from which they are removed on the other side into another, entirely separate room B, so that the articles which have been cleansed and disinfected are not again exposed to infection.

The disinfecting oven is made of boiler iron and has double shells, into the space between which steam may be forced through a pipe. It is also arranged for the admission of live steam into the interior of the chamber. It is of an elliptical section, the longer diameter being perpendicular, and is 7 feet 2 inches long, 7 feet 5 inches high, and 5 feet 4 inches wide. It is lined with wood, and has wooden forms on which the mattresses, clothing, etc., to be disinfected may be hung or placed. The ends revolve on vertical axes, and are supported on castings running on circular tracks. The ends are locked by screw handles. There is also placed in the wall, between the chamber which receives infected articles and that in which cleansed articles are delivered, a large iron kettle or caldron, 3 feet 3 inches in diameter, with a capacity of 90 gallons. This is a double-jacketed boiler, heated by steam, and has two hemispherical covers, one opening into the outer and one into the inner room. In this articles of clothing or bedding of small quantity can be steamed or boiled, thus avoiding the necessity of heating up the large disinfecting chamber when but few articles are to be treated.

PART IX.—ARRANGEMENT OF BATH BOILERS AND PIPING AND HOT-AIR CHAMBERS IN BASEMENT OF OCTAGON WARD, DETAIL OF THERMOMETER FITTINGS AND SPECIAL VELOCITY APPARATUS FOR HOT-WATER PIPES.

FIGURE 46 is a perspective view of one of the basement rooms under the general rooms belonging to the octagon ward. (See Figs. 29 and 30.) A is a galvanized-iron closet, whose door is removed to show within the boiler B, about 3x6 feet, supplying hot water for the bathtubs, washbasins, and nurses' use. Cold water is supplied by the 1-inch pipe C, and is heated by an interior steam coil which receives live steam through ¾-inch pipe D. Hot water is delivered through 1-inch pipe H; E is a hot-water circulation pipe; H' and C' are distribution branches of hot and cold supply pipes H and C respectively; G is the return steam pipe; F is an emptying cock; I I are steam connections to steam trap J; K, L, and M are soil pipes receiving the branches from water-closet, urinal, and washbasin lines respectively; N is a 3-inch riser receiving safe wastes, etc.; O is a hot-air chamber, which contains a radiator coil that receives hot water from main P through branch Q, and returns it to main R through branch S.

Figure 47 is a section through the hot-air chamber O, Fig. 46; T is a cast-iron hot-water radiator supported on iron beams U; X is an adjustable damper controlling the admission of fresh, cold external air through inlet W, which is protected by an iron screen Y; V V V are cast-iron doors.

For purposes of experimental observation, thermometers are fixed at various points in the flow and return pipes of the hot-water system, in order to determine the temperature of the water at various distances from the source of heat, and before and after it has passed through the heating coils and given off some of its heat to the air passing up between the heating surfaces. The thermometers T, Fig. 48, are attached to special connections A; they were designed by Bartlett & Hayward to provide a position for the thermometer bulb and would keep it in contact with the flowing water and not in the cooler water of a dead end.

Figure 48 is an elevation of connection A; Fig. 49 is a section at Z Z, Fig. 48; and Fig. 50 is a section at Y Y, Fig. 49. Water enters at B, and passing around diaphragm E, impinges against the thermometer bulb, which is close to opening F, and circulates on through pipe C. D is a plug for clearing out the chamber. By reversing positions of plug D and pipe B, the fitting A may be used in a straight pipe, instead of at an angle, as here shown.

Two pieces of apparatus have been inserted in the heating system for the purpose of determining the velocity of the current of hot water in the pipes under various circumstances of external temperature, and thus obtaining data as to the amount of water producing a given heating effect in a given time. One of these is placed in the basement of the octagon ward; the other near the point most distant from the boiler in the isolating ward. The plan of this apparatus is shown in Fig. 51. It consists essentially of a by-pass connected with one of the smaller supply pipes P in such a way that all the water coming through this pipe can be sent through a glass tube having the same diameter as the pipe. In this tube the velocity of the stream can be measured by injecting from A a small quantity of colored fluid, such as a solution of carmine, and noting the time required for it to pass a measured distance in the glass tube. Ordinarily valve B is open, and valves C C are closed, and the water flows directly through pipe P without entering the by-pass; but by reversing the valves the velocity apparatus can be put in operation whenever desired. With a temperature of 92.6° Fahr. in the flow pipe, and 85.4° Fahr. in the return, the rate of flow as determined by this apparatus is 13.5 feet per minute. With a temperature of 134.8° Fahr. in the flow pipe and 129.7° Fahr. in the return the velocity was found to be 16 feet per minute.

THE HEATING AND VENTILATION OF THE WILLIAM J. SYMS OPERATING THEATER, WITH TEST OF EFFICIENCY OF THE HEATING COILS.*

The problem of heating and ventilating the large number of hospitals constantly being constructed and remodeled, in civilized communities, is one of the greatest questions that engage the study of the sanitary and mechanical engineer.

Hospital ventilation, unlike that of small buildings, requires large quantities of pure air, and, if the indirect system of heating is used, air must consequently be introduced at a low temperature. These

*Graduating thesis of Henry D. Whitcomb, Jr., and Henry C. Meyer, Jr., of the Stevens Institute of Technology, class of 1892.

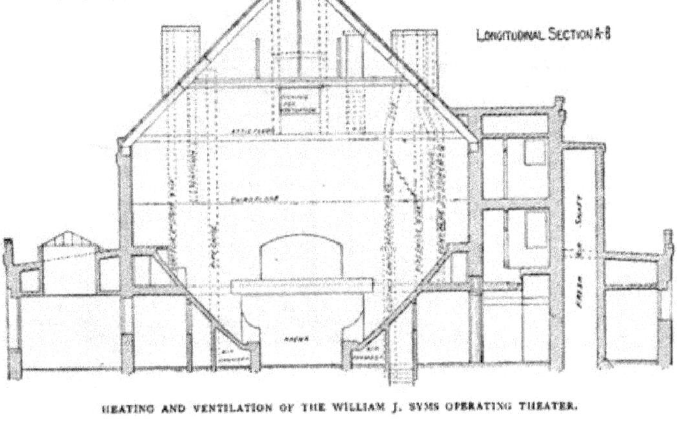

HEATING AND VENTILATION OF THE WILLIAM J. SYMS OPERATING THEATER.

conditions necessitate the use of fans and fan engines to force the fresh air to various parts of the building, and also steam and hot-water coils to heat the air at some point in its passage. Not only is the initial cost of such a plant large, but the cost of fuel is also. Plans and specifications for such a plant might be easily prepared by an engineer posted in such matters, but to do the same with economy is the question.

Good ventilation, according to Dr. John S. Billings, requires the admission of as much pure air as is necessary to "keep the vitiated air constantly diluted to a certain standard." The number of cubic feet of air admitted to maintain the standard varies with the use to which the building is to be put. If it be a hospital, the nature of the diseases of the patients effects this figure. The volume of air admitted must

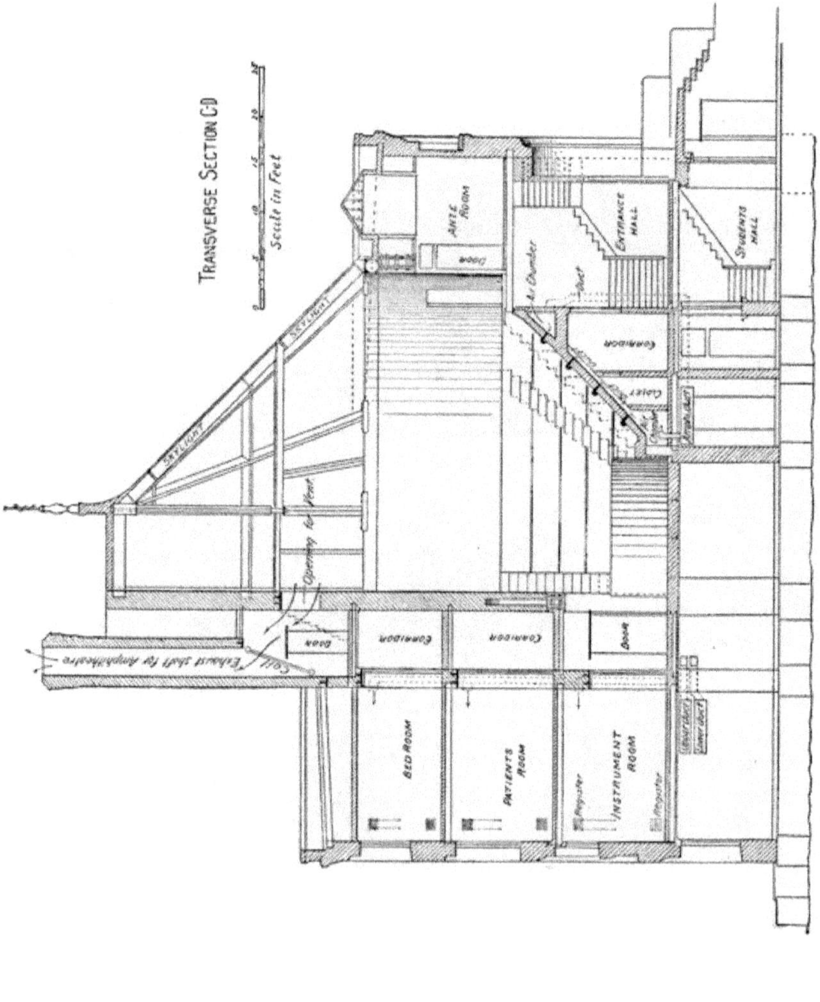

HEATING AND VENTILATION OF THE WILLIAM J. SYMS OPERATING THEATRE.

be very large, the air must be at a moderate temperature, and have a low velocity, in order that drafts may not be perceptible. It must not be supposed that large quantities of air will protect a patient from disease germs floating in the air, but it will lessen their quantity per unit volume, and hence diminish the chances of an infection. As to number of cubic feet of air to be delivered per head per hour, authorities differ. General Morin, whose figures are often quoted, estimating it at 3,530 cubic feet for hospitals for wounded, and 5,300 in times of epidemic.

As a notable example of hospital ventilation, recently constructed, we have selected the William J. Syms Operating Theater of the Roosevelt Hospital, New York City. The structure was designed by Mr. W. Wheeler Smith, Mr. William J. Baldwin being the mechanical engineer, both of New York City. The building occupies an area of 120x80 feet, and consists of a large amphitheater for operating, private operating-room, instrument-rooms, etherizing-rooms, photographic-rooms, bedrooms for patients about to be operated upon, bath, coat, janitors' rooms, etc. The operations are performed in the amphitheater before large classes of medical students, and the absolute necessity of having pure air throughout the operations makes the problem of the heating and ventilating of great importance. Owing to the many novel features contained in the plant of this building it was decided to make the description of these the subject of this thesis.

The indirect system is used. The cold air is taken in by a shaft 5x5½ feet, running from the roof to the basement and then passed through a single coil of heating pipes into the fan chamber. These primary coils are only used in extremely cold weather. At this point the current of air divides, one part, to which we will refer later, going to the bedrooms, etc. The supply for the amphitheater is drawn through a circular opening in the wall into the heating chamber by means of a 66-inch Sturtevant fan. The heating chamber, shown in Fig. 3, contains six coils of pipes, presenting in all a heating surface of 546.3 square feet. Each coil has its own service pipes and valves, thus enabling the engineer to turn on as many coils as is necessary to raise the incoming air to the required temperature. The coils are inclosed on the top, bottom, and sides with galvanized sheet iron. One end is open for the admission of air, and the other contracted into a duct through which it is led to the chambers under the amphitheater. The air is finally delivered to the amphitheater through the goose-necks shown in detail in Fig. 1. These are 4 inches in diameter and are placed under the seats. The curved form is used in order that the air may be diffused upon the floor, and so not place the occupants of the seats in a draft. One hundred and one of the goose-necks are employed.

The entire flooring of the amphitheater is of stone and asphalt, thus permitting it to be washed down with a hose when deemed necessary. The curved form of the goose-neck also prevents the water from being carried into the air chamber. The foul air is removed by an aspirating chimney 6x3 feet, running from the top of the side wall to the roof. A single coil of pipes is placed in the shaft to stimulate the circulation of air. They are heated by the exhaust steam from the fan engine.

Mention was made of the air for the remaining parts of the building. This is taken from the fan chamber in the same way as the air for the amphitheater; it also passes through a circular opening into a heating chamber. It is driven by a fan similar to the one before mentioned which is run by a separate engine. The heating chamber is the same as the one for the amphitheater, except that it has two sets of coils, each one being the same size as coil No. 3.

A separate set of ducts, however, running parallel to each other, lead from each set of inclosed coils to the various register heads in the different rooms and parts of the building. The intention of the double duct and duplex coil system is to supply two currents of air at different temperatures to each register head, the sections of the coil in one cluster being all turned on or nearly so, giving the temperature of about 110° Fahr., while only one or two sections of the other have steam within them giving a temperature of about 60° Fahr.

The coils in each cluster are of equal power and surface, so that should the hot one fail or be out of order the cold ducts can be made to supply warmed air to the hospital rooms, thus minimizing the chances of leaving the building without heat through damage to the coils by frost or otherwise.

The two currents flow through the parallel ducts to the registers in the rooms, and are there mixed by a device shown in Fig. 4. A movable slide, held in position by grooves, is placed over the ends of the ducts. This is controlled by a shaft of square section, connected by a double lever. To the outer end of the shaft a lever is keyed. The slide is of such a size as to entirely cover the mouth of one duct, and by moving this the two currents of air are mixed to the desired temperature. By the use of this register any room may have any temperature within the limits of the air in the two ducts. It will also be seen that whatever this temperature may be, the volume of air admitted will be constant. The registers are placed in the side walls about 7 feet from the floor. The outlet register is placed as close to the floor as possible, there being but 2 inches between the lower edge of the register and the floor.

The fans are run by two independent Metropolitan engines with 7x8-inch cylinders. The boilers of the Roosevelt Hospital supply the coils and fan engines with steam; the steam pressure for the coils being regulated by a Davis reducing valve. A steam gauge is placed upon each side of this valve. The condensed water is pumped into the hospital receiving tank by a Worthington duplex pump controlled by a pump governor. A Davis steam trap is placed between the pump governor and the coils, to prevent the live steam from passing through the pump and governor, when high pressures are used. The cost of the building was $175,000, and of this $10,500 was expended in the heating and ventilating apparatus described.

THE TEST OF THE HEATING COILS.

A test of value of each successive heating coil was made, the object being to find the rise in temperature of the air when 1,000,000 cubic feet were forced past the coils per hour; this being the figure sought by the designer of the plant. The test was made on the coils for the amphitheater, communication to the other being shut off. Five tests in all were made at 2, 5, 10, 15, and 20 pounds steam pressure in the coils. Unfortunately, the condensed water from the coils ran in with the drips and condensed water from other parts of the hospital, and its temperature and volume could not be ascertained. We were thus unable to check the results shown by the thermometers.

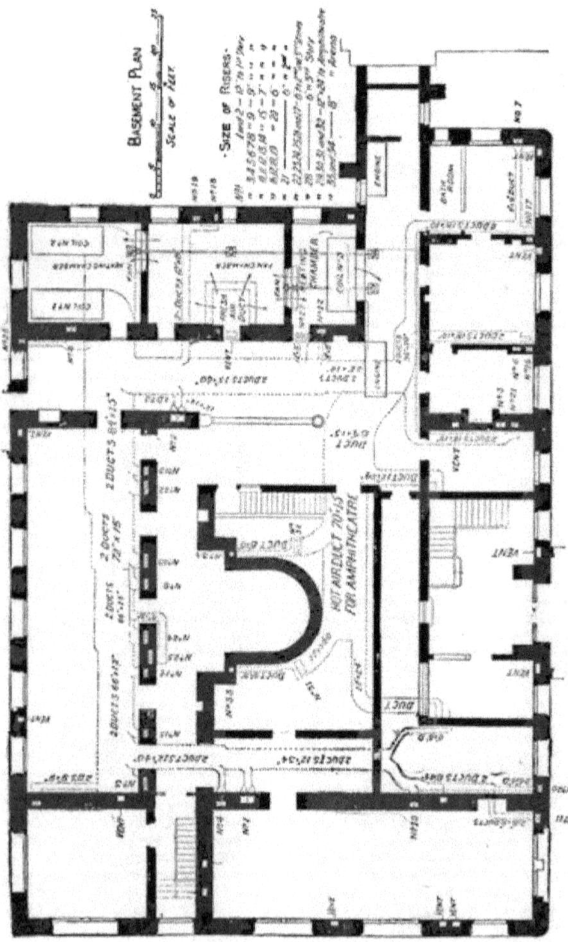

HEATING AND VENTILATION OF THE WILLIAM J. SYMS OPERATING THEATER.

A thermometer was placed in the fan chamber, about 2 feet from the circular opening containing the fan, and directly in the line of the shaft. A second one was placed in the duct leading to the amphitheater, at A, Fig. 3, about 3 feet beyond the coil. This was done by punching a hole in the duct and inserting a cylinder of pasteboard. The thermometer, wrapped in cotton waste, was placed in this. To prevent radiation from the coils, the bulb and all of the thermometer inside of the duct was protected on the top, bottom, and side next to the coils by thin boards, nailed together, and held in place by strings from the top of the duct.

The thermometer in front of the fan was unprotected during the first three tests, and then it was found that the service pipes for the primary coils, which were about 4 feet away, were quite hot, and undoubtedly caused too high a reading for the air at that point. The results were therefore thrown out and the tests made over. The valves to these coils were tightly closed, and the feed pipe covered with a board. The thermometer was placed in the pasteboard

Fig. 2

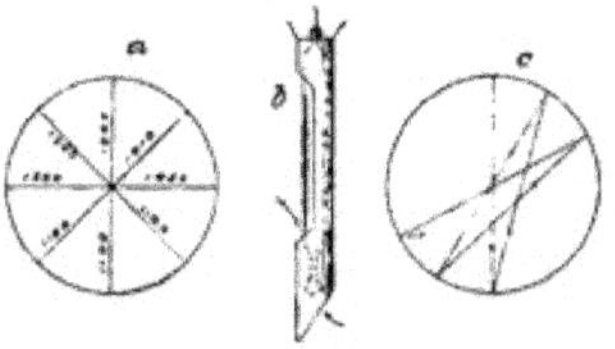

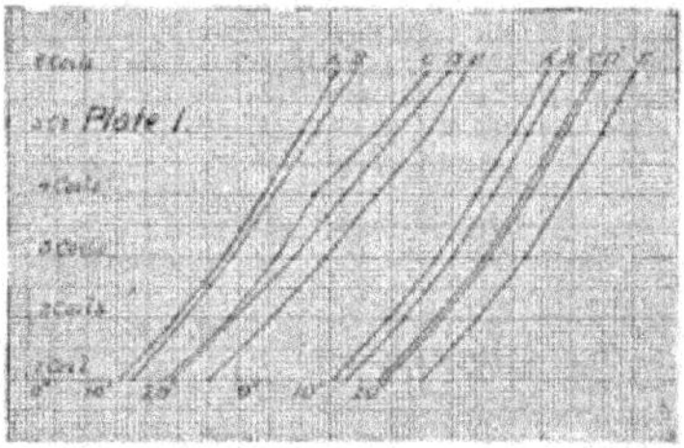

Fig. I

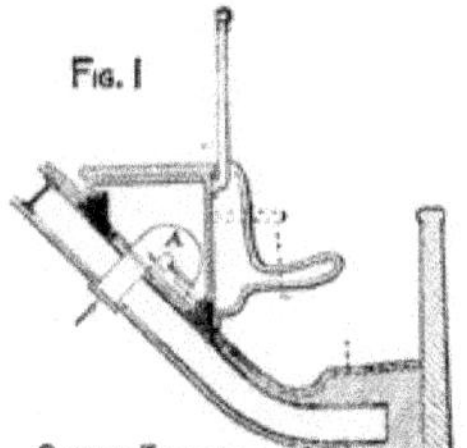

Section Through Amphitheatre Floor.

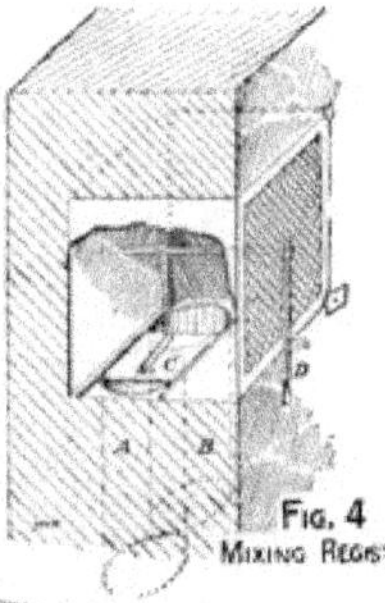

Fig. 4
Mixing Register.

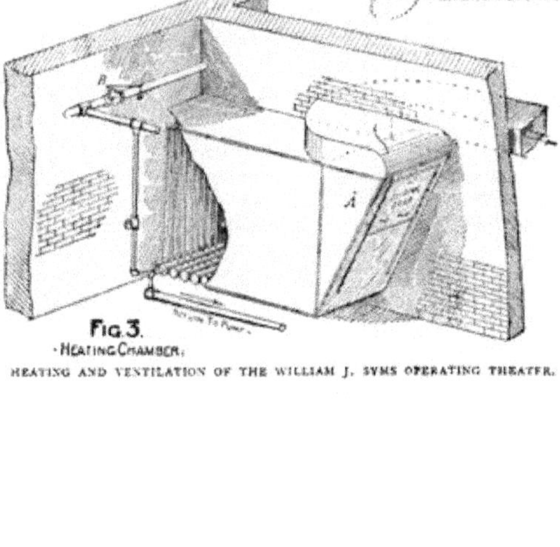

Fig. 3.
Heating Chamber.

HEATING AND VENTILATION OF THE WILLIAM J. SYMS OPERATING THEATER.

cylinder, which was cut as shown at *b*, Fig. 2. Hygrometric tests were made at the middle of each test, the results of which are shown in Table No. 3. Readings of the thermometers were taken every 10 minutes throughout the test, until their difference was constant. The revolutions of the fan were also taken at the same time.

To determine the linear velocity of the air in passing through opening into the heating chamber, measurements were made with an anemometer. On taking readings it was found that the person holding the instrument interfered greatly with the current of air, and it was afterward placed upon the end of a rod about 6 feet in length. There is a spider bearing in the opening, consequently the velocities at all points equally distant from the center are not equal, nor are they at different distances from the center. Measurements were made in several different ways. The first consisted in moving the anemometer, with as uniform a speed as possible, back and forth upon different radii of the circular orifice, the object being to start at a certain time and move at a uniform speed, back and forth, on the radii, and at end of trial arrive at the starting point. Eight of these readings were taken on eight different radii, giving a mean linear velocity of 1,307 feet per minute. Another trial was made by moving the instrument in the manner shown by *c*, Fig. 2. This was done for five minutes, at the end of this time again arriving at the starting point. A mean of two trials gave a velocity of 1,325 feet.

The opening is 4 feet in diameter. One million cubic feet of air per hour passing through this would have a linear velocity of 1,344 feet. One thousand three hundred and sixteen linear feet, the mean of the trials, is 0.979 of what it should be when actually passing 1,000,000 cubic feet, and taking into account the friction of the anemometer, it was supposed, during the tests, that the fan was passing the required amount.

TABLE NO. 1.

Actual Rise in Temperature for Each Coil Section of 91.05 Square Feet in Degrees Fahr.

	Number of Pounds Pressure				
	2 Lbs.	5 Lbs.	10 Lbs.	15 Lbs.	20 Lbs.
One coil section.......	15.5	17.8	23.0	23.0	30.0
Two coil sections.....	24.7	28.2	31.5	34.0	40.3
Three coil sections ...	32.7	34.0	41.4	44.0	40.5
Four coil sections	39.2	40.7	47.5	52.4	57.5
Five coil sections.....	45.4	47.5	57.4	61.8	68.6
Six coil sections	51.9	54.3	65.0	69.5	71.2

TABLE NO. 2.

Theoretical Rise in Temperature for Each Coil Section of 91.05 Square Feet in Degrees Fahr.

	Number of Pounds Pressure.				
	2 Lbs.	5 Lbs.	10 Lbs.	15 Lbs.	20 Lbs.
One coil section.......	13.5	18.0	23.2	24.1	30.2
Two coil sections.....	24.7	27.2	32.4	33.3	39.4
Three coil sections ...	32.7	35.2	40.4	41.3	47.4
Four coil sections.....	39.2	41.7	46.9	47.8	53.9
Five coil sections.....	45.4	47.9	51.1	54.0	60.1
Six coil sections	51.2	53.7	58.9	59.8	65.9

TABLE NO. 3.

Reading of Thermometer Fahr.		Weight of Vapor in 1 Cubic Foot of Air.	Per Cent. of Moisture as Calculated.	Weight of Vapor in 1,000,000 Cubic Feet of Air.	Humidity Saturation. 1.000.
Dry.	Wet.	Grains.		Pounds.	
62.5	58.2	3.357	53.5	426.45	.536
71.0	55.8	3.850	46.6	501.50	.466
67.5	55.2	3.852	52.7	501.51	.524
68.4	50.0	3.480	63.1	453.10	.632
70.5	62.5	4.680	44.5	609.30	.443

Table 1 shows the value of the heating coils for 91.05 square feet of heating surface. Plate 1 shows Table 1 plotted on cross-section paper, the abscissas showing the increments of temperature. The curves marked A, B, C, D, and E show these increments for 2, 5, 10, 15, and 20 pounds pressure. The tests at two pounds seem to be the most accurate, each increment of temperature being a little less than the one before it. At the higher pressures the temperature seems to be too high after turning on the fourth coil. The temperature of the sheet-iron covering for the coils at this point seemed to be higher than that of the air, it being undoubtedly due to direct radiation from the coils themselves. This seemed to affect the readings of the thermometer. What was thought to be the correct curves for the higher pressures, assuming that the rise in temperature for the first coil to be correct, is shown on the right side of Plate 1, and marked A^1, B^1, C^1, D^1, E^1. A^1 is simply A moved over to the right.

Table 2 is made from these curves. The same percentage of moisture is understood to be present in A^1 as in A, B^1 as in B, etc.

We are indebted to Mr. W. Wheeler Smith, the architect, for access to the plans of the building; to Mr. William J. Baldwin, M. E., for details of the ventilating plant; and to Mr. P. A. Sullivan, engineer in charge of the Roosevelt Hospital, for courtesies extended while making the test.

HEATING AND VENTILATION OF THE ROYAL VICTORIA HOSPITAL AT MONTREAL.

THE grounds of the Royal Victoria Hospital at Montreal, Canada, which are situated on the eastern slope of Mount Royal, are somewhat irregular in shape, extending for a distance of 825 feet along the street and 1,425 feet back in a direction up the mountain. The main building comprises 13 distinct blocks connected by corridors. The administration building, forming the center of this group, is flanked on each side by the medical and surgical wings, both of the latter extending outward toward the street, thus inclosing the driveway and lawns that make up the approach to the hospital.

In describing the ventilating and heating apparatus only that pertaining to the medical wing of the building will be taken up, as the apparatus in the other buildings is entirely independent of and similar to the one about to be described. The medical block is composed of five floors known as the basement, ground, first, second, and third floors. The ground floor, first and second floors each contain a ward for 30 patients. Attached to each large ward are the nurses' room, day rooms for convalescent patients, the ward kitchen and a separation ward. The bath-rooms and ward offices are contained in the round towers at the ends of the wards, and are so designed as not to interfere appreciably with the outlook from the large end windows and balconies provided for the use of the patients.

Adjoining each ward block and connected with them by cross ventilating corridors is the staircase block containing, besides a broad staircase, patients' lift, patients' clothing and linen closets. In the rear of this block is another containing private and children's wards and medical officers' rooms. The last block contains a theater capable of seating 250 students. The third floor contains male and female private wards.

The buildings were designed by Mr. H. Saxon Snell, F. R. I. B. A., of London, England, and the building erected under the superintendence of Mr. J. R. Rhind, of Montreal. The heating and ventilating plant, which was designed by Mr. Snell, was modified considerably by the contractors of the plant, Messrs. Garth & Co., of Montreal, to whom we are indebted for the details of this description. The building is heated throughout by hot water, partly by the direct and partly by the indirect systems. The building, which is 250 feet in length by 45 feet wide, contains 370,000 cubic feet, which is warmed by 5,636 square feet of surface in direct radiators, 1,997 square feet in box coils, and 620 square feet in circulations, making a total of 8,253 square feet of heating surface, this giving a ratio of about 1 to 45.

Figure 1 is the basement plan and Fig. 2 is a plan of the ground floor showing the various wards, etc. The water is heated in three sets of twin boilers and one single boiler located in the basement as shown on the plan. A section through one of these boilers is shown in the sketch Fig. 3. The flow and return pipe from each boiler are connected to an 8-inch header. Three 5-inch and one 2½-inch branches from this header lead to the various parts of the building for the supply of the direct and indirect radiators. Two 6-inch branches from the header mentioned are carried into the administration building. On each of these branches a valve is placed and close to the valve is a 1¼-inch pipe with valve for draining the system of water.

The air supply for the hospital, after passing through a fine wire screen, enters through a 5x6-foot opening into a large heating chamber in the rear of the boilers in which the air is slightly heated by the smoke flue from the boilers which passes through this chamber. The air is carried from the heating chamber through a large duct under the floor until it reaches a point (see Fig. 1) where it branches, each branch running on one side of a large extraction trunk to the farthest end of the basement. At certain intervals indirect coils are placed on top of the fresh-air duct or trunk, the opening to it being controlled by a sliding damper.

Figure 3 represents a section on the line F G of Fig. 1, this line passing through the boilers, their connections, ducts, and indirect stacks. Each indirect coil contains 172 square feet of surface, the coil being constructed of 500 feet of 1-inch pipe. The coils are encased with brick at the ends, and on the front by wood lined with tin. Three ducts in a cluster, each 14x14 inches in size, lead from each coil box to vertical flues of the same size, which are built in the side walls of the building. The flues terminate in the wards above, the register by which the air enters being 12x12 inches, and located at a point about 13 feet above the floor.

Figure 4 is a sketch showing a section through the building walls, extraction trunk, etc. The two foul-air registers on each upright flue are 2 and 5 feet above the floor level, and each is connected to the extraction trunk in the basement by a flue and duct. The smaller wards of the building are ventilated in a somewhat similar manner by using recesses in the corner of the closet-room. A vent register is placed in these rooms, and is connected to a ventilating shaft. The main extraction trunk in the basement terminates at the base of a vertical vent shaft 8 feet in diameter. In this shaft a smoke pipe, 42 inches in diameter, from the boilers is carried up to the roof. The extraction trunk, which measures 6x6 feet on the inside, is built of brick and lined with Keen's cement. The vertical flues are lined with the same material. A 3-inch air space separates the back of the flues and the stone wall of the building.

Each boiler (Fig. 3) contains 500 feet of 1¼ inch lap-welded pipe inclosed in brickwork. A 12-inch smoke pipe from each boiler is connected as shown to an iron flue 31 inches in diameter, cast in several sections, flanged and bolted together. One end of this flue is provided with a blank flange which may be removed for cleaning purposes. Each 12-inch flue is provided with a sliding damper. The main flue is carried to and up the main vent shaft before mentioned. One side and the bottom of the heating chamber are built of brick, while the other side and top are of ⅜-inch boiler-plate. The inside of the chamber contains four vertical diaphragms of boiler-plate extending half-way across the heating chamber, so that the air in passing through is obliged to pass across instead of parallel to the main smoke flue.

Two Heine Safety boilers, each of 75 horse-power, fed by a 6"x4"x7" Blake duplex steam pump and Penberthy injector, supply steam for the electric light plant and steam coils in two cylindrical boilers for heating the water for the house supply. The cylindrical boilers are 48 inches in diameter and 11 feet long. At one end of each of these boilers a cast-iron circular frame with a clear opening of 2 feet 6 inches is riveted onto the boiler head. A cover is held on the frame by hinged bolts. This cover is bored to receive two 3-inch brass pipes which are connected on the inside to a coil containing 500 feet of seamless brass pipe 1 inch in diameter. The coil is drained by a 1-inch pipe connected to a No. 2 Nason steam trap, the discharge of which is led to a 4'6"x2'6"x2" receiving tank, from which the water is pumped back into the Heine boilers.

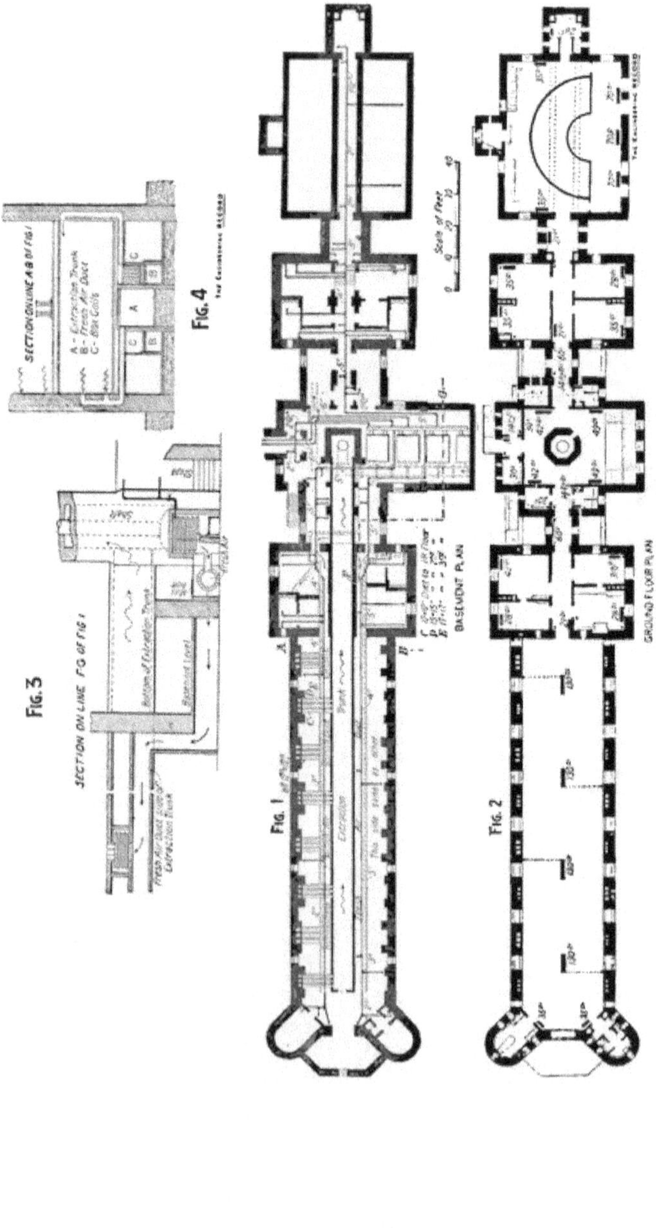

HEATING AND VENTILATION OF THE ROYAL VICTORIA HOSPITAL, MONTREAL.

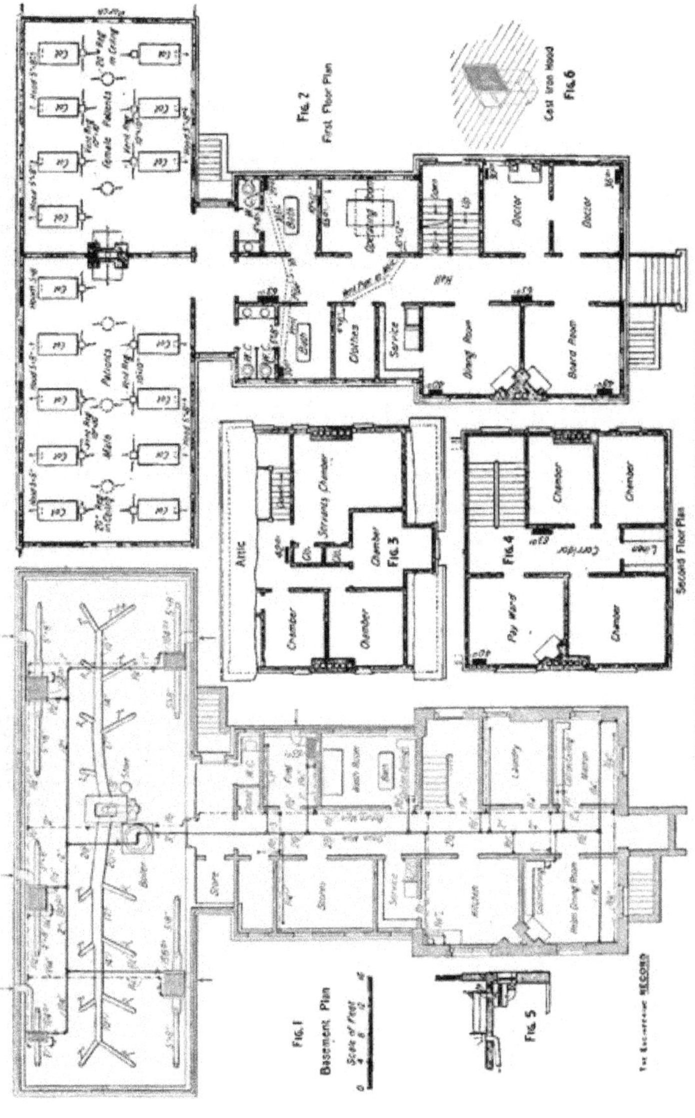

HEATING AND VENTILATION OF THE MOUNT VERNON, N. Y., HOSPITAL.

The exhaust steam from the electric-light engine and pumps can be turned into this brass coil, a grease separator being used to collect the oil in the exhaust steam. The buildings are lighted altogether by electricity from three direct-acting dynamos and twin compound engines, made by Messrs. W. H. Allen & Co., London, England, and a 58-cell Crompton-Howell storage battery.

HEATING AND VENTILATION OF THE MT. VERNON, N. Y., HOSPITAL.

The accompanying drawings show the ventilating and heating apparatus of the Mt. Vernon Hospital at Mt. Vernon, N. Y., of which Messrs. Boring, Tilton & Melen, of New York City, were the architects, Mr. E. Rutzler, of New York City, was the engineer and contractor for the heating plant. The building, as will be seen, is in the shape of the letter T. It is about 35 feet on the front by 65 feet deep; the rear wing is about 75x25 feet.

The rear basement, Fig. 1, contains the heating apparatus, while the front basement contains a kitchen, laundry, storehouse, and dining-room for the servants. The first story, Fig. 2, contains two large wards and an operating-room, besides doctors' rooms. The second story and attic, Figs. 4 and 3, which cover only the front wing, contain bedrooms, pay wards, and the servants' quarters.

The heating is partly by the indirect and partly by the direct system. A vertical tubular steel boiler is used. The boiler is 42 inches in diameter and 5 feet in height, and contains 110 2-inch tubes. The shell of the boiler is encased in wrought iron, allowing a space of 4 inches around the outside of the shell. A casing of galvanized iron is also placed outside of the flue to prevent radiation. The boiler is located in the center of the rear basement, and from this pipes lead to the indirect stacks in the basement and to the direct radiators in the different parts of the building. The large wards and the operating-room on the first floor are the only rooms that are heated by the indirect system. This is accomplished by the use of six indirect stacks distributed throughout the basement, each being located as near as possible to a window with the exception of the stack supplying heated air to the operating-room. From each of these stacks horizontal ducts lead in opposite directions to vertical flues which lead to the ward above. There are 16 beds in the two wards, and a 5x8-inch vertical supply duct leads to the head of each bed, the air being discharged toward the adjacent wall.

Figure 5 is a section through the wall of the building showing the cold-air duct, indirect stack, and supply duct in detail. Each cold-air duct has its separate damper with brass lever weight, quadrant and pin, so that the damper may be set at any desired angle. Figure 6 is a sketch of the cast-iron hood over the supply ducts.

The air is drawn from the wards by means of a 10x10-inch vent register set in the floor at the foot of each bed. A duct leads from each of these registers to a main duct in the basement, which finally empties into a vertical vent shaft located as shown on the plan. A movement of air is maintained in this shaft by carrying the chimney from the heater up through the shaft. In the summer-time a fire is kept burning in a stove in the basement, the smoke pipe of which is also carried up in the vent shaft, and this creates the draft in the vent shaft. Five circular registers, each 20 inches in diameter, also serve to ventilate the large wards. Ceiling registers and open fireplaces ventilate the other rooms in the hospital.

The building contains 11 direct radiators, distributed as shown, presenting a total radiating surface of 508 square feet. The basement is provided with three ceiling coils constructed of 1¼-inch pipe. The coils are suspended from the ceiling beams by cast-iron saddles.

A WET AIR-SCREEN FOR VENTILATING PURPOSES.

In a paper describing the ventilating and heating system at the Victoria Hospital, at Glasgow, recently presented to the British Association for the Advancement of Science by Mr. William Key, an account was given of the method of air filtration there employed.

The fresh air is drawn by propellers down an inlet 16x4 feet feet; the mouth of which is placed 10 feet above ground level, so as to escape dust. The air admitted is washed by passing through an air-washing screen of cords formed of horsehair and hemp, closely wound over a top rail of wood, and under the bottom rail, forming a close screen 16 feet long by 12 feet high. There is a constant stream of water trickling down the screen, by which dust and soot particles are removed and carried away. By an automatic flushing tank 20 gallons of water are instantaneously discharged over the screen every hour.

HEATING OF RAILWAY SHOPS.

STEAM-HEATING PLANT FOR NORTHERN PACIFIC RAILROAD SHOPS.

PART I.—GENERAL DESCRIPTION, MAP AND DIAGRAM OF MAINS.

The Northern Pacific Railroad has recently established extensive repair, machine, and car-building shops at Tacoma, Wash., which comprise a complete plant for construction and maintenance of rolling stock. The shops occupy a large area, afford employment to about 500 men, and were constructed at a total cost of more than $1,000,000. The designs were in accordance with very careful and comprehensive plans of Chief Engineer J. W. Kendrick, who aimed at a thorough and perfected equipment. Especial attention was given that the heating system should be adequate for the severe conditions imposed by the climate and the location. The system adopted was of a steam supply from a central battery of boilers through a principal delivery and return main to the branch mains of all the principal departments, except a section which is heated by engine exhausts, and the isolated round-house, which has an independent battery of boilers. Excepting the office and machine shops, which have cast-iron radiators of the Joy pattern, the radiation is chiefly from wall and ceiling coils of 1-inch wrought-iron pipe in from six to 12 branches. The distributing main is carried in a stone-walled trench, and the submains from it to the different buildings are chiefly carried beneath the floor. The total contract price for the heating was about $26,000 exclusive of boilers, and the work was installed by W. F. Porter Company, of Minneapolis, Minn.

Figure 1 is a general plan showing relative size and position of the principal buildings and indicating the position of the distributing mains. The leading dimensions are as follows: Coach repair and erecting shop, two stories; cabinet and upholstering departments in second story, 100x243 feet; wood-working shop, 90x152 feet; engine-house and steam-heating room, 42x74 feet; boiler-house, 42x76 feet; coal-house, shaving and dust tower, 19x76 feet; chimney (6-foot flue), height 150 feet; paint shop, 90x242 feet; paint-shop storehouse (two stories), 35x90 feet; freight repair shop, 90x302 feet; two lavatories, 26x42 feet; double dry kiln, 40x72 feet, machine shop, 120x244 feet; engine-house for machine shop, 40x40 feet; boiler, tank, tin and copper shop, 80x321 feet; blacksmith shop, 80x192 feet; coal and iron storehouse, 28x150 feet; office and storehouse, 43x156 feet; oil-house, 43x60 feet; round-house, first-class, 22 stalls; turntable, length, 65 feet; two water tanks, each 49,000 gallons; ashpit, length, 100 feet.

The system consists of high-pressure distributing mains through all the buildings except the coach shop, round-house, planing mill and engine-rooms. The steam is taken from these mains at each of the other buildings and passed through a pressure-reducing valve to the system of piping of each building, which is arranged so that any pressure more than required to return the water and less than pressure on mains can be carried. All pipes in separate buildings were arranged to properly drain and to discharge all water of condensation formed in heating surface through proper traps into return mains except in the oil-house, where they discharge into the sewer. The return main delivers to the 42'x10' receiving tank in the tank-room.

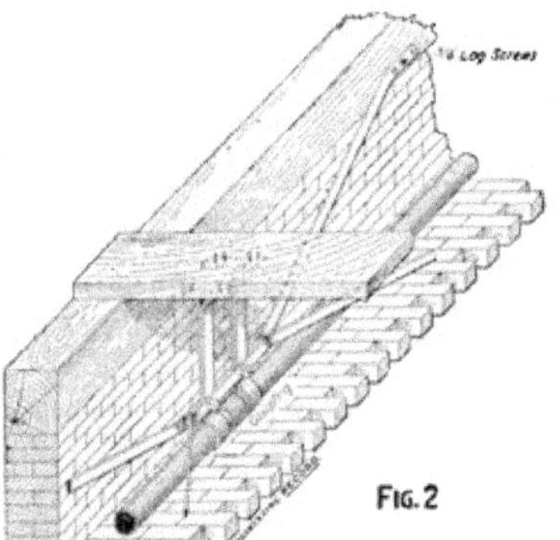

FIG. 2

The location of the shop buildings proper having been changed, while the round-house was left on the ground originally intended for the shops, it was found necessary to construct two separate heating plants. A boiler 56 inches in diameter and 22 feet over all, of the locomotive type, was furnished by the railroad company and fitted by the contractor to furnish steam heat for the round-house. The main taken from this boiler has a pressure-reducing valve and is large enough to supply steam for the house when its walls are extended to a full circle. The water of condensation is taken to a receiving tank and thence pumped back to the boilers. The boiler connection is made in such a manner that a second boiler can be coupled on to it. The round-house now contains 22 stalls and is provided with 5,280 square feet of radiation and piping arranged with 1¼-inch pipe coils in pits four pipes high. Besides the round-house a small oil-house, about 20x30 feet, is heated

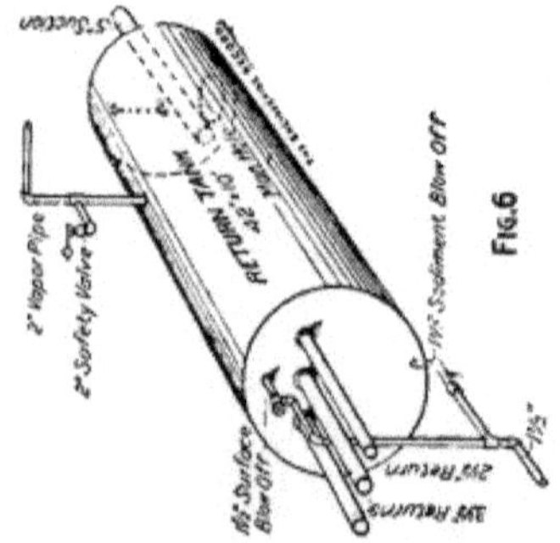

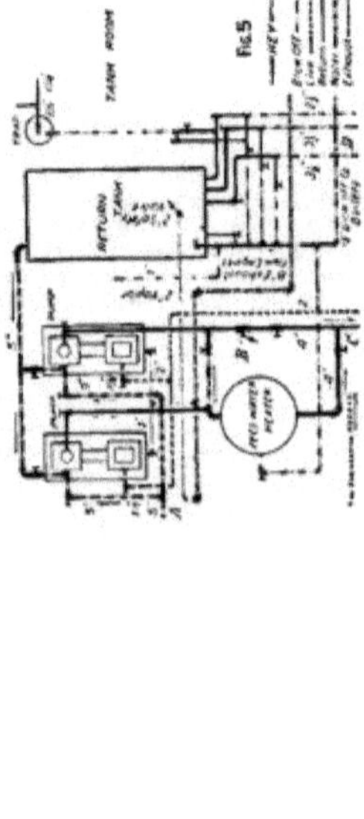

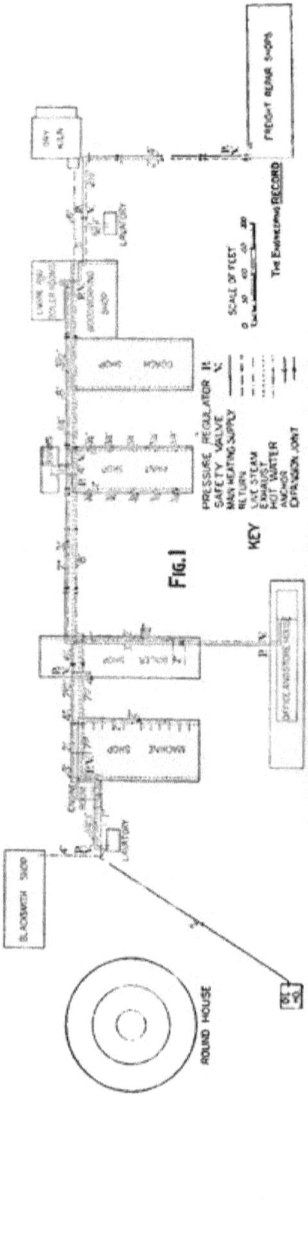

STEAM-HEATING PLANT IN THE NORTHERN PACIFIC RAILROAD SHOPS, TACOMA, WASH.

by this system. There are to be four batteries of two Babcock & Wilcox boilers each (only three batteries are now set) in the main system. The boilers are of the locomotive type, furnished by the railroad, each having 84 tubes, of No. 10 wire gauge, 3 inches in diameter and 12 feet 6 inches long. The battery of boilers is furnished with two 10"x6"x10" Worthington type duplex steam pumps set on brick foundations with cut-stone caps, and having direct live steam pipe connection with the boilers and arranged so as to feed the boilers with hot or cold water.

Live steam pipes, separate from the heating pipes, are run from the boilers to the engines in the boiler shops and machine shops, with live branches to the pits as shown on plans; also a steam pipe to the blacksmith shops to connect with steam hammers. A steam storage tank of sufficient size is provided close to the steam hammers. Live steam connections are made for the Corliss engine and a dynamo engine in the engine-room. The exhaust steam from these engines is used for heating the water in the Berryman feed-water heater. The steam hammers exhaust into the open air. The exhaust from the engines in the machine shops and boiler shop is utilized for heating these buildings. When exhaust steam is used for heating, the water of condensation is not returned to the general receiving tank. The apparatus is arranged, however, so that live or reduced-pressure steam direct from the boilers can be used in the whole or part of these buildings, and the condensation water returned to the boilers.

The heating contractors provided a Hughes Brothers' pump, set on a brick foundation with cut-stone cap in the pump-room of the boiler-house, and made steam and water connections in such a manner as to furnish water to a pressure tank for the general water supply. All high-pressure steam pipes are insulated with sheet iron and air space (Flegle patent) and manilla paper covered with eight-ounce duck and painted two coats where steam pipes are conveyed inside of buildings. Steam pipes outside of building are carried in stone trenches covered with plank. Figure 2 shows the method of anchoring the mains in the trenches as indicated at various points in Fig. 1.

PART II.—PLAN OF BOILER-ROOM, ELEVATIONS OF BOILER CONNECTIONS, TANK AND PUMP-ROOM CONNECTIONS AND DETAILS OF RETURN TANK.

FIGURE 3 is a plan of the boiler-room containing the main battery of boilers, which furnish steam both for general heating and power purposes.

Figure 4 is a smaller scale diagram of different elevations of the steam main and boiler connections. The boiler-room is intended to receive eight boilers, six of which, B B, etc., are now set, and room is left at C C for two more. They are supplied through the 2-inch pipe A with water from the feed-water heater H or directly from pumps P P. Each boiler has a steam connection N, controlled by an angle valve M, to the main D, which terminates in a 12-inch header F, from which the different supplies G G, etc., controlled by valves L L, are taken to required points. T is the return tank, which receives the water of condensation and is connected to the pumps by pipes, which are omitted in the drawing. I is an exhaust main. E is a blow-off. S is the 150-foot smokestack. Pipes A D and E have plugged tees for future connections with the two boilers yet to be set.

Figure 5 is a diagram not drawn to exact scale or position, but intended to show clearly the principal connections in the tank-room and how the returns enter the main tank. This tank is 42 inches in diameter by 10 feet long, and is provided with a gauge glass, manhole, etc. The pumps may draw directly from this tank or through the independent suction pipe A from the city mains, and then delivery may pass through the heater or through the by-pass and back-pressure valve B to the boiler supply pipe C. Figure 6 shows the connections to the return tank.

PART III.—SYSTEM IN THE OFFICE, RISERS, FLOOR PLANS, RADIATORS, AND CONNECTIONS OF TRAPS.

THE buildings are heated when the outside temperature is at zero to the following temperatures: Wood-working shop, 55 degrees; coach shop, 55 degrees; boiler, tank and copper shop, 55 degrees; machine shop, 55 degrees; lavatories, 50 degrees; oil-house, 50 degrees; paint shop, 60 degrees; paint shop stock-room, 60 degrees; round-house, 60 degrees; office, 70 degrees. All pipes are valved both at the source of supply and at the point of delivery with angle or globe valves of standard make. Each coil of circulation is provided with steam supply, return, by-pass and air valves.

The office building is heated with upright tube radiators with oval tops, the valves having nickel-plated trimmings and rosewood handles. There are 1,000 square feet of radiation in the office, piped so it can be run at high or low pressure with pressure regulator, traps, etc.

Figure 7 is a basement plan of the office building showing the steam mains and risers, the vertical pipes being conventionally indicated by oblique lines. The supply pipes are shown in full black lines and the return pipes are throughout lines broken with one dot. Figures 8 and 9 show the arrangement of radiators on the first and second office floors respectively. The radiators are indicated by black rectangles and their size and superficial area and the size of pipe is marked on each. One steam trap is placed on the return from the office system as indicated at T, Fig. 7. When the return pipe is over 2 inches in diameter, two traps are placed as shown in Fig. 10.

There are in all 18 traps used, three of which are located in the machine shop, two in the lavatories, one in the oil-house, two in the freight repair shops, three in the paint shop, two in the storeroom three in the boiler shop, one in the office, and one in the engine-room.

PART IV.—SYSTEM IN THE PAINT SHOP, PAINT STOREROOM, MACHINE SHOP, AND BOILER SHOP.

THE paint shop has 4,400 square feet of radiation in coils. No exhaust steam is utilized, any pressure can be carried, and all water is returned to the tank. The paint storeroom has 1,160 square feet of radiation arranged in coils 10 pipes high on the first floor and eight on the second floor and piped the same as the paint shop. Figure 11 is a diagram plan of the first floor of the shop and storeroom showing size and location of mains and the arrangement of coils and radiators. The radiators are indicated by solid black squares. A is a 10-branch vertical coil of 1-inch pipe on the wall below the windows, one branch being outside the valves so as to serve for circulation. The supply and return pipes are laid in trenches, accessible by removing loose sections of the floor. In the storehouse, B is a 10-branch vertical coil of 1-inch pipes on the wall below the first-story windows; C is an eight-branch vertical coil of 1-inch pipe on the wall below the first-floor windows, and D is an eight-branch horizontal gridiron coil of 1-inch pipe below the sinks on the first floor. These coils, together with similar eight-branch 1-inch vertical wall coils E E, in the second story, are shown in perspective in Fig. 12. Figure 12 is a view from Z Z, Fig. 11. In all coils one branch is left outside the valves for the purpose already stated. Figure 13 is a vertical elevation from X X, Fig. 11, of coil A. Figure 14 is a plan of the machine shop showing the size and location of mains and the arrangement of radiators and coils. A is a 10-branch vertical coil of 1-inch pipe on the wall under the windows, B is a similar eight-branch coil, D is a back-pressure valve on connection to the engines. Figure 15 is a vertical section of the coil A, Fig. 14. All the radiators contain 84 feet of surface each.

The machine shops and boiler shops are arranged so that the exhaust steam of the engines can be used for warming and supplemented by live steam

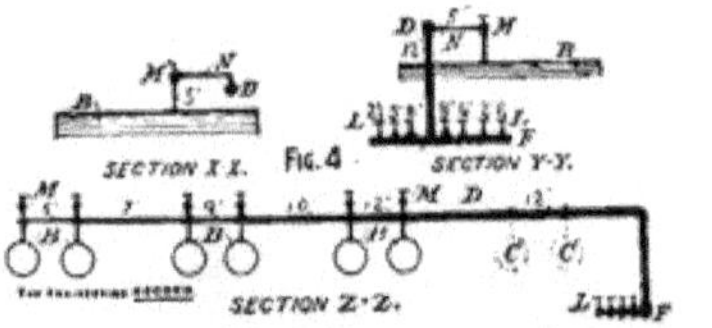

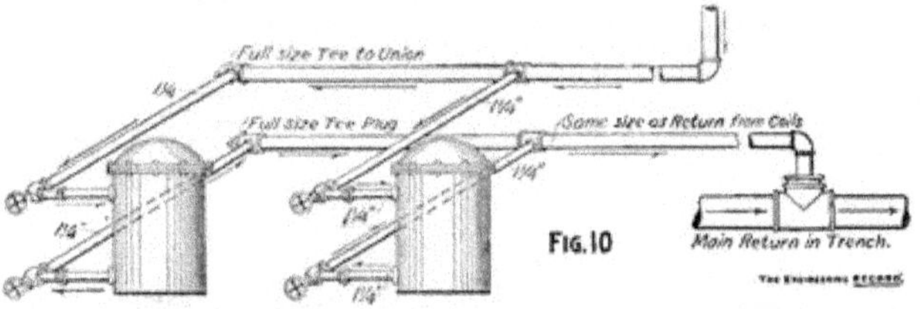

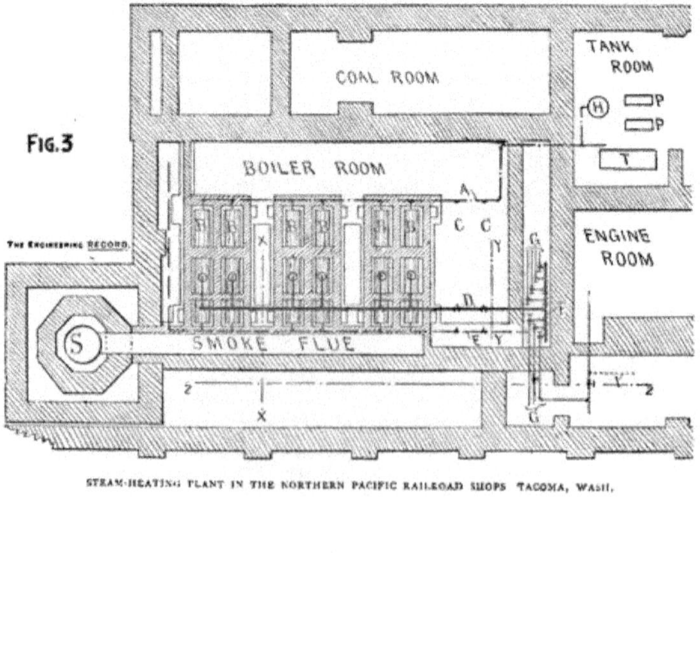

STEAM-HEATING PLANT IN THE NORTHERN PACIFIC RAILROAD SHOPS TACOMA, WASH.

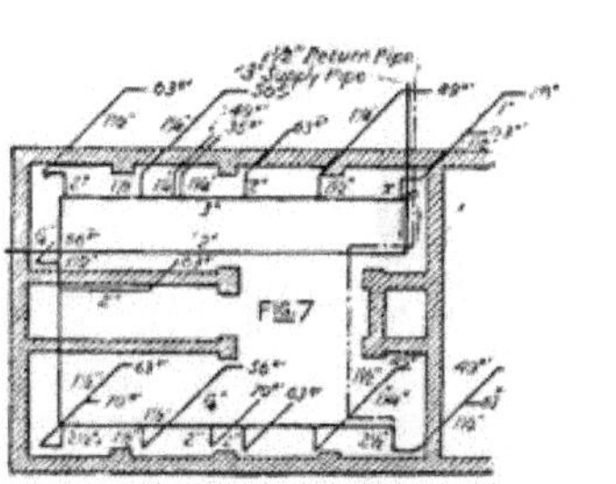

through a pressure-reducing valve if necessary. When exhaust steam is used, the condensation can be discharged into the sewer through traps. The piping is so laid out that all the steam can be taken through a pressure-reducing valve when it is not desirable to utilize the exhaust, but in case that is done all water is returned to the receiving tank through traps and return mains.

The machine shop has 4,300 square feet of radiation consisting of 1,600 square feet of coils on three sides and 2,700 square feet of gridiron coils arranged overhead. The supply mains are run overhead and the returns are brought back above the floor. The boiler shop has 4,200 square feet of radiation divided into two horizontal overhead coils as shown

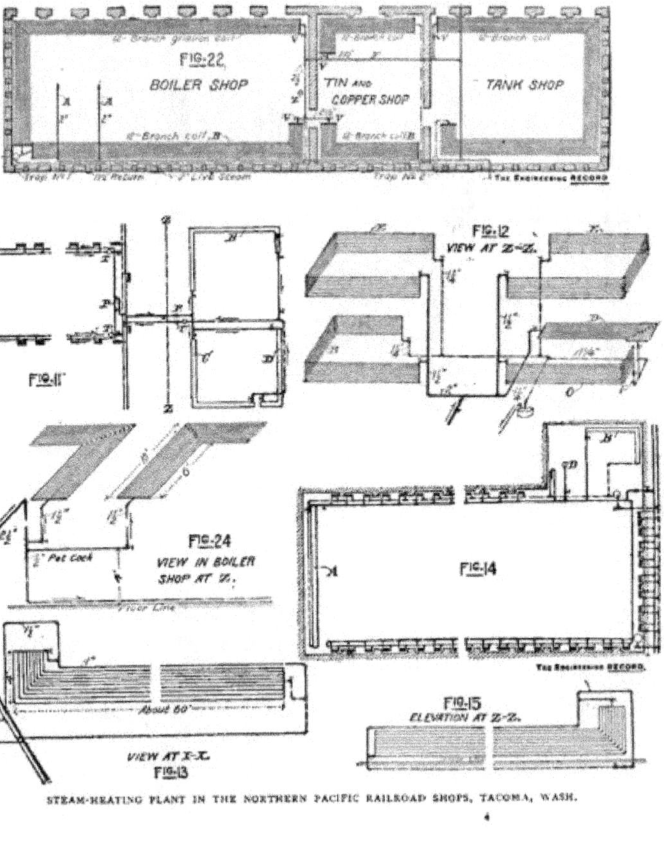

STEAM-HEATING PLANT IN THE NORTHERN PACIFIC RAILROAD SHOPS, TACOMA, WASH.

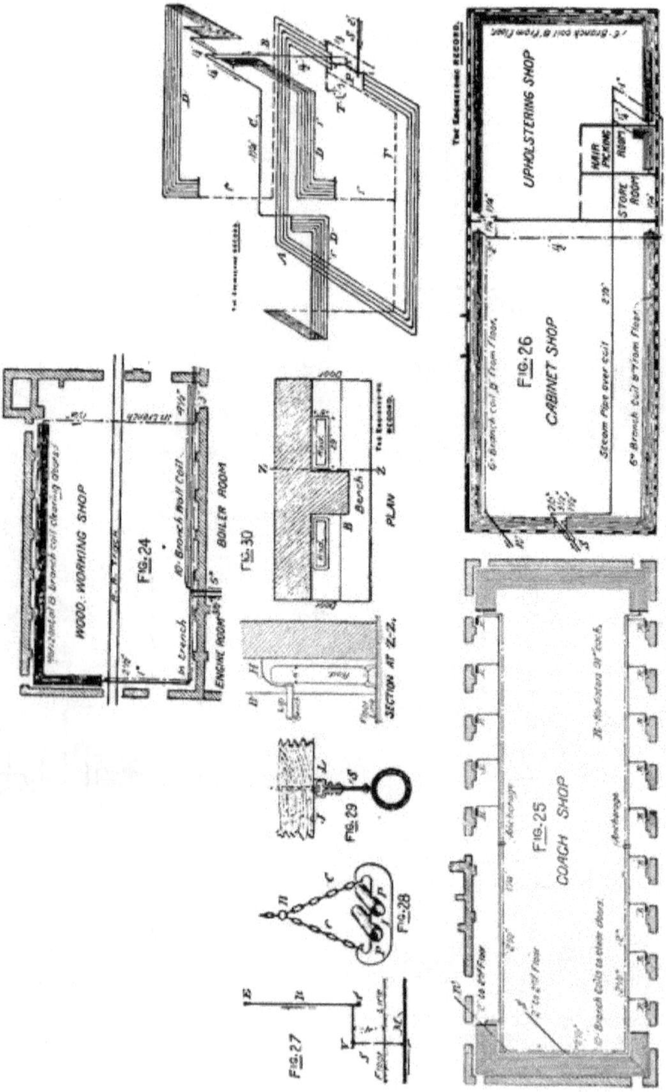

STEAM-HEATING PLANT IN THE NORTHERN PACIFIC RAILROAD SHOPS, TACOMA, WASH.

by the general plan, Fig. 22, where A A are globe valves with hose nipples for supplying steam for tests, etc., and V V, etc., are globe valves with extension vertical stems. All the valves on the returns are tapped for 1¼-inch petcocks. The traps are set on the floor and the coils are connected as shown in Fig. 24, which is a perspective diagram from Z of the supply ends of coils B B.

PART V.—FREIGHT REPAIR SHOPS AND LAVATORIES.

The freight repair house is 302x90 feet in size, one story high. The specification provided that it should "be warmed by 4,100 square feet of radiation arranged in coils, pipes to be run at high or low pressure and return water through traps and returns to a receiving tank." In accordance with these requirements, the arrangement shown in Fig. 16 was devised.

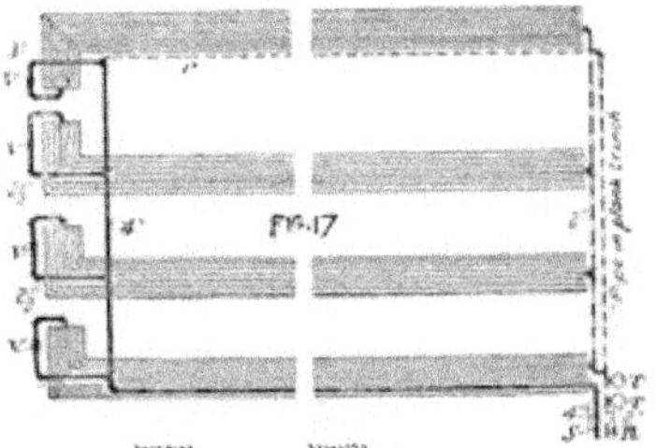

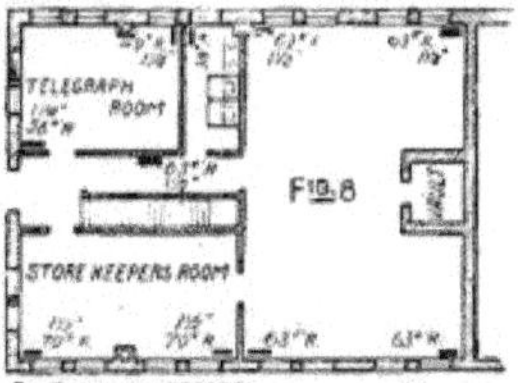

THE ENGINEERING RECORD.

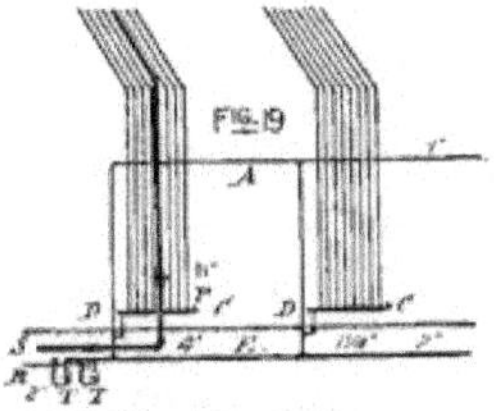

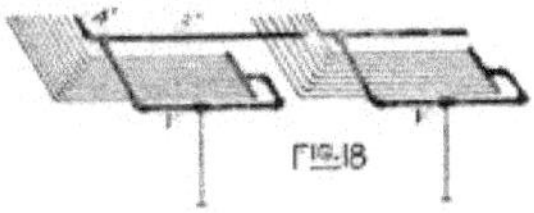

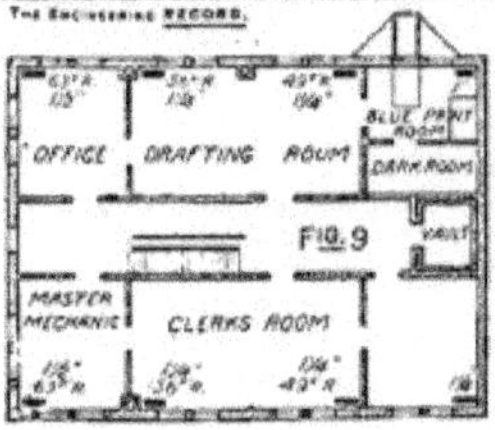

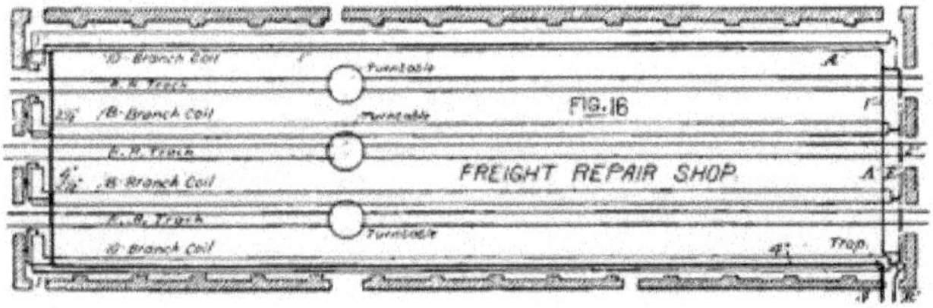

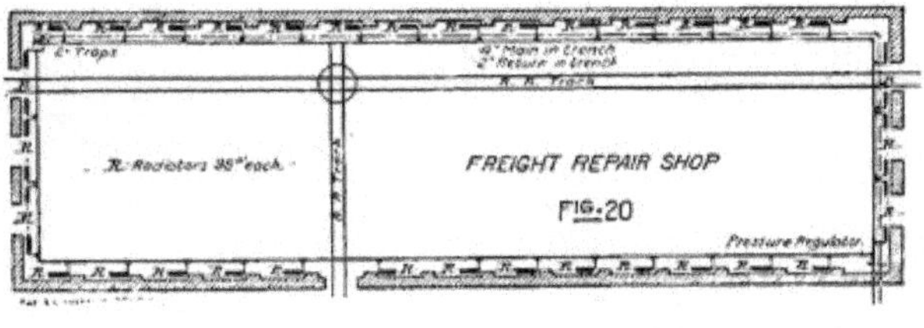

STEAM-HEATING PLANT IN THE NORTHERN PACIFIC RAILROAD SHOPS, TACOMA, WASH.

Figure 17 shows the arrangement of the coils, and Figs. 18 and 19 show the connections at the supply and return ends respectively. S is the 4-inch main supply pipe, and R is the 2 inch main return. A is a 1-inch drip pipe from supply branches, and E is the 2-inch return pipe from the ceiling coils. T T are traps. W is the 4-inch main gate valve. P is a 3-inch pressure regulator. D D are 1¼-inch angle valves 4 feet from the floor. C C are ½-inch petcocks, and V V are 3-inch globe valves with extension stems on the wall.

Figure 20 is a diagram showing the arrangement of the radiator system finally adopted in place of the overhead coils. Each radiator is composed of 14 sections of standard height, and is connected to the supply with a 2-inch arm and a 1¼-inch stub and valve. The variations in the length of the pipe, nearly 300 feet long, through temperature changes are so great that the radiators were made free to conform to the different positions by being each

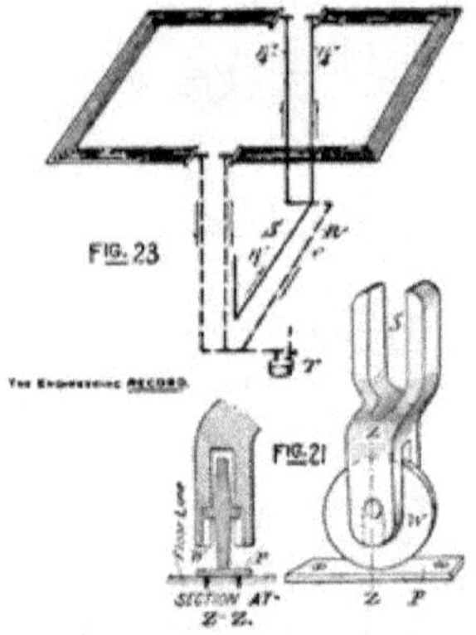

mounted on four iron casters, shown in Fig. 21. The casters had slots S, just fitting the bottom of a radiator section, and the wheel W ran on an iron bed-plate P screwed to the floor.

Figure 23 is an isometric diagram of the horizontal radiator coils in the lavatories. They are overhead, and the supply and return mains S and R are carried in a trench beneath the floor. The trap T and a pressure reducing valve, not here shown, are located in an accessible brick well. The lavatories have each 250 square feet of radiator service.

PART VI.—WOOD-WORKING, COACH, CABINET, AND UPHOLSTERY SHOPS, AND OIL-HOUSE.

Figure 24 is a plan of the radiators and mains in the wood-working shops, which are heated by low-pressure steam and the condensation returned to a receiving tank. The coach shop, planing mill, and engine-room are also heated by low pressure provided either through a pressure reducing valve or by the exhaust steam of the Corliss and dynamo engines, and return all water of condensation to the receiving tank.

The coach shop is provided with 3,600 square feet in coils, 2,200 on the first floor and 1,400 on the second floor. The planing-room or wood shop is provided with 1,200 square feet in coils. Figure 25 is a diagram plan of the first floor, showing horizontal overhead radiator coils at each end of the coach shop and the location of the radiators, each of which is 34½x10 inches. The cabinet and upholstering shops are in the story above the coach shop, and are heated by overhead horizontal coils as shown in Fig. 26, which shows them in plan with an isometric projection of the supply and return mains S and R from the lower floor, Fig. 25.

The oil-house is warmed with 550 square feet of direct radiation placed in the form of horizontals or gridiron coils run about 8 feet above the floor. Figure 31 is an isometric diagram showing the arrangement of coils and mains in the basement and first floor. The supply from main S passes through pressure regulator P, and is delivered by branches B and C to the coils, of which D D D are vertical ones on the walls of the first story, and A is a horizontal one in the basement. The return and condensation water is discharged through trap T to the sewer.

Figure 27 is a diagram of the general method of taking a supply S for the radiation in any building from the main M in the trench. It is controlled by an angle valve V, beyond which a ½-inch petcock C drains the riser R, which extends upward along the wall to a point above the ceiling coils, where its tee E receives the branches to the coils.

Figure 28 shows the wrought-iron plate I which supports overhead pipes P P, so as to swing freely in chains C C which unite in a ring R secured to the roof or floor joist above.

Figure 29 shows another convenient method used to support the suspender S by a 2x½-inch strap L screwed directly to the timber above.

Figure 30 shows how, in the machine shop, the radiators are set behind the work bench, in the corners by the buttresses B, and have their tops covered by a galvanized-iron hood H. Throughout the buildings each coil is provided with by-pass and supply and return valves as well as each main line of piping. The dry kiln is heated by high-pressure steam, and the condensation water is returned to the receiving tank. The Corliss engine-room is warmed by 400 square feet of radiators. There are provided 11 ½-inch branches with ½-inch compression bibbs in the paint shop, and a ½-inch bibb for each sink in the stock-room.

STEAM HEATING IN THE BOSTON AND ALBANY RAILROAD STATIONS AT SPRINGFIELD, MASS.

We show, in the accompanying cuts, foundation plans of the adjoining stations at Lyman and Liberty Streets, Springfield, Mass., of the Boston and Albany Railroad Company, illustrating the main features of the steam-heating system with which they were equipped by Messrs. Norcross Brothers, of Spring-

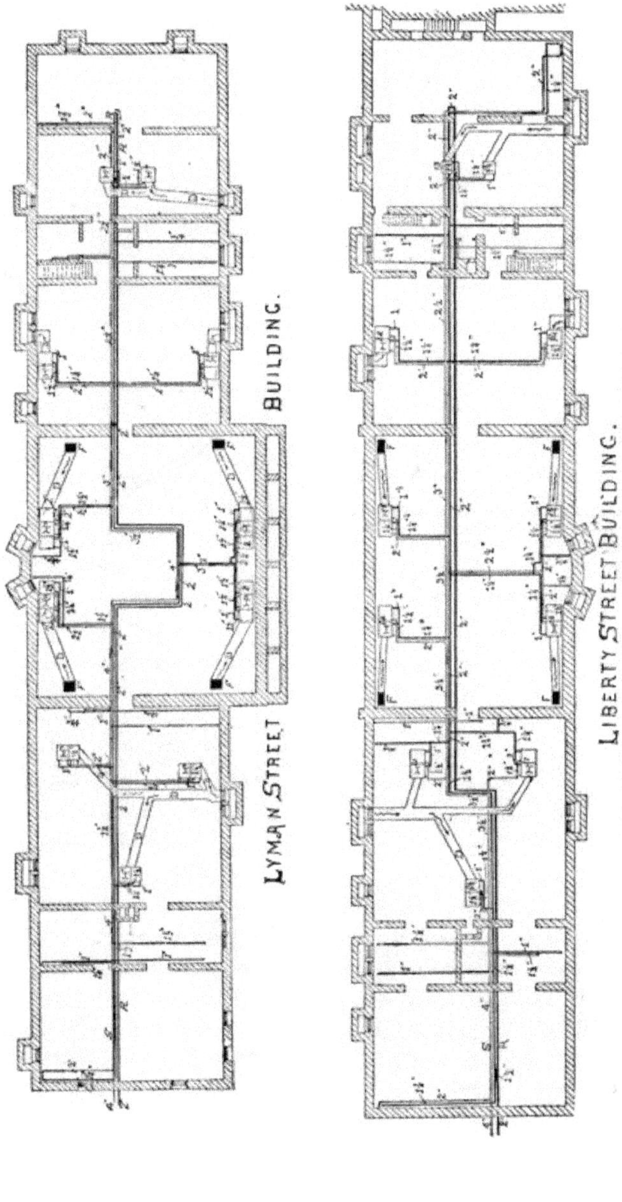

STEAM HEATING IN THE BOSTON AND ALBANY RAILROAD'S SPRINGFIELD, MASS., STATIONS.

field, under the superintendence of Mr. E. M. Harding, now with the United States Heating and Plumbing Company, of Boston, Mass.

The boilers, which are of the Hennessy triple-draft pattern, designed for using oil fuel on the plan of the Aerated Fuel Company, of Springfield, are located in a detached boiler-house, about 1,000 feet west of the two stations, and carry a pressure of 80 pounds, supplying, as they do, steam for electric light engines also. The boilers, four in number, are all connected with one 10-inch steam drum, from which the supply pipes for heating and power are taken.

The steam for heating is used at reduced pressure—not over five pounds—both direct and indirect radiation being employed, with gravity return. The stacks of indirect radiators are designated by the letters H, the numerals at the right indicating the number of sections. These are of cast iron, with extended surface suspended from the ceiling by heavy wrought-iron hangers. Each stack is incased in two thicknesses of wood, consisting, first, of a box of sound flooring lined with tin, the bottom being removable; and second, of an outer covering of matched and beaded stuff, with a layer of heavy building paper between, put together with screws. The air supply to the stacks for the smoking-rooms, restaurants, and women's waiting-rooms, on each side of the middle portions of the buildings, is taken direct from the outside; for the main waiting-rooms in the middle, however, a system of interior circulation has been adopted, the air being taken from the rooms through registers F in the floor at convenient points, passed through the radiator stacks, and discharged again into the rooms. It is thought that a sufficient supply of fresh air to these rooms is maintained by the continual opening and closing of the doors leading to the outside.

The fresh air supply is led through ducts D, along the basement ceiling. Each duct, where connecting with the outside, has a galvanized wire screen, and near the inlet to the stacks H, has a balanced damper arranged to work automatically by the pressure of the steam. The ducts consist of matched and beaded stuff. The dampers have the obvious advantage of stopping the supply of fresh, cold air to the stacks as soon as the steam is cut off, and there is thus no danger of freezing of the radiators. Under each floor register, between it and the coil, is suspended a pan of galvanized iron, as a receptacle for any dirt that may drop through the grating. These pans can be removed for cleaning.

The steam main from the boilers to the buildings is 6 inches in diameter, underground, and has a branch with a reducing valve and by-pass. From it two 4-inch mains are taken to the two buildings.

The piping in both buildings is practically the same. The steam-supply pipes are shown by double lines in the illustrations, and are marked S, while the return pipes R are in full black. The steam mains, starting with a 4-inch diameter, continue along the basement ceilings, reducing gradually in size, as indicated, as branches are taken off for the various coils and radiators, to 2 inches at the far ends of the buildings. The return mains, in both cases, start at the left-hand ends, 1¼ inches in diameter, and also run along the basement ceiling, increasing gradually as connections are made to them, to 2 inches at the other ends. From these ends the two main returns are brought back, retaining that size, and are joined in one 2½-inch pipe, which is carried to the boiler-room independently on the same line followed by the steam-supply pipe, and is trapped into a tank.

The steam and return mains from the boiler-house are carried in masonry trenches. At intervals of 200 feet along the line of these trenches are manholes of masonry with stone coping and cast-iron covers. Drainage of water from the trenches is secured by means of earthenware tile pipe in the bottom, with proper outlet connections.

Expansion joints are provided in the pipes at every 200 feet. Expansion of the mains in the buildings is allowed for by offsets, as shown. All the underground pipes are covered with asbestos cement, and all steam and return pipes in the buildings, above ¾-inch in diameter, are also provided with non-conducting covering.

The direct radiation consists of 1-inch wrought-iron pipe wall coils, and of Bundy loops.

HEATING AND VENTILATING A ROUND-HOUSE AND RAILROAD SHOP.

The accompanying illustrations show the system just installed for heating and ventilating some large buildings of the Chicago and Grand Trunk Railway at Port Huron, Mich. Figure 1 is a plan of the 26-stall round-house, Fig. 2 is a plan of the paint shop, and Fig. 3 is a cross-section at Z Z Z, Fig. 2. The paint shop, Fig. 2, contains 894,000 cubic feet and has a No. 80 heater N, with a 72-inch fan, which will deliver from 34,000 cubic feet of air at 1,200 feet velocity per minute to 146,000 cubic feet at 5,175 feet velocity (one ounce pressure). The round-house C, Fig. 1, contains 828 cubic feet and has a No. 60 heater E with 60-inch fan, which will deliver from 23,000 cubic feet of air at a velocity of 1,200 feet per minute to 101,500 cubic feet at 5,175 feet velocity (one ounce pressure). In connection with the round-house is a machine shop B, which contains 220,100 cubic feet and an office building A, containing 90,500 cubic feet, making a total of 310,600 cubic feet. These (the machine shop and the office building) have a No. 30 heater F, with a 42-inch fan, which will deliver from 11,500 cubic feet of air at 1,200 feet velocity per minute to 49,700 cubic feet at 5,175 feet velocity (one ounce pressure). The No. 80 apparatus contains 8,000 lineal feet of 1-inch pipe, and its 72-inch fan is capable of delivering 60,000 cubic feet of air every minute at an expense of about 12 horse-power, which is furnished by a directly-attached vertical engine, manufactured by the Huyett & Smith Manufacturing Company, Detroit.

This shop was entirely completed before the apparatus was installed, but has no piping save what is necessary to carry the air across the end of the room at which it enters, and the variation of the heat in the different ends of the room is said to seldom exceed 3° Fahr.

In Fig. 1 the fresh air is taken from outdoors at K, and, passing through the heater E, is distributed through the mains H H, laid in a trench, to the outlets L L, etc., which deliver it around the circumference of the building and are intended to prove equally efficient at all points by the expedient of making those most remote from the fan of larger diameter.

Figure 4 is an enlarged section at Z Z, Fig. 1, showing the arrangement in the pit P. The outlet L has a movable joint Q, which may be set in any position so as to discharge the hot air in any direction underneath an engine that may be over the pit, and thus melt off snow and ice and facilitate the cleaning of the engines.

In Fig. 1 additional branches, not here shown, are connected to the main G to distribute the hot air in the separate rooms of the office A.

In Fig. 2 fresh air is admitted at M, and is delivered by heater N to the main conduit R and distribution branches O O O.

This apparatus was built and erected by the Huyett & Smith Manufacturing Company, of Detroit, Mich., according to plans approved by Mr. H. Roberts, Superintendent of Motive Power, Detroit, Mich.

HOT-WATER HEATING OF AN ELEVATED RAILROAD STATION.

The conditions attaching to the heating of the station buildings of the New York Suburban Rapid Transit Elevated Railroad are severe in that the buildings are entirely detached, are elevated, and exposed to wind and weather beneath as well as on all sides and on the top, and each station has at certain hours of the day large number of persons passing in and out. The stations are generally heated by a hot-water system similar to that of the Wendover Avenue station, shown in the accompanying illustrations, and installed by Hitchings & Co., contractors, of New York City.

The isolated position and general arrangement of the passenger stations are indicated in Fig. 1. The station A is warmed from a boiler placed in an under room C. The open platform B is not warmed. The plan of the station is shown in Fig. 2, and a vertical

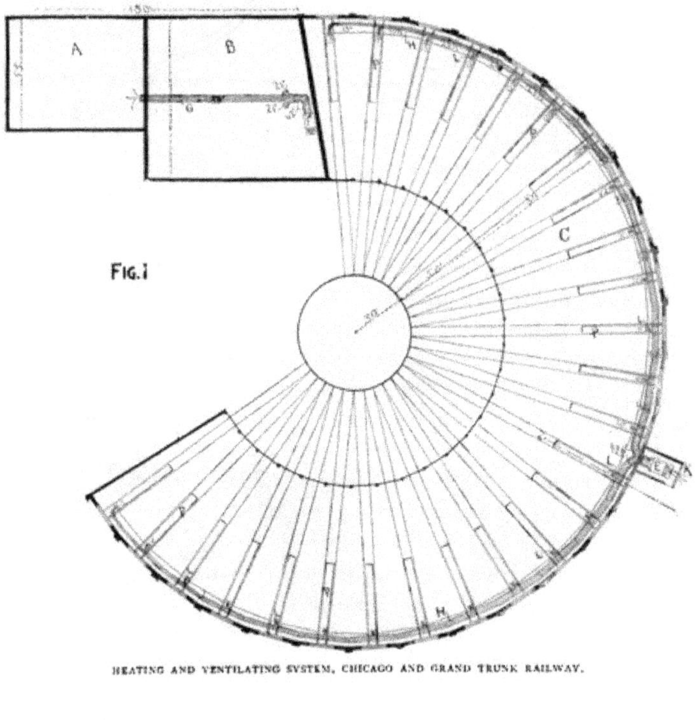

HEATING AND VENTILATING SYSTEM, CHICAGO AND GRAND TRUNK RAILWAY.

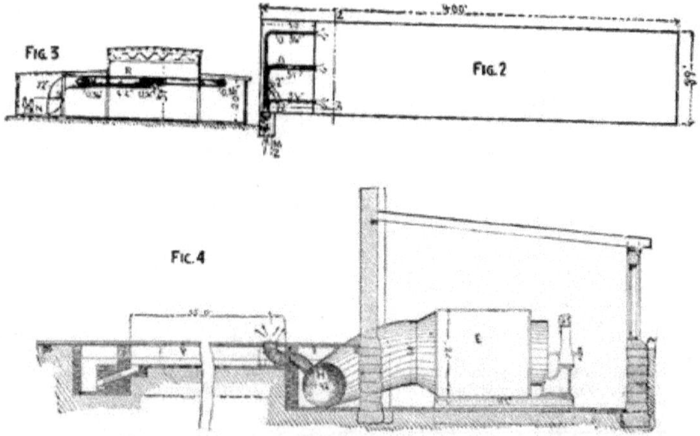

HEATING AND VENTILATING SYSTEM, CHICAGO AND GRAND TRUNK RAILWAY. (See page 187.)

elevation and sectional diagram at Z Z is indicated in Fig. 3. The radiators M and N, Fig. 4, in waiting-room R are operated from the boiler E, Fig. 3. Radiator N has an additional coil L to increase the heat in the ticket office K, which is exposed to cold air from the adjacent entrance door.

The radiators are connected with the boiler by risers at G and H, and with each other by an air pipe W, which opens into the 10-gallon cast-iron closed expansion tank V, which may be filled by opening a key valve O, in the water-supply pipe Y, or may be emptied through the petcock Z. It overflows through the open pipe X into the cistern of an adjacent water-closet. Circulation in it is maintained and danger of freezing averted by connecting to it a riser I from the boiler and a return J, which serves a flow pipe for radiator M. The connections of the flow and return pipes to the boiler E are shown in Fig. 5. The boiler may be emptied through the key valve P. Circulation 16 feet below the boiler is effected by extending the return pipe S from coil N, vertically downwards to form the loop Q, from which the water returns to the boiler through branch T. The water, gas, and soil pipes are carried to below the frost line

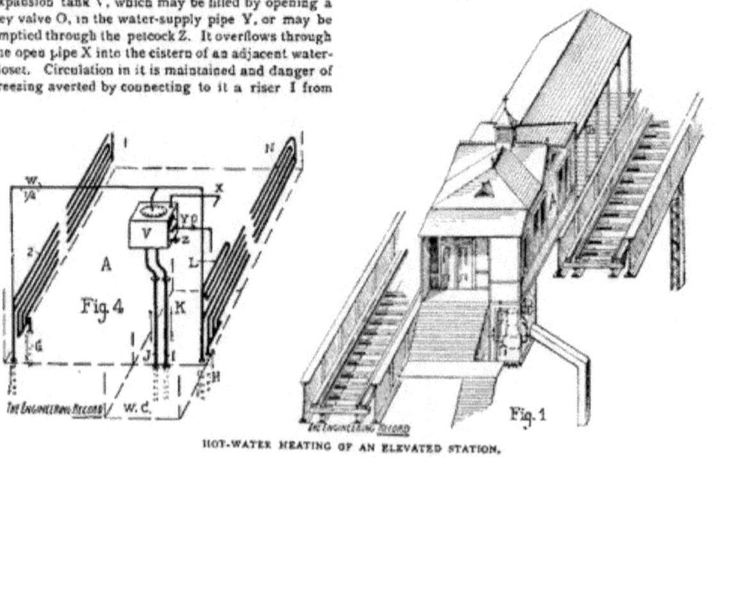

HOT-WATER HEATING OF AN ELEVATED STATION.

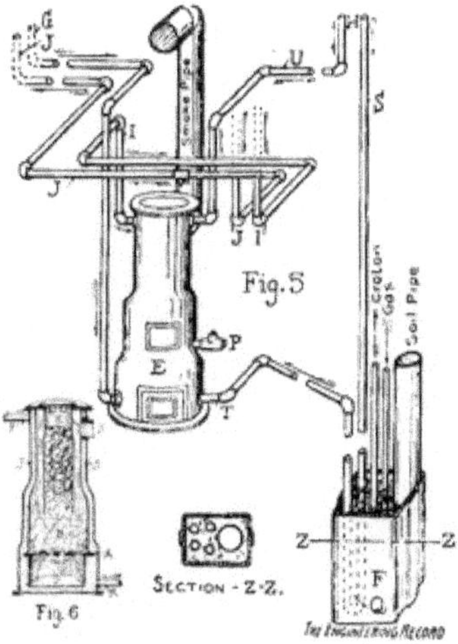

in a cast-iron box F, Fig. 3, packed with hair, etc., and having the coil Q as an additional precaution against freezing.

Figure 6 is a sectional view of the boiler E, Figs. 3 and 5; B is the fire box; G, the grate; K, the coal reservoir; P, the ash-box; and J J, the water jacket surrounding them all. S is the smoke pipe, and F and R are respectively flow and return pipes. The boiler is Hitchings' self-feeding heater No. 23, designed to heat a room with 8,600 cubic feet of contents.

A VACUUM CIRCULATION STEAM-HEATING SYSTEM.

THE Thirty-first Street plant of the Minneapolis Electric Street Railway Company consists of a group of brick buildings, mostly one story high, covering an area about 225x175 feet in extreme dimensions. The buildings comprise car storage, and receiving and repairing rooms, boiler-room, engine-room, round-house for motor cars, and various shops and storerooms, the whole being heated by the exhaust steam from the engines, supplied at a pressure substantially not above atmospheric to long pipe coils and a few cast-iron radiators. The steam plant consists of four Babcock & Wilcox boilers and 10 vertical

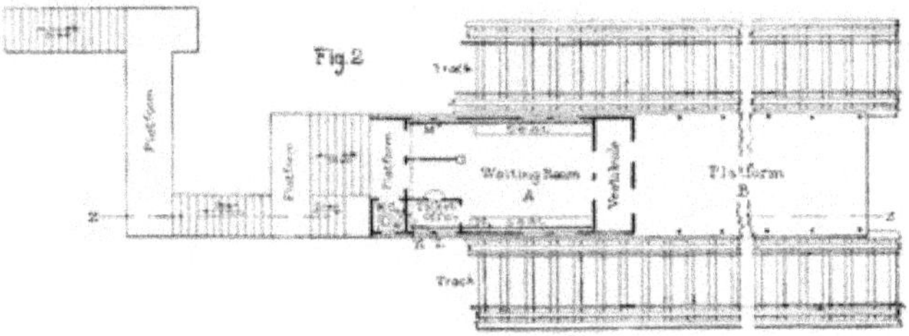

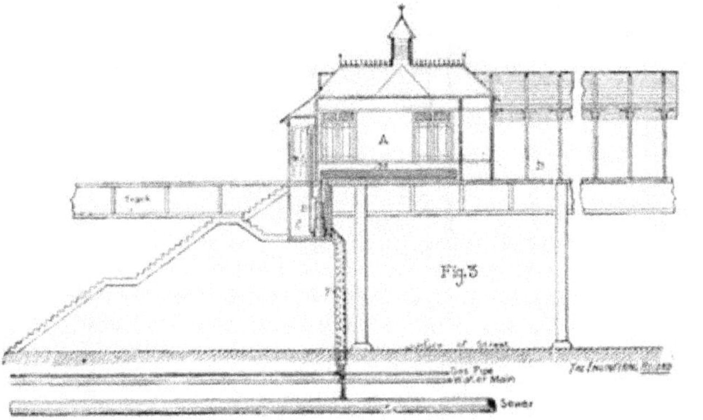

HOT-WATER HEATING OF AN ELEVATED RAILWAY STATION.

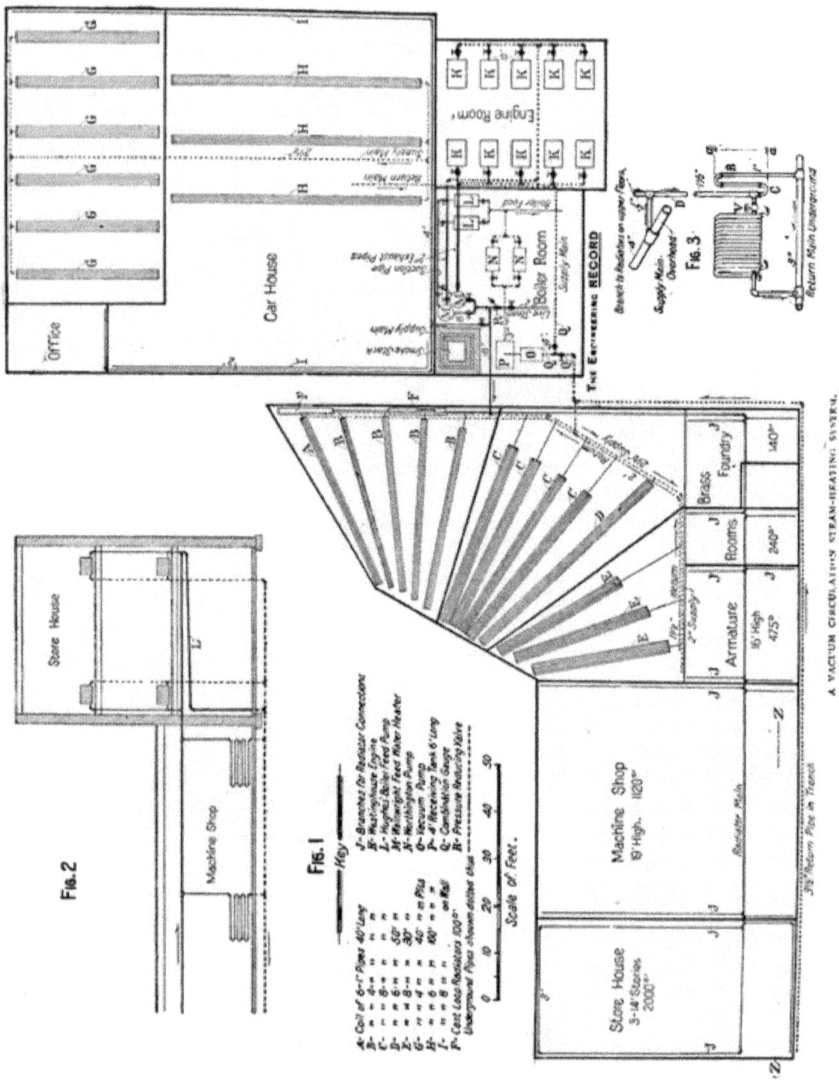

A VACUUM CIRCULATING STEAM-HEATING SYSTEM.

engines. The exhaust steam from the engine first passes through two Wainwright feed-water heaters and then into the heating mains. The exhaust is separately received from each of the two sets of five engines, and is of course provided with suitable valves for free discharge to the open air and with by-passes around the heaters to enable them to be cut out if occasion requires. The feed water is injected by two 10"x12"x6" Hughes' boiler-feed pumps, which are interchangeably connected with two Worthington pumps set under the floor to take the condensation water out of the return tank into which the contents of the mains, coils, and radiators are delivered by a vacuum pump which produces circulation.

Figure 1 is a diagram, not drawn to exact scale or position, but from a sketch made from a description given to a member of our staff, to indicate the arrangement of the buildings and location of coils and showing the special features of distribution and the characteristic essentials of the connections. Only the principal valves are indicated and the pipe lines and coils are shown conventionally. Most of the return branches are parallel to the supplies and one size smaller and are omitted to avoid confusion. The numerals marked on the different rooms show the number of square feet of radiation provided. The radiators in the storehouse connected to the "radiator main" are of the Detroit type and in some instances are composed of 1-inch pipe coils on the walls. In general they are connected up without valves, and a typical arrangement of pipes is indicated in Fig. 2, which is an elevation of Z Z, Fig. 1, where the return loop L is introduced merely to give extra radiating surface. Figure 3 shows the connections of the radiators on the first floor of the storehouse. The valve V is normally open and some condensation water collects at C, which is pulled up into the return bend B by the action of the vacuum pump exhausting the return main and thus forms a seal whose height *a a* of about 17 inches is sufficient to prevent the pump from taking the contents of the branch D before it passes through the radiator. The system was designed and installed by Archambo & Morse, of Minneapolis, to whom we are indebted for the data upon which this description is based. They state that the circulation of the system is satisfactory and its operation efficient and very economical under a pressure of less than one-half pound with the exhaust open freely to the air.

HEATING OF HOTELS.

STEAM HEATING IN THE PLAZA HOTEL, NEW YORK.

PART I.—GENERAL DESCRIPTION, CELLAR PLAN, BOILER-ROOM, BOILER SETTING, AND BOILER FEED-WATER HEATER.

THE Plaza Hotel, New York City, is a seven-story building, situated at the Fifth Avenue entrance to Central Park, about 150 feet front and 20 feet deep, intended for a family hotel to accomodate about 600 guests. It contains about 100 suites of rooms, each of from three to six apartments, besides single-guest chambers, parlors, offices, etc., for general hotel service, and the kitchen laundry, etc., etc. Gillis & Geoghegan are the designers and contractors for the steam engineering.

Steam from one boiler plant is supplied for the general domestic purposes of the building in cooking, lifting, pumping, laundrying, ventilating, and heating. It supplies two 150 horse-power Corliss engines that drive the electric lighting plant, an ice machine and compressor, and an engine for laundry purposes.

All the heating is by direct steam radiators that are ordinarily supplied by the exhaust from the numerous pumps, engines, etc., but can at will be supplied with live steam.

Figure 1 is a general diagram illustrating the system of steam mains and regulating valves, and showing the return mains and steam risers. The heating mains are indicated by single heavy lines, exhaust mains by full double lines, return mains by broken heavy lines, steam-power mains by heavy lines broken with one dot, branch pipes are, in general, omitted, but the principal rising lines are indicated by small circles.

A A are brine pumps connected with the ice machine engine; B B, etc., are the steam boilers; C is an automatic governor for the tanks D D, that receive water from the return mains; E is an automatic governor for the boiler feed pumps F F; G G, etc., are steam engines; H H are dynamos; P is a pump; I is a feed-water heater; J is the laundry engine; K K are house pumps; L L are elevator pumps; M M are hot-water boilers; N N are tanks, one to receive drips and the other for a blow-off tank; O O O are suction tanks; Q is the vertical exhaust pipe from the exhaust main; R is the regulating valve, controlling the communication between steam main W and exhaust main X; U is a back pressure valve; V is a common gate valve; S S, etc., are risers from the exhaust main, designed to promote upward current in ventilating shafts, and T T, etc., are the risers from the heating and return mains to which the radiators are connected.

Ordinarily all the exhaust steam passes through the feed-water heater I, and after returning to the exhaust main is delivered through various branches to the dry-room, the hot-water tanks, the ventilating shafts, and to the general heating system, but any one of these can be cut out and the exhaust sent through the others, or directly up the stack at Q. The house pumps may also be cut off from the exhaust main and send their exhaust directly to the hot-water tanks or the dry-room.

Portions of the building calculated to be in constant use, such as dining-room, restaurant, billiard-room, café, etc., are each supplied with direct, separate steam heat, controlled from the boiler-room.

The heating apparatus circulates at less than a pound indicated pressure of steam, and is arranged to operate at a pressure not exceeding five pounds at any time. Back-pressure valves are placed in the exhaust mains and set at this limit of five pounds, so that any excess of back pressure on the engines will cause them to open and thus give free escape to the roof for the exhaust. The system is designated a "combination system," and in order to supply any deficiency in the quantity of exhaust steam, or to make up a sufficient quantity for the entire apparatus running at one time, live steam is let in through differential pressure valves at any pressure under the five pound limit set on the confined exhaust.

Figure 2 shows the arrangement of steam mains and valves in the boiler-room (see also Fig. 1). B B, etc., is the battery of Babcock & Wilcox double boilers of 628 total horse-power. These boilers, though not originally designed to be operated by forced draft, are fitted with the McClave argand steam blower. Buckwheat coal is now burned under them.

A A are branches to the power main; C C are branches to the heating main; D D are expansion joints; E is an 8-inch main to dynamos, etc.; F is a 3-inch main to boiler pumps, ice machine, and kitchen; G is a 20-inch main supplying the heating system.

H is a 6 inch main to power engine, elevator pumps, and house pumps; I is a 4-inch main to the boiler feed-water heater; J is a back-pressure valve (see also Fig. 1); U is a differential valve set at 80–5 pounds. Usually valve M is closed and L and L are open, admitting boiler pressure to valve K, but the by-pass is arranged so that by closing L and L and opening M steam will be admitted direct to G from the boilers (and likewise permits of the pressure regulator being removed for repairs without interfering with use of steam); O O, etc., are the coal bins.

Figure 3 is a plan showing the boiler setting and smoke flues, and Fig. 4 is a vertical section of the same.

Figure 5 shows the Berryman heater for boiler feed water (see I, Fig. 1). A is the main exhaust

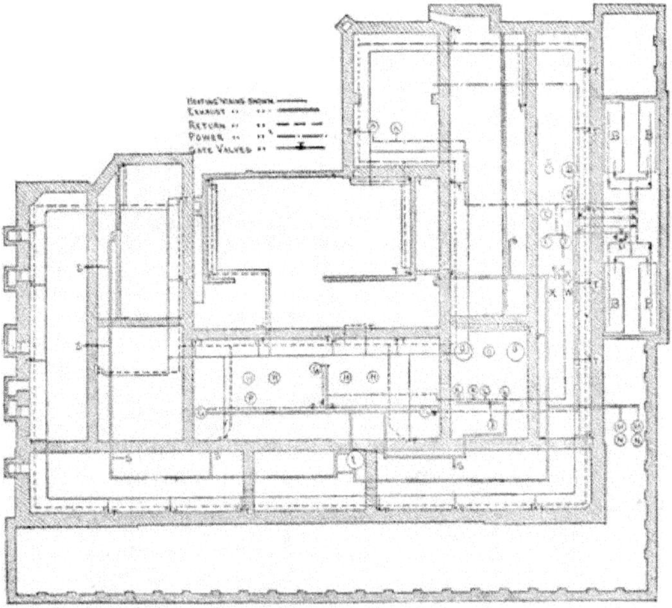

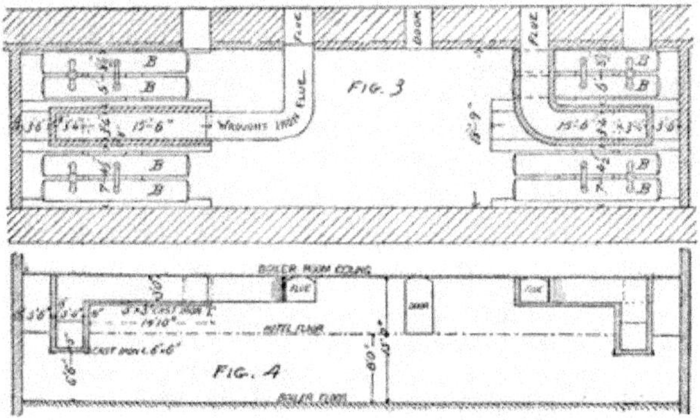

STEAM HEATING IN THE PLAZA HOTEL, NEW YORK.

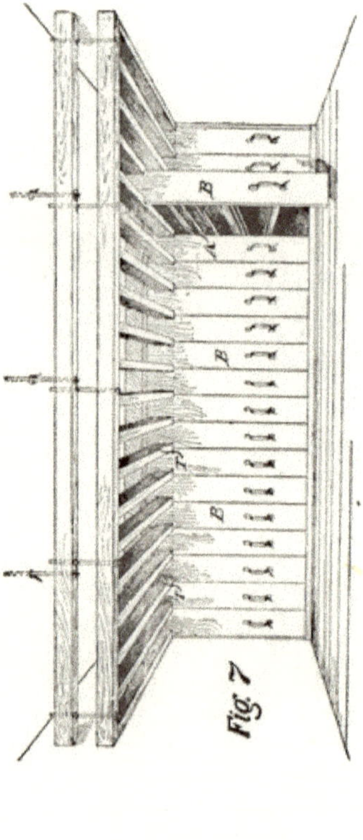

Fig. 7

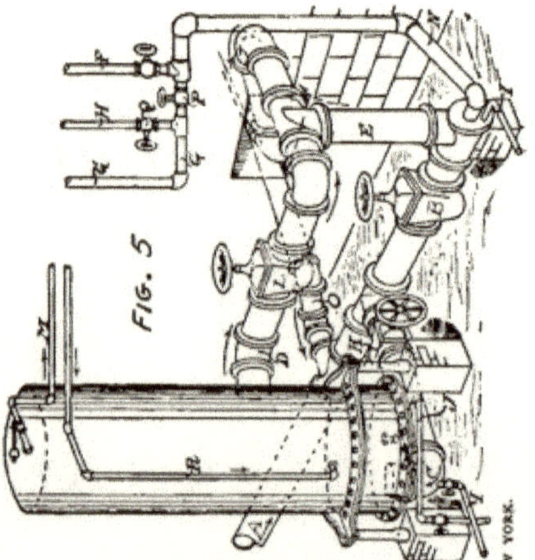

FIG. 5

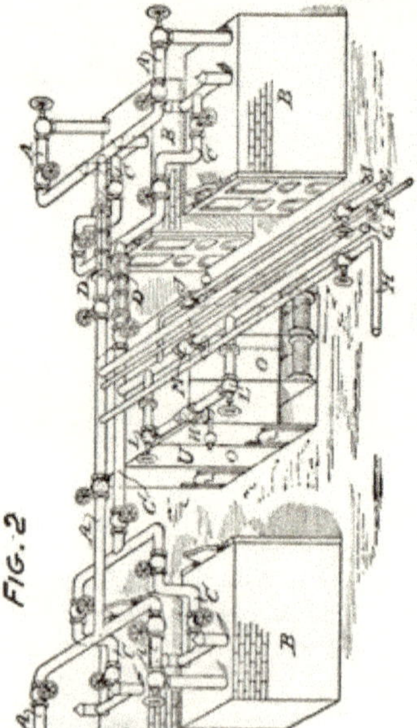

FIG. 2

STEAM HEATING IN THE PLAZA HOTEL, NEW YORK.

(12-inch) from engine-room; O is the 8-inch exhaust from house and elevator pumps and laundry engine.

Ordinarily valve B is closed and K and L are open, and the exhaust steam passes through branch C to the heater, and out through D to heating main and main exhaust pipe, as indicated by the arrows; but by reversing valves B, K, and L the heater is cut out and the exhaust steam passes directly through by-pass E, as shown by broken arrows. The 5-inch branch N supplies exhaust steam to 3 inch lines F and G, the former to promote circulation in ventilation shafts, and the latter to heat the coils in the laundry dry-room.

If valve P be closed and Q be opened, live steam from 1½-inch pipe H will be supplied to the dry-room; I I are 2 inch drip pipes delivering the condensation water to a drip tank (N, Fig. 1); J is a 1-inch blow-off; L is the 2-inch cold-water supply to the heater; and M is the 2-inch pipe delivering hot water to the boilers.

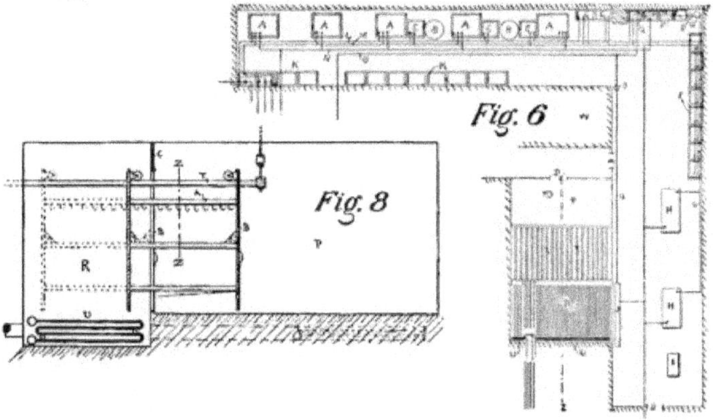

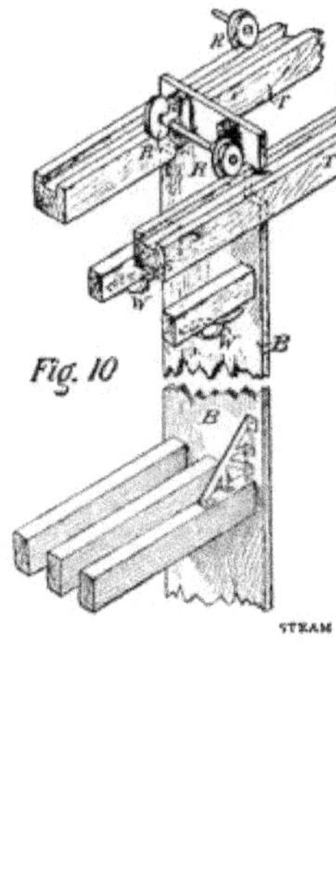

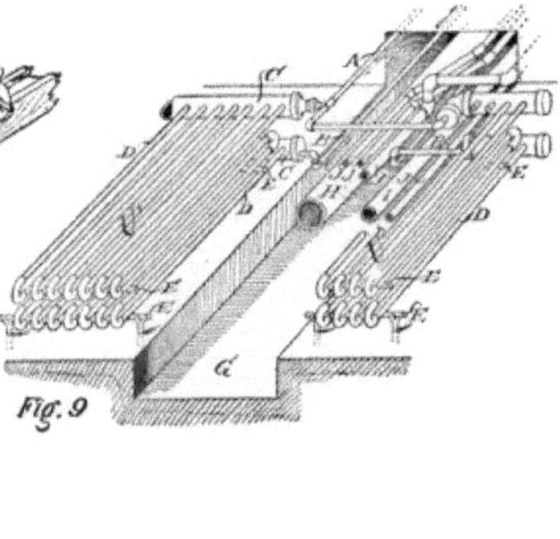

STEAM HEATING IN THE PLAZA HOTEL, NEW YORK.

PART II.—SYSTEM IN LAUNDRY AND DRY ROOM, DETAILS IN DRY-ROOM, AND STEAM COILS IN HOT CHAMBER.

FIGURE 6 is a diagram plan, not to scale, of the laundry and adjacent dry-room (L and M, Fig. 1,) in the basement.

Hot and cold water is supplied through pipes L and M respectively, live steam through N and exhaust steam through O. All pipes except branches Q Q are overhead, suspended from the ceiling, and the risers from the different machines are indicated by solid black circles.

A A, etc., are nonpareil power washing machines. B B power centrifugal wringers. C C, etc., rinsing tubs. F F, etc., are common kitchen laundry tubs,

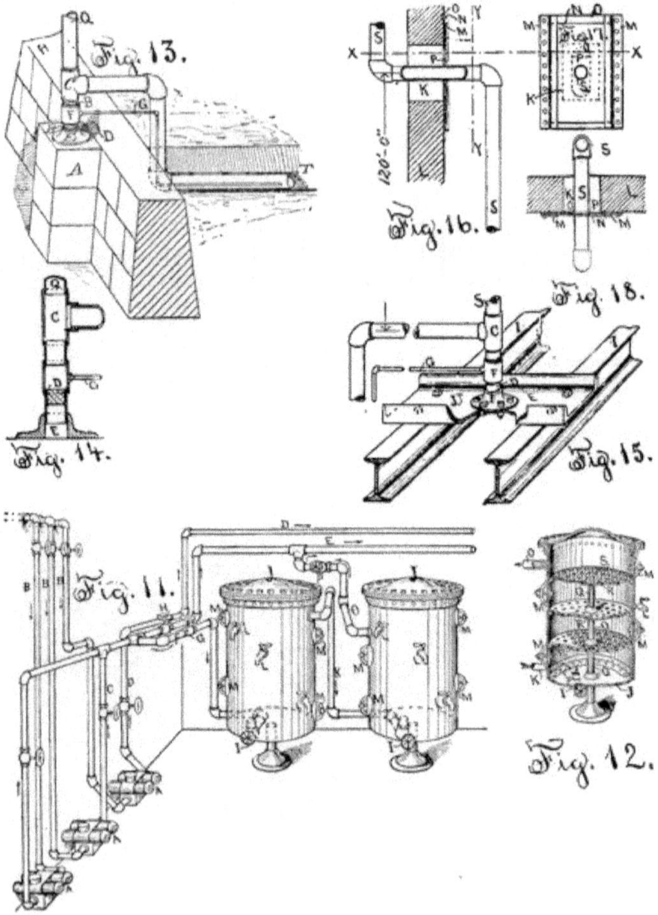

STEAM HEATING IN THE PLAZA HOTEL, NEW YORK.

half of which are provided with perforated steam pipes, etc. K K, etc., are bins for the reception of soiled linen.

H H are steam-heated French power mangles. I is a collar and cuff ironing machine. J is a lace curtain ironer, which consists essentially of a large hollow iron cylinder mounted on hollow trunnions through which steam is received to heat the cylinder.

G is a starch kettle and D D are steam traps. The machines in the laundry are driven from counter-shafts that are operated by a special steam engine. Adjacent to the laundry is the dry-room, fitted with patent racks (omitted in Fig. 6) that are suspended from overhead tracks T T, etc., on which they travel from the room P to the hot chamber R, where the clothes are dried over the large steam coils U. These racks are 12 feet long by 7 feet 6 inches high and 1? inches and 18 inches wide.

Figure 7 is a perspective from Y, Fig. 6, showing position of overhead tracks T T, etc., and the 20 racks A A, etc., one of which is shown partly withdrawn into room P, Fig. 6, to be filled or emptied; while the rest are in the hot chamber R, for which the panels B form a tight partition from room P. The ventilating flues are omitted.

Figure 8 is a section at Z Z, Figs. 6 and 7, showing a section of the coil radiator U and the pit in which it is contained, and the sheathing C that, with the panels B, partitions off hot chamber Q from room P.

Figure 9 is a general view of the radiator coils U U', Figs. 6 and 8. They receive live or exhaust steam through A from pipes G and H, Fig. 3, and it returns, together with condensation water, through pipes B B. U' is a small radiator with only four coils. U has headers C C of 5-inch pipe about 12 feet long, tapped to receive about 50 coils, D, of 1¼-inch pipes about 10 feet long connected by three return bends each. The coils are separated and sup-

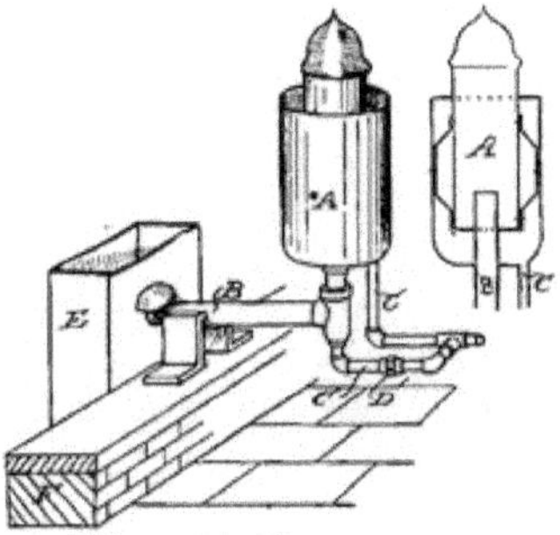

Fig. 19.

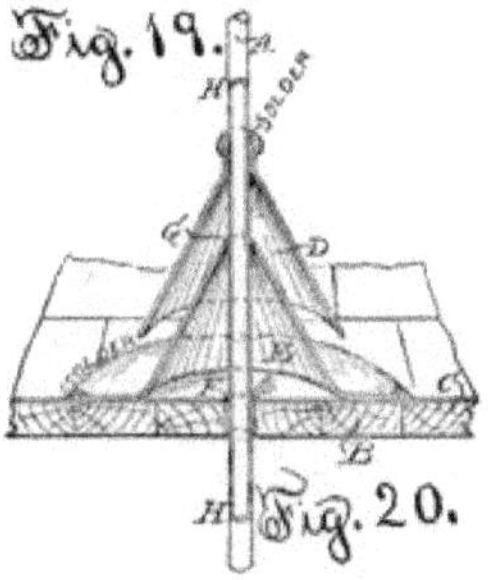

Fig. 20.

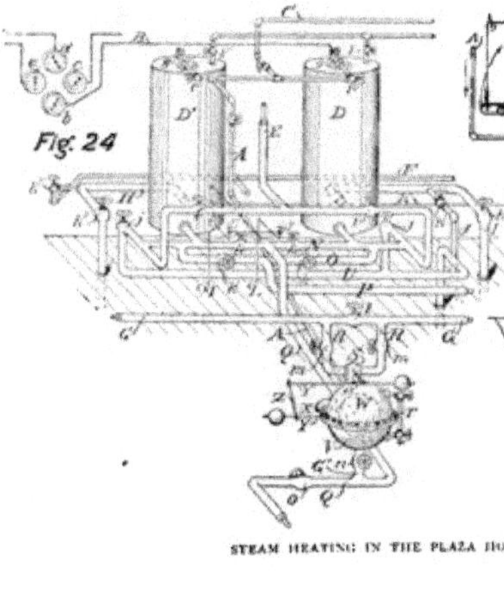

Fig. 24

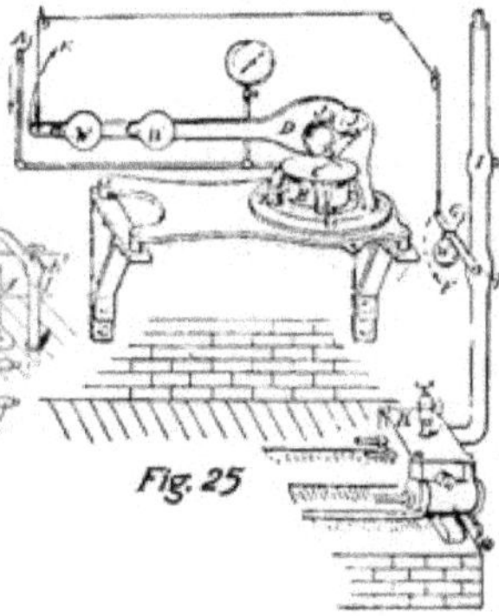

Fig. 25

STEAM HEATING IN THE PLAZA HOTEL, NEW YORK.

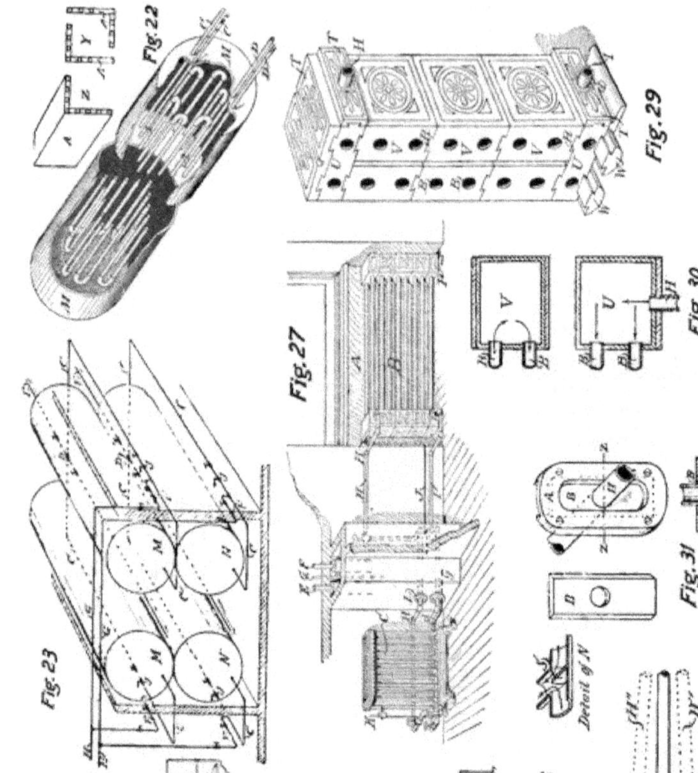

STEAM HEATING IN THE PLAZA HOTEL, NEW YORK.

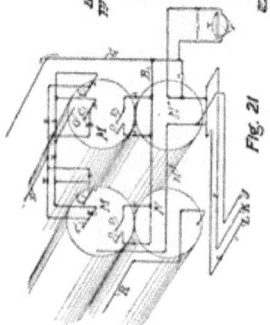

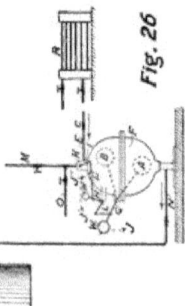

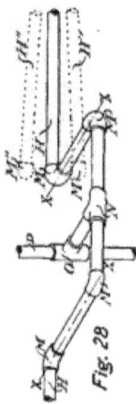

ported by the 1½-inch pipe rollers E E, etc., which allow temperature movements; the lowest roller is supported by a pedestal bar set in the ground and having a saddle head. G is the trench in which the main pipe lines are carried through the cellar. I is a 6-inch exhaust main *to* and H a 12-inch *from* the boiler feed-water heater. J J J, etc., are relief pipes, etc., not connected with this portion of the system.

Figure 10 is a section at X X, Fig. 9 and shows the details of construction of frames A, Figs. 6, 7, and 8, and the arrangement of rollers R R, etc., on the tracks T. W W, etc., are the galvanized hooks from which the clothes are suspended.

The dry-room and laundry are fitted with machinery supplied by Oakley & Keating, New York.

PART III.—SPECIAL FILTERS, DETAILS OF MAIN EXHAUST STACK AND RISERS IN VENTILATION SHAFTS.

FIGURE 11 shows one of the two sets of special filters, designed by Mr. F. Nagle, Chief Engineer for the lessees of the hotel, and built by Gillis & Geoghegan. All the croton water used for domestic purposes must pass through these filters, which may be so connected as to filter the boiler feed water also.

Croton water from the Fifty-ninth Street main is supplied through branches B B B to the Worthington meters A A A, which discharge it through pipes C C and D. Pipes C C generally deliver through branch F to the twin tanks J J, which, connected by the pipe K, form one filter whose outlet is through branch O into main pipe E. By closing valves G and P and opening valve H, the filters are cut out and the water passes directly from the meters to pipe E.

I I are valves controlling the tanks J J, which are each about 3 feet in diameter by 4 feet high, constructed as shown in section by Fig. 12, where R R R are diaphragms with numerous ½-inch perforations. Each diaphragm is supported by a nickel-plated piece of 1-inch brass pipe Q Q Q, and is covered by a $\frac{1}{16}$-inch mesh screen S S (shown removed at the middle diaphragm). Between the diaphragms is packed the filtering material, which can be removed, or washed, through the handholes M M, etc., when valve I is closed and the water is drawn off through cocks L L, etc.

The main exhaust stack Q (see also Fig. 1) is of cast iron, 12 inches in diameter, about 1 inch thick, and 130 feet high; it is supported along the external wall H by bands under each hub, and rests on a special foundation pier A, Fig. 13. The pipe comes through the building in trench T, and rising enters the stack by tee C, that is connected downwards to the cast-iron pedestal D and bearing flange E. B B are screwed joints. F is a tee with a solid bottom plug D and a branch G just above D, through which all condensation water is drained off from the stack.

Figure 14 is a section through the center of the stack and pedestal to show the connections more clearly. Figures 15 and 16 show respectively the supports for the bottom and top of one of the exhaust risers S (Fig. 1), which are arranged for ventilating purposes. The arrangement of the bottom is similar to that illustrated in Fig. 13, the pedestal E here resting on a channel bar J, supported by floor beams I I.

At the top pipe S passes out through a slot K, in the vertical wall L. Figs. 17 and 18 are sections and elevations from Y Y and X X, Fig. 16. The slot K is covered by the galvanized sheet-iron plates O and N, the former being fixed and having a slot P to permit vertical movements of the pipe, and the latter having a round hole just large enough to receive the pipe, with which it moves up and down, sliding under the cleats M M, which hold it close to O.

Figure 19 shows a section and elevation of the Hoey & Conrow's patent exhaust head A, that is made of copper and fitted to the top of every exhaust riser.

The steam impinges on the dome, and is deflected to the bottom, whence it escapes through the external annular space. In the sketch E is the ventilating flue from the swill-room, and the 3-inch exhaust pipe B is to promote circulation in it.

C C are 1-inch drip pipes for condensation water, and D is a union inserted to enable the pipes to be readily disconnected. F is the parapet wall around the roof.

Figure 20 shows the expansion arrangement for carrying the exhaust pipes through the roof. A is an exhaust pipe (shown in section from H to H) passing loosely through hole F in roof boards B and tin sheathing C. E is a flanged galvanized funnel soldered tightly to the roof tin, and permitting pipe A to pass through with clearance at G. D is a similar funnel, without a bottom flange, that is soldered fast to pipe A.

The pipe is thus free to rise and fall with temperature variations, while the holes F and G are protected from all leakage by the funnels E and D respectively.

PART IV.—RETURN TANKS AND AUTOMATIC PUMP GOVERNORS.

FIGURE 24 shows the 3x6-foot tanks D D' (Fig. 1) which receive the condensation water through the return pipes of the heating system. B is a ½-inch pressure pipe to the gauge *b*, which indicates the steam pressure in the receiving tanks and return pipes; C is a 1½-inch equalizing pipe between the heating main and the tanks; *g g* are petcocks for air vents; A is a 1-inch equalizing pipe between the pump governor W and the tanks. E is a 1¼-inch return pipe from the billiard and dining rooms; F is a 2-inch relief pipe from the principal heating main; H and I are 3 inch return pipes from the main system of the hotel; J is a 2-inch relief pipe from the main system of the hotel; K is a ¾-inch relief pipe from the pipe supplying the heating apparatus for the café and dining, lounging and billiard rooms.

L is a 2-inch main return pipe from the café; M is a 2-inch main return pipe from the restaurant; Q is a 1½-inch suction pipe from the boiler pump, and is connected by branch N to both tanks, from either or both of which it can draw; P is a 1½-inch suction pipe from the kitchen pump, and is similarly connected to draw, through branch O, from either or both of the tanks; U is the return pipe from the cooking apparatus; T is a gauge tube, showing the level of the water in the tanks; *c*, *d*, and *e* are gauges

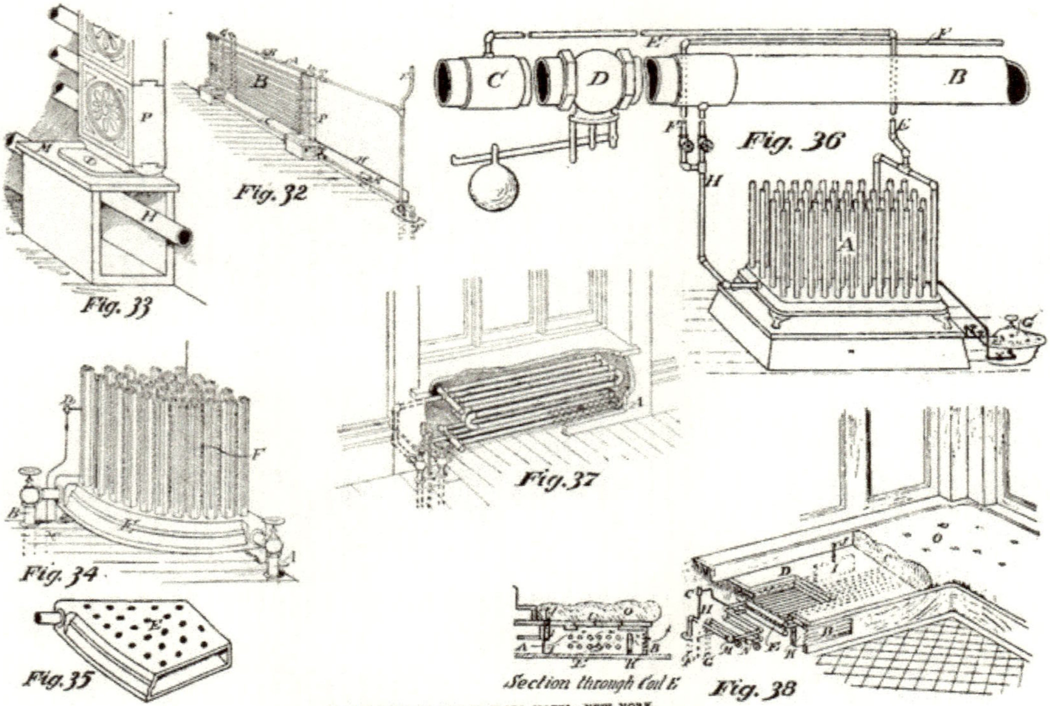

STEAM HEATING IN THE PLAZA HOTEL, NEW YORK.

showing respectively the pressures in the heating main, the exhaust main, and in the steam boilers; W is a Kieley automatic pump governor, connected by branch G' to suction pipe Q of the boiler pump which receives steam through pipe G. Ordinarily valve *t* is closed and valves *m m* open, so that the steam goes through the by-pass R R which is controlled by the valve S, operated by the automatic governor W. The latter, when valve *n* is open, is in free communication with the suction pipe Q. The water therefore rises to the same height as in tanks D D, and, raising the large internal float V which is pivoted on axis X, operates through link Z, and counterweighted levers Y Y, the stem *a* of throttle valve S. The latter thus lets on steam and starts the pump as soon as the water reaches a certain height in the tanks D D. If the water level is lowered the float V falls, shuts off steam, and stops the pumps until accumulating water in the tank again raises the float, and so on, but the adjustment has been so made that just enough steam is admitted to keep up a very slow, practically continuous motion of the pump. The governor can be cut out by closing valve *n*, or the steam may be controlled independently of valve S by opening valve *t* and closing *m m*.

The valves and connections shown in the figure enable the two tanks to be used either independently or together, and to be emptied by either the boiler feed pump or the pump belonging to the cooking apparatus.

If, as is likely to be the case, tank D must be used for kitchen system and tank D' for the main heating system, then for D, valves *h*, h_2, h_3, *i'*, *r*, *k'*, *w'*, and *j'* must be closed, and valves *j*, *k*, *i*, and *w* must be open; and for D', valves *h'*, h'_2, h'_3, *n*, *o*, *l*, *m*, *n*, *i*, *r'*, *w'*, *s*, and *s'* must be open, and valves *l'*, *j'*, *i'*, *r*, and *w'* must be closed.

Figure 25 shows the Kieley automatic pump controller C, attached to the house pumps (K K, Fig. 1), which deliver into the roof tanks through specially made valves. These are closed when tank is full. The delivery pipe is then subjected to pressure, which is transmitted through the communicating ½-inch pipe A to a diaphragm in cylinder B, and causes a movement which raises lever D in direction E, and permits a corresponding descent of connected lever G toward F, closes valve H, and thus cuts off steam from pump K, and immediately stops it. I is a throttle valve to control the pump independently by hand; W W W are adjustable weights; J J are knife edges. As soon as the pressure in A is relieved, levers D and G resume their original position, as shown, and the pump again commences to work, thus keeping the tank always full. Similar governors are attached to the fire pumps.

Among the trade articles used in the steam systems above described are six back-pressure valves, eight pressure regulators, six damper regulators, several return steam traps, and four tank pump controllers, some of which are here illustrated, or have been before shown in these columns, and were manufactured by Timothy Kieley, New York City.

PART V.—HOT WATER, DRIP, AND BLOW OFF TANKS, RETURN FROM LOW RADIATORS, GENERAL ARRANGEMENT OF RADIATORS, CONNECTION OF RADIATOR BRANCHES TO STEAM RISERS, SPECIAL RADIATOR CONNECTIONS AND SPECIAL WALL PLATE.

FIGURE 21 is a perspective showing steam connections at front end of boilers M M, Fig. 1, which furnish the hot water required throughout the house, except for part of the laundry work, where steam is used.

Each boiler M has two 70-foot internal coils of 3-inch brass pipe A A (Fig. 22). The pipes of each coil were arranged in perpendicular planes, as shown at Z, and these were put together in pairs so that the two coils formed a square cross-section, as shown at Y in each boiler, the inlet and outlet ends C C and D D alone being exposed. Live steam is received through E, Fig. 21, or exhaust steam through O, and after passing through the coils A A, etc., Fig. 22, returns through pipes D D, etc., to branch B. N is a tank intended to receive the discharge from the drip pipes, etc. N' is a blow-off tank. Neither of these tanks is shown completely connected up.

G is a pipe to the feed-water heater; F is a Kieley steam trap; R is a drip pipe; H, J, K, and L are relief pipes.

Figure 23 is a rear view of the boilers and tanks shown in Fig. 21. C C, etc., are the sediment or blow-off pipes, which are connected by crossheads, D D, etc., to pipes E E. These discharge into a special sewer built expressly for this purpose, and running direct to the street. G G are equalizing pipes. They have relief branches H H terminating with open ends above the roof. F F are check valves and J J are stop-cocks.

The condensation water from nearly all pipes and radiators drains by gravity into the tanks D D, Fig. 24, but it was necessary to place some radiators at a lower level than that of these tanks, and their drainage is effected as shown in Fig. 26, where R is a radiator set too low to discharge into tank D. Its drip is therefore received in the Kieley steam trap F, which is set at a convenient lower level. As the water accumulates in trap F, the internal float A rises to position B and turns levers G, H, and I in the direction indicated by arrows J J' J'.

L L are connecting links and W W are counterweights. As soon as lever H passes the vertical its weight W carries it further toward J' and through link L', and lever I quickly opens valve K, admitting live steam through pipe M to the surface of the water in trap F and forcing it out through pipe N into tank D. When the float falls to position A the valve K is closed, shutting of the steam, and water is again received through pipe E, and so on.

O is an exhaust pipe. C is a check valve closing with pressure from the trap and so prevents the steam from pipe M entering the radiator R. D is another check valve closing with pressure toward the trap, and thus prevents steam pressure from tank D entering trap F.

Figure 27 shows the arrangement, in many of the suites, of radiators in adjacent rooms. Wherever there was a window, as at A, located near the riser shaft D, the horizontal coils B were made to fit the wall recess, and were preferably used, but when, as often occurred, there was no adjacent window recess, the vertical coil C was considered more desirable.

E is the 2½-inch supply, and F the 2-inch return steam riser, with 1-inch branches, H and I respectively; G is the ½-inch air escape pipe, with ⅛-inch branches J J; K K are air valves.

Figure 28 shows the connection of branches H and L to risers D and F. The elbows M M, etc., are arranged to permit the vertical displacements of the tee O, due to temperature elongations and contractions of D, to be taken up by bending in the length of branch H, which is free to swing about axis X X. The latter moves either up or down, and carries it to the (exaggerated) positions M' H', or M' H". N is a special tee, made, as shown in the enlarged section, with a central diaphragm S, designed to prevent the possibility of water backing up in one radiator by reason of an unbalanced pressure in the other. This device was invented by Edward Noonan, New York, and is placed on return riser connections only. C is an ordinary Bundy radiator, and on the first floors and in all corridors these radiators are provided with an ornamental screen.

Figure 29 shows the patent connection details of this radiator, which was invented by J. L. Wells, to avoid the use of a connecting strap across the pipes. Each element V V V, etc., is a simple cast box, constituting a return band, which slides freely on the dovetailed joints R R. Any desired number of elements can be used, and are supported on detached pedestals W W. Steam is received at H and discharged at I, after passing through the radiated pipes B B, etc. T T, etc., are ornamental face plates. The connections at the other end of the radiator are the same, except for the omission of elements U U. Figure 30 shows a horizontal section through U and a vertical section through V.

Figure 31 shows the details of expansion wall plate L, Fig. 27. The branch H passes through a slot in plate A, which permits it to rise and fall with the expansion and contraction of the riser D, and through a close-fitting hole in plate B, which latter slides on the face of A and always closes the slotted hole.

PART VI.—DETAILS AND ARRANGEMENT OF SPECIAL RADIATORS IN THE SERVANTS' CORRIDOR, IN THE DINING-ROOM, IN THE REFRIGERATOR ENGINE-ROOM, IN THE PARLOR, AND IN THE LOUNGING-ROOM.

FIGURE 32 shows a radiator in the ninth-floor corridor of the servants' wing. The radiator B is there connected to the risers E and F by very long branches, A and C, to allow for the 2⅜-inch vertical movement which must be there provided for in the risers. The radiator is similar to B, Fig. 27, except that it is a single instead of a double coil; E is the 1¼-inch supply, and F the ¾-inch return steam riser, with 1-inch branch H and ¾-inch branch A respectively; G is an air valve with ⅛ inch pipe H; K K are supporting brackets.

Figure 33 shows the detail of a connection block P, exactly the same as those shown in Figs. 27, 29, and 30. The pedestal M is made hollow to permit the movement of branch H.

Figure 34 shows the special arrangement of Bundy elements in the dining-room radiators, which are designed to occupy the corners and have marble cover slabs on top (not shown here) to serve for waiters' table. They will also be provided with curved brass screens. There are six of these radiators, each having about 225 square feet of surface; A is the 1¼-inch supply, and B the 1¼-inch return steam pipe; D is the air valve with ⅛-inch pipe C.

Figure 35 is a sectional view of the cast base E, which forms a single chamber, establishing communication between the steam pipes A and B and the radiator elements F F, etc., the latter being here removed for clearness.

Figure 36 shows the arrangement of the radiator A in the refrigerator-room; B C is the exhaust main with back-pressure valve D, set to five pounds. Steam can be supplied from the stack end C at a pressure of five pounds or less, through 1¼-inch pipe E, or from B under a greater pressure through 1¼-inch pipe H; F is a 1¼-inch live-steam pipe, and G is a steam trap.

Figure 37 shows the coils under the parlor windows, the warm air passing out freely through the brass screen A.

The lounging-room has a divan D, Fig. 38, surrounding all the sides, and as it was not desirable to have the radiators in the middle of the floor, nor to have them interrupt the continuity of the divan, they were arranged underneath it, as shown in Fig. 36. In this E is the radiator coil, F the 1¼-inch steam supply, and G the 1¼-inch return, C is the air valve and H its ⅛-inch pipe.

The panel L gives access to the valves C, M, and N. Cold air from outdoors is supplied through duct A. Its delivery is regulated by gate I, operated by handle J, which is always concealed by the cushions O. The dotted lines show its closed position I'. The arrows indicate the course which the cold air must take, passing through the radiator coil E and over bridge K, and thus being tempered before gaining admission to the room through the register B. The dining-room ceiling is of papier-maché, filled with small perforations, concealed by the pattern. Above the ceiling is a free space, the air in which is exhausted by a 36-inch fan, driven by an electric motor.

The system of ventilation was designed and executed by J. L. Wells, of Gillis & Geoghegan, New York.

STEAM PLANT IN THE NEW NETHERLAND HOTEL.

PART I.—GENERAL DESCRIPTION, ARRANGEMENT OF MAINS IN BOILER-ROOM, AND DIAGRAM OF HEATING MAINS.

THE New Netherland is one of the recently built hotel structures of New York City. It is a 17-story building standing on the corner of Fifth Avenue and Fifty ninth Street. The power, heating, lighting, ventilating, and refrigerating plants are located in the cellar, the boilers being in the area under the sidewalk and the main part of the apparatus being assembled in what is known as machinery hall. The cellar floor is 25 feet below street grade and the cellar has a head room of 11 feet. Steam for all purposes is generated in four Babcock & Wilcox boilers, each a "double-decker" of 160 horse-power. They are set two in a battery, head on, the firing being either way from the center. They are fitted with the Kirkwood rocking grate and the Spencer draft regulator. By using the "double-deckers" it was possible to get this 640 horse-power, a 12x13-foot center firing-room and the combustion uptakes in a floor space of 13x70 feet. The boilers are under the sidewalk at the corner of the street and the avenue, and there is sufficient head room over them to allow of inspection and repairs. The boiler space is well ventilated and is lighted in the daytime by the open and illuminated sections of the sidewalk and at night by incandescent electric lamps. A 4x4-foot No. 8 sheet-steel smoke flue is suspended above the boilers, connecting with the uptakes B B B B, Fig. 2, and descending into the arched 16-inch walled horizontal 4x5-foot brick flue C which is built upon the floor and passes to the end of the area, and then turning to the left continues to and connects with the 5-foot square chimney flue D which rises 247 feet, clearing the highest point of the roof. The boilers and all the live and exhaust supply and return mains and all other steam pipes, connections, traps, etc., and the entire heating and ventilating system was installed by Messrs. Gillis & Geoghegan, of New York City, in conformity to the plans and specifications of William H. Hume, architect.

The boilers furnish steam for all the power which is used in the hotel, house heating, laundry work, water heating, kitchen work and steam cooking, and are so connected that any one boiler or all of them may be used for any one or all of these services. The arrangement of dome and cross-connecting is such that all expansion is taken care of without the use of slip joints.

Figure 1 is an isometric general diagram of the steam mains only rising from the boilers settings, and shows the general arrangement and connections, the reference letters corresponding with those on the heating and power plans. Figure 2 is a plan of the basement showing a diagram of the heating supply main only, and omitting all other pipes. For heating purposes the steam leaves the boilers through the 6-inch dome connections E, which start from beneath the 4-inch "consolidated" safety valves F, located midway on the boilers and the mains are laid to the

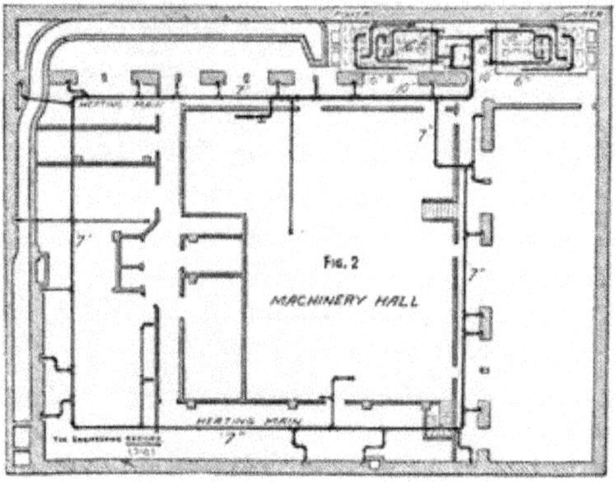

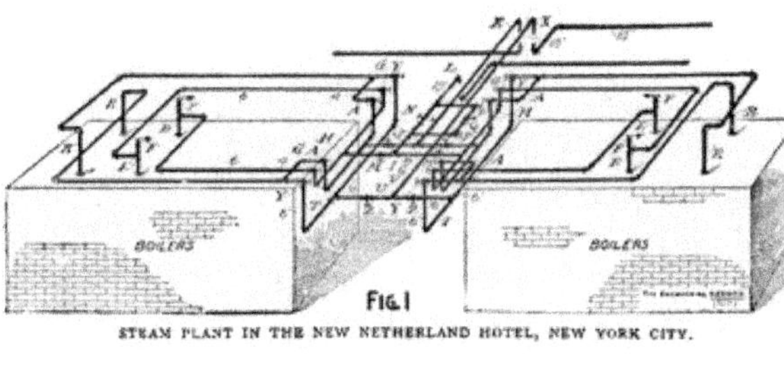

FIG. I

STEAM PLANT IN THE NEW NETHERLAND HOTEL, NEW YORK CITY.

fronts of the boilers and to the angle valves G. Turning down it passes by an 18-inch section and an elbow into the 8-inch crossmains H. These two crossmains are connected by the 8-inch connecting main I to the 10-inch main J, which continues to the point L, where it drops sufficiently to pass to points of distribution on the cellar ceiling. By closing its valve G any boiler may be shut out, or if it is desired to shut off either battery its valve M is closed. If it is desired to use the full boiler pressure for heating, the by-pass valve N is opened, but in ordinary service it is kept closed, the steam passing by the loop and valves P through the equalizing valve Q again to the main J. The valves P are provided to control the valve Q when repairs are being made. The heating main G is connected to the exhaust main which serves all the principal engines by a vertical down pipe at T, and the supply is ordinarily received from this source, passing through a grease extractor which is arranged with a by-pass and globe valves, between which and the boilers the main is 10 inches in diameter. Beyond these points the circuit around the outer edges of the cellar ceiling is used by a main of 7 inches uniform diameter, most of whose branches to the risers are 1½, 2, and 2½ inches. Parallel with supply main G, but a little nearer the wall in a trench underground, runs a corresponding 2-inch return main, and receives drips from all the risers and empties into the receiving tank. When exhaust steam is not used for heating, or when an additional supply is needed, it is taken direct from the boilers through branches E E, and enters the system through the reducing valve Q, which is ordinarily set to five pounds pressure. If the system is being operated by a direct live-steam supply from the boilers it can be changed to a supply from the exhaust main by simply closing valves P P and N, Figs 1 and 2, opening

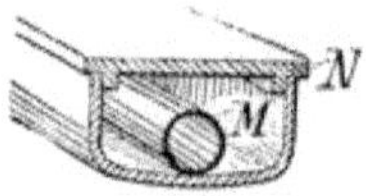

valve W, Fig. 2, and adjusting valve X to the required pressure, when the surplus exhaust steam, if there is any, will escape through the main stack in the vent flue to the open atmosphere above the roof.

PART II.—DETAILS OF HEATING APPARATUS, RETURN TRAPS ON RISERS, EXPANSION JOINTS IN LONG VERTICAL PIPES, CONNECTIONS OF RADIATORS, AND ARRANGEMENT OF INDIRECT RADIATOR STACKS.

FIGURE 3 is a general plan of the ninth story, showing arrangement and location of the direct radiators R R, etc., which is substantially the same on all of the guest floors. All the risers on the outside walls are run in built-in recesses, and each riser is dripped into a 1¼-inch pipe into the main return pipe, which is run in an accessible brick-walled iron-covered trench just below the cellar floor. The height of the heating risers, and the fact that the living-rooms were almost duplicates of each other, allowed a uniformity in locating the radiators and coils. As

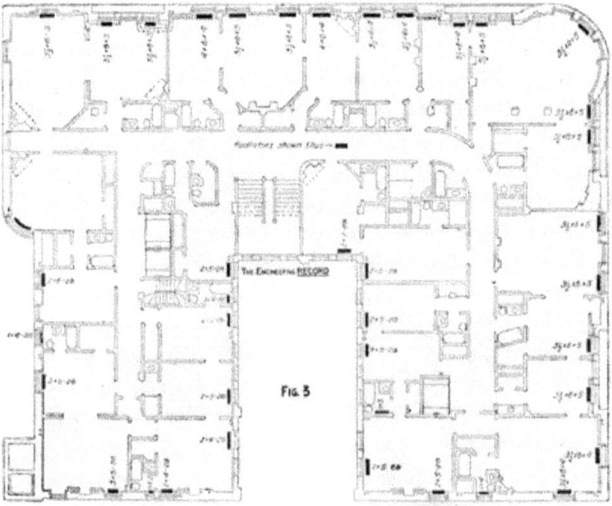

STEAM PLANT IN THE NEW NETHERLAND HOTEL, NEW YORK CITY.

shown on the plan Fig. 3, it was desired to run the risers without horizontal leads, but to do so it became necessary to provide for the expansion which, caused by high-pressure steam, might at times reach a considerable maximum. The height of some of those risers is 230 feet, running to the seventeenth floor. Four inches expansion was allowed for in three sections. Each was suspended at the top as shown by Fig. 5, and expanded downwards. Each section A of the riser was hung from its elbow H at the upper end, the rim of which rested on the I beam J. The lower end had a spring of 8 feet as shown with the horizontal return bend K. All of those fittings were specially tapped to give the desired spread.

Excepting in the halls, bathrooms, and throughout the third and thirteenth stories, where the window sills were too low to admit sufficient heating surface, all the radiators on the guest floors were set in the deep recesses of the window cases as indicated in Fig. 6, and were inclosed by an iron perforated screen S, through which a small opening gave access to the regulating angle valve V that served as a connection for the long supply main M, which was pitched as indicated by the arrows to insure drainage of the condensation water into the riser H, and was made long enough for its spring to take up the expansion and contraction of the riser. The height of the radiator was such that its top B formed a convenient seat or bench.

The fresh air that is forced into the halls and public rooms in the lower stories is drawn from outdoors through the covered sidewalk louver M on the Fifth Avnue side, and passed through a 43x43-inch duct V, Fig. 7, made of No. 20 galvanized iron, to the chamber N, also made of No. 20 galvanized iron, and about 3'x4'x8' square, which contains 52 Bundy Climax sections put together in four stacks or coils, each containing 13 sections, each section containing 13 square feet of heating surface, and each stack having independent connections to the 2-inch supply pipe A and the 1-inch return pipe B. The air is forced to pass through and around these radiators, and is thus warmed before it is delivered by fan E to duct E.

Where it was necessary to lay the horizontal pipes M on the floors, they were placed in cast-iron trenches,

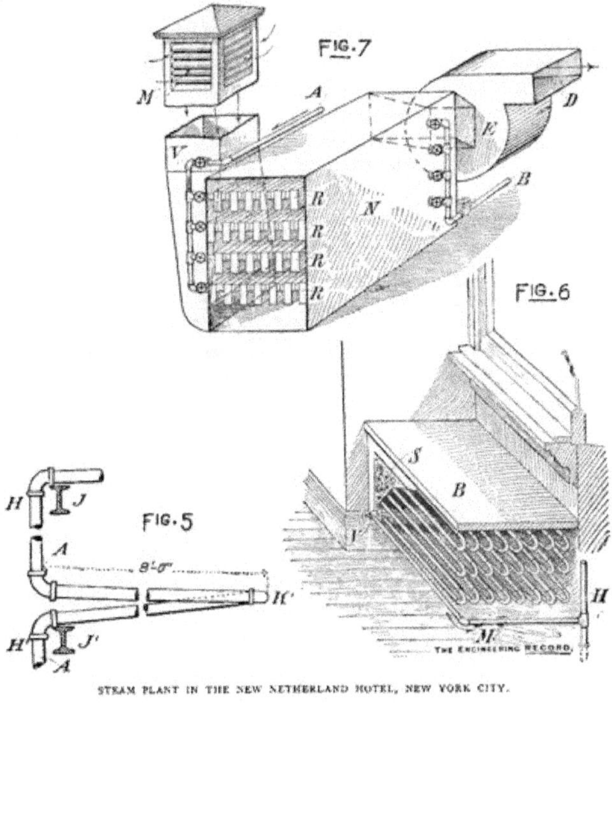

STEAM PLANT IN THE NEW NETHERLAND HOTEL, NEW YORK CITY.

Fig. 4, made to suit the section, and having loose-fitting cast-iron covers N as shown in detail, to give ready access for repairs or inspection. All the heaters are not connected on the one-pipe system, but are connected to two risers, one for conveying steam and the other for conveying water of condensation to the main return in a trench under the cellar floor, the main return emptying into a receiving tank. From this tank the return water passes to pumps, each under control of the Kieley pump governors, and is by them pumped as feed water to the boilers.

PART III.—PLAN OF POWER MAINS, DETAILS OF EXHAUST STACK, LOCATION AND OPERATION OF ENGINES, PUMPS, AND DETAILS OF AUTOMATIC GOVERNOR.

FIGURE 8 is a general plan of the basement floor, showing the size and position of the live and exhaust steam mains, and the location and arrangement of the engines, pumps, tanks, ice plant, and other machinery. The steam-heating mains and the main E and its connections, which supply steam for the boiler pump, kitchen laundry, and cooking apparatus, are omitted here to avoid confusion, but are shown in the diagram, Fig. 1, and in plan in Fig. 2. The 6-inch dome connections R for the power service start from the rear of the boilers, passing to the front. They have the angle valves Y, and, like the heating connection E, drop about 3 feet, so that the connection of this system may be below those of the heating system. The 6-inch crossmains T are united by the 8-inch connecting main U, from which the 10-inch power main V branches and continues to the point X, where it also drops to pass into machinery hall. The closing of the valve Y will shut out any one boiler from the system, or the closing of a valve Z will shut off either battery of boilers. From the boiler sides of the valves Y were taken the 4-inch connections A, which were cross-connected and controlled somewhat on the lines of the power connections. This passes between the heating pipes above and the power pipes below, as indicated by a center line broken with two dots, to a line E, which supplies steam to the boiler pump, kitchen laundry, and cooking apparatus. All these several connections are dripped into the drip tank F, which is below the floor, and is also used as a blow-off tank for the boilers. The main exhaust stack Y rises through the chimney Z, and is topped by a Hoey condenser head, the drip from which passes back to the blow-off outside the valve of the drip tank F. This 12-inch exhaust pipe, which is 250 feet high, rests on a lower section, the detail of which is shown in Fig. 14. The head G is welded into the end of the sup-

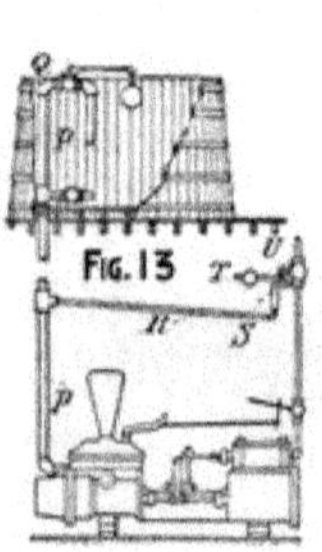

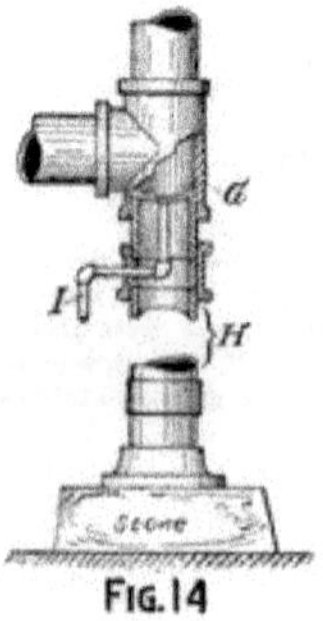

porting section H, thus giving a flush internal face, which is tapped and dripped as shown by the pipe and valve I.

The 4-inch pump main E supplies steam to one 5½"x3½"x5" duplex pump B for emptying the drip tank F, two 7¼"x4½"x10" duplex pumps Q' Q' for feeding boilers, and two 12"x6"x10" duplex pumps H H for domestic and fire service. All of those pumps are of the Worthington make, and are dripped into the tank F. This pump main E also supplies steam to the cooking apparatus in the kitchen, and for heating water in the heating boilers I I I, when the exhaust steam, a connection for which is provided from the main exhaust pipe, is not available. The

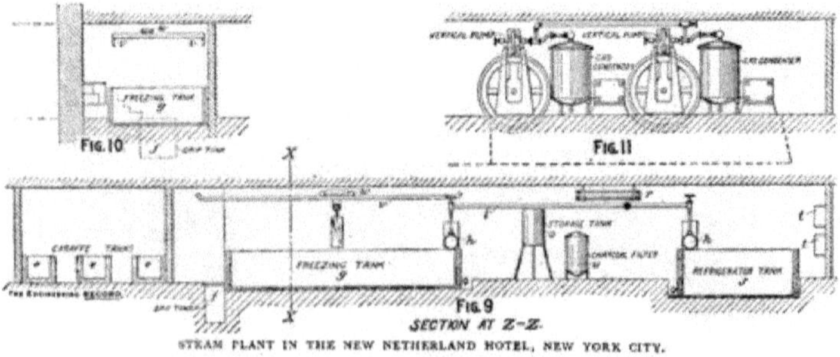

STEAM PLANT IN THE NEW NETHERLAND HOTEL, NEW YORK CITY.

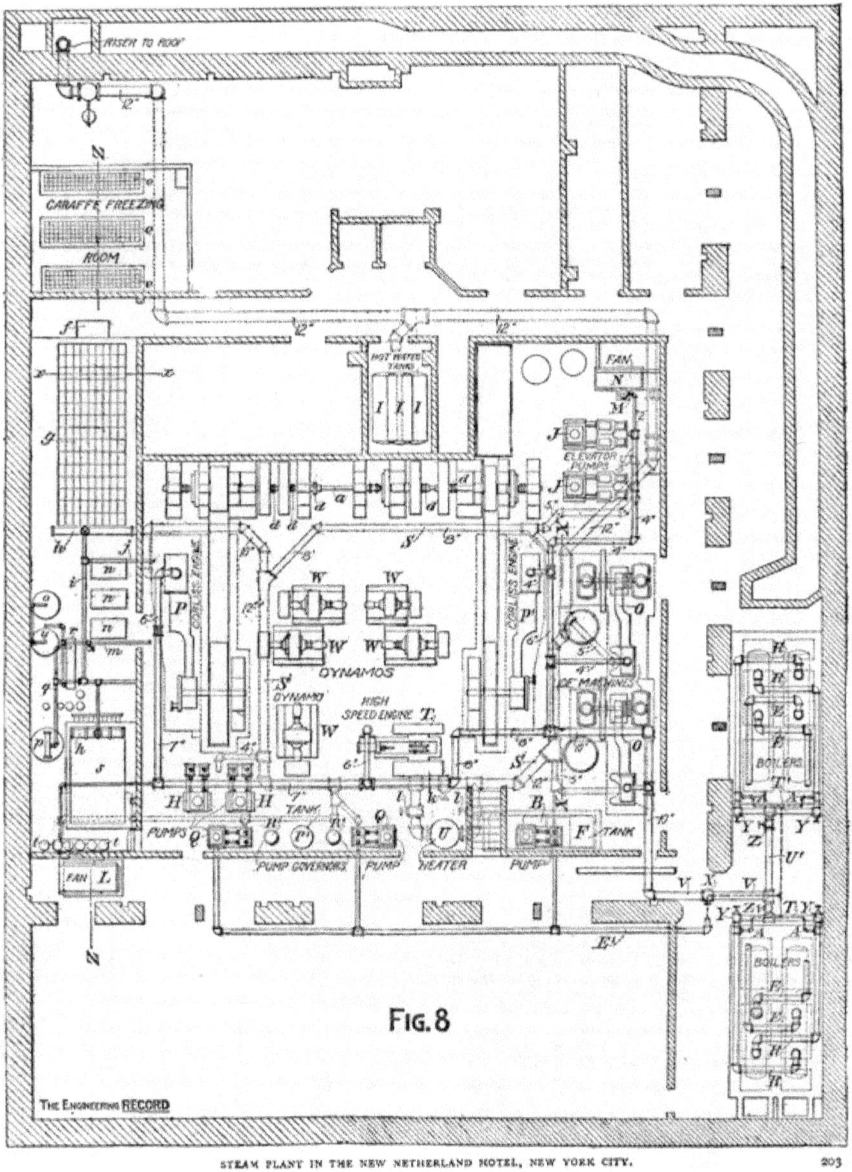

STEAM PLANT IN THE NEW NETHERLAND HOTEL, NEW YORK CITY.

pumps H H, which are used for filling the house tanks, one performing service for the upper tank and the other for the intermediate tank, are fitted with the improved Kieley tank pump controller, which operates as follows: The delivery end of the tank supply pipe P, Fig. 13, has upon its end the counterbalanced lever float valve Q. When the float is down this gives a clear waterway from the pump to the tank, and when the float has been raised by the inflowing water it serves to close the valve. The shutting off of the discharge from the pump brings an immediate back pressure in the pipe P, which is transmitted instantly through the pipe R to the cylinder S, whose piston is then forced back by this increased water pressure, and the piston-rod being attached to the lever T of the steam throttle U raises it, shuts off the steam, and immediately stops the pump. The action is almost instantaneous, the air chamber on the pump being sufficient to prevent water hammer and relieve the strain. When the water in the tank has been sufficiently lowered to allow the ball valve Q to open and relieve the dead pressure which is at all times on the discharge P when the pump is not working, the counterweight on the lever T falls, opening the steam throttle, and starting the pump, which runs until again stopped by the action of the valves Q and U.

The 10-inch power main V, Fig. 8, supplies steam to two 14" and 20"x12"x10" compound duplex Worthington pumps J J for elevator service, one 6x9-inch engine K, which runs the 72-inch heating fan L, and a 4½x6¾-inch engine M which drives the 60-inch fan N, the two Whitehall engines O O for the refrigerating and ice-making machinery, one 22x42-inch and one 16x42-inch Watts-Campbell Corliss engines P P, and one 15x16-inch McIntosh & Seymour high-speed engine T for the electrical department. The exhaust from all pumps and engines is carried in the pipe S and its feeders to the Berryman heater U heating the boiler feed water, and then to the heating system or the external air as described. The valve *k* acts as a by-pass if so desired, and the valves *l l* control the heater.

The electric current which drives the motors attached to the ventilating fans, laundry machinery, pneumatic blowers, and one house elevator which is used for baggage, etc., and which furnishes the electric lighting for the entire building, some 6,000 incandescent lamps, is furnished from three 100 and one 60-kilowatt standard Edison compound wound dynamos W W W, which are driven by ¾-inch cotton rope drives from the 5¾"x54' hammered steel shaft *a*. Either dynamo can be instantly thrown in or out of service without interfering with the rest of the service by means of the Lane clutch *d* which is attached to each driving pulley L. To one end of this shaft is connected the 22x42-inch Watts-Campbell Corliss engine P of 200 horse-power and to the other end the 16x42-inch 160 horse-power engine P' of the same make. Both engines are connected to the driving wheel of the countershaft by a 1-inch cotton-rope drive and with those of the dynamos have gravity tighteners. The 15x16-inch 125 horse-power McIntosh & Seymour high-speed engine T is independently connected to the 100-kilowatt dynamo W' by a rope drive, and is to be used in case of an emergency or for any special service. All those engines, dynamos, and pillow blocks are laid on heavy granite block work imbedded 6 feet below and forming part of the concrete mass which composes the entire floor of the machinery hall.

To guard against injury to the main rope drives from water seepage entering at the lower side of the 14-foot main engine driving wheels, water-tight ¼-inch wrought-iron plate troughs, conforming to the lines of the main drives and the extended lower sides of the rope drives, were set in the concrete floor and given a 24-inch clearance below the wheel for examination, etc.

The entire system of electric circuits is controlled by switches on the 9'x8'x1¼" thick polished slate switchboard, upon which are the fuse blocks, registers, etc. There were about 35 miles of wire used in wiring this job. The whole electrical contract was with the General Electric Company, 44 Broad Street, New York, and cost $75,000.

In the refrigerating department three pumps *n n n* draw cold brine from machines *o o* through 2½-inch suction pipe *j* and circulate it through the 3-inch pipe *i* to the different points required—viz., the refrigerator tank *s*, the freezing tank *g*, with a capacity of 135 150-pound cans, and the tanks *e e e* in the carafe-room, each of a capacity of 100 bottles. In this figure *f* is a dip tank where the cans of ice are momentarily immersed in hot water to enable their contents to be easily slipped out; *h h* are distributing headers for the brine; *u* is a charcoal filter to purify the water used in ice-making, which is drawn from storage tank *o*; *p* is a Wheeler condenser communicating with vapor tank *p* by the pipe *q*; *m* is a 2½-inch waste-water pipe, and *t t* are two batteries of Buckring filters arranged on the wall. Figure 9 is a general section at *z z*; *w* is a trolley hoist running on elevated track *v* to handle the ice cans. Figure 10 is a section at *x x*, Fig. 9, and Fig. 11 is an elevation at *z z*, Fig. 8, of the ice machines.

Figure 10 is a section at *z z*, Fig. 8, through the refrigerator-rooms. Figure 11 is a section at *x x*, Figs. 8 and 10, and Fig. 12 is an elevation of the brine pumps. This refrigerating plant was installed by the Whitehill Pictet Company. It maintains a temperature of about 30° Fahr. in the ice refrigerators, and operates one cooling coil of pipe for reducing the temperature of the fresh-air supply in summer.

VENTILATION OF THE NEW NETHERLAND HOTEL.

This hotel is steam-heated throughout by direct and indirect systems, and has mechanical ventilation of the principal public rooms and all toilet-rooms by a system installed by Messrs. Gillis & Geoghegan, in accordance with the plans and specifications of William H. Hume, of New York City, architect.

Figure 1 is a basement plan showing the location of fans, chimneys, etc., and by distinguishing conventions the different hot and cold-air and ventilation main ducts. The 6x3½-foot fan E is driven by a 6x9-inch engine, and is designed to force heated air to the several points shown on the plan, Fig. 1, through a rectangular No. 20 galvanized sheet-iron duct A, the delivery ends of which are controlled by graduating wing registers, so that the persons in charge of the rooms supplied may check or increase the flow of warmed air at will. The delivery to the basement and first floors is indicated by vertical flues shown with diagonal marks across them, while the delivery to the cellar rooms is through registers indicated merely by arrows emerging from the side of the duct. The fresh cold air for fan E is drawn from a low louvered cast-iron shaft M close to the building line on the Fifth Avenue sidewalk, and is taken thence to the fan through the No. 20 galvanized sheet-iron duct V, which connects to the heating chamber N, which is also made of No. 20 galvanized sheet iron. Within the chamber are connected the indirect heating stacks. This same fan and system of ducts is to be used for cooling in the warm season. In a similar metal chamber, located in the rear of the heating chamber, are placed numerous closely-nested coils of 1¼-inch wrought-iron pipes, through which can be circulated brine at a very low temperature from the refrigerating apparatus. The chilled surfaces of those pipes condense moisture from the air passing over them. This congeals as frost or ice on the pipes, making a very thick mass which both cools and dries the air. Air entering by sliding valves, the intake M can easily be diverted through the hot or cold chamber at will.

The 4½x2½-foot Sturtevant fan F is driven by a 4½x6¾-inch engine, and is intended to supply fresh cold air only to the points of delivery indicated by broken arrows. This supply of air is brought from the Fifty-ninth Street side of the house through another louvered intake, whose galvanized sheet-iron duct is marked W. The same conventions are used here as in the hot-air system to indicate the position of the outlets in the duct B. The main ventilation duct C is also made of galvanized sheet iron, and, like the others, is suspended overhead and run in the most convenient places. Its lateral branches have direct inlets and vertical flue openings controlled by graduating dampers. The trunk is gradually increased in size as each lateral is added, having at its point of entrance to the vent shaft Z' a cross-sectional area of 20 square feet. This vent shaft Z' has a cross-sectional area of 30½ square feet, and rises to a height of 250 feet above the point at which the vent duct enters. The main steam exhaust pipe Y passes up and through this shaft, which has the additional accelerating influence of 200 feet of 1¼-inch steam-heating coils. The heat from the exhaust stack and the length of the shaft give a powerful draft in favorable weather. The steam coil is provided to

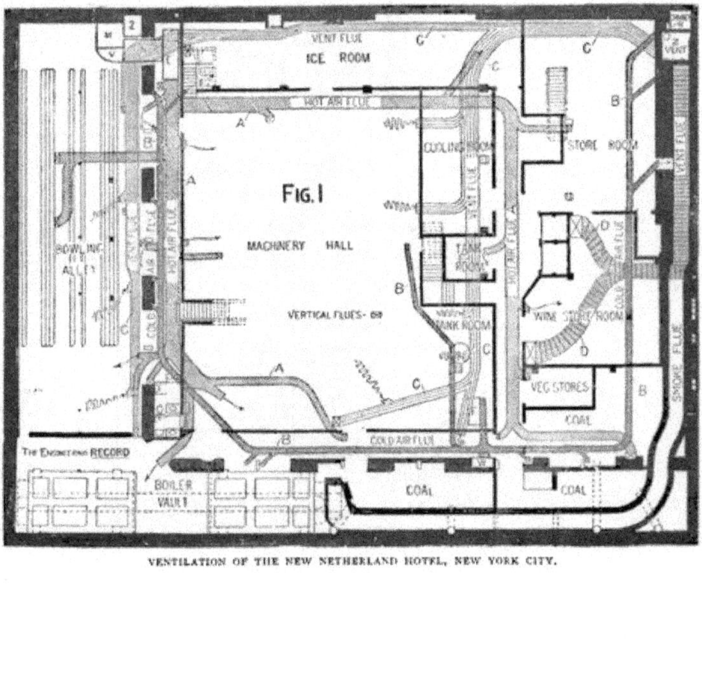

VENTILATION OF THE NEW NETHERLAND HOTEL, NEW YORK CITY.

assist if necessary, and to assume a sufficient draft under all conditions of atmosphere. There was placed upon top of this shaft, as shown in Figs. 2, 3, and 4, a 72-inch Blackman fan, to which is belted a 10 horse-power Edison electric motor, receiving its current from the house electric plant. These conditions are intended to assure at all times a strong, regular current, withdrawing foul air from the machinery hall, kitchen, bowling alley, closets, and other apartments in the cellar and basements which are connected to this system.

The basement rooms are connected into the duct formed by a false ceiling hung below the true one in the halls and communicating with the main duct. Special care has been given to securing the removal of odors arising from the cooking, and to preventing their entrance to the dining-rooms, business offices, parlors, or living apartments, by the erection of ample canopies covering all cooking or steaming apparatus, and connecting by the galvanized duct D, Fig. 1, to the special vent shaft R, which was built into the wall inclosing the boiler flue.

To secure a change of air in all of the living apartments on the several floors there were built 11 local vent shafts S, Figs. 5 and 6. Beginning at the right

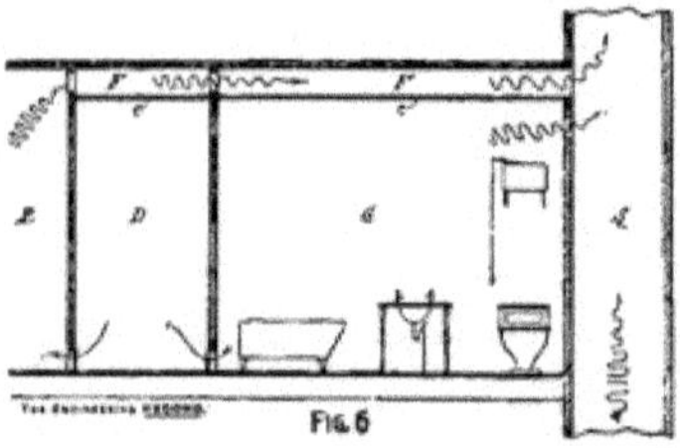

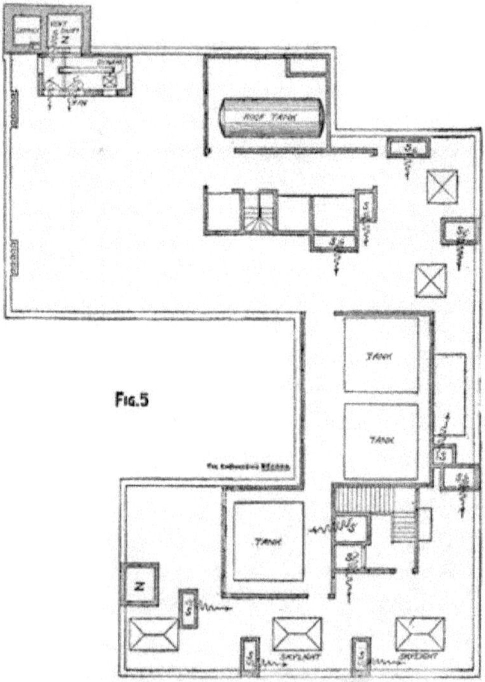

VENTILATION OF THE NEW NETHERLAND HOTEL, NEW YORK CITY.

and left of the main stairway hall the false ceiling C, Fig. 6, was built, dropping 10 inches below the ceiling proper, closed at the starting lines. Following and covering all the side halls D, the shafts S were so distributed as to give each its proportionate amount of service, and each sleeping apartment and parlor E were connected to this ceiling duct F. At the top of each shaft S S, etc., a Blackman fan was set, six of them being 36-inch diameter, and driven

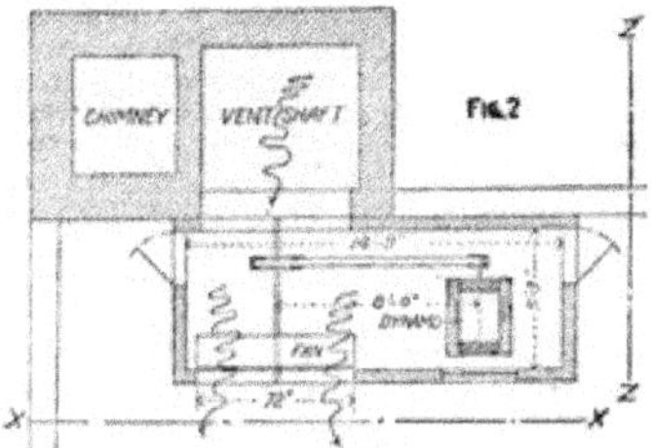

PLAN

FIG. 3
ELEVATION X-X.

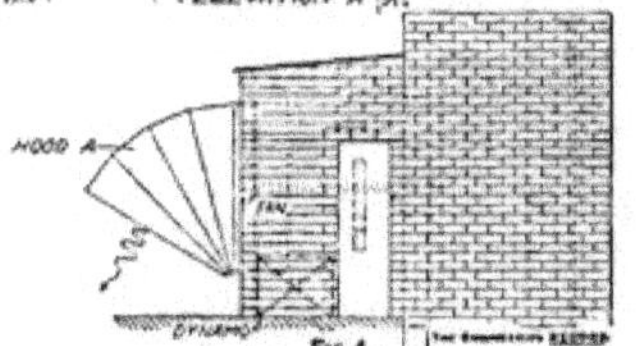

FIG. 4
ELEVATION Z-Z.

by three horse-power Edison electric motors, both working on the same shaft. Five were 30-inch fans, driven by two horse-power motors, all of the same make. The contract for this part of the work was in the hands of H. Ward Leonard & Co., 136 Liberty Street, New York. The housing and general arrangement of these fans and motors were on the same general plan as that shown by Figs. 2, 3, and 4, excepting that they are connected direct, while that in Figs. 2, 3, and 4 is driven by a belt. There is also a steam-heating coil set in each of those shafts near their upper ends, and designed to be sufficient to create a strong upward current in ordinary weather. At other times the electric fans are used, their speed being controlled by the engineer from the switchboard in machinery hall. The currents of air drawn through those shafts are designed to effect a general movement of air in the several rooms connected with them, fresh air being admitted to the rooms from the halls through a protected opening through the walls near the floor. The exhaust air passes through another protected opening, into and through the ducts F, or directly into the shaft S, Fig. 6.

Figures 2, 3, and 4 are diagrams showing the plan and elevations of the fan house on top of the vent shaft, the dynamo and fan being conventionally indicated. A special feature is the 6½-foot copper elbow-hood A which protects the fan from the severity of driving winds and storms while permitting its unobstructed delivery of foul air outwards. Figure 5 is a plan of the roof showing the location of tank-house, skylights, main shafts Z Z, chimney, etc., and the local vent shafts S S, etc. Figure 6 is a typical section through bathroom, etc., showing the double-hung ceiling and connection to shaft S.

STEAM HEATING IN THE HOLLAND HOUSE.

PART I.—EXHAUST-STEAM ARRANGEMENTS AND VENTILATION SYSTEM.

The Holland House is an 11-story marble hotel at Thirtieth Street and Fifth Avenue, New York City. The appointments throughout are modern and the mechanical work is extensive and interesting. The building is heated by direct and indirect steam radiation and the rooms are ventilated by registers and wall flues delivering into ducts which lead to an exhaust stack extending above the roof. The system and details of the work conform in general to current practice and comprise essentially the features that have been shown in these columns. All heating is by exhaust steam, though connections have been made to permit the use of live steam if necessary. The radiator main has a pressure regulator and a pressure gauge near the place where the exhaust is received. With the regulator open and the most remote radiators well heated, no pressure was indicated by the gauge index.

The system is what is known as the one-pipe system, with single pipe risers and branches, having special pipes for returning condensation water parallel to the horizontal distributing mains only. The arrangement and exhaust-steam connections in the engine-room are shown in Fig. 1, where A is a 6x3-foot iron tank receiving all exhaust steam, that from the large and small engines respectively through the 4-inch and 3-inch pipes B and C, and from the elevator pump through the 5-inch pipe D. From tank A the oil overflows at a level of about 18 inches from the bottom through a vertical stand-pipe trapped into the main sewer, and the steam passes through the 12-inch pipe E to the grease extractor F and is discharged through pipe G, which can deliver either to the boiler-feed water heater H, to the radiator system, or to a direct exhaust above the roof. When valves N, I, and J are closed and P and K are open, the steam exhausts directly through pipe M. If K is closed and J opened, the steam enters the radiator main L, which has a pressure-regulating valve, not here shown, to reduce the pressure to any desired maximum. If valves K and P are closed and N, E, and J are opened, the steam passes through the feed-water heater and enters main L; or if J be closed and K open, it then exhausts above the roof. O is a pipe commanded by valves, not here shown, which connects with the coils in the hot-water tanks. Q is a cold-water supply from the pumps, and R is the hot supply for the boiler feed. S is a surface pipe, and T T are blow-offs. U U are steam traps, V V, etc., are drip pipes, W W W are check valves, X is a water glass, Y is a live-steam pipe, and Z is an air cock.

The living rooms are ventilated by separate wall flues extending to the top of the building. A low chamber or air space exists between the ceiling of the upper floor and the cement roof, and to this level the flues are brought by 36 risers, each of which was intended to be separately connected with the accelerating ventilating shaft. But owing to urgent haste in the construction, the plan was slightly modified by allowing some of the 36 risers which were nearest to the shaft to open freely into the closed space under the roof which then became an exhaust chamber connected by special ducts F F, etc., Fig. 3, with the ventilation shafts. All the other and more remote risers were connected with separate individual ducts F F, etc., Fig. 3, to the vent shaft. The circulation in the ventilating shaft is continually promoted by the heat from the smoke and exhaust pipes which passes through it. The steam supply risers have a vertical height of about 116 feet and are firmly anchored and supported near their centers, so that the total linear expansion of about 1¾ inches is equally divided at the ends and causes not more than 1 inch maximum movement at any point. Bad effect upon the radiators is prevented by connecting them with long horizontal branches which have sufficient pitch to preserve a fall to the riser at its maximum or minimum extension and which will readily spring enough to accompany the rise and fall.

Figure 2 is a view of the galvanized-iron top of the exhaust shaft on the roof. Figure 3 is a section at Z Z, and Fig. 4 is a section at X X. A is the 40-inch smoke pipe from the steam boilers, B is the 20-inch smoke pipe from the kitchen range, C is the 12-inch main exhaust pipe, D is the drip pipe. F F, etc., are the ends of foul-air ducts, from which the air is discharged against deflector G. The steam-heating system provides for about 600 radiators, and the work was done by J. S. Haley & Co., of New York City, in conformity to the requirements of the architects, G. E. Harding and Gooch, of New York.

PART II.—VENTILATION STACK, CONDENSATION TANK, WATER FILTER, SUPPLY MAINS, SPECIAL RADIATORS, AND DETAILS.

FIGURE 5 shows the construction of the ventilation shaft, which also serves as a smokestack. The main smoke flue B enters the upper part of chamber C upon the walls of which rests a flanged cast cap K, which forms a base for the steel smokestack S, which is inclosed by brick walls and anchored to them by clamps J J, etc., at every floor. The smoke and hot gases rise freely from B, enter S, and are discharged beneath cowl G, as indicated by the full arrows. Soot falls freely to the bottom and is removed through door D, thus preventing the accumulation of obstructions and the trouble and expense of elbow connections, doors, etc., at B. A duct A from the

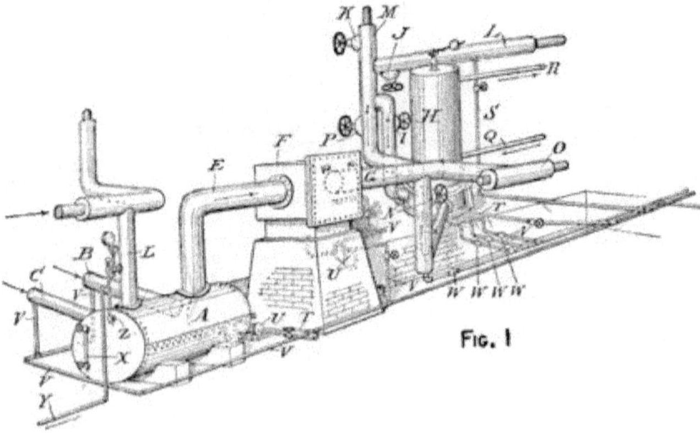

FIG. 1

STEAM HEATING IN THE HOLLAND HOUSE, NEW YORK CITY.

kitchen ranges communicates with the space between the smokestack and the outside walls where the air is heated and induces an upward circulation, discharging the kitchen vapors and foul air from the ventilation ducts F F F, etc., Fig. 3, at the top of the stack L, above the roof as indicated by dotted arrows.

Figure 6 shows the return condensation tank and connections by which all the water of condensation is collected and automatically pumped into the boilers. A is a steel tank about 3x6 feet in size, which receives condensation water from the different divisions of the system through the 2½-inch pipes B B, etc. C is a ¾-inch circulation pipe connected with the upper radiators, and D is a ¾-inch vent pipe, open above the roof. E is a 3-inch suction pipe to the pump F, which delivers to the boilers through the 3-inch pipe G. H is a ¾-inch escape pipe for air that may be drawn into the pump. I is a connection to the regulator J, in which the water rises to the same level as in A. When it reaches a certain height it operates an internal float, which commands the counterweighted levers K K K, and opens the valve L, which admits steam from the supply pipe to branch N to the pump F. This pump is thus driven until the water level falls in A and J, and the descending float reverses valve L, and cuts off steam to the pump.

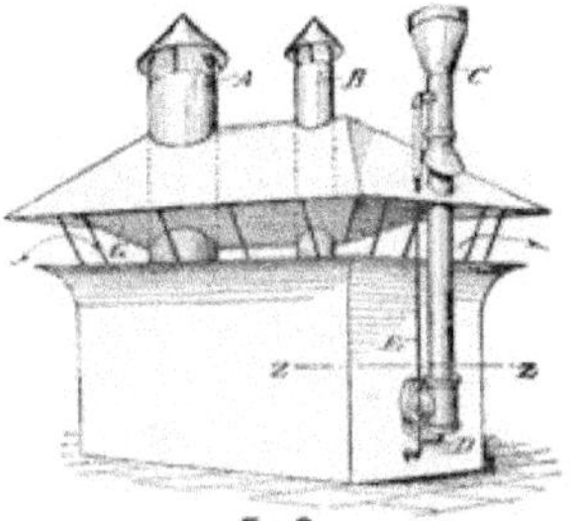

Fig. 2

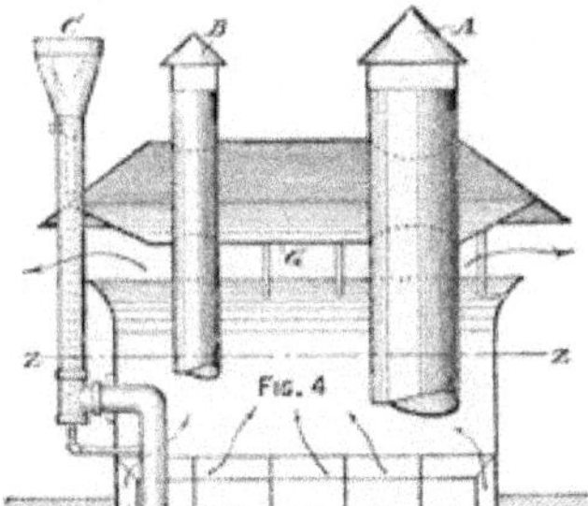

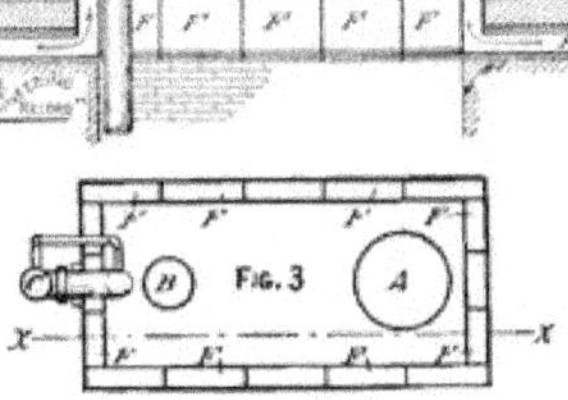

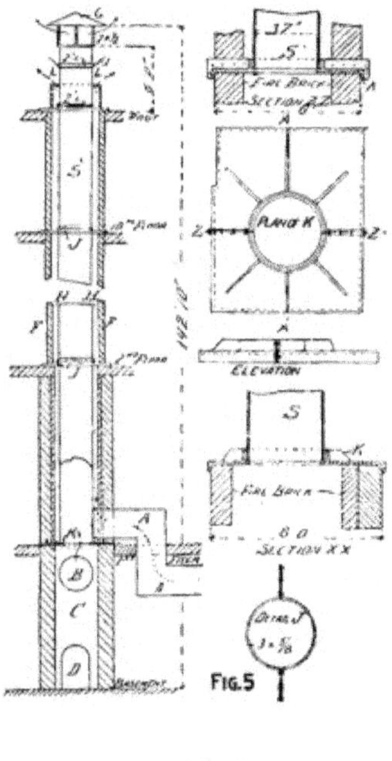

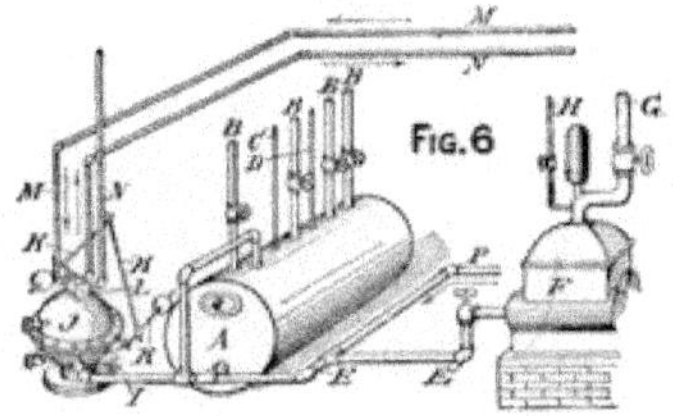

STEAM HEATING IN THE HOLLAND HOUSE, NEW YORK CITY.

All the water used in the building, either for domestic purposes or for the steam boilers, is filtered through a pair of Potter filters, made by J. S. Haley & Co. The filter is about 2'x3'x8' deep; and consists of two cast-iron half-shells A A, flange-bolted together, as shown in the vertical longitudinal section, Fig. 7, with pipe connections B B C and C. The filter is divided into two equal chambers F and G by a center partition composed of a sheet of ½-inch felt D, confined between two sheets of galvanized iron, with ¼-inch mesh wire netting E E, all of which are tightly bolted in between the flanges. Ordinarily, valves I and J are closed and H and N are open, and the course of the water is indicated by the black arrows. It enters from the city pressure pipe L, passing through branches C C to the chamber G, thence filtering through the felt D to the chamber F and into the pipe N to the suction tank. When it is desired to wash the filter, valves H and K are closed and J and I are opened, admitting water under tank pressure through pipe O. The water follows the course indicated by the dotted arrows, and is delivered to the sewer through pipe M. Each filter has a capacity of about 1,000 gallons an hour under 15 pounds pressure.

Figure 8 shows the connection of the main horizontal branches A to the basement exhaust main B, supplying steam to the radiators. Branch A is taken

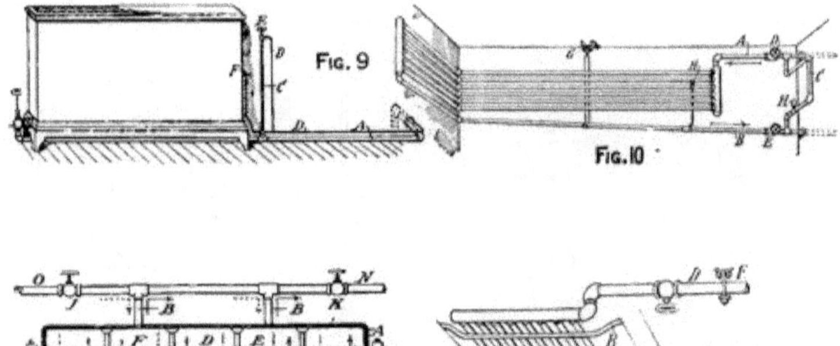

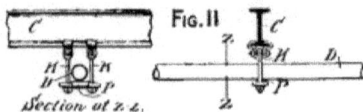

STEAM HEATING IN THE HOLLAND HOUSE, NEW YORK CITY.

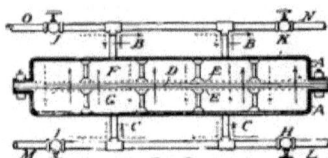

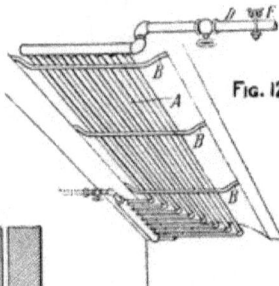

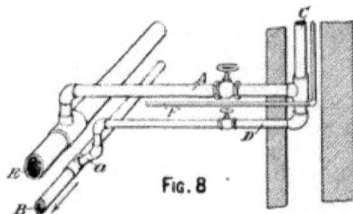

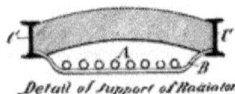

from the top of the main E, and supplies the riser C, to which the radiators are connected. Riser C terminates at the foot in a reducing elbow from which a smaller pipe D drains the water of condensation back to the return pipe B, which delivers into the tank A, Fig. 6.

Figure 9 shows the connections of the most remote radiators on the top floor. The steam pipe A is pitched about 1¼ inches away from the radiators, and is commanded by valve B, which has an extra long stem for convenience of operation. Near the radiator a branch C is carried up to an air valve E, and returns by pipe D to the tank A, Fig. 6, so that when the radiator valve B is closed the steam will still circulate as indicated by the arrow. F is an air valve.

Figure 10 shows the arrangement of a basement radiator which is located about 200 feet from the exhaust steam tank A, Fig. 1, and at a slight elevation above it. The radiator is composed of 1-inch angle pipes, with branches 4 feet and 9 feet long, and is hung on the wall near the ceiling by hanger G. The 2-inch supply pipe A and the 1-inch return pipe B are commanded by valves D and E and connected by a 1-inch by-pass C which permits circulation when valves D and E are closed. H is an air valve. On the upper floor Bundy radiators are used, with Detroit radiators elsewhere, except for indirect radiation in the basement, where ordinary pipe coils are used.

Figure 11 shows the location of one of these pipe coils in the hollow of an arch between the iron floor beams in the ceiling of the basement toilet-room where there was not room to set it on the walls or floor. The radiator A is simply supported on pieces of gas pipe B B, etc., the ends of which are flattened to rest on the lower flanges of the iron beams C C. The supply pipe D is supported from the floor beams C C, etc., by hangers E, shown in Fig. 12. These clamp over the lower flanges and are adjustable by means of the screw rods K K, and have loose gas-pipe thimbles P to serve as expansion rollers. Most of the main exhaust steam pipes are jacketed with wood pulp, which is said to be satisfactory here.

HEATING OF OFFICE BUILDINGS.

POWER AND HEATING PLANT, MANHATTAN LIFE INSURANCE BUILDING.

PART I.—GENERAL DESCRIPTION OF BOILERS, ENGINES, ELEVATORS, STEAM AND EXHAUST PIPING.

THE power and heating plant of the Manhattan Life Insurance Company's building at 64 to 68 Broadway, New York, presents an interesting opportunity for study. The building has a frontage on Broadway of 67 feet 3 inches and a depth back to New Street of 122 feet. It contains besides 17 stories a basement and pipe cellar, and measures 350 feet from the sidewalk to the top of the tower. The offices of the company are on the sixth and seventh floors, while the remainder of the building is devoted to the uses of an office building.

The steam plant is located in the pipe cellar and there will be found the engines, boilers, elevator pumps, etc., which are necessary to the warming and lighting of a large office building. The boiler plant consists of three internally fired boilers of the Scotch marine type placed under the Broadway sidewalk. The boilers, which were built by the Quintard Iron Works, contain about 1,620 square feet of heating surface each and are 11 feet 6 inches in diameter by 10 feet long. Each is fitted with two Purves ribbed steel furnace flues 45 inches in diameter by 7 feet 5 inches in length. This is connected at the real end of the boilers to the usual combustion chamber from which the gases pass to the breeching through 146 Servés ribbed steel tubes 3½ inches in diameter. The furnaces and tubes were supplied by Charles W. Whitney, of New York City. Figure 2 shows a longitudinal section through one of the boilers.

Steam is taken from a 7-inch nozzle on each boiler through the necessary piping to a 12-inch main, which drops into a short 12-inch drum lying in a horizontal position. Three steam pipes are connected to this drum, and each is provided with a stop valve placed near the drum. One of these pipes, 8 inches in diameter, supplies the electric light engines, elevator pumps, etc., with steam at 100 pounds pressure. A second 8-inch main containing an Acton reducing valve is used to carry low-pressure steam for heating the building. Exhaust steam is usually used for this purpose and boiler steam at reduced pressure is only supplied when the exhaust steam is insufficient. The third main leaving the drum before referred to, is 5 inches in diameter and supplies the entire pumping plant with steam. It will be noticed on the plan that the pipe supplying steam to the fire and boiler feed pumps is connected to this 5-inch main at a point near the elevator pumps. It is also connected to a 3-inch steam main in the boiler-room, which is fed by a 2¾-inch feeder from each boiler. This 3-inch main is known as the "Sunday main," that only being used to run the boiler feeds, etc., on Sunday or a holiday when the greater part of the plant is shut down.

The large 8-inch mains run back toward the rear of the building, gradually decreasing in size as the branches to the right or left draw their supply from them. The exhaust steam from the engines and elevator pumps is collected in a 10-inch main and led toward the heater, where the exhaust from the pumping plant and fan engine enters it. From that point a short length of 13-inch pipe leads to a 700 horse-power Berryman feed-water heater, the heater being so connected as to be thrown out of use if necessary. A connection is made between the heater and the atmosphere by the 12-inch free exhaust containing an Acton back-pressure valve. The exhaust pipe is carried up a vent shaft to the roof. A 12-inch pipe provided with a stop valve and a Hussey grease extractor leads to the heating main.

The electric plant consists of four Armington & Sims vertical engines directly coupled to the General Electric Company's iron-clad dynamos. Three of these dynamos are for lighting purposes, while the remaining one is for power, supplying current, as it does, for two electric elevators and three exhaust fans. Each lighting engine has a 12x12-inch cylinder, makes 275 revolutions per minute, and is rated at 80 horse-power. Each drives a 50-kilowatt multipolar generator working against a pressure of 120 volts. The engine driving the dynamos for electric power has a 9½x10-inch cylinder, makes 300 revolutions per minute, and drives a 30-kilowatt multipolar generator, the voltage in this case being 220.

The building is wired on the three-wire system for 3,236 16 candle power lamps. Nine different circuits are controlled from the switchboard in the engine-room. These circuits comprise the circuit for the pipe cellar and basement, the halls from the ground floor to the mezzanine, from the mezzanine to the roof, the dark-room circuit supplying those rooms below the fifth floor that need light in the daytime, and five circuits supplying the remaining rooms in the building. Independent of the main cut-out box on each floor are placed above the hanging ceiling four distribution boxes so that each room may be cut out from the mains separately. The dark-room circuit on each floor is also controllable by a distribution box for the same reason. A double throw switch at the main switchboard serves to connect the whole system with the Edison Electric Illuminating Company's streets mains in case of any accident to the lighting plant in the building.

The elevator plant consists of five hydraulic elevators, and two run by electric power, all furnished by Otis Brothers & Co. Four of the five hydraulic elevators are designed for passenger service only, while the fifth has special heavy parts for lifting safes when needed, and at other times is in passenger service. All of the elevators have vertical machines run by water furnished by two Worthington compound pumps, with cylinders 16" and 25"x13¾"x15". Beside these is an Otis electric pump discharging into the pressure tank, it being possible to use this when there is no steam for supplying the larger pumps. The pressure tank for this system, which is 6 feet 6 inches in diameter and 20 feet long, is located on the roof to economize space in the pipe cellar. The safe elevator when run as such is operated by a 10"x3¾"x10" hydraulic pressure pump, the delivery pipe from the pump discharging directly into the elevator cylinder. One of the electric elevators before referred to connects the two floors occupied by the insurance company's offices. The other electric elevator is in the tower.

boilers. The last two pumps are also cross-connected. A 4½"x2¾"x4" pump is used to pump the contents of the blow-off tank into the sewer. The water connections between the feed pumps and boilers are so arranged that the water may pass through the heater or not as desired. Still another 6"x4"x6" pump is used to pump the contents of the cesspool into the sewer, the cesspool containing the drips from the engines and elevator cylinders.

The cylinders of every pump and engine are relieved by the required drip pipes connected to Nason traps, the discharge from these leading to the blow-off tank. The blow-off tank is 3 feet in diameter and 9 feet long. The discharge from the pump emptying this tank connects with the sewer outside of the plumber's trap. The steam pipes throughout the plant are covered with magnesia sectional covering, furnished by Robert A. Keasbey, of New York City.

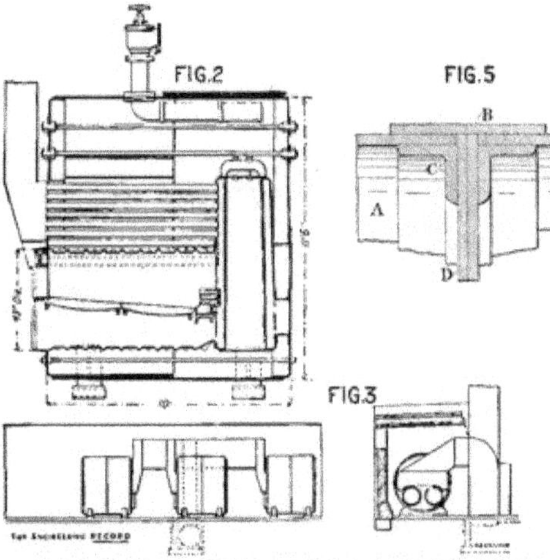

POWER AND HEATING PLANT, MANHATTAN LIFE INSURANCE BUILDING.

Outside of the elevator plant there are six Worthington duplex pumps and one Knowles deep-well pump for drawing water from a 1,056-foot artesian well driven by P. H. and J. Conlon, of Newark, N. J. The water from this well will be used at a later date for condensing purposes and flushing water-closets set in the building. Of the six duplex pumps the largest is a 14"x7½"x12" Underwriters' fire pump, and the next in size is a 7½"x4¼"x10" for supplying the house tanks. The boiler feed pump is a 6"x4"x6" and is so piped with the house tank pump that either may be used for either purpose. A second 6"x4"x6" pump operated by a Kieley pump governor is used to return the condensation from the heating system to the

The engine-room is ventilated by a 48-inch Sturtevant blower operated by a vertical engine. The air is forced through a system of ducts, shown by the dotted lines, to various parts of the pipe cellar. The boiler-room is ventilated by a different but somewhat similar system, a Seymour fan in this instance discharging the air into the boiler-room and at the same time ventilating the three front rooms in the basement of the building, the fan drawing its supply from those rooms.

Figure 3 shows an elevation of the breeching and smoke flue in the boiler-room and the manner in which it drops into an underground flue, which is shown by dotted lines in Fig. 1, and which connects

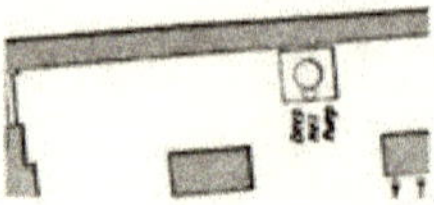

to a large sheet-iron stack running to the roof. The construction of the underground smoke flue presents one of the most interesting pieces of work in the steam plant. Because of the position of the caisson piers supporting the cantilevers upon which the building rests, and the vast network of pipes, and of the desire to keep the boiler-room as cool as possible, it was next to impossible to run a smoke flue from the boilers to the stack above the floor of the pipe cellar. The fact that the subsoil under the pipe cellar was considerably below the level of the surface water prevented, in the minds of the designers of the plant, the use of a brick flue, and for that reason recourse was had to a series of connected sheet-iron

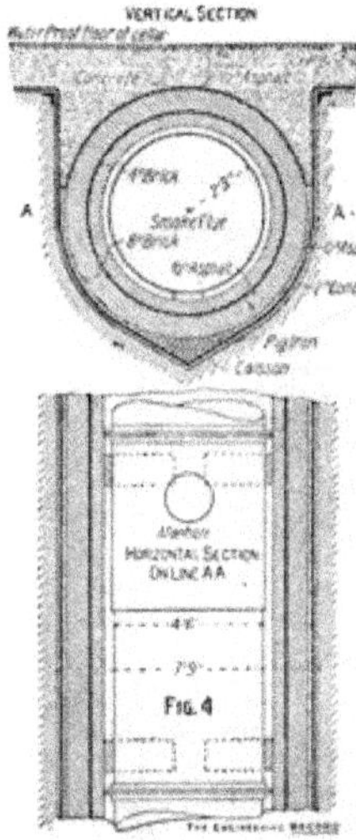

FIG. 4

caissons in which the flue is laid. The caissons were of the section shown by Fig. 4. Each caisson was about 11 feet in length by 7 feet in width. At four different points in the bottom of each caisson were cut four 5-inch holes, and a flange riveted on the inside from which a 5-inch pipe extended upward for a few feet above the top of the caisson. The sinking of the caissons was done separately, each being put in its proper location, when sheet piling was driven around the place in which it was to be sunk. After a temporary bulkhead of sheet iron was bolted on to each end of the caisson, it was heavily weighted with pig iron and steam syphons connecting to one or more of the 5-inch pipes referred to, the latter sucking out a sufficient amount of the subsoil to allow the caisson to sink a few inches. The other end of the caisson was then sunk in the same manner, and by this means it was possible to lower it uniformly.

Figure 5 shows the manner of connecting the caissons. In this instance the caisson A, to which the strap B and angle iron C were first riveted, would be sunk first, the bulkhead D being temporarily bolted to the angle iron. The next caisson would then be sunk, this also having a bulkhead similarly bolted in place. When the second caisson had reached the required depth the bulkheads would be removed and the two caissons drawn together and secured by rivets through the angle irons.

Figure 4 shows a horizontal and vertical section through the caissons, flue, and insulating materials. Inside of the caissons comes a coating of asphalt, and at the bottom is allowed to remain a part of the

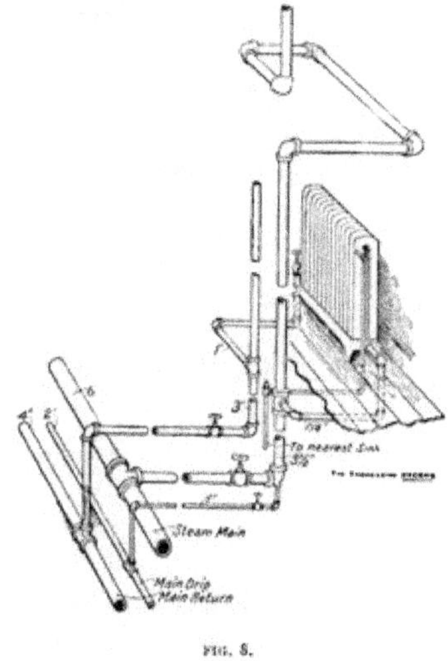

FIG. 5.

pig iron to weight the caisson while the flue was being put in. The spaces between the pig iron are then filled with cement to make a smooth surface on which to lay the 10-inch brickwork. A second layer of bricks, separated from the first by a layer of asphalt, lines the chamber that contains the flue. The flue is supported by the brick collars, which are cut away at the bottom to allow any water that might collect in the air space around the flue to drain toward one end, where it can be syphoned out through a pipe provided for the purpose.

We are indebted to the architects of the building, Messrs. Kimball & Thompson, and to Messrs. Gillis & Geoghegan, the contractors of the heating plant, beside the several parties before mentioned, for blueprints and data from which this description was prepared.

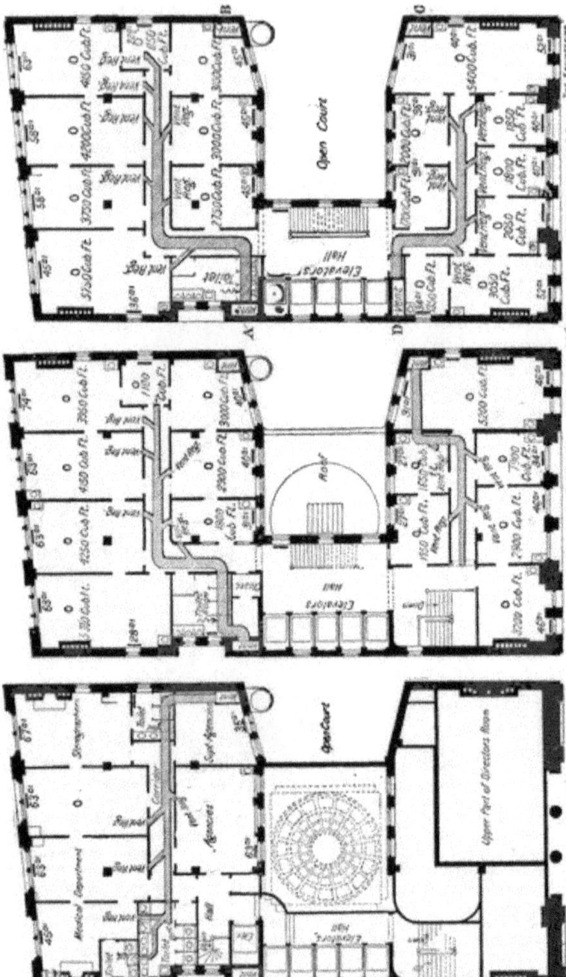

MEZZANINE FLOOR
FIG. 5.

EIGHTH FLOOR
FIG. 6.

ELEVENTH FLOOR
FIG. 7.

POWER AND HEATING PLANT, MANHATTAN LIFE INSURANCE BUILDING, NEW YORK CITY

PART II.—DESCRIPTION OF THE HEATING AND VENTILATING SYSTEM.

NINETEEN riser lines, starting from various points in the pipe cellar, supply steam to a large number of radiators placed on the 17 stories of the building. The building is heated by the direct system, containing in all about 500 radiators, presenting a total heating surface of about 19,000 square feet.

Figures 5, 6, and 7 have been made from the plans of three of the floors, these being selected to show the manner in which the building is warmed and ventilated. It will be noticed that many of the radiators are located under the window, and where that is the case the window sash contains a metal bronze register with a 2½x20-inch opening provided with a slide and handle. The fresh cold air entering the rooms at this point meets and mingles with the warm current rising from the radiators.

There may be seen on Fig. 7, the eleventh floor, four vent flues lettered A, B, C, and D. The flues A and B extend from the pipe cellar to the roof, flue C from the ceiling of the second floor to the roof, and flue D from the ninth floor to the roof. The plan adopted in ventilating the various rooms was, when possible, to connect each room by means of registers directly to the main vent shaft, and where the rooms could not be vented in this way they were connected to the nearest vent shaft by galvanized ducts built in the corridors in a space between the ceiling and the floor above. All of the floors of the building are ventilated in a manner similar, in a general way, to one of the three floors shown.

To insure a constant movement of air from the rooms into and out of the flues A, C, and D, three Blackman exhaust fans were erected in the space between the sixteenth and seventeenth floors. A fan was not provided for the flue B, as that contains the 12-inch exhaust pipe that runs from the pipe cellar to the roof, as it was thought that the heat radiated from that pipe would be sufficient to insure a movement of air in the flue. The fans for the flues A and D are both 42 inches in diameter, while that for C is 48 inches. Each of the three fans is driven by Lundell electric motor.

The radiators are supplied, as has been said, by 19 riser and return lines. The sizes of the risers are 3½ inches from the subbasement to the third floor, 3 inches to the sixth floor, 2½ inches to the ninth floor, 2 inches to the twelfth floor, 1½ inches to the fourteenth floor, and 1¼ inches to the sixteenth and seventeenth floors. The returns are one size smaller.

Figure 8 shows in a general way the manner in which the pipe lines are run, the method of taking care of expansion, the connection to the radiators, etc. All branches to the radiators are made between the floors and ceilings and inclosed in galvanized iron. The air valve on each radiator is connected by a ¼-inch drip pipe into a drip main running to the pipe cellar, where it is led to the nearest sink.

The grade of the steam mains and branches in the basement is such that the pipes drain toward the riser lines. Drip connections at their extremities, or where they connect with the riser, are provided, which connect with an independent drip pipe which carries any water of condensation by a main drip pipe back to the return tank. The connection for the radiators is made by a short length of pipe, an elbow being introduced forming a right angle in the run of the pipe so that the turning of the elbow on the short length of pipe will allow the riser to expand without moving the radiator or straining the pipe. The loop shown above the connection to the radiator shows the method of taking care of the expansion of the risers.

TEST OF A STEAM-HEATING PLANT IN THE CARTER BUILDING.

BOSTON, MASS., May 22, 1894.

To the Editor of THE ENGINEERING RECORD.

SIR: It may be of interest to your readers to be made acquainted with the results of a test which I made last March on the heating plant of the Carter Building, which has recently been erected at the corner of Water and Washington Streets, Boston, and I take pleasure in submitting the principal facts concerning the same, as follows:

The building is one of nine stories and a basement, and the accompanying plan of the basement shows the form of the building, together with the general arrangement of the mains and returns, as well as the risers. The building is heated on the gravity system, the water of condensation returning into a Webster heater, which seals the returns, and from which the water is pumped automatically to the boiler. The boiler is one of the Babcock & Wilcox type, rated at 100 horse-power, and it supplies steam to the main heating pipe through a reducing valve. The intended pressure in the system is limited to five pounds, and arrangements are made so that the exhaust steam from the elevator pump of the building can be employed in connection with live steam. The main heating pipe has a diameter at its largest point of 7 inches. There are 19 steam risers and 19 return risers, and these, with one or two exceptions, extend from the basement to the ninth floor. The radiators in the halls on the first floor are of the National type. The remaining radiators on the first floor and all on the second floor are of the Perfection flue type. Those on the remaining floors are of the National type, 26 inches high, all being manufactured by the American Radiator Company. The total amount of heating surface contained in the radiators is 5,633 square feet; that in the risers and their branches, 1,927 square feet. The covered mains in the basement present a surface of 388 square feet; the uncovered mains, 262 square feet, and the return pipes beneath the floor of the basement, 229 square feet, making a total uncovered radiating surface amounting to 8,051 square feet, and a total covered surface of 388 square feet. The area of the window surface of the building is, in round numbers, 14,000 square feet, and the total volume of the building, including halls and basements, 430,000 cubic feet.

The test was made in comparatively warm weather, the temperature of the atmosphere outside being 60 degrees. For this reason the results are not of much value in showing the capabilities of the plant in warming the building in cold weather, but the test does show the rate of condensation in the entire building, and the quantity of steam required to maintain it in a working condition under the circumstances which then existed, and this furnishes data which are of unusual valve owing to the infrequency of such tests. The quantity of steam consumed was determined by observations upon a meter properly verified which was attached to the feed pipe of the boiler. All the radiating surface was set to work, and when this had become thoroughly heated the test was continued for a period of two hours. Observations were made of the temperature of the air in each room, together with the pressure of the steam, the temperature of the water, and the record of the meter. The results were as follows:

1. Duration	hours	2
2. Weight of steam used per hour	lbs.	1,049
3. Temperature of return water	degs.	175
4. Pressure of steam in system above the atmosphere	lbs.	3.7
5. Average temperature of the air in the various rooms of the building	degs.	84.5
6. Rate of condensation per square foot of total uncovered surface	lbs.	0.24
7. Rate of transmission of heat per square foot of uncovered surface per degree difference of temperature between that of the steam and that of the surrounding air	B. T. U.	1.74

GEORGE H. BARRUS.

HEATING OF THE COLUMBUS BUILDING IN CHICAGO.

THE Columbus Building is a 13-story structure covering a 100x90 lot located on the corner of State and Washington Streets, Chicago, Ill. The first floors of the building are occupied by the two firms of Hyman, Berg & Co. and Ritkin & Brooks. W. W. Boyington & Co. were the architects of the building, and the contractors for the heating plant were F. W. Lamb & Co., of Chicago, Ill.

We have only given a plan of the basement floor, Fig. 1, and of what is known as the pipe chamber, Fig. 2, and a vertical section, Fig. 3, these being sufficient, together with the text, to describe the heating system. The heating system of this building, as is customary with most buildings of its character, is supplied with the exhaust steam from the elevator pumps and engines, and when this supply is insufficient live steam is turned into the system through a reducing valve. The piping is arranged on the overhead system with single-pipe supply. Starting from the low-pressure header in the boiler-room, a 10-inch main supply pipe with valve is carried up through a pipe chase to the pipe chamber before referred to. The pipe chamber is a space specially provided between the ceiling of the twelfth story and floor of the thirteenth story. Here the 10-inch riser divides into two 8-inch horizontal main supply pipes, which are run around the building, one taking in the north and part of the west side of the building and the other main the balance, as will be seen by Fig. 2. Each branch supplies the same amount of heating surface. These mains are reduced in size as the descending risers are taken off from them, eccentric fittings being used when such reduction takes place,

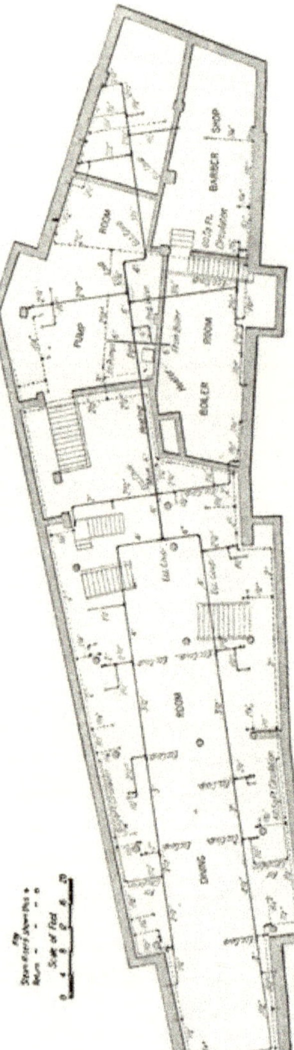

STEAM-HEATING PLANT IN BASEMENT OF THE CARTER BUILDING, BOSTON, MASS.

thus keeping the entire bottom of the pipe on a line, varying only with the pitch of the same. The descending risers are taken out of the bottom of these mains in the manner shown by Fig. 4, this being an enlarged view of the pipe chamber in Fig. 3. The pipes that lead to the thirteenth story and the attic are also shown. These are shown by broken lines in Fig. 2 to distinguish them from the descending risers. Each riser is provided with a valve close to the main and one in the basement, where it enters a return main. The descending risers are 3½ inches in diameter from the pipe chamber to the seventh floor, where they reduce to 3 inches and so continue to the third floor, where they reduce to 2½ inches in size and as such continue to the return in the basement. At the seventh floor each riser is fitted with a special expansion joint shown by Fig. 7. Each riser is anchored in two places, half-way above and below the expansion joint. This by reason of the expansion brings a vertical movement at the top and bottom of each riser, and to permit this the horizontal steam main at the top and the return main at the bottom are suspended by pipe hangers shown in Fig. 6. A helical spring is introduced to allow a movement of the pipe. The return main, which is shown in Fig. 1, is 4 inches in diameter. This main leads the condensation in the heating system back to a 300 horse-power Excelsior feed-water heater of the open type.

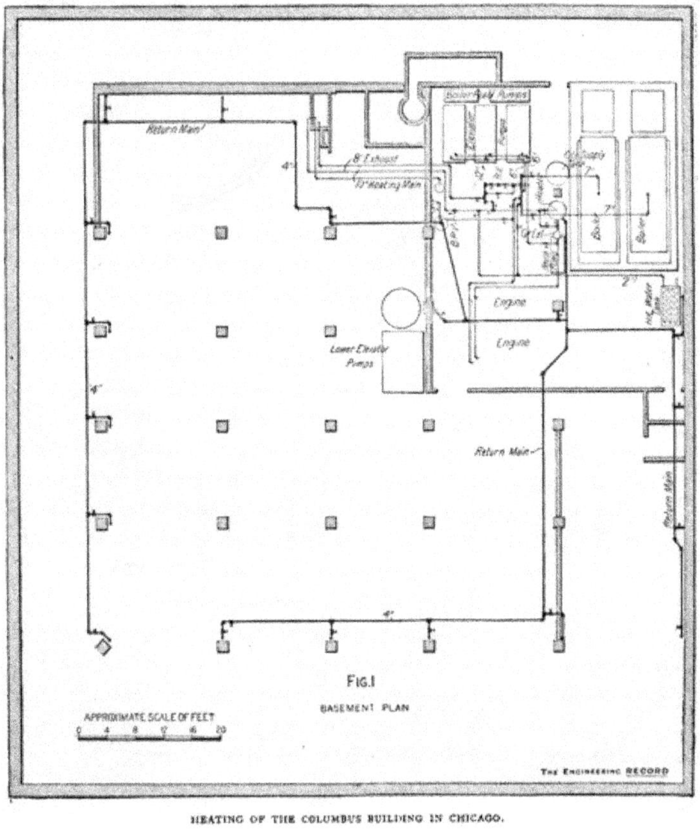

FIG. I

BASEMENT PLAN

HEATING OF THE COLUMBUS BUILDING IN CHICAGO.

The total amount of radiation in 375 radiators in the building is estimated at 13,725 square feet. The radiators are located under the windows, and are connected to the risers by a single pipe run under the floor. Figure 5 shows in plan the manner of making the connections, a special T being provided as shown. The boilers are two in number, of the return-tubular type, 7 feet in diameter and 21 feet long, each containing about 2,400 square feet of heating surface. Each is rated at 200 horse-power. A 7-inch pipe from each boiler is connected into a 12-inch drum, and from this a 6-inch pipe containing a 6-inch reducing valve with by-pass leads to a 12-inch drum for the low-pressure heating system. From this drum the 10-inch pipe before mentioned leads to the pipe chamber.

The boiler supplies with steam, besides the various pumps, two 75 horse-power Ideal engines, which drive dynamos for the incandescent lights in the building. Two 7½"x4½"x6" Blake boiler feed pumps are provided. As the heater receives the returns from the heating system, the boiler feed pumps are so connected that they may draw their supply from the heater and return it to the boilers, or when the heating system is shut off, as in the summer months, the pumps can take their supply from the city pressure tank and force the water through the heater and then into the boilers. The house pump, which is 14"x6"x10" in size, has its suction connected with the city pressure tank. The house pump is controlled by a Fisher pump governor. The house and boiler feed pumps are cross-connected in such a manner that either may be used for either duty. In determining the size of boilers to be used the designer of the plant had also to provide for three elevator pumps; two compound pumps with 14"x20"x12"x12" cylinders and one 20"x12"x12" in size.

A 10-inch pipe receives the exhaust and carries it to a Kieley grease extractor, from which it passes into the heater before mentioned. Just beyond the heater an 8-inch connection is made with the heating system, while at a point beyond in the exhaust pipe is an 8-inch Kieley back-pressure valve with a by-pass as shown on Fig. 1. The exhaust pipe is 8 inches in diameter from the heater to Lyman exhaust head on the roof, it being reduced from 10 inches, as a part of the steam would be condensed in the heater as well as in the heating system.

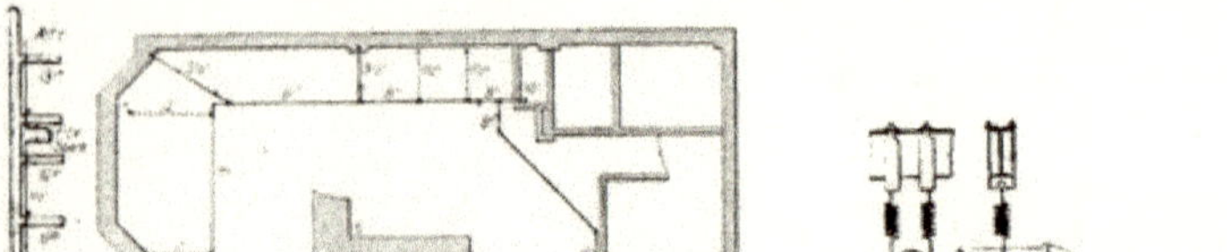

HEATING OF THE COLUMBUS BUILDING IN CHICAGO.

THE HEATING AND VENTILATION OF A NEW HAVEN OFFICE BUILDING.

THE accompanying plans show the heating and ventilating apparatus of the new office building that has recently been erected for the New York, New Haven, and Hartford Railroad Company at New Haven, Conn. The building is in the shape of the letter L, extending along two streets nearly at right angles for distances of 275 and 235 feet. The building is about 60 feet deep on both streets. A plan of the first floor and basement is given, and for convenience a continuous part of the wing is shown removed to one side. The building is ventilated and heated entirely by the hot blast or blower system. A Sturtevant heater, containing the equivalent of 14,500 lineal feet of 1-inch pipe, warms the air, which is driven to the various parts of the building by two steel-plate blowers with blast wheels 7 feet in diameter and 4 feet wide. The heater is so subdivided that the temperature of the entering air may be varied by shutting off the desired portion. The fan is rated to deliver 85,000 cubic feet of air per minute when making 235 revolutions in that time. Each fan is driven by a 7x10½-inch Sturtevant horizontal engine, connected to the shaft of its respective blower by belt. The exhaust from these engines is utilized in the heater. The fans are set on brick foundations with a long shaft connecting the two, couplings being provided so that they may be disconnected if necessary, and so that both fans may be driven by either or both engines.

Galvanized iron ducts lead from the fans to the bases of the vertical flues. The latter are lined with galvanized iron, and are all 8x12 inches in size, unless otherwise marked on the basement plan. The inlet registers, which are placed about 8 or 9 feet above the floor level, are about 65 per cent. greater in net area than the flues leading to them. A vent duct draws off the foul air from the rooms. This duct is the same size as the inlet, unless marked to the contrary on the plan.

Steam for the plant is generated in three horizontal return-tubular boilers, each 54 inches in diameter, 16 feet long, and containing 58 3-inch tubes, each 15 feet in length. The condensed steam from the heating coils is collected in a receiving tank, from which it is pumped back in boilers by a 6"x4"x6" Worthington pump.

The plant was designed by the B. F. Sturtevant Company, Boston, Mass., who were also the contractors for the heating and ventilating plant. The New Haven Heating and Plumbing Company were

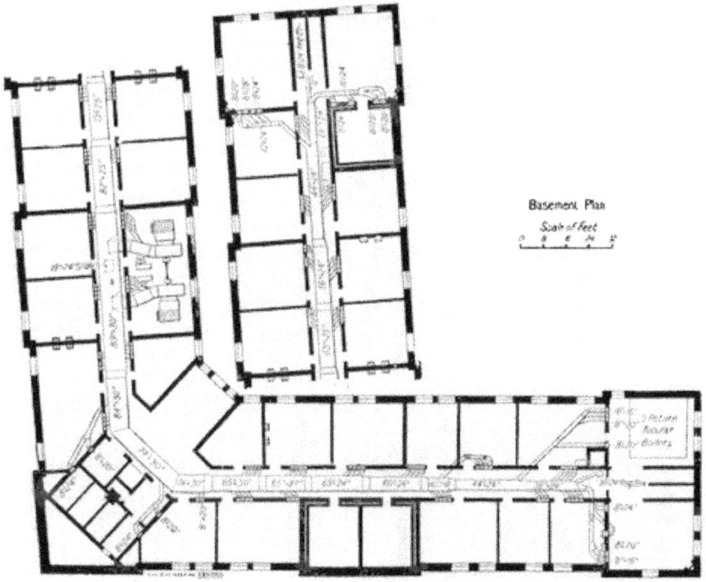

HEATING AND VENTILATION FOR THE N. Y. N. H. & H. R. R. CO.'S OFFICE BUILDING, NEW HAVEN, CONN.

the contractors for the steam plant proper in the building.

HEATING AND VENTILATING THE WALBRIDGE OFFICE BUILDING, TOLEDO, O.

The plant recently installed in the Walbridge office building, Toledo, O., is an illustration of the manner in which a complete system of indirect radiation, fresh-air supply, fan circulation and ventilation has been carefully applied to a large office building constructed without regard to the location or operation of such a plant and affording no special opportunities for the convenient arrangement of the ducts, flues, etc., which were, nevertheless, arranged simply and directly without essentially obstructing or defacing the rooms. The building is about 44x130 feet, six stories high throughout and eight stories high at one end. It is of iron and brick construction and has large exposed window surfaces on all sides except the narrow rear wall. The building is owned by Shaw, Kendall & Co., Mr. E. O. Fallis, of Toledo, is the architect, and the heating and ventilating apparatus was installed by the Buffalo Forge Company, from whose working drawings and data we illustrate (see pages 224 and 225) and describe the general features.

The main requirements were that practically the total volume of air in the building should be withdrawn and replaced every 10 minutes by fresh air at a temperature of 70° Fahr. in the coldest weather, and it was designed to accomplish this with a uniform velocity of 700 feet in the small branch ducts. Fresh cold air from the cellar intake is delivered to the heater by a 120-inch fan of the steel-plate type, full housing, with a blast wheel 85 inches in diameter and 42½ inches wide. The fan is propelled by means of a Buffalo direct-connected 10x8-inch horizontal engine. The discharge of the fan is right-hand, bottom horizontal discharge. The heater contains nine coils of 1-inch steam pipe, four rows in each manifold, 4 feet 6 inches long by 7 feet high, a total of 4,500 feet. Dampers are so arranged that the fresh air may either pass through the heater or around it and be delivered hot or cold or mixed in any desired proportion, in the foot of the vertical brick hot-air flue F, Figs. 1, 2, 4, 5, 7, 8, and 9, whence it is distributed throughout the different floors through rectangular galvanized-iron conduits T T, etc., which are placed just under the ceiling and are made 12 inches deep (not 9 inches as first designed and as marked on the drawings) and wide enough to reach within 1½ inches from each wall, thus practically forming an inconspicuous new ceiling 1 foot below the plas-

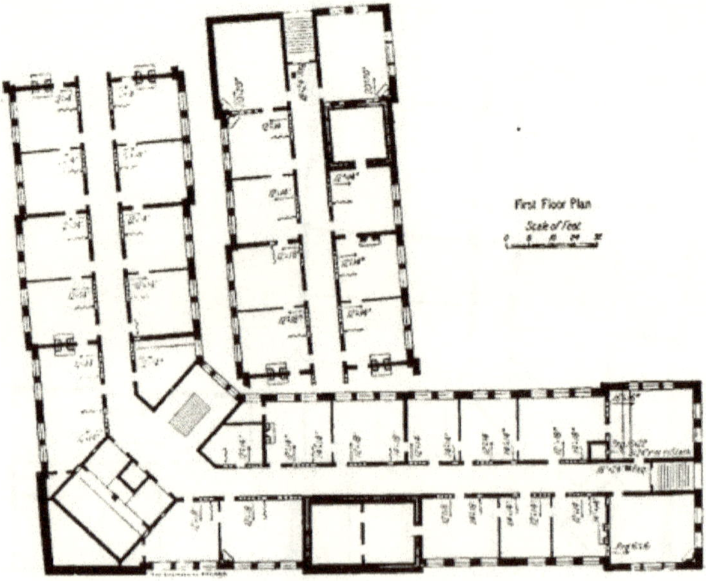

HEATING AND VENTILATION FOR THE N. Y. N. H. & H. R. R. CO.'S OFFICE BUILDING, NEW HAVEN, CONN.

ter. The conduits T T, etc., enter the flue F by flaring openings A A, etc., sloped upward to promote easy flow of the air. From the main conduits T T, etc., side branches B B, etc., deliver the hot air to the various separate rooms, where they terminate in registers R R, etc. These registers are all just below the ceilings, and just above the floors other registers V V, etc., each command separate flues which existed in the thickness of the wall or have been built specially as flat rectangular pipes D D D, etc., on the wall surfaces, and carry the foul air up to the ventilation chamber U between the upper-story ceiling and the roof. The currents of fresh air and of foul air throughout are indicated by straight and crooked arrows respectively, and sections of fresh and foul-air ducts or conduits by solid black rectangles.

The 88 rooms heated are from 9 to 15 feet in height, most of them being 11 feet high, and have a total contents of 251,813 cubic feet of air to be warmed, besides 54,400 cubic feet in the halls. This air is intended to be entirely changed every 10 minutes, thus requiring a supply of from 193 cubic feet per minute for the smallest rooms, 11'x16'x11', to 418 feet per minute for the largest rooms, 19'x20'x11'. The average size of the rooms is about 16'x16'x11', and requires about 300 cubic feet per minute. One exceptional room, No. 88, on the seventh floor (Fig. 9), is 18'x48'x9', and requires 776 cubic feet per minute, which is supplied by two 11-inch ducts. The ducts are all cylindrical, 8, 9, 10, 11, or 12 inches in diameter, their areas being respectively 50, 63, 78, 95, and 113 square inches. The area of cross-section required to furnish the necessary amount of air at a uniform velocity throughout of 700 feet per minute having been determined, the nearest larger corresponding size of duct was generally used. For instance, for room 28 58-inch section was required, and a 9-inch duct with 63-inch area was used. In some cases, however, the second larger size was thought suitable so as to give an excess, as in room 16, Fig. 1, where 48 square inches was required, and a 9-inch duct with an area of 63 inches was used. The hot-air registers were 9"x12", 8"x15", 9"x14", 10"x16", 10"x14", 14"x22", 12"x15", and 12"x19", their areas being always considerably in excess of those of the ducts they commanded. The vent flues are rectangular, 5"x8", 5"x11," 5"x9", 5"x13", and 5"x16", with registers 6"x10", 8"x10", 8"x10", 8"x12", and 8"x15". In most instances only one heating and one ventilating register was used in any one room, but in a few cases they were doubled.

Figure 1 is a plan of the first floor, and all the other stories except the second correspond to this in the lower part of the building. Figures 2 and 3 are vertical sections at Z Z and X X, Fig. 1, respectively. Figure 4 is a part plan of the basement, showing the position of the engine, fan, heater, etc. Figure 5 is a section at X X, Figs. 1 and 4, showing the location of the heater in the bottom of the flue F, and the connection of one of the main ceiling conduits T, which have a handhole H to permit the adjustment of the movable inlet damper K. Figure 7 is a plan of a part of the third floor, the fourth and fifth floors immediately above it being the same. Register P is in the side of main hot-air flue F to heat the hall. Figures 8 and 9 are respectively sixth and seventh

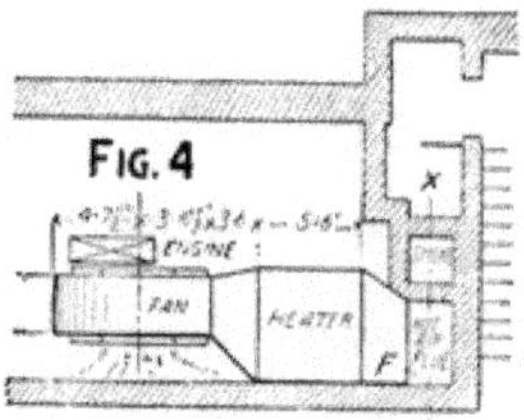

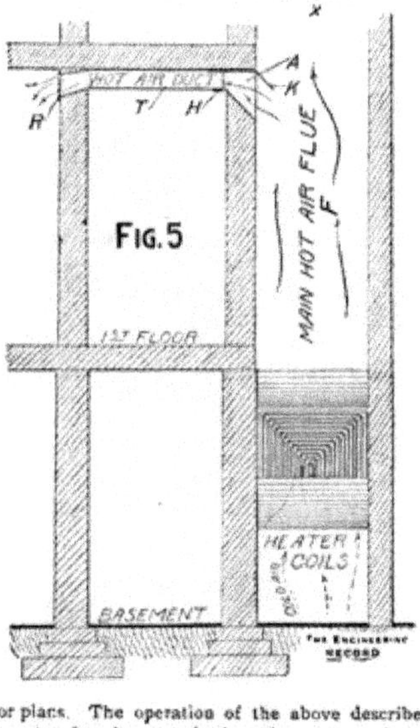

floor plans. The operation of the above described apparatus has been such that the owners of this building are having the same system put in another building under similar conditions.

HEATING IN THE WAINWRIGHT BUILDING, ST. LOUIS, MO.

PART I.—GENERAL DESCRIPTION AND OPERATION OF SYSTEM, AND BASEMENT PLANS SHOWING ARRANGEMENT OF PLANT AND SIZE AND LOCATION OF PIPES, ETC.

The Wainwright Building is a 10-story structure of modern fireproof construction, situated on the corner of Seventh and Chestnut Streets, St. Louis, Mo., and is intended for use as an office building. The architects are Charles K. Ramsay, of St. Louis, Adler &

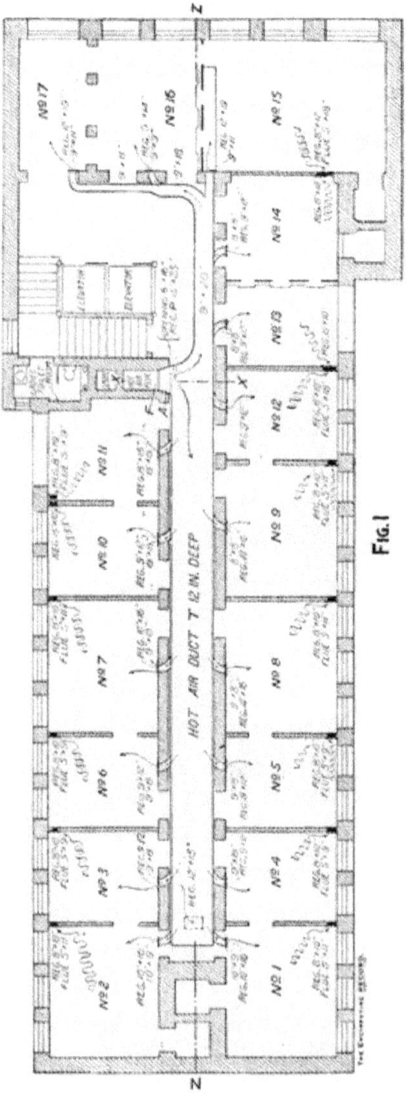

FIG. 1

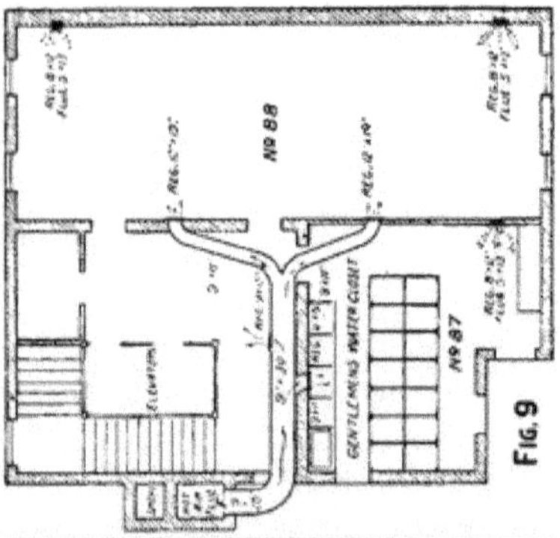

FIG. 9

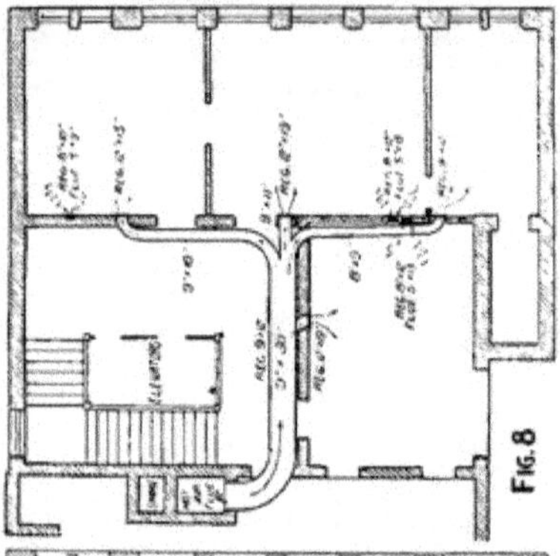

FIG. 8

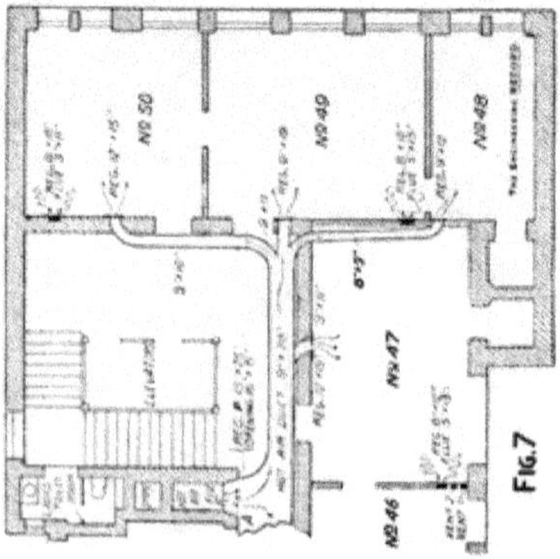

FIG. 7

HEATING AND VENTILATING THE WALBRIDGE OFFICE BUILDING, TOLEDO, O.

FIG. 2

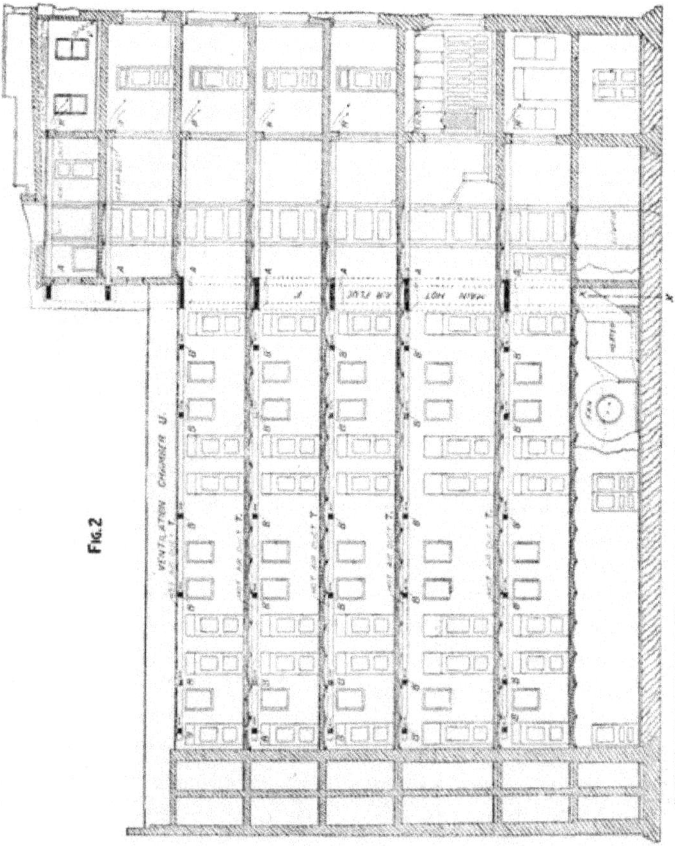

FIG. 3

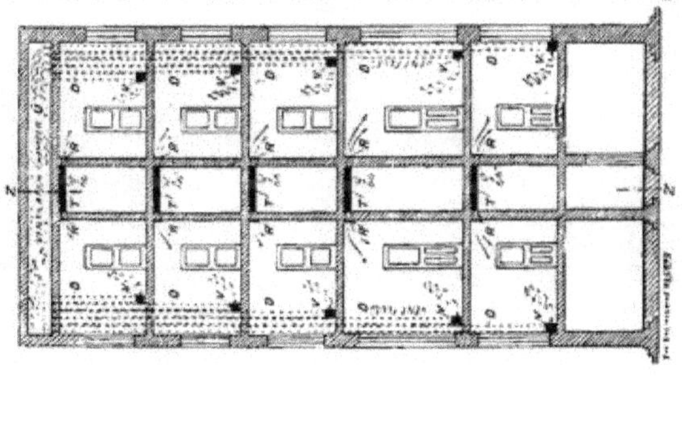

HEATING AND VENTILATING THE WALBRIDGE OFFICE BUILDING, TOLEDO, O.

Sullivan, associated, and the building is 117¾x127¼ feet in ground plan. A court 30 feet wide at the rear divides that part of the building into two wings 40 and 50 feet wide respectively, and 75 feet deep. Every room is provided with direct light and air, and four fast-running elevators afford access to the several floors. There are in all 210 offices in the building.

The system of heating is of the low-pressure type, having all drips, drains and return "sealed" before intersection, and was installed by the Missouri Steam Heating Company, St. Louis, Mo., to whose Secretary, Mr. H. W. Stone, we are indebted for the data and for descriptions and explanations to a member of of our staff. All the exhaust steam from the several engines, of which there are three of 65 horse-power and one of 35 horse-power, pumps and other mechanism, is utilized for heating purposes, as required. The operation of the system is in general as follows: The exhaust from the several engines in the building is gathered into a 12-inch header, and passed through a Stewart grease extractor; then to a 500 horse-power "Hoppes" feed-water heater and purifier, from which it goes to the main receiving tank. There is a by-pass around the heater and exhaust connection to the atmosphere at this point, the use of which is controlled by valves. The exhaust pipe, which is 12 inches in diameter, runs to the roof, where it is capped by a "Lyman" "A" exhaust head. This arrangement allows the steam to be used in the heating system, passed through the feed-water heater, or both, or discharged directly into the open air.

The main receiving tank has a 3½-inch high-pressure steam connection to the boilers for the purpose of supplying such additional steam to the heating system as may be required. The flow of the steam is regulated by a pressure-reducing valve, controlled by the pressure in the tank. The distribution of steam for heating purposes throughout the building is made from the main receiving tank. The water of condensation from the entire apparatus is returned to the bottom of this main receiving tank, from which it escapes continuously through the medium of a "Fitch" chronometer balanced valve, operated by a lever and ball float, so adjusted as to maintain a nearly constant water line in the tank. This valve is suitably by-passed for use in case of emergency. The escaping water is led to the hot-water discharge tank through a valved 4-inch pipe. Arrangements are so made that the water is discharged from the tank at a point midway between the bottom of the tank and the surface of the water, when discharging either through the balance valve or the 4-inch by-pass outlet.

The receiving tank is placed as near the flow as possible in order to give it the utmost hydraulic head in connection with the heating apparatus. The water of condensation from the entire heating apparatus is thus returned to the hot-water discharge tank, and from there returned to the boilers by two Smith-Vaile direct-acting, outside-packed plunger pumps, 7"x5"x10". The boilers carry a minimum steam pressure of 110 pounds, while that of the heating apparatus does not exceed two pounds, gauge pressure.

A catch-basin 24 inches in diameter by 48 inches deep, with the top set flush with the floor, is provided and has a 4-inch trapped overflow to the sewer. This catch-basin receives the blow-off from the boilers and the various drip and drain pipes from the several pumps, engines, tanks, etc. It has a 3-inch vapor pipe discharging 4 feet above the roof. The main receiving tank sets on brick piers with stone caps. It is 4 feet in diameter and 12 feet long and is made of C. H. No. 1 wrought iron, the shell one-fourth inch thick, the heads dished and three-eighths inch in thickness. The tank was tested to a hydraulic pressure of 100 pounds. This tank is provided with a glass water gauge, a low-pressure steam gauge, and has in the center of one of the heads a No. 2 Eclipse manhole and fittings. Attached to the inside of this tank is a copper ball float with a lever made of ¾-inch steam pipe, having a small hole drilled into it near the float. The other end is fastened to a piece of brass tube, fitted to a stuffing-box screwed into the side of a tank. The outer end of this tube is supplied with a compression cock, and the whole operates so as to remove the air from the inside of this tank at a point near the water line, when at its extreme positions, or at any point between the high and low-water mark as desired.

Attached to this tank is also a 12-inch back-pressure valve designed to "cushion" before seating so as to prevent noise. All return connections from the heating apparatus, the feed-water purifier, and the hot-water boiler used in connection with the house supply are made to the bottom of the receiving tank, and all discharge connections from traps used in connection with high-pressure steam are made at the top. Each return-pipe connection to this tank at the bottom has a valved outlet, just back of the return valve, with a connection to the catch-basin, so arranged that each division or appliance may be drained without interfering with the working of the others; suitable connections are also made for draining this tank into the catch-basin.

The hot-water discharge tank is 36 inches in diameter by 6 feet long, and of similar design and construction to the main receiving tank. This tank has one 4-inch and one 6-inch threaded flange, riveted on to the bottom, and a 1½-inch threaded opening, re-enforced at the top. The 6-inch opening is commanded by an inside arrangement which is secured from syphoning, and draws off the water from a point half-way between its surface and the bottom.

Within this tank is a copper ball float, the stem of which works through a stuffing-box, and is connected by chains to a "Fitch" chronometer valve on the steam supply to the boiler feed pumps, for regulating the steam supply to same, according as the level of the water rises or lowers in the tank. This tank is also fitted with a glass water gauge and suitable drainage connections to the catch-basin. This tank is set at such an elevation that its low-water line is 6 inches above the upper set of pump valves, and at the same elevation as the water line of the main receiving tank. This tank lies horizontally and has its normal water line established at a point about two-fifths of the diameter up from the bottom.

The feed-water heater and purifier receives steam from the 12-inch exhaust header and discharges it into the main receiving tank. It draws feed water from the surge tank and delivers it to the main receiving tank and is fitted with an automatic air valve and a compression air cock for exhausting the air from the interior. A hot-water boiler 4 feet in diameter and 10 feet long, of similar design and construction to the main receiving tank is provided and contains 36 lineal feet of 3½-inch brass pipe in two coils fitted up with both live and exhaust steam connections so that either series may be used to heat the water in the tank for the house supply. This tank is set at such an elevation that the return outlet of the coils is 36 inches above the high-water line of the main receiving tank. There are for boiler feed service two boiler feed pumps of the size previously stated and two duplex pumps of the same make, size 10"x6"x12", for house tank service. The boiler feed pumps are controlled by automatic regulating valves operated by float attachments and the house pumps are controlled by Fisher gravity governors, both automatically regulating the supply of steam so that they will begin and stop pumping when the water level in the tanks reaches the desired limits. The suction and discharge headers of the tank pumps are so cross-connected with the suction and discharge headers of the elevator pumps that either one or both may be called into service when extra service is required on elevators. There is a tank in the basement, 5½ feet wide, 8 feet long and 6 feet high, to which the condensed water from the exhaust condensing head on top of main exhaust pipe is drained. This water is used entirely in the elevator system, and when there is more than is required for this purpose, as is very frequently the case, it is used in the boilers also. A 4½"x2¾"x5" "Hooker" steam pump, omitted in the drawing, Fig. 1, to avoid confusion, is used for this purpose. There is also one 6"x6"x8" "Hooker" single-acting air pump for forc-

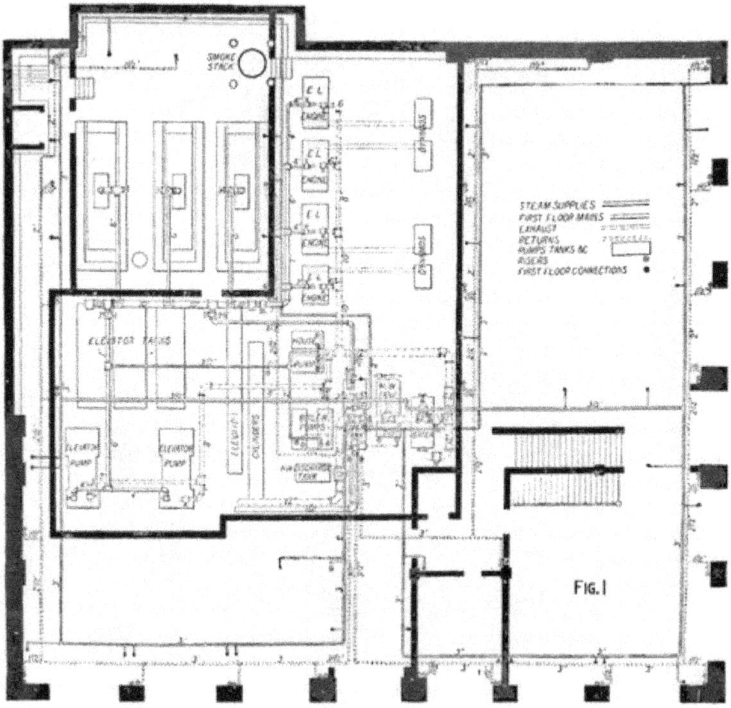

FIG. I

HEATING IN THE WAINWRIGHT BUILDING, ST. LOUIS, MO.

ing air into the compression tanks for the elevator service.

In addition to the four passenger elevators, there is an ash hoist, from the boiler-room to alley, with a capacity of 2,500 pounds, operated from the same tanks as the other elevators.

The elevator service is operated by two "Blake" compound duplex pumps 18½ inches and 20'x16'x16'. A marble gauge board in engine-room contains seven "Marsh Tripod" gauges, indicating the pressures in the different steam and water services. The house supply pumps are connected so that they may be used in connection with one of the elevators, when additional pressure is required for lifting unusual weights, as safes, etc.

Figure 1 is a general plan showing the arrangement of pumps, boilers, and other apparatus, the principal valves and connections and the sizes and location of pipe mains in the basement. The basement is spacious and well lighted and ventilated, affording exceptionally ample and convenient space for the apparatus, which is well set and is notable for its neat and attractive appearance.

PART II.—OPERATION AND PROPORTIONING OF RADIATOR PIPES, CONNECTIONS OF RADIATORS AND RISER LINES, AND PLAN OF DISTRIBUTION MAINS.

The main steam supply pipe for radiators on the first floor and basement is 4 inches, starting from the valved outlet on the main receiving tank, extending to the outer wall of the building, as shown on the plans. It is there divided into two 3½-inch branches, extending around the outer walls of the building to a point of intersection. This pipe is pitched to drain in the direction of the steam currents, and is supplied with all the necessary drain and drip pipes, connected with the return pipe. The main supply pipe for the rest of the building is 10 inches, starting from the valved outlet on the main receiving tank. It leads to the pipe shaft, up which it is extended to the attic space. It there divides into two branches, each one running along the center line of the building to the extremes of each L. From these branches arms are run over to the outer walls and extend down through the several stories to the basement, where they are connected into the main return pipe of this system leading to and connected with the receiving

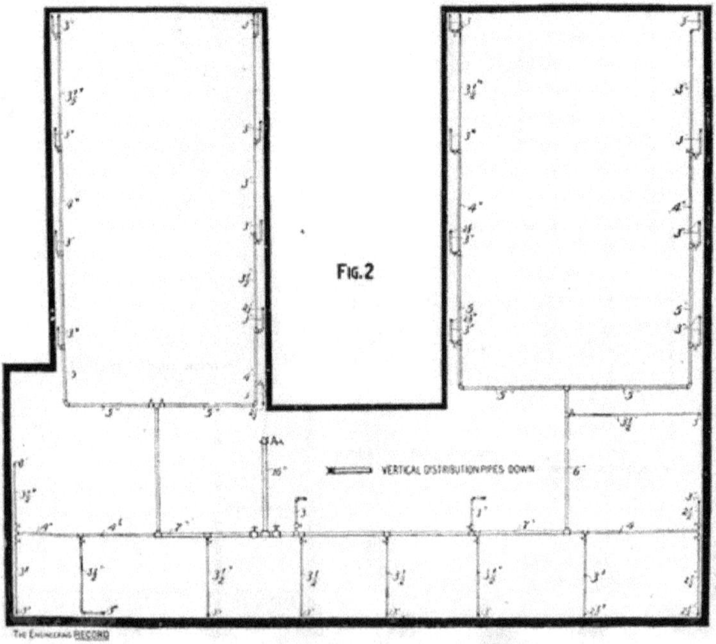

The Engineering Record

HEATING IN THE WAINWRIGHT BUILDING, ST. LOUIS, MO.

tank. The descending lines are of sizes according to the following requirements: Descending lines supplying 225 feet or less to start, 1½ inches; supplying 235 to 320 feet to start, 2 inches; supplying 320 to 480 feet to start, 2½ inches; supplying 480 to 800 feet to start, 3 inches; supplying 800 to 1,200 feet to start, 3½ inches. The horizontal arms in the attic connecting the descending lines to the main supply pipe are made of pipe one size larger than the top section

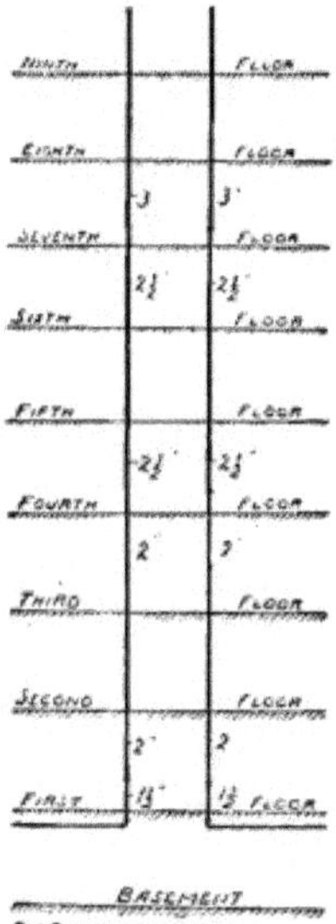

THE ENGINEERING RECORD.

FIG. 3.

of the descending line. In all horizontal steam pipes, where they are reduced in size, they are provided with eccentric fittings to avoid pockets.

Each of the descending lines is connected at a point near the ceiling and below the shut-off valve, by a ¾-inch connection to a 1½-inch air pipe extending around the basement near the ceiling and continued to the tank-room, where it is connected with the main return pipe, back of its cut-out valve. An automatic air valve and a common compression air valve are attached to this air pipe for drawing off the air in the system. The connections to each descending line are made so that none of the condensed water in them can enter this air pipe. All the radiators are connected to the mains and the descending lines by a single feed connection, and where this connection is more than 18 inches long it is made of pipe one size larger than that called for by the opening in the radiator, and this has a heavy pitch toward the riser. The total heating surface in the building is 18,000 square feet, this quantity being based upon a unit having an efficiency equal to that of a two-row vertical tube radiator. All radiators are tapped right hand for single-feed connections of the following sizes: Radiators containing 20 feet or less, tapped 1 inch; containing 20 to 40 feet, 1¼ inches; containing 40 to 60 feet, 1½ inches; containing 60 to 120 feet, 2 inches. The center of the feed opening is always 6½ inches above the floor.

All steam traps are automatic, continuous discharge float traps, by-passed and drained to the catch-basin. These traps are designed of a size to discharge double the quantity of water received. All steam valves are of globe or angle pattern and all water valves are gate pattern. All valves are fitted with soft removable disks. All radiator valves have finished trimmings and rough body wooden wheel and composition disk, and have a lift so that the disk rises entirely out of the steam current. Each radiator has an automatic air valve provided with a float so that in the event of the radiators becoming filled with water through a wrong manipulation of the supply valve the operation of the air valve will stop until the error has been corrected.

Wherever any of the pipes are supported from an iron girder, pillar, column or pier, incombustible insulation is interposed between the pipe and hanger, so as to avoid, to the greatest possible extent, the transmission of sound from the pipes to the structure. All high-pressure pipes were tested to 135 pounds steam pressure, and all low-pressure pipes to 10 pounds steam pressure. During the cold winter that has elapsed since the installation of this work the radiators have heated all the rooms to a temperature of 70° Fahr., with a pressure in the receiving tank not exceeding 60 inches hydraulic head.

Figure 2 is a plan of the attic showing the arrangement near the ceiling of the steam distribution pipes that are supplied from main riser A and branched to feed the vertical pipes going down around the outer walls. A riser which is located near the middle of the end of each wing and is connected to the attic distributing mains by a 3-inch branch is accidentally omitted here.

Figure 3 is a diagram of a vertical elevation of a pair of vertical supply pipes. The connected radiators on every floor are not shown.

PART III.—LOCATION OF RISERS AND RADIATORS ON OFFICE FLOORS, DETAILS OF CONNECTIONS AND AUTOMATIC PUMP GOVERNOR.

FIGURE 4 is a typical plan showing arrangement of offices and location of risers and radiators on one of the office floors.

Figure 5 is a sketch showing connection of riser to distribution main in the attic.

Figure 6 shows the connection of a radiator to the vertical riser by means of a branch crossing behind it and provided with ells to allow the rise and fall due to temperature variations.

A Fisher steam pump governor performs the dual service of controlling the water supply to the elevated

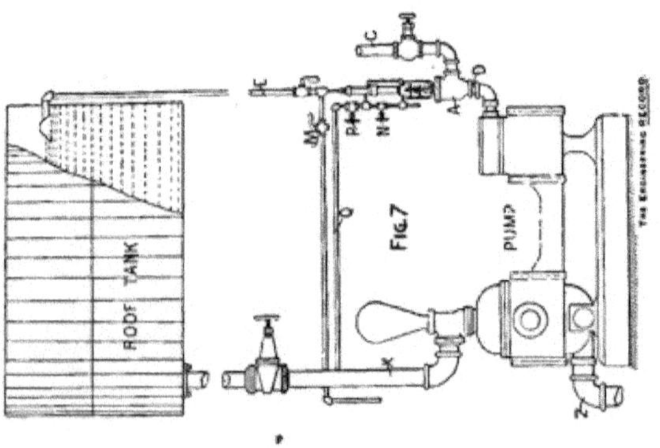

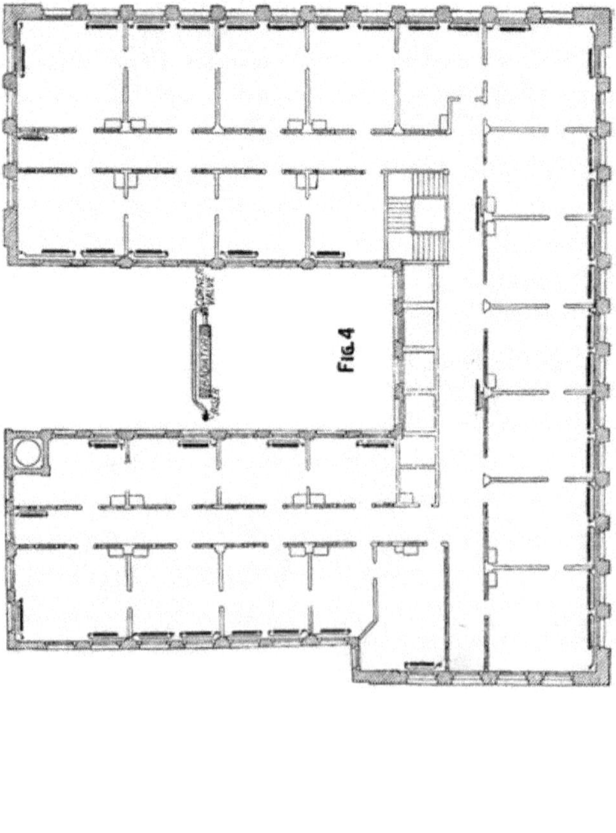

HEATING IN THE WAINWRIGHT BUILDING, ST. LOUIS, MO.

tank system and to the high or fire service system from the same pump, so that either may act independently of the other. Figure 7 shows the connections to pump and tank. Figure 8 is a vertical section showing the construction and position of the governor. The casing A, with its double-disk unbalanced valve B, is set between the cylinder chest and throttle valve. Steam enters at C, lifts valve B, because the upper disk is larger than the lower one, filling casing or shell A, and passing out at D into the engine cylinders and operating the pump. When the elevated tank has been filled to the desired level, at which point the ¾-inch overflow or down pipe E is connected, the water quickly fills this pipe, and the weight of this column of water on the top of the piston F in the upper cylinder G presses it down. The piston-rod H is an extension of the stem of valve B. All move together, thus gradually closing valve B, shutting off the steam, slowing up, and finally stopping the pump. The column of water

FIG. 5

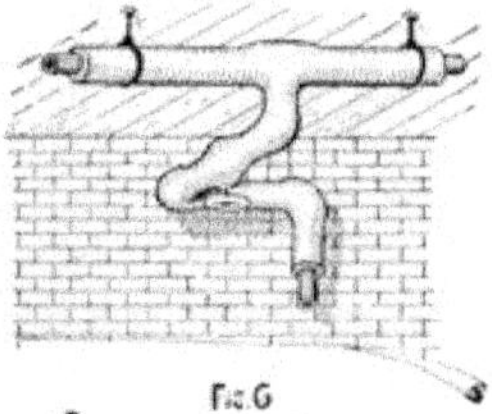

FIG. 6

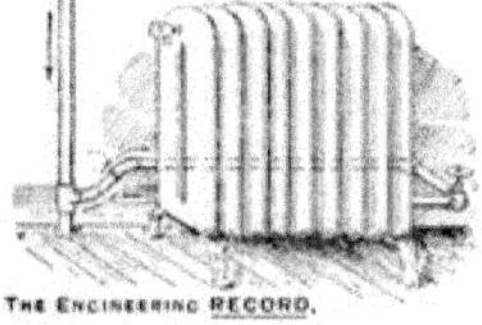

THE ENGINEERING RECORD.

in overflow pipe E runs out through stop-cock M—which is always partly open—to lower tank or sewer, releasing the pressure from the top of piston F, when the steam, aided by spring U, raises valve B and starts the pump again, thus automatically starting and stopping the steam pump as may be necessary

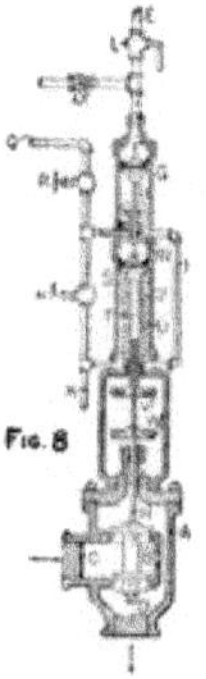

FIG. 8

to keep the water in the elevated open tank at practically the same level all the time. The pressure in the discharge pipe from the pump is practically constant, and this pressure is brought through the ½-inch pipe Q and valve P on to top of piston R in lower cylinder and controls the ordinary running of the pump. Wheel W in yoke is fixed immovably to the valve stem. Turning it to the right the stem is screwed up into the bottom of the piston-rod T, compressing spring U and raising valve B and admitting steam, and when the steam pump is moving at speed desired the upper wheel in yoke is turned to the right until screwed up against the bottom end of piston-rod locking the parts in place. The whole action of the device is automatic when once set and regulated.

The entire basement, including boiler and engine room, the toilet-rooms on every floor and barber shop in attic, are ventilated by the exhaust system, two 90-inch "Buffalo" steel-cased blowers being used for this purpose, and operated by one 25 horse-power "Eddy" electric motor, placed in the attic, and discharging the foul air through the roof.

MISCELLANEOUS HEATING INSTALLATIONS.

UTILIZATION OF LOW-PRESSURE STEAM FOR HEATING AND ELEVATOR SERVICE.

The San José apartment house at the corner of Madison Avenue and Ninetieth Street, New York City, is heated by steam at 10 pounds pressure, and the elevator pumps are run by steam of the same pressure. The steam for all purposes is generated in two boilers made by the Bigelow Company, of New Haven, Conn. Each boiler is 13 feet long and 36 inches in diameter, with 34 3½-inch tubes, and is provided with a steam dome from which steam is taken. The steam pipe from each boiler runs into a tee from which it branches, one line going to the heating system and the other going to the pumps, hot-water tank, etc., which require steam the entire year.

Figure 1 is a general basement plan showing the arrangement of the apparatus and the size and location of steam pipes and risers, etc. The supply pipes for the heating system are shown by full black lines, their corresponding returns by broken lines, and the supply to pumps, tank etc., by a line broken with one dot. Vertical riser lines are indicated by open circles. Figures 2 and 3 are side and front elevations showing the arrangement and general connections of the elevator pumps.

The elevator, which was put in by the Whittier Elevator Company, is run by a vertical hydraulic machine. One hundred and forty pounds pressure is carried in the pressure tank. In order to force the water into the pressure tank under such a pressure with only 10 pounds pressure in the steam end of the

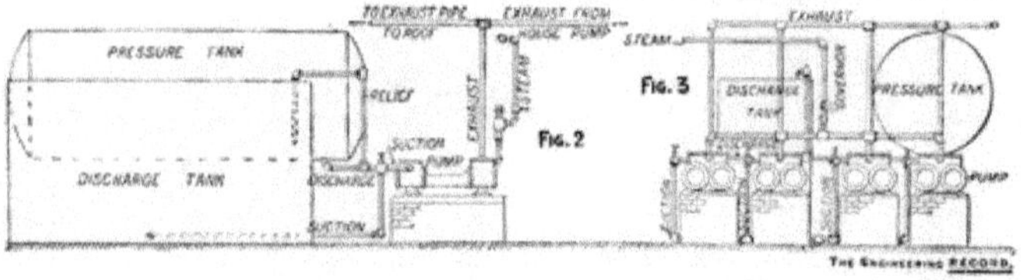

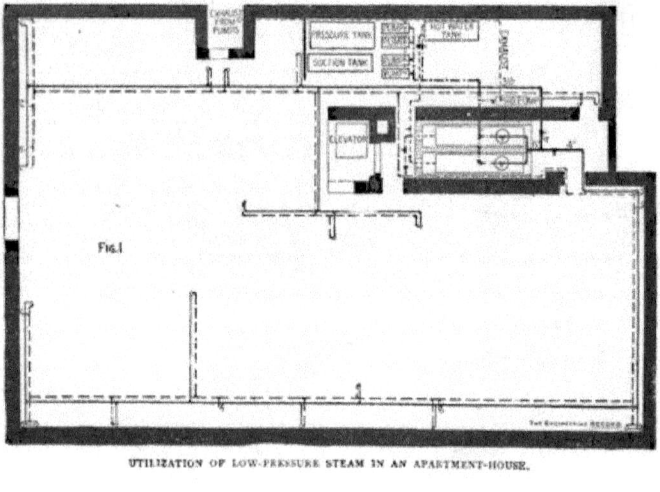

UTILIZATION OF LOW-PRESSURE STEAM IN AN APARTMENT-HOUSE.

pump, a comparatively small piston was used in the water end. There are four Snow pumps, each 6"x1¾"x6", provided, but three pumps only are necessary to do the work, the extra one being for use in case repairs are needed. It is claimed by the contractors that the four pumps of that size were cheaper than two of larger size with a total capacity equal to that of the four pumps. The pumps take steam directly from the boilers and the exhaust is through a vertical pipe to a point above the roof.

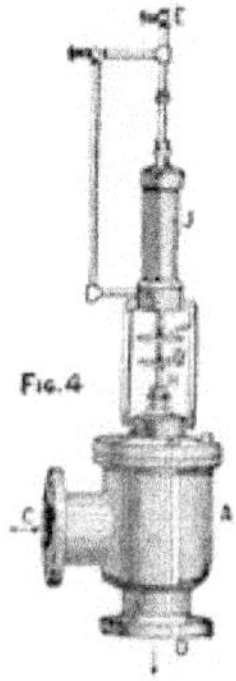

Fig. 4

The pumps are automatically regulated by a Fisher patent governor, Fig. 4. When there is not the required water pressure in the tank, the steam supply entering at C passes freely through valve A and is delivered at D to the pump. When, however, the water attains a sufficient pressure in the tank it is exerted through pipe E, which is connected to the receiving tank, and forcing down a piston in cylinder J, causes its rod H to close the valve in A and stop the pump until the tank pressure dimishes sufficiently to enable the counterspring in cylinder J to raise its piston and open the valve in A. W W are adjustment wheels to regulate the countersprings in J.

There are about 1,948 square feet of radiating surface in the building, the radiators all being on the one-pipe system. The return water is led back to the boilers and enters through the blow-off connection. The boilers are fed by a pipe supplying water at city pressure, its flow being controlled by a float of standard pattern. The boilers require but little feeding, as all the steam evaporated is returned to them as water in the return pipes with the exception of the little required for the elevator pumps. An automatic damper regulator controls the draft in the chimney and consequently maintains a constant steam pressure in the boiler. The hot-water tank shown in the drawing is heated by a coil of pipes supplied with steam from the boilers. The house pump, which has to raise water to a tank on the roof through a distance of 75 feet, is also of the Snow make and is 6"x2"x6" in size.

The plant was designed by Mr. J. T. Mulhern, of Mulhern, Piatti & Kirk, of New York. They were also the contractors for the plant.

HEATING OF A MINNEAPOLIS STORE.

The large dry-goods store of S. F. Olsen & Co. in Minneapolis, Minn., is housed in a building covering a lot 196x158 feet in size, and containing three floors beside the basement. The building is warmed by steam, partly by direct and partly by indirect radiation. The accompanying cuts, which were made from sketches, show the heating arrangements. The piping is laid out on the one-pipe principle. All pipes are pitched down toward the boiler, and the direction of flow of the condensation water is indicated throughout by broken arrows and that of the steam by full arrows.

The plan shown is that of the basement, and for convenience of representation the power plant, which is in reality located in a sub-basement, is shown on that floor. The plant contains four return-tubular boilers, each 16 feet long by 4 feet in diameter, beside the necessary pumps, return tank, etc., that go to make up a plant of that class. The upper floors of the building are heated by direct radiation, the ground floor by a combination of direct and indirect systems, and the basement by direct radiation from the steam mains which are carried around its ceiling. These mains are shown in the plan, and it will be noticed that they are carried around each side of the building until they meet at a point opposite the boilers. At D these pipes drop, and are carried under the floor from that point to

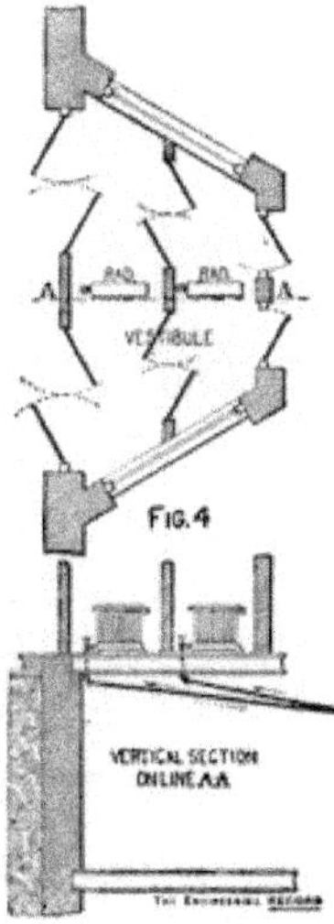

Fig. 4

the return tank. Especial pains was taken to introduce the bends at E to allow swing for the temperature expansions and contractions in the long mains.

The ground floor of the store is not provided with any ventilation, as it was supposed a considerable quantity of air would enter through the constantly opening and closing doors leading to the street. The main marked A on the basement plan, Fig. 1, supplies 15 indirect stacks located about as shown. These stacks are shown more in detail by Fig. 2, which shows them to consist of two clusters of indirect stacks suspended from the ceiling and inclosed by galvanized iron. Figure 3 shows the pipe connections. The air for these stacks is taken from the store that is to be warmed, passes down through register face A and around a baffle-plate B, so as to come up between the coil sections. The air after being thus warmed passes up into the store through a second register C. If the heating thus obtained proves to be insufficient, steam can be turned into eight large wall coils and about 20 direct radiators of 100 square feet of surface each that are placed about the walls and windows of the store. The steam for these radiators is supplied by the main B, and is at

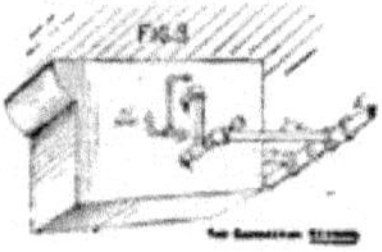

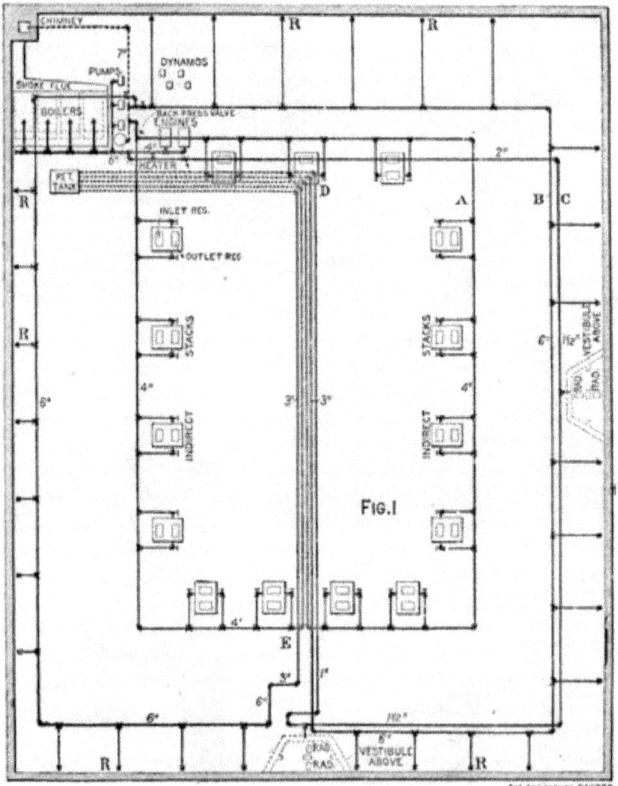

THE ENGINEERING RECORD

HEATING OF A MINNEAPOLIS STORE.

about five pounds pressure, as is the steam for the indirect stacks. The main B also supplies about 3,500 square feet of radiation on the second and third floors.

As before stated, the doors of the store are opening and closing constantly and this might cause discomfort to those near the doors, and especially so in a locality of so severe a climate as that of Minneapolis. To prevent cold drafts double doors were provided in the vestibule beside the main doors to the store, and a radiator containing 120 square feet of surface was placed between each pair of doors. A plan of this is shown by Fig. 4, as well as a sectional elevation showing the pipe connections to the main. As the steam for these radiators is supplied by a main that is independent of the others, it can be supplied by steam at any pressure below the pressure carried in the boilers.

The proprietor of the building states that the plant proved satisfactory in every way during a severe winter. Mr. W. H. Dennis was the architect of the building and Messrs. Archambo & Morse, of Minneapolis, the heating contractors, and to the latter we are indebted for data from which this description was made.

HEATING OF THE HORTICULTURAL BUILDING, WORLD'S COLUMBIAN EXPOSITION.

The Horticultural Building at the World's Columbian Exposition is a conservatory on a large scale, consisting of a central and two end pavilions, connected by front and rear curtains as is shown by the accompanying drawing. The building is about 1,000 feet in length and about 280 feet in width at its widest part. Quite an area, however, is deducted from this by two open courts. The central pavilion is covered by a dome about 190 feet in diameter by 113 feet high, entirely of glass. This central dome and the two curtains, which are finely shaded on the drawing, are the only parts of the building that are heated by the system about to be described.

Each of the curtains is about 273 feet in length by 73 feet in width. The total floor space covered by the central pavilion and the front curtains is about 71,000 square feet. The roof and sides of the curtains are of glass.

The building is warmed by the hot-blast system. The boilers which supply the heating coils with steam are located in a separate building about 400 feet away. The boiler plant consisted of three water-tube boilers, each of 200 horse-power, made by Wicks Brothers, of Saginaw, Mich. These boilers, beside furnishing the heating coils and fan engines with steam, also furnished steam to heat the Service Building and a number of adjacent greenhouses. The fans and about 36,000 square feet of heating surface are located beneath the central dome of the Horticultural Building. The fans and heating plant were entirely concealed by a pyramid of palms and semi-tropical plants which reached nearly to the top of the dome. The fans are all inclosed in a fan chamber and they discharge the heated air through ducts which cross from the center of the dome each way to the curtains. On reaching the nearest end of the curtain they rise, following the line of the side wall.

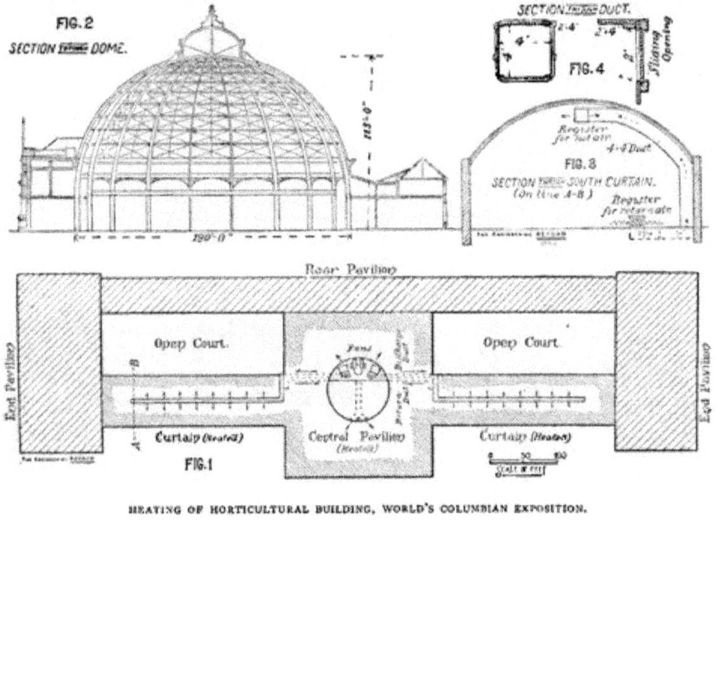

HEATING OF HORTICULTURAL BUILDING, WORLD'S COLUMBIAN EXPOSITION.

to the center and top of the building, and then run to the ends of the curtain, decreasing in size as they near the end. The ducts are pierced with openings for registers which allow the hot air to be discharged against the glass roof. Figure 4 shows the cross-section of the duct and a detail of the sliding opening. The air supply for the fans comes from a register in the curtains placed near the floor, and at the end that is nearest the fan (Fig. 3). A duct connects this register with the fan chamber. The air is then passed through the fan chamber and discharged as before, thus making a continuous movement of the same air.

In addition to these two ducts for heating the curtains, the dome is heated by the same system, ducts from the central fan leading to four points in different parts of the pavilion, each discharging at the side wall of the fan chamber as shown by the arrows in Fig. 1. The fans were capable of delivering 329,000 cubic feet of air per minute.

The apparatus was in use during the winter of 1892–93, while the building was filled with the most delicate palms, and it is said the plant gave excellent satisfaction. The plant was designed and installed by the Samuel I. Pope Company, of Chicago, Ill., by whom it will be removed in the spring of 1894.

HEATING A FLORIST'S DELIVERY VAN.

Although the heating of cars by hot water has been practiced for a long time we believe the application here described is somewhat of a novelty in the development of hot-water heating. This is a warming apparatus designed by Mr. John A. Scollay, of Brooklyn, for a delivery van for James Weir's Sons, florists of the same city. This firm conducts a store on Fulton Street, Brooklyn, which is supplied by its extensive greenhouses at Bay Ridge, a distance of some 5 miles from the store. Not only in making this long trip, but in delivering the goods, the flowers would perish from frost if in an ordinary delivery wagon during excessive cold weather, and it was for the protection of the flowers that the van was designed.

Figure 1, which is a rough sketch only, shows the van to be entirely inclosed on the top, sides, and front. Two large doors, which, however, are not shown, serve to cover the back. Underneath the wagon, and suspended from it by straps, is a specially designed hot-water heater, and from this a 1-inch flow pipe extends up into the van and around the inside, gradually rising until it terminates in a header. The header is carried up for some distance, and the top is blanked off. A petcock is inserted in the pipe for relieving the apparatus from air when it is being filled. It is filled up to the level of the cock, which is then closed, and the water on expanding compresses the air in the expansion pipe above it. Two 1-inch returns are carried back from the header as shown dropping as they go toward the boiler so that there will be no chance for the lodgment of air in them. The return pipe drops under the boiler, and an upward branch from a tee in it is connected into the boiler. It must frequently happen that a wagon of this kind is left out-of-doors for a considerable length of time when it is not in use and when the fire is allowed to go out. This of course in cold weather would mean the freezing of water in the pipes and their consequent rupture. To prevent this water is drawn off through the plug cock B, which has a screw thread on the outer end enabling a hose to be attached to it and the apparatus filled with water.

Figure 2 shows a longitudinal section of a specially designed boiler. It consists of two cast-iron boxes, both open at one end and provided with flanges as shown. The larger box is about 15 inches long, 11 inches high, and 9 inches in width. The smaller casting is about 1 inch smaller in each dimension and the space between the two shells contains the water to be heated. Both are cast with a slight taper so as to allow the pattern to be readily withdrawn from the mold. To prevent radiation of heat from the boiler the

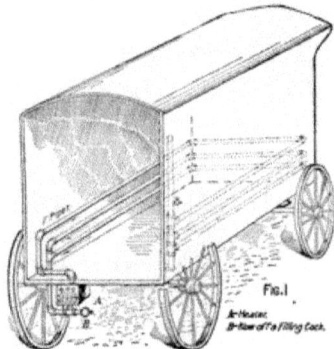

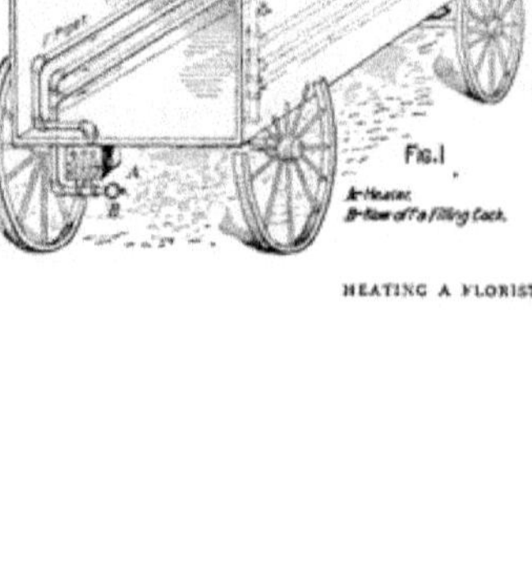

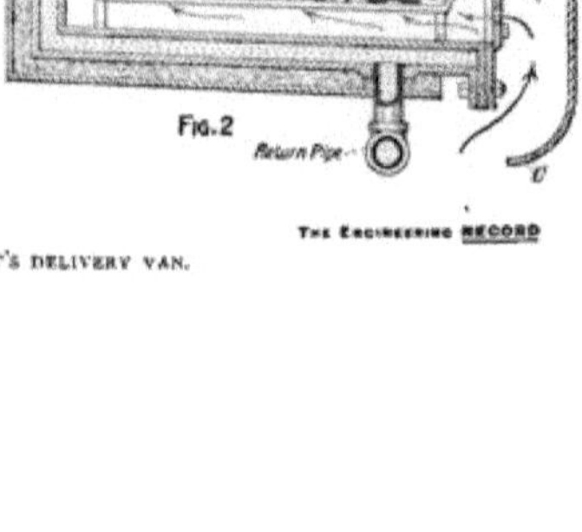

HEATING A FLORIST'S DELIVERY VAN.

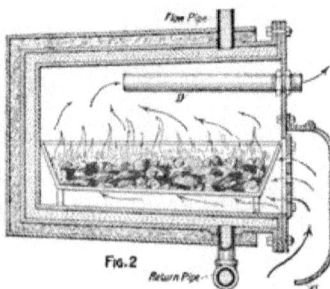

whole is covered with 1 inch of mineral wool incased in a wooden jacket. A cast-iron front is bolted onto the front end of the boiler and this is pierced for three 1-inch tubes and a firedoor. The back of the boiler is turned toward the front of the wagon, and to obtain a sufficient draft the deflector C is provided so as to cause a current of air to flow through the dampers under the fire. The lower part of the deflector turns upon a hinge so that it may be swung out of the way if desired. A charcoal fire is made in a basket made of wire netting and the basket is put in the boiler. To insure a more uniform distribution of the air as it flows through the bottom of the basket to the fire the pipes D were introduced, each one being fastened to the front by lock nuts. They are of such length as to extend well back over the fire so as to prevent the air from taking a short cut on entering the furnace and only going through that part of the fire that is nearest the furnace door. Ordinarily the charcoal fire will last about an hour without having to be replenished. Starting with cold water it takes about a half an hour in extreme weather to heat the van up to such a temperature as will permit the flowers to be carried safely.

A STEAM PIPE CONDUIT.

As the buildings of the University of Minnesota are somewhat scattered it is necessary to carry steam for a considerable distance to warm some of them. The accompanying sketches show the details of the steam pipe conduit that connects the boiler-house with the chemical and physical laboratories and with the building known as Pillsbury Hall.

Figure 1 is a plan of the buildings and conduit, and Fig. 2 shows a sectional view of the same. The conduit rests upon 23 sewer brick laid on edge, and on this is placed a layer of fireproof tiling, each tile be

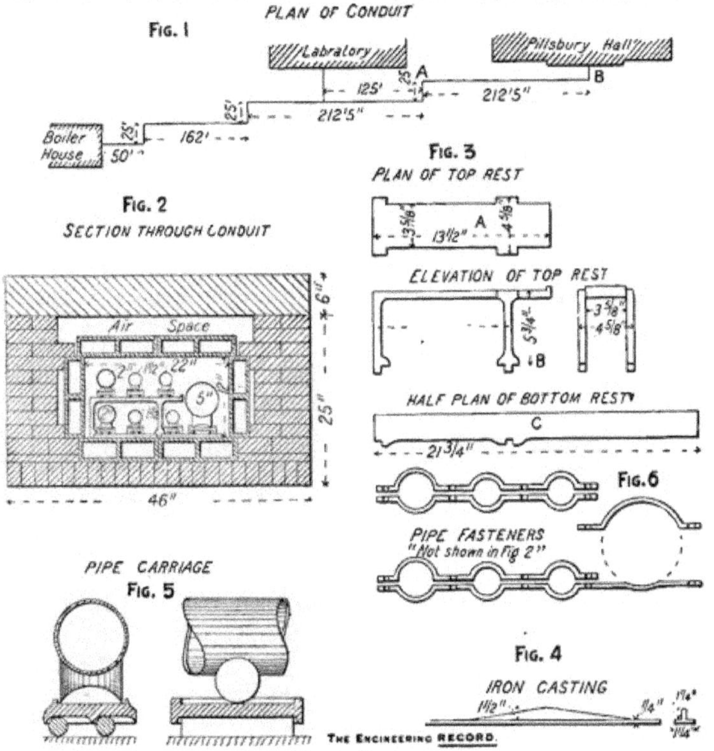

A STEAM PIPE CONDUIT.

ing 3'x12'x12". A single tile placed on edge forms each side of the conduit, while the top is also made of tiles, supported by the casting shown in Fig. 4, which are put in at intervals of 12 inches. The sides are inclosed by brick laid in cement, leaving the air spaces as shown. A stone over 4 feet 6 inches in width by 6 inches deep completes the conduit, as far as the masonry is concerned.

Three pipes are carried from the boiler-house to the laboratory, while four are carried over to Pillsbury Hall beyond. The former run in a branch conduit to the laboratory. Expansion is provided for by means of long bends. The pipe line rises 9 inches at each bend as it leaves the boiler-house so that the return water will flow back more readily.

Figure 3 shows the specially designed pipe rests which were used in the conduit. The pipes are carried on two levels. The pipe rests are in two castings—the lower one, which is a cast-iron plate, shown in plan by C, and the top rest, as it is called, is shown in plan by A and in elevation by B without the pipes in position. Each pipe is supported on a spool to allow for longitudinal expansion of the pipe, the spool in turn rolling upon a carriage which moves laterally across the conduit on rollers. This is shown in detail by Fig. 5. The pipes are held together by the fasteners shown in Fig. 6.

As the laboratory is not yet finished, Pillsbury Hall is the only building which is heated by pipes in the conduit. The lower row of pipes are the only ones now in use, the 5-inch supplying steam to that building, the two 1½-inch pipes and the 2 inch serving as returns from the different heating systems in the building. When the pipes were put in place the air space around them was filled with mineral wool.

The conduit was built by W. F. Porter & Co., of Minneapolis, and we have obtained from Mr. George C. Andrews the bill of material and labor for laying the conduit, which is as follows:

Pipes and fittings	$745.96
Sewer brick, 67,500 in number	540.00
Mineral wool, 15,080 pounds	488.18
Cement (Milwaukee), 152 barrels	176.80
Fireproofing, 5,100 pieces, 3"x12"x12"	442.02
Stone, 84 perch	212.10
Labor bill for steamfitting and digging ditch	448.07
Sundries	73.50
Patterns	22.85
Additional pipes and fittings	3.25
Castings	71.73
Labor bill, bricklaying, etc	667.00
Sundries	60.00
Total cost	$3,830.38

Prof. J. H. Barr, now of Cornell University, at one time made some experiments on the 210.7-foot length of 5-inch pipe, or from A to B in the plan, and found that there was an expansion of 4.45 inches when steam at 55 pounds pressure was admitted, the temperature being 43.1° Fahr. before, with the pipes exposed to the air, and 300 degrees after steam was admitted in the pipe.

FIRE UNDER A BOILER-ROOM FLOOR.

A RECENT fire in New York City may prove instructive to erecters of heating apparatus as emphasizing the necessity of providing sheet-metal protection for adjoining wood floors of so generous and ample design that it will prevent the possibility of small coals getting under or against the floor when hot ashes are being scraped from the ashpit. In a large store on Broadway, New York, a strong smell of smoke gave warning of a fire. After fruitless search by those in charge of the premises, the firemen were called in, but their thorough inspection, aided by cutting through where the fire might be suspected, revealed no source. As the smoke and odor remained the firemen established a watch on the premises. The evidences of fire increased in volume and pungency, until 24 hours after they were first noted the cause was located beneath a wooden floor in the basement. When the heating boiler A was set the flooring with its wooden joists was removed; a space 2 feet on all sides of the location of the heater was filled with sand C, and bricks set on edge B, upon which the heater was erected. The sheet-iron covering E was nailed on the floor F for several feet outside the brickwork. Later, another flooring E

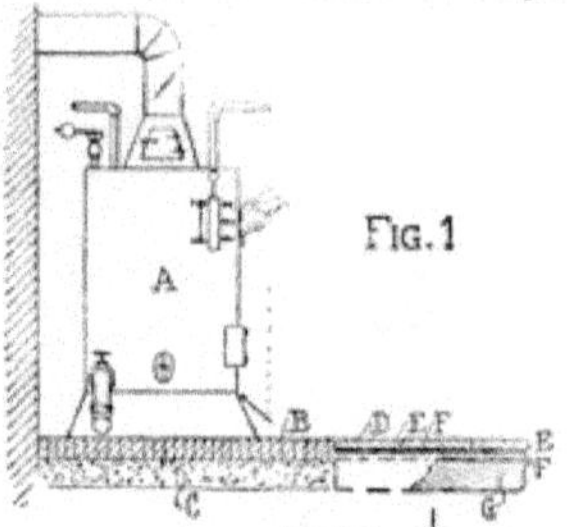

FIG. 1

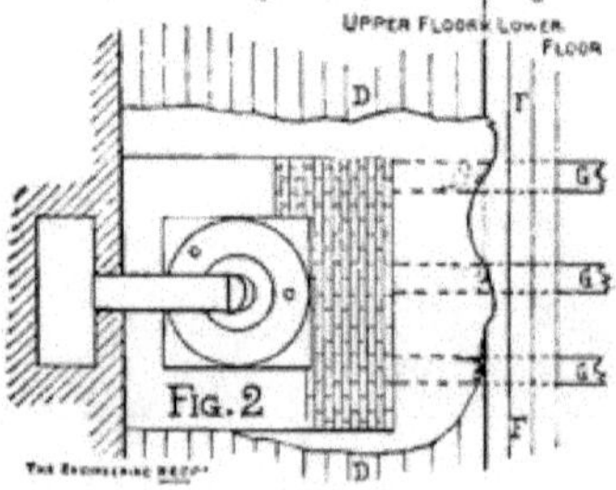

FIG. 2

was laid over the old one, covering the sheet iron. From some cause the floor boards F and the joists G took fire. As there was no draft under the sheet iron and the brickwork in front helped to exclude the air, a process of charring was set up, which, in the 24 hours that elapsed before its discovery, had eaten its way around as shown in the plan, Fig. 2. It had not burst into flame at any point until the firemen cut through the sheet iron, although there is little doubt that it would have burned upward as soon as it passed the confines of the iron sheeting. The sheet metal should obviously have been carried down in front of the flooring far enough to remove any chance of the communication of fire.

STEAM HEATING—NOTES AND QUERIES.

BOILER PROPORTIONS.

HORSE-POWER OF HEATING BOILER

E. H. Adams, Fitchburg, Mass., writes:

"Will you inform me of the number of horse-power required to heat 4,000 square feet of direct radiation?"

[One square foot of direct radiation will condense from one-quarter to one-half a pound of steam per hour, as the pressure varies from three to four pounds to 50 or 60 pounds. Assuming you are to heat with low-pressure steam you will condense one-quarter of a pound per square foot of surface; your 4,000 square feet would then condense about 1,000 pounds per hour. By horse-power we suppose you mean 30 pounds of water evaporated per hour into steam in the neighborhood of 70 pounds pressure. Neglecting in the calculation the slight differences of pressure, you would then require 1,000 ÷ 30 = 33 horse-power in boiler capacity.]

PROPORTIONS OF BOILER AND RADIATING SURFACE.

B. G. Carpenter & Co., Wilkesbarre, Pa., write:

"Supposing we have a building to heat requiring 2,500 square feet of radiation. To heat same with low-pressure boiler we generally rate 1 foot boiler surface to 5 or 6 square feet of radiation. What size boiler would be required if we wished to use high-pressure boiler using reducing valve; also if water is put back into boiler by use of pump or trap. Also what size if we wish to run a 10 horse-power engine from same boiler, and again if exhaust from engine were used as far as it would go towards heating building, condensation being pumped back into boiler?

"Supposing we have a schoolroom of 10,000 cubic feet seating 50 scholars; we wish to heat the room by indirect radiation giving each scholar 30 cubic feet of fresh air per minute. What amount of radiation would be necessary to do the work?

"Can you suggest any good book from which we can make calculations similar to above?"

[The boiler of an apparatus is proportional to the work done, and as a low-pressure radiator of, say two to five pounds pressure, will condense no more steam whether the pressure is reduced by a valve or otherwise, the boiler will remain the same, provided the water is returned by any means and not lost.

If you give each scholar 30 cubic feet of fresh air per minute it is equal to 1,800 cubic feet per hour, and for 50 scholars it will require 90,000 cubic feet per hour for the room. This air may be at zero or 10 degrees below outside, and may have to enter the room at from 80° to 90° Fahr., in zero weather, so that the maximum increase of temperature of the air will be about 100 degrees. Then 90,000 cubic feet $\times 100^\circ = 9,000,000$, or the total number of cubic feet of air warmed 1° Fahr. This divided by 50 gives the heat units required to warm so much dry air (or 180,000 heat units), and this again divided by 1,000 gives the number of pounds weight of steam that must be condensed in an hour in the radiators or coils, or 180 pounds of steam condensed to water. A good indirect radiator with ample flues, etc., will condense one-half pound of water per hour, so that about 360 square feet of radiator with natural draft in the flues will be required for such a room. With a fan, about half the amount, or 180 square feet, will do. Although this particular question is not discussed the rules applying are given in Baldwin's "Hot-Water Heating and Fitting."]

FIGURING THE CAPACITY OF STEAM-HEATING BOILERS.

J. F. Kelly, Butte City, Mont., writes:

"Will you, through the columns devoted to notes and queries in your valuable paper, show the usual method for figuring the capacity of horizontal tubular boilers for heating? The boilers are each 42 inches in diameter by 12 feet long, and contains 25 3½-inch flues. What would be the nominal horse-power of such a boiler, and what the amount of direct radiating surface such a boiler would supply? Where two such boilers are designed to be used together for a single system of steam heating, is it necessary that they should be cross-connected with steam and water-equalizing pipes in addition to the steam and return headers? An uninterested engineer has suggested an arrangement for the same."

[The term horse-power is very misleading, especially in connection with steam-heating boilers, and in power boilers it should be considered only after deciding what pressure is to be carried, the service and action of the engine, the fuel and water, etc. In common practice the superficial surface of all the flues, one-half of the shell and one head, are considered in figuring the heating surface of a boiler. Fifteen square feet of boiler surface is considered a boiler-maker's horse power, and 30 pounds of water evaporated per hour per nominal horse-power of boiler the Centennial standard. The proper method of finding the amount of surface a boiler will make steam for is that the average horse-power (15 square feet of surface) will evaporate without trouble in average practice 30 pounds of water per hour. The average radiator will condense from one-quarter to four-tenths of a pound of water per hour, varying with the pressure, the former with very low-pressure steam and the latter with steam at about 50 pounds; so that we have a ratio of about 1 of boiler to 8 of radiator for low-pressure steam. This will fall to 1 to 5 or 6 for steam at high pressure. It may be that there is some local reason why the steam-equalizing pipe you mention should be used, but in ordinary practice we should not endorse it, as the main steam connections, if sufficiently large, are in themselves equalizers. There is an objection to them on the score of expense and the additional complication of so many more valves, the improper use of which might result in serious consequences. We cannot endorse the water-equalizing pipe, as it would nullify the action of your check valves in retaining the return water in each separate boiler, in fact making the two practically one. The return pipes should be large enough to perform the functions of a water equalizer, and having valves on steam and return water mains only you are enabled to use either boiler separately or in battery.]

HOW TO FIND THE BOILER SURFACE WHEN THE RADIATING SURFACE IS KNOWN.

FOREMAN writes:

"Will you kindly give a simple rule through the columns of your valuable journal for finding the heating surface of a boiler when the radiating surface of the building is known?

"What I require is to be able to reason the subject out for myself when I have new conditions.

"I am very well aware that the books say 'about 7 to 10 of radiating surface to 1 of boiler,' but this is taking the matter blindly. I know of one case where there is 15 or 16 of radiating surface to 1 of boiler and good results are obtained."

[The proper way for a beginner in engineering to reason on this subject is to find how much water a given radiator surface will condense, and then consider the amount of boiler surface that will evaporate the same amount of water in the same time.

Usually in these questions an hour is the unit of time that the engineer should familiarize himself with, and when other units are used such as the minute, etc., if he desires to commit things to memory he had better to reduce it to the hour for easy memorization.

The experiments made by the Nason Manufacturing Company, the Walworth Manufacturing Company, William J. Baldwin, and George H. Barrus with vertical steam radiators in this country, and by Tredgold and Hood in England on pipes, all prove that the units of heat given off by a square foot of radiating surface exposed to the action of the inclosed air of a room, is between 1.25 heat units for large diameter horizontal pipes (2½ to 4 inches) and 2.25 heat units per hour for best vertical radiator pipes of ordinary height, for each degree the temperature of the surface of the radiator is warmer than the air of the room.

Nearly all vertical pipe radiators of smooth vertical surfaces have been found to give off as much as 2 heat units per degree difference between pipe and air per hour, but to be safe under all conditions when finding boiler surface 2¼ or 2½ heat units may be taken as the basis of calculation.

Therefore, if we have a radiator of 100 square feet in a room at 70° Fahr. with a temperature of steam corresponding to one pound pressure, say 212 degrees at the outside surface of the pipe, then we have $212° - 70° = 142 \times 100 \square = 14{,}200 \times 2.5$ heat units $= 35{,}500$ total heat units given off by 100 square feet of surface under these average conditions.

Having found the total units of heat divide them by the latent heat of steam for the pressure assumed (one pound above atmosphere), say 962, and you have the pounds of water that must be evaporated into steam for each 100 square feet of radiator, which is

$$\frac{35.500}{962} = 37 \text{ pounds.}$$

Now 30 pounds of water evaporated in a boiler is ordinarily considered a horse-power, and if instead of taking 2.5 heat units as at first, we take an average of 2, we will find that an average horse-power of boiler furnishes steam to 100 square feet of surface. This is well to remember for rough mental calculations.

To consider the matter carefully, however, we have to find the square feet of average boiler surface that will evaporate 37 pounds of water in an hour, and in this we are likely to find a variety of conditions that are not met with in condensing steam. The conditions to condense steam by air currents and radiation are nearly always the same. The conditions to make steam from water depend on the fuel, the draft, the form of the boiler, and other circumstances, and therefore only averages can be considered here.

Ordinarily, however, 1 average square foot of either horizontal or upright fire-tube boilers will evaporate two pounds of water per hour, or if it will not do it it is not well set and has a badly proportioned grate and chimney. Under circumstances more favorable than ordinary, boilers have done double this duty, and three pounds of water is not uncommon. Still we would never advise a designer to count on more than two pounds unless he is sure from actual experience with a similar plant that he is able to get a greater evaporation.

Under such conditions then all we have to do is to divide the pounds of water to be evaporated by 2, and we have the heating surface of the boiler, which is 18.5 for the case we have been assuming for the 100 square feet of surface, and which gives a ratio of 1. square foot of boiler to 5.45 square feet of radiating surface.

It must be remembered, however, we have assumed the greatest possible condensation for radiators in finished rooms, and it may be that we have exceeded it nearly one-fifth. This fifth, however, is a good factor to remain, as in our calculations the condensation in main pipes and in branches has not been considered, and unless carefully considered and treated separately we had better let it remain.

In the question of the boiler, however, we have taken the lowest duty, so that our answer gives us the maximum boiler surface that is ever required with proper boilers.

If the mains and the branches in recesses of walls or under floors are all carefully covered, 2.1 heat units per degree difference per hour is ample to consider, and if you are able to get three pounds of evaporation per square foot of boiler, we will then have 212° — 70° = 142° × 2.1 heat units = 292.2 × 100 □ + 962 = 30.98 pounds of water to the 100 of radiating surface, which latter divided by three pounds of evaporation = 10.33 square feet of boiler surface, or a ratio of 1 of boiler to 9.68.

Thus 1 of boiler to 5.45 of radiating surface for direct radiators forms the maximum for boiler surface, while 1 to 9.67 should form the minimum.]

THE STEAM HEATING OF A PUBLIC BUILDING.

J. K., New Orleans, La., writes:

"There has been some discussion about the heating of a public building, and the decision has been left for your opinion. The cut shows a rough plan of the buildings, which are exposed on all sides. The questions are (1) the size and horse-power of the boiler required to furnish the necessary steam to warm the buildings; (2) the size of main steam pipe

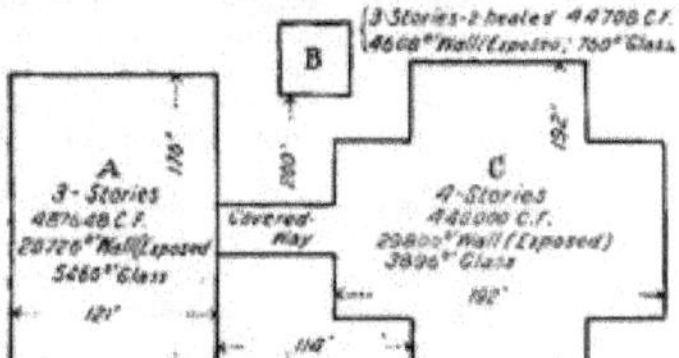

for high-pressure heating; (3) the amount of radiation required at 30° Fahr. outside and 70 degrees inside. I would also like to know if steam traps are used in a well-designed apparatus. The houses stand on level ground and the condensation is pumped back into the boiler. The cubical contents of the houses, their distances apart, and their wall and glass surfaces, are given on the plan."

[The first thing to do will be to calculate the amount of radiating surface required. To find the amount of radiating surface in square feet to balance the radiation from 1 square foot of glass, Baldwin gives the rule: Divide the differences between the temperature at which the room is to be kept and the coldest outside temperature by the difference between the temperature of the steam and that of the room. The room being at 70° Fahr., the coldest outside temperature 30 degrees, the temperature of the steam at 10 pounds pressure would be about 240 degrees; hen

$$\frac{70 - 30}{240 - 70} = 0.235.$$

You will have to add 50 per cent. more to this to balance the effect of the inleakage of air at the windows. This would give you 0.352 square foot of radiation to every square foot of glass in the buildings. On this basis, 1 square foot of glass being taken of the equivalent of 10 square feet of wall surface, the building A would require 2,651 square feet, the building B 429 square feet, and C 2,420 square feet of surface. This will give a total of 5,500 square feet of surface. Allowing that each square foot of surface condenses 0.3 pound of steam per hour you would have to have a boiler of sufficient size to evaporate 5,500 × 0.3 = 1,650 pounds of water per hour. The evaporation of 30 pounds of water being taken as a horse-power, you would then require a 55-horse-power boiler to do this work. A boiler-maker usually allows 15 square feet of heating surface to the horse-power. The 55 horse-power boiler is estimated solely upon the amount of water condensed in the radiators. Condensation in the steam mains and steam used by the pump, if you have one, has not been allowed for. A 65 horse-power boiler would cover this and give you some power in reserve. As to the size of steam pipe carrying the steam from the power-house for, say 10 pounds pressure, we think one 6 inches in diameter would be sufficient.

The question as to whether or not it is good practice to use steam traps depends entirely upon the system you are to install. In a dwelling-house or a building where you would probably have a uniform pressure throughout your heating system, a gravity apparatus is used. This of course does not require a trap. If, however, you have scattered buildings and a variable pressure, the returns would be led to steam traps, the discharge from them being carried to a receiving tank from which the water is returned to the boiler by an automatically governed pump.

If you have high pressure to drive engines, etc., and use the exhaust steam for the heating steam, you would of course have to return your condensation in the heating system to a receiving tank, and pump it from there into your boilers. You could, if your pressures be constant, return directly to this tank, but as a safeguard to prevent water from backing up into the return which is under the least pressure, it is getting to be the common practice to run each return into a steam trap and discharge from that into the receiving tank.]

HEATING BOILER PROPORTIONS.

Dom, Lancaster, Pa., writes:

"I would like to have your opinion as to the best rules for proportioning boiler surface on both steam and hot-water heating apparatus: 1. How many square feet of grate surface should be allowed to a square foot of boiler heating surface? 2. How many square feet, or fractions thereof, of grate surface should be allowed to a square foot of radiating surface? 3. How many square feet of boiler surface should be allowed to 1 square foot of radiating surface in a building? I would like to know the proportions of steam as well as hot water."

[We will try to answer the questions in the order in which they appear, but as the questions all relate to the same subject, and probably will not admit of a direct reply to each, the order may become mixed.

First you ask, "How many square feet of grate surface should be allowed to a square foot of boiler surface?" The reply to this is, that it will vary from 1 of grate to 10 of boiler, to 1 of grate and 40 of boiler, all depending on the style of boiler used. Your second question, "How many square feet of grate surface to allow to a square foot of radiation?" admits of a more satisfactory reply, and is really the key to the whole situation. Chapter X. of Baldwin's book on "Hot-Water Heating and Fitting" shows pretty clearly that the heat given off by hot-water radiators under ordinary conditions of practice is twice heat units per degree difference of temperature per hour, and the heat units per square foot of surface per degree difference of temperature between the water in the coils and the air of the room to be warmed, is a proper basis on which to start in making the necessary calculations pertaining to this subject.

For hot water, assume the temperature of the radiator to be 170 degrees, and the temperature of the room 70 degrees, the difference will be 100 degrees; multiply this by twice the heat units, and you have the number of heat units given off per square foot of surface of an ordinary radiator for the average conditions just assumed. If you have, say 1,000 square feet of surface in your house, the total heat units required for that house for an hour will be about 200,000 heat units. In a pound of coal there are about 15,000 heat units; 10,000 of these, however, are about all that are available in ordinary practice, so that by dividing the 200,000 heat units by 10,000 heat units, the practical value of one pound of coal, you have 20, the number of pounds of coal it is necessary to burn in an hour to do this work. You can burn this 20 pounds of coal on 2 square feet of grate in an hour with good draft and clean fires, or you can burn it on 4 square feet of grate, under ordinary conditions of house practice. If your grate is larger, say 5 to 6 square feet, it makes no material difference when you have a large and thick bed of fire, as is generally used in house-heating boilers. This gives you the data from which you can figure the answer to your second question when you know the conditions. The heating surface in the house and the grate surface, as above, can be fixed with some limit of certainty.

The ratio of the boiler surface to the grate surface of the boiler surface to the radiating surface, is the variable quantity, and cannot be fixed with any degree of accuracy for boilers in general. If a man has a hemisphere directly over a fire, concave side down, he will have a boiler of great efficiency per square foot of grate, and 1 of grate to 10 of boiler and to 250 of radiator, may give very good results, whereas, with a very complicated pipe boiler it may stand 1 of grate, 40 of boiler, and 250 of radiator, all for hot water. Ordinary heating boilers, of course, will come within those extremes, and, in our judgment, about half-way between. The same rules apply to steam. The direct reply to No. 2 will be 4.56 square feet of grate to 1,000 square feet of radiation.]

HOW TO PROPORTION RADIATING SURFACE.

W. R. CRITTENDEN, Bucyrus, O., writes:

"Will you, either by personal letter or through the columns of your journal, give me a comprehensive rule for determining the proportion of radiating surface to cubic contents in rooms to be heated by steam, together with allowance usually made for exposed situations, etc.? We are building new offices, and would be glad to have some information of this kind with which to work."

[There can be no very accurate rule for determining the heating surface by the cubic contents of rooms or buildings. The method followed by many of allowing 1 square foot of radiating surface to 25 cubic feet of air space as a maximum, to 1 square foot to 200 cubic feet as a minimum, is of no service except in the hands of a man of considerable practical experience, and then serious blunders are made with it.

A room of 1,000 cubic feet in a tower with windows on all sides would probably require "1 to 25," and in windy weather that might not do.

The same room on a corner of a building, with windows and outside walls on two sides, would probably require 1 to 40; a middle room of same size, with windows on one side, 1 to 55 or thereabouts. Now if the latter room is made twice as large, by increasing its front measurement only, it will require just the same proportion (1 to 55) as above, because the cooling surfaces increase in the same ratio as the cubic contents, whereas, if this room was made twice as large by increasing its depth only it would require very little more surface than it did when it had only 1,000 cubic feet.

Now take a room of 10'x10'x10', or 1,000 cubic feet, on a corner, with windows and outside walls on two sides, and it is evident it will require as much surface as a room 20 feet on the front by 10 feet deep and 10 feet high.

The radiating surface in a room must vary in the proportion of the outside walls and windows, and not in the ratio of its cubic contents.

There is no accurate rule that we know of, but the most comprehensive on the wall and window surface is found on pages 26 and 27 of "Steam Heating for Buildings," the summary of which is to allow three-quarters of a square foot of surface to each square foot of glass or window opening, and the same for each 7 to 10 square feet of outside wall surface. This is for low-pressure steam, with good efficient radiators or coils. With some radiators that are sold, however, three-quarters of a square foot would not be sufficient.]

HEATING A SWIMMING BATH.

A. T. ROGERS, of New York, writes:

"This problem has been put to me: 'How much coal is required to heat the water in a swimming bath, containing, say, 85,000 gallons of water, by hot-water circulation?' The system proposed is to connect the flow pipe from the heater at one end of the tank, near the top, and the return at the bottom at the other end, thus causing all the water of the bath to pass through the heater. From data which I have obtained I have calculated that to raise this amount of water from, say 40° to 90° Fahr., would require the consumption of about 2,800 pounds of coal. Allowing a consumption of five pounds of coal per hour per square foot of grate surface, a grate containing 56 square feet of surface would be required to do the work in 10 hours. I should like to hear from the experience of others in this kind of work."

[The 85,000 gallons of water would weigh approximately 708,333 pounds. To raise the temperature of this weight of water from 40° to 90° Fahr., or through 50° Fahr., would call for the expenditure of about 708,333×50, or 35,416,650 heat units. Assuming that 10 pounds of water can be evaporated by one pound of coal (good ordinary practice), it will require about 3,541 pounds of coal to do the work required in 10 hours, or at the rate of 354 pounds of coal burned per hour. Allowing, say 10 pounds of coal to be burnt per square foot of grate surface per hour, which is a fair figure, 35.4 square feet of grate area would be required to do the work.]

CONDENSATION NECESSARY.

AMOUNT OF RADIATION IN INDIRECT STACKS.

F. W. J. writes:

"I have a schoolroom 25'x32'x14' and must supply it with 1,650 cubic feet of hot air per minute, the temperature of room to be maintained at 70° Fahr. by natural ventilation. What will be the size of the indirect radiator required, also the hot-air duct, cold-air duct, and foul-air or exhaust duct? Would it be better to have two hot-air ducts to maintain the temperature at 70 degrees when the outside temperature is at zero?"

[According to the data sent, each room requires 99,000 cubic feet of fresh per hour—say 100,000 cubic feet. The warming of 100,000 cubic feet of air to 100 degrees requires the condensation of 200 pounds of water in the same time, and 400 square feet of good indirect radiation will condense this amount of water when properly boxed. You can get a velocity of about 7 feet per second in a good smooth flue with natural draft, so that one flue of 2x2 feet will be about the proper thing, or two flues, each 2'x16" in sectional area. Cold-air inlets and foul-air outlets should be of equal size and should have long easy turns. A flue larger than 5 square feet in area for the room mentioned is not necessary, nor should the area be less than 4 square feet.]

PROPORTIONING OF RADIATION.

Percival H. Seward, Syracuse, N. Y., writes:

"I will be greatly obliged if you will indicate the amount of direct radiation necessary to heat the store described with low-temperature steam, say five pounds per square inch. The store is 190'x40'x17' containing 129,200 cubic feet. A building adjoins the store for the entire distance along one side and for more than half the distance on the other. The front is all of glass. There are six windows 3x8 feet at the rear of the store and five 2x3 feet at the side, these five windows being set at the ceiling line. There is a skylight about 35x40 feet over the store floor.

"I am aware that this will seem like a rather urgent request, but some of our fellow fitters and myself got into an argument about the amount of radiation necessary, and we decided to leave it to you."

[The glass front has an area of about 690 square feet, five 2x3-foot windows on the side have 30 square feet, and six 3x8-foot windows in the rear 144 square feet. The total area of exposed wall minus window area, is about 1,356 square feet, or the equivalent of about 135 square feet of glass, making the total cooling surface 999 square feet of glass, or its equivalent in cooling surface. Now, the radiation to warm the store must be proportional to the cooling surfaces. One square foot of average radiation in good coils or radiator will offset the cooling done by 2 square feet of glass, provided there is nothing else to cool the air. This would call, then, for 499.5 square feet of radiation. Air, however, is cooled in other ways, and it has been found that one-half as much more radiation as will offset the cooling of the windows is usually ample to provide for all further contingencies in ordinary buildings. This would call for 749.25 square feet of radiation. This is 1 square foot of radiation to about 172 cubic feet. The heating surface should be near the front and rear of the store—about two-thirds of it near the door and front windows and one-third in the rear.]

HEATING SURFACE REQUIRED TO HEAT WATER IN TANK.

J. M. & S., New York, writes:

"Will you kindly inform us how many square feet of heating surface in a tank will be required to raise 21,000 gallons of water from 160 degrees to 320 degrees in 10 hours, using steam at 80 pounds pressure?"

[To warm 21,000 gallons in 10 hours through the range of 160 degrees, between 160 degrees and 320 degrees, is the equivalent of warming 17,125 pounds of water in an hour through the same number of degrees, or 2,740,000 heat units, and it calls for the condensation of 3,009 pounds weight of steam at 80 pounds pressure to water at the same temperature.

This steam must pass into the coil at a difference of something less than five pounds pressure; in other words, a pressure must be maintained in the coil of something over 75 pounds pressure if the water is to be made 320° Fahr. This determines the minimum diameter of the pipe that supplies the coil; and with a volume of about 5 cubic feet to the pound weight, and a velocity of, say 600 feet per second, it will require a pipe of just about 1 square inch of area in cross-section to pass the steam required at 80 pounds.]

RULES FOR ESTIMATING RADIATING SURFACE FOR HEATING BUILDINGS.

L. F. BELLINGER, Northfield, Vt., writes:

"In my estimate I used a rule for single rooms, which is given on page 37 of Babcock & Wilcox's catalogue. The heating firms have used the cubic-foot method so far; taking no account of windows evidently. I counted the halls as outside surface and fell about one-third under the heating firms in heating surface. Is B. & W.'s formulas applicable to single rooms? I have books of the two Billings published by you and like them."

[The formula which you site for calculating radiating surface for buildings is ample; it is:

"*Add together* the square feet of glass in the windows, the cubic feet of air required to be changed per minute and one-twentieth the surface of the external walls; then multiply this sum by the difference between the required temperature of the room and that of the external air at the lowest point it is likely to reach and divide the product by the difference in temperature between the steam in the pipes and the required temperature of the room."

Take the case of a corner room 14x14 feet by 10 feet high with four windows, each of 24 square feet of glass and changing its air once in 15 minutes. Thus by this rule we have 96 square feet of glass plus 130 cubic feet air plus (one-twentieth the outside wall) 9.2 square feet, or 245.2 as a total. Then assuming the room to be kept at 70° Fahr., the outside temperature zero, and the steam pipe 212° Fahr., we find the radiating surface thus:

$$\frac{245.2 \times (70^\circ - 0^\circ)}{212^\circ - 70^\circ} = 120.8 \text{ square feet.}$$

This, under the cubic-foot rule, gives a ratio of 1 square foot of surface to each 16.2 cubic, which is a much higher ratio than is obtained under any other rule we know of.

The contents of this room being 1,960 cubic feet, 65 square feet, or 1 to 30, would be called good for direct radiation, with accidental ventilation.

The rules laid down by "Thermus" in our columns for steam and hot-water heating give different results, and as he is an engineer of experience his rules are worth considering.

Rule 1 is

$$\frac{\text{Temp. room} - \text{outside temp.}}{\text{temp. steam pipes} - \text{temp. room}} = \text{the square foot}$$

of surface of radiator to counteract a square foot of glass, which give very nearly one-half square foot of radiator to 1 square foot of glass with steam pipe at 212, room 70, and outside air zero.

Then for unventilated buildings, or ones with only accidental ventilation, he adds from one-fourth to one-half more surface, as the judgment of the engineer suggests, to cover the contingencies of ordinary construction. He also considers that a square yard of wall—brick—cools about the same amount of air as a square foot of glass.

He treats the question of the air admitted for ventilation differently. Taking the same room again, its cubic contents being 1,960 cubic feet, with a change every 15 minutes he would have 7,840 cubic feet to warm per hour from zero to 70° Fahr., and proceeds as follows:

$$\frac{\text{Cubic feet of air admitted per hour} \times \text{rise of temperature}}{50} = \text{heat units.}$$

Then

$$\frac{\text{heat units}}{1,000} \times 4 = \text{heating surface in square feet.}$$

Then by Rule 1 we have 58.2 square feet, and by the rule for air admitted we have 43.9, or a total of 101.1 square feet, which by the cubic contents rule is in the ratio of 1 of heating surface to 19 cubic, which is very liberal.

The rule for finding the diameter of mains which you allude to is Baldwin's rule for gravity apparatus. It first appeared on page 129 of "Steam Heating for Buildings," and his words are: "The increase of the diameter of a steam pipe is directly as the square root of the heating surface, and according to the arbitrary unit adopted (the 1-inch pipe to 100 feet of surface) the diameter of the pipe in inches is one-tenth the square root of the heating surface in feet."

This rule is ample for very low pressures, and will cover all ranges of pressure. It is based on the idea, however, that as pipes enlarge in diameter they also increase in length; and whereas the rule for constant lengths, but increased diameters, would be in the proportion of the fifth root of the square of the surface. The rule for lengths increasing from 50 to 100 feet within the ranges of diameter ordinarily used, the first rule—viz., the ratio of the square root of the surface, is about right.]

RULES FOR FIGURING STEAM-HEATING SURFACE.

W., Honesdale, Pa., writes:

"Since using Baldwin's 'Steam Heating for Buildings' for computing heating surfaces, I have had several encounters with steam-heating men, and in every case I have found by his rules less steam-heating surfaces than they claim to use. Will you kindly figure the following for me by his rules and see if I am correct?

"I want to warm a well-built brick house situated on the northeast corner of —— Street. It has hard-finished walls, papered. The sitting-room is 16'x20'x10', has 160 square feet of wall exposed to west wind, 200 square feet of wall exposed to north wind, 63 square feet of glass surface; desired temperature 70 degrees, 10 degrees below being about our lowest outside.

"I figured that 57 square feet of direct heating surface was necessary and would answer. Am I right? This is taking it for granted that the house is properly warmed in other rooms and hall. The room mentioned is on the first floor, 2 feet above the level of the sidewalk, and is unusually tight, so far as joiner-work is concerned."

[To figure for a room 16'x20'x10' high, hard-finished walls, Baldwin's rules (pages 26 and 27, eighth edition), we have 16' × 10' = 160 □ west wall,

20' × 10' = 200 □ north wall,

360 □ total for both outside walls, less the window area, which you give at 63 square feet = 297 square feet of cold walls.

He considers the cooling value of a square foot of a hard-plastered wall (one without furring and lathing) to be one-fifth the value of a square foot of glass, which places the total value of the wall in cooling powers as equal to $59\frac{6}{13}$ square feet of glass. To this, according to his rule, we must add the glass and treat all as glass Thus we have $59\frac{4}{13} + 63$ square feet of glass $= 122\frac{4}{13}$ square feet of glass, or the equivalent thereof if it were all glass.

Then, to find the amount of average pipe surface that will warm as much air as a square foot of glass will cool, he says: "Divide the difference in temperature between that at which the room is to be kept and the coldest outside atmosphere by the difference between the temperature of the steam pipes and the air of the room," and the product will be the square foot of pipe surface that will offset the cooling of a square foot of glass.

According to this, then, we have room 70° Fahr., outside temperature 10° Fahr. (below) = (difference) 80° Fahr. Again, temperature of steam pipe 212° Fahr., less temperature of air of room = 142° Fahr. Then $\frac{80}{142} = 0.563$ as the plate or pipe surface, in square feet, that will offset a square foot of glass.

According to this, then, we have $122\ 4 \times 0.563 = 68.91$ square feet of pipe or radiator surface to offset the cooling done by the walls and windows. To this must be added the amount of radiator surface necessary to warm the air admitted to the room in an hour. For instance, your room has a cubic contents of 3,200 cubic feet, and say it changes twice in an hour and is warmed from 10 degrees below to 70 degrees above. Then we have $\frac{3.200 \times 2 \times 80}{50} = 10,240$ heat units; in which the (50) is the approximate number of cubic feet of air that a heat unit will warm 1° Fahr. This, say, is the equivalent of 10 pounds weight of steam and requires about 30 square feet additional of radiator surface to condense it. Thus we have $68.9 + 30 = 98.9$ as the total quantity of pipe in square feet for 10 degrees below zero.

If your room is a tight box, or nearly so, 70 to 75 square feet will do, but should air be admitted accidentally or otherwise it must be provided for on the above basis.

Although the mercury may reach 10 degrees below at an odd time in your neighborhood, we think 10 degrees above as amply low to figure on as a basis of calculation. This will lessen the surface by one-fourth, or to about 75 square feet for your room. With 98.9 square feet of radiator in such a room it is 1 square foot to about 33 cubic, and with a radiator of 75 square feet it is about as 1 to 43, and, as the steam man usually figures it, either of them appears pretty ample.]

RELATIVE CONDENSATION IN HEATING APPARATUS.

"CHIEF ENGINEER," from Maine, writes:

"I have 14 separate buildings to be heated by steam, the average distance of each from boilers being nearly 300 feet. Steam is carried to, and condensation returned from, them by cast-iron pipes inclosed in brick trenches or ducts laid in cement, although all are not so protected. Some are carried in ex. hy. C. I. soil pipe, with calked lead joints. None of these ducts or sleeves are below frost (from 4 to 5 feet here some winters). Thus, the aggregate distance is over 4,000 feet for each supply and return. The "Williams system" is in use here. Aside from this, I have seven buildings warmed by small L. P. boilers (gravity system), averaging about 10 horse-power each, warming 264,598 cubic feet with 3,654 square feet of radiation, about evenly divided, direct and indirect. The buildings are all exposed on all sides, two brick and five wood (one a greenhouse). What, I would ask, is the probable average weight of water per hour per square foot of heating surface in both systems, also the probable evaporation per pound of coal in each? C. A. Williams system, 1,253,398 feet; gravity, 264,598 feet; heating surface, Williams, 9,471 square feet; gravity, 3,654 square feet."

[In reply, we will say, *the pressure being the same*, there will be *no* difference in condensation between the Williams system and any other system.

The condensation per square foot of surface for ordinary direct coils and radiators will be equal to from 1.5 to 2.25 heat units per hour per degree (Fahr.) of difference between the air of the rooms and the surface of the coils or radiators. Average vertical radiators condense an average of 2 heat units, so that for a 100 square-foot radiator at 2 pounds steam (218° Fahr.) in a room at 70° Fahr., you will have $100 \times 2 \times (218° - 70°) = 29,600$ heat units as the equivalent work of the radiator, which, divided by the latent heat of steam at 2 pounds (960), gives the pounds of water per hour $\frac{29,500}{960} = 30.8$ pounds of steam or water condensed.

The condensation for indirect radiation depends on the draft of the flue. The better the draft, the more the work; though the ratio of condensation is not quite equal to the increased quantity of air. In common practice (without fans) 3 heat units per degree of difference will cover the case.

The condensation in mains is an unknown quantity, depending on the covering, etc. It is not safe to put it at less than one-fourth the condensation in an uncovered pipe.

The evaporation in both cases will probably be found to be between 8 and 10 pounds water per pound of fuel. As to which system does the best, we are unwilling to hazard an opinion on such insufficient data. Weigh the water of condensation and compare it with the fuel burned and the heating surface, and reasonably approximate answers will be obtained to your questions]

STEAM-HEATING ESTIMATE WANTED.

GEORGE E. ROBERTS, of Providence, R. I., writes:

"Will you state what allowance in cubic feet the current practice will estimate can be heated per horse-power; steam at an average of 30 pounds pressure?"

[The accepted horse-power of the present day is 30 pounds in weight of water evaporated to steam, and of course the same steam condensed to water is a horse-power. The pressure is not an important element when heating only is the object; it is simply the weight of the steam used. One hundred square feet of heating surface will condense just about 30 pounds of water in an hour, when the pressure is about 30 pounds per square inch, so that the value of 100 square feet of surface is the equivalent of one horse-power. One hundred square feet of surface will warm from 5,000 to 10,000 cubic feet of average air space. If the rooms or chambers are very large 1 square foot to 100 will do. If the rooms are small an average of 1 to 50 is generally ample in the coldest weather in this latitude.

The above are approximations to the truth, and are as near as it is possible to come without a careful scientific review of the whole subject.]

STEAM CONSUMPTION FOR HEATING IN NEW YORK IN DIFFERENT MONTHS.

PRACTICAL tests on a large scale in New York City have shown that the steam required for heating buildings is represented by the following percentage during the different months of the heating season:

	Per Cent.
October	5
November	5
December	15
January	25
February	20
March	20
April	10
	100

STEAM-HEATING SURFACE FOR DRYING-ROOMS.

"STEAM GAUGE," Kingston, Ont., writes:

"Will you please state what radiating surface at a steam pressure of 100 pounds will be required to heat a drying tunnel 6x7 feet in cross-section by 75 feet long to a temperature of 300 degrees?

"Also, what amount of radiating surface will heat an unplastered brick building 70'x40'x13' to a temperature of 160 degrees with steam at 70 pounds pressure?"

[You have omitted a very important item of data in asking the first question. You make no mention of the amount of air that is to be changed in any given time. With the tunnel absolutely closed up and well protected we have no doubt but that a coil of 10 1-inch pipes run the whole length will be sufficient. This surface, however, being only about 37 degrees hotter, at 100 pounds pressure of steam, than the temperature at which you wish to keep the air of the tunnel, will not more than make up the heat that is carried off through the brickwork, and when steam is first turned on, if the brickwork is fresh and contains much moisture, it may take a considerable time before it will bring it to this temperature. This does not take into consideration the moisture to be driven off from the materials to be dried. The moisture in the air of a drying kiln plays an important part in the rise of temperature. If the air can be kept absolutely dry or nearly so, it can be warmed considerably above 212° Fahr. When moisture is present, however, evaporation first goes on, and afterwards the moisture held in the air and the air itself may readily advance to a temperature of about 212 degrees (the temperature of the vapor of water at atmospheric pressure). But to warm the vapor beyond this point it requires to be superheated and kept from contact with walls, or elsewhere from which it can draw additional moisture. If this is done the temperature of the drying kiln or tunnel may be advanced by a further application of heat, and we have no doubt that temperatures of 300 may be obtained, if moisture and the passage of air be cut off. High temperatures are best obtained within metal-lined chambers backed with some non-conducting surface. Air at 300 degrees can also be obtained by forcing it between steam coils at high temperatures, but not allowing it to come in contact with the moisture afterwards. The specific heat of air is low, and though it is easily warmed on that account, the same reason accounts for its being unable to evaporate moisture to any considerable extent without materially lessening its own temperature. To make a drying-room effective, therefore, large quantities of dry air must be moved through it, and our experience is, that when a temperature of 300 is required to be maintained, and drying by evaporation is required, the only way is to force hot dry air through the tunnel, the air being heated outside. If a baking oven only is required to harden varnishes, then it may be done with direct steam heat, and we are of the opinion that fully 700 square feet of surface will then be required.

In reply to the second question, you give us no data in regard to the window surface or the amount of air to be moved, so that it becomes almost guesswork to reply to this question. According to Baldwin's rule it would require about 360 square feet of surface, at 70 pounds steam, provided there were no windows in the building, and no air move, or no great amount of moisture to evaporate, and that, with ordinary proportioned windows, it would require from two to three times as much surface, say 700 to 1,000 square feet.]

COST OF STEAM HEATING.

ESTIMATING COST OF STEAM.

THE ECONOMY STEAM HEAT COMPANY, of St. Paul, Minn., writes:

"Having been referred to you by Henry Carey Baird & Co., of Philadelphia, would like to know the following: We have the largest steam-heating plant in St. Paul. During the last few years we have added an electric-light and power plant. Owing to your experience in figuring on heating, I would like you to mention the best practical book containing information for estimating and charging for heating buildings. We want a book, not with formula, stating experience, etc., but a book which has derived as far as possible, the rules and charges for heat. Also, if there is any practical meter. Any information in regard to the New York Steam Company, or any other information will be thankfully received."

[We do not know of any book that gives the information necessary for estimating and charging the cost of supplying heat to buildings. The New York Steam Company meters all the steam which it supplies. The meter is the invention of Dr. Charles E. Emery, the first engineer and superintendent of the company. The steam flows with a constant difference of pressure of two pounds through a variable opening, the flow of steam being closely proportional to the size of the opening, which is recorded on a strip of paper moved by clockwork. The area between the base and record lines is integrated by a planimeter and the result interpreted into "kals," forms the basis of the bill for supplying steam. The kal is a convenient commercial term coined by Dr. Emery to express a unit of caloric or heat, and denotes the equivalent of one pound of water evaporated into steam.

In estimating the cost of supplying steam you must commence by assuming a price for fuel, say $5 per ton for anthracite coal, which gives you the basis for a sliding scale of cost, depending on the current price of coal.

Either 30 pounds of water evaporated to steam or 30 pounds of steam condensed to water per hour may be taken as the equivalent of one horse-power, the first for power and the second for heating. Ten pounds of water evaporated in a boiler by a pound of coal is considered good in common practice. Therefore we will assume in your case that you are able to produce a horse-power for three pounds of coal per hour, and as coal at $5 per ton costs one-quarter of a cent per pound, the cost of the horse-power to you, without wear and tear of plant, engineer's hire, etc., is three-quarters of a cent per hour. Thus, if you either buy or sell power or steam at the rate of one horse-power you have, for 182½ days of cold weather, at 24 hours per day, 75 cents × 24 hours × 182.5 days = $32.85, as the cost of the fuel necessary to produce a horse-power for a winter for heating; or, for a whole year, at 12 hours per day (Sundays included), for power in a good engine. To estimate the total cost of any particular case we may figure it on, say 100 horse-power for one year, for 12 hours a day, thus:

Fuel, 100 horse-power at $32.85 equals	$3,285
Depreciation 10 per cent. on cost of plant ($5,000)	500
Engineer's wages	1,000
Fireman's wages	600
Sundries	100
Total	$5,485

Dividing this by 100 horse-power you have $54.85 per year as the cost of producing one horse-power if you manage everything carefully.

In small plants the cost may reach $60 to $65 per horse-power, but in well-regulated plants, supplying over 100 horse-power, it should be kept to about $50, with coal at $5 per ton.

Radiators condense from one-fourth to one-half pound of steam per hour per square foot of surface according to exposure and the pressure of the steam. Under the usual conditions three-tenths pound is probably a fair average, in which case a radiator of 100 square feet uses steam at the rate of one horse-power, and at $50 per horse-power for a season of 182½ days at 24 hours per day, the cost of supplying 1 square foot of radiating surface for that time would be 50 cents. Profit must be added to the above when you are dealing with a consumer.

As to the cost of supplying steam in New York, F. H. Prentiss, Superintendent of the New York Steam Company, says:

Before our meter was developed our rates were from $2.50 to $5 per 1,000 cubic feet (of space) per season. Hence on basis of 1 foot heating surface per 100 cubic feet, the price would be 25 to 50 cents per square foot.

The Boston Heating Company charges about double our price for steam, and I think the Denver Company charges $1 per square foot per season.

Mr. Prentiss inclosed a set of regulations for the supply of steam by his company, from which it appears that the present charge by meter varies from 80 cents per 1,000 kals, for a consumption of 12,000 kals per month, down to 45 cents for 1,000 kals for a consumption of 400,000 kals per month, so that the charges of the steam company will range from about 1.35 cents to 2.4 cents per horse-power per hour. There is also a minimum charge, varying from $10 a month on a 1-inch supply up to $45 a month on a 6-inch supply. The cost of introducing the steam is paid by the consumer; service pipes from street mains to buildings cost about $75 and remain the property of the company; the connections to the meter, etc., cost about $25 and belong to the consumer; the meter combination costs from $30 for 1-inch to $160 for 4-inch, ownership not stated; the cost of fitting up the house trap, to remain the property of the company, varies from $10 to $30. The usual pressure is 75 pounds, which has been on continuously since April, 1882.

E. E. Magovern, formerly one of the engineers of the New York Steam Company, also writes as follows:

The New York Steam Company began supplying steam about August, 1882, and continued supplying on other than a meter basis until 1884. The rates were based on the number of cubic feet of space heated, and varied from $2.50 minimum to $4.50 per 1,000 cubic feet of space heated per season. Experiment had shown that the condensation per square foot of surface in radiators, under varying conditions of exposure and draft, varied from 0.25 to 0.45 kals per hour. Hence, following a general ratio of heating surface to space heated of 1 to 100, gave the kals condensed per 1,000 cubic feet of space per hour, as 2.5 to 4.5. The contract was based upon a heating season of 2,000 hours giving 5,000 to 9,000 kals per 1,000 cubic feet of space heated. With coal at $5 per ton, allowing for profit, attendance, interest, depreciation, etc., it was found that for an ordinary sized plant the cost of furnishing steam was not far from 50 cents per 1,000 kals, hence the rates per season for 1,000 cubic feet of space varied from $2.50 to $4.50 as stated.

When the meters were introduced it was found that the rates, as shown by meter, varied from $1.44 to $4.56 per 1,000 cubic feet of space heated per season.

Mr. Magovern accompanies his reply by a table, given on page 248, containing the cost of heating several buildings, under various conditions, taken from actual measurement of the steam used.

Exposure.	Material.	Period Heat is Used.	Hours Used per Day.	Pressure, Pounds.	Cost of Heating 1,000 Cubic Feet per Season.	Remarks.
Exposed on all sides.	Stone and iron.	Oct. 1 to May 30.	24	4 to 20	$2.56	Large windows.
Corner.	Iron.	Oct. 15 to May 15.	20	30 to 45	5.78	Large windows.
Corner.	Stone and brick.	Oct. 15 to May 15.	12	6 to 11	3.74	Large windows.
Corner.	Brick.	Oct. 15 to May 15.	24	5 to 25	3.45	
Corner.	Stone and brick.	Oct. 15 to May 15.	12	25	1.80	
Corner.	Stone and brick.	Oct. 1 to May 30.	15	10 to 30	3.20	
Ordinary.	Stone and brick.	Oct. 15 to May 1.	24	17 to 24	4.47	
Ordinary.	Iron.	Oct. 15 to June 15.	24	6 to 15	4.38	
Ordinary.	Marble.	Nov. 1 to April 30.	10	20	2.50	
Ordinary.	Stone and brick.	Oct. 15 to May 15.	8	15	1.50	

The information given above from Messrs. Prentiss and Magovern was furnished in March, 1889, about which time A. V. Abbott, Chief Engineer of the Boston Heating Company, wrote concerning a similar matter:

We charge $1 per thousand pounds of steam without any regard to the pressure at which the steam is furnished, finding that that price averages very well for our service. We also find that here in Boston it is customary to charge 60 cents per square foot of radiating surface for the heating season, which is considered to be 200 days of 10 hours each, and that the prices for power, that is to say, the prices for steam supplied to engines by small plants throughout the city, vary from $100 to $150 per year of 300 days of 10 hours each. Our own charge for steam for engines is $90 for the same time.

Station B, the first steam station erected by the New York Steam Company, and designed to hold boilers aggregating 16,000 horse-power, was described in THE ENGINEERING RECORD on January 25 and February 1, 1883, and Station J, of the same company, was described on April 7 and 14, 1888. The plant of the Boston Heating Company was described in issues from April 28 to May 26 inclusive, 1888.

THE CHARGE FOR HEATING SERVICE.

F. K. S., Boston, writes:

"I am supplying heat to an adjacent store where they have about 350 square feet of direct steam-heating surface (pressure about five pounds), and want to know what to charge for it. The steam is turned on for about eight hours each day."

[We do not know whether or not your radiating surface is properly proportioned, but assuming that it is, you will condense about a third of a pound of water per square foot of surface per hour, or about 116 pounds per 350 feet, and 116 × 8 = 928 pounds in a day of eight hours. If your boilers are working properly you will probably evaporate, the condensed steam being returned to the boiler, 10 pounds of water per pound of coal, and hence 928 pounds would require the consumption of 92.8 pounds of coal per day. You know what your coal costs, and from this you can get the cost of fuel for heating. You can rightly add to this an additional amount to pay for a part of your fireman's wages.]

COAL REQUIRED.

STEAM REQUIRED FOR HEATING A RAILWAY TRAIN.

JAMES EMERSON, Williamsport, Mass., writes:

"In your journal of 29th of January, 1887, is an article relative to the heating of cars by steam, in which an estimate of the maximum quantity of steam for the purpose that possibly could be required is the first *definite estimate* that has met my notice. The estimate, however, is far too high. For five years my attention has been devoted to the subject of car-heating, and experiments have been made at different times in order to determine the best quantity of heating surface per car. All of the cars piped on the Connecticut River road are 70-seat cars, 55-foot sills, 18 windows each side. One-inch, 1¼, 1½, and 2-inch pipe has been tried, and I prefer 1¼-inch, so arranged as to get 125 square feet of heating surface per car, which is found to be abundant; the only complaint with passengers being that the cars are kept too hot. There are 20 or more trains on this road now each way per day warmed with steam, beginning with one and showing a steady increase, and as trains are added they are piped and heated by steam, from the engine, though in case of accident to the engine there is a small auxiliary boiler under each car; or rather, that is my system. This little boiler has a 12-inch firebox, in which a fire about 6 inches deep can be kept, and that makes the car so warm that it has been found necessary to put pipes on top of the car in which the steam could be condensed and returned to the boiler, to keep up the water supply in boiler. In the first trial the boiler, holding 35 gallons of water, was run for 10 days without renewal. About four ordinary coal-hods full of coal were used in running from Springfield to St. Albans, a distance of 250 miles, and return, or 500 miles the trip. There is less trouble in heating the cars by steam from the engine or from the auxiliary boiler than by stoves. It takes just about the same steam to heat as to operate the brakes, certainly less than a horse-power per car, and actually less than where stoves are used, for it saves eight seats in a car, so that seven cars heated by steam from the locomotive are equal in seating capacity to eight heated by stoves. They are far safer, more comfortable, and of course far better ventilated, for the ventilators at the top are never closed. I don't believe that 17 cars of ordinary length are ever taken in one train, unless in some special case, for that number could be better run divided in two, one a half or a whole hour later than the other, and would accommodate the traveling public better; and, still further, no engine of 17x24-inch cylinder could draw 17 cars and make express time, unless they have engines on the N. Y. Central unknown upon the Eastern roads."

[The estimate of the steam required for warming a rain to which our correspondent alludes above, was made with the view of being greater than actual practice could demonstrate, as it was our object to show how small a percentage of the steam made by a locomotive was required for the purpose of warming the train. We also agree with our correspondent that no, or at least very few, passenger trains ever have 17 coaches, for such a train would of necessity be about a quarter of a mile long, but here we again took the extreme condition mentioned by the President of the N. Y. Central Railway, so that our estimate would not be open to the criticism of not being ample or of being unfair to the railroad companies.]

AMOUNT OF COAL REQUIRED TO HEAT WATER FROM 40° TO 200°.

ENGINEER, Togus, Me., writes:

"Will you kindly inform me how many pounds of coal would be required to heat water from 40° up to 190° or 200°? Boilers are evaporating eight pounds of water to one pound of coal, and the desire is to apply steam, at 40 pounds pressure, to heating water, passing it through a brass coil in a hot-water boiler, the water of condensation being saved. The amount to be heated to 190° or 200° is about 300 gallons per hour (average), or 7,200 gallons in 24 hours, and the question is, How much will it cost per gallon to heat it?"

[To warm one pound of water from 40° to 200° Fahr. will require 160 heat-units, and, as 300 gallons of water at 40° Fahr. weigh very nearly 2,500 pounds, it is evident that it will require 400 000 heat-units per hour to be taken in form of steam from the boilers. This is the equivalent of 440 pounds weight of steam per hour at 40 pounds pressure condensed to water at the same temperature, and if cooled to the atmospheric pressure and temperature will require about 110 pounds weight. Then, if you get eight pounds weight of steam per pound of coal, it is plain it will cost you the value of 55 pounds per hour to warm the water (300 gallons); or in other words, 5½ gallons of water can be warmed by one pound of coal in the case you cite.]

METHODS OF HEATING.

HEATING BY THE GRAVITY SYSTEM.

LEWIS, Saginaw, Mich., writes:

"Some time ago I noticed an inquiry in a scientific paper as to heating a building by the gravity system, the answer being that it could not be done successfully if 40 to 50 pounds pressure was carried. Is this correct? As the water of condensation returns by gravity to the boiler, why would it not return as well with 50 pounds pressure in the system as with one or five pounds?"

[There is no valid reason why a gravity system should not work successfully and return its water to the boiler, provided all pipes and connections are properly proportioned and care is taken that no air or water trapping exists in the system. Water, whether hot or cold, seeks its level under all natural conditions of atmosphere. The same statement holds good with water under pressure of air or steam, provided the pressure is the same in the boiler and return pipes, which condition practically exists, where the pipes and connections are sufficiently large to furnish steam as rapidly as condensation takes place, thus keeping up an equal pressure above the water in return pipes. Should the steam-supply pipes be insufficient in size, then in that proportion, and in addition to the friction generated, the return water will rise in the return pipes. Coils, radiators, heating or return pipes are after all but a part of the steam boiler, no matter how far removed, and in a great measure water in return pipes is influenced the same as the water in a gauge glass may be. If it departs from the natural laws of level a sufficient mechanical reason will be found for it.]

HIGH AND LOW-PRESSURE HEATING.

J. E. L., Athol, Mass., writes:

"One of the questions most commonly asked by young apprentices to the steam-heating business is, What is the difference between high and low-pressure heating? Will you kindly explain through your valuable journal, and oblige many inquiring minds?"

[There is no definite point at which low-pressure heating ends or high-pressure begins. If we assume that there is sufficient steam pressure for power service and apply its full head in the heating system this would properly be called high-pressure heating. Now, if a steam "reducing valve" is introduced between this high steam pressure and the heating system, and the heating pressure is reduced to, say five or 10 pounds, then we have what is known as low-pressure heating, as applied to power plants. In domestic heating "low pressure" is the form commonly used; but, unlike the assumed case cited, there is no high-pressure combination. Such apparatus seldom exceeds 10 pounds pressure, and if properly proportioned and constructed may operate well at much less. Many satisfactory jobs are running at from one to three pounds. Good results are attained at even less than this figure, and we have known perfect heating where a partial vacuum existed in the heaters; but in such cases the radiating surfaces and boilers were sufficiently large, and all joints were perfectly tight. It is possible that your discussion has taken you into the field of exhaust heating. While this does not properly belong to the classes of high or low-pressure heating it has within itself "high," "low," and "open" heating.]

DIRECT OR INDIRECT RADIATION FOR SCHOOLHOUSES.

A Reader, Milwaukee, Wis., writes:

"We are building a new schoolhouse which will be heated by steam, containing four classrooms each on first and second floors, and amusement hall on third. The architect of this building has specified direct radiation for the entire building; he claims that as long as the foul-air outlet ventilating shaft is of sufficient size there will be pure air in the classrooms (there are no fresh-air inlets); he claims that we get all the fresh air desired through the crevices of windows and doors. I have visited several of the public school buildings and find that they contain at least one-half indirect radiation for ventilation; please state your opinion on the direct system."

[A direct system of radiation without systematic ventilation is not suitable for a school building. It requires much care so to design the heating arrangements of a schoolhouse in a cold country as to properly warm it and ventilate it at the same time.

When a schoolroom is warmed altogether by indirect radiation it is seldom properly warmed—though it may be well ventilated. On the other hand, when it is warmed altogether by direct radiation, it is very rarely sufficiently ventilated.

It is necessary to admit into a schoolroom about 2,000 cubic feet of air per child per hour to have good ventilation, so that in a school of 60 scholars 120,000 cubic feet should enter each hour.

This amount cannot enter by accident, through cracks around doors or windows, and no increase in size or number of exhaust shafts will draw it in, however that amount of air could be sucked in from outside. If it did it would be impossible to live in the room, as 120,000 cubic feet of air at zero, or even 20 above, cannot be drawn into a room in an hour without making it too cold to remain in, much less sit and study in.

When schoolrooms are warmed altogether by indirect radiation and supplying sufficient air to produce proper ventilation, very fair results are obtained in the way of heating when the temperature is not excessively cold outside.

In the climate of our Northern and Eastern States, however, it is necessary to supplement the indirect radiation by direct radiation under the windows and at the outside walls of the room, otherwise the air chilled by the cold outer walls and windows will fall to the floor and flow towards the center of the rooms or towards the foul-air outlets, causing cold drafts on the children who are in its course.

When air is admitted in sufficient quantities to produce abundant ventilation, its temperature as it enters the room cannot be much above the living temperature of the room, or the room will become insufferably warm. The low temperature, therefore, at which it must enter makes a little direct radiation necessary to warm the colder parts of the room.]

DIRECT-INDIRECT VERSUS INDIRECT HEATING FOR LARGE BUILDINGS.

Earlham, Earlham College, Richmond, Ind., writes:

"Will you please be kind enough to inform me on the following subject: Is not the 'direct-indirect' system of heating much better for large college and public buildings than the 'indirect,' and will it not obtain as good, if not better, results in the way of ventilation? I mean the 'direct-indirect' system as given by Baldwin on page 30 of his valuable book 'Steam Heating.'

My reason for asking is that the above college is now erecting two new college buildings, and the architect recommends the indirect system, with radiator coils in the cellar. The buildings will be entirely exposed, and often the thermometer is as low as 26 degrees below zero. One building is 56'4"x35'6" and two stories high; the other is in its extreme dimensions 174'4"x156'4" and three stories high. If you would like to see the plans I can send them to you in a week or so. Should not the warm-air registers in case of the 'indirect' system and the radiators in case of the 'direct-indirect' system be *always* placed immediately under the windows with the vent registers on the opposite side of the room at the floor?

"P. S.—Our old plant is 'direct' and we are going to do all of the heating from the one boiler-house and the same set of boilers."

[The system of warming and ventilating known as the "direct-indirect" system, in which the radiators are under the windows with the fresh air taken to them through air passages under the sills leading to box bases, should never be depended upon for ventilating schoolrooms. Baldwin describes this system in his book as one of the usual methods of warming with which ventilation or the admission of air is combined; but he does not place it ahead of "indirect" heating as a means of ventilation, nor should it be so considered, all other things being equal, for one moment.

With "direct-indirect" radiation enough air may pass through the inlets to supply air for two or three persons in a room, and therefore it may do for office rooms or residences, provided it is properly done; but for schools or crowded academic rooms or auditoriums it is wholly inadequate.

Take, for instance, an inlet 4x12 (an average size for such work) or the third of a square foot, and assume it is passing air at 5 feet per second—an unusually high velocity for such work—6,000 cubic feet of air per hour is all that can possibly pass through it; and even if you have *three* such inlets to six rooms, 18,000 cubic feet, or sufficient for 10 or 12 persons, is all that can pass. But in ordinary practice a velocity of 5 feet per second is not obtained, and the writer has known several of such apparatus that have worked the wrong way—i. e., passed the warm air out-of-doors instead of drawing cold fresh air in.

Again, such apparatus are supplied with dampers to keep the coils from freezing, etc., and to prevent the passage of cold air when the coil or radiator is not in use. The habit is to neglect this damper, and in nine cases out of 10 it remains permanently closed.

Baldwin evidently considers it a good plan to have some direct radiation, in addition to the indirect, in all schoolrooms, the former to be used in very cold weather. He says this direct heating surface should be on the cold walls of a building or under the windows, so as to counteract the loss of heat from the bodies of the scholars, which radiates to the cold walls and glass, and also to prevent the fall of very cold currents of air from the glass to the floor, along which it usually flows to the outlets, etc., keeping the children's feet in a lower stratum of air a foot or so high, and 10 or 15 degrees colder than the air at the breathing or head line.

Schoolrooms for small children should have an ample system of indirect radiation that will admit from 500 to 1,000 cubic feet of air per hour per capita, and it should be so arranged that the regulation of the temperature will not lessen the air supply. There are several good methods of accomplishing this known to heating engineers. The air should enter the room averaging in temperature from 80 degrees to 100 degrees, according to the outside temperature. In addition to this there should be the direct coils before mentioned.

For healthy youths or adults, as in your case, the air supply should not be less than 1,500 cubic feet per capita, and if they are not confined to one position too long the addition of the direct coils is not so important.

The place for air to leave a room depends largely on the system of heating and ventilation used. When air enters through one or more large registers they may be near the floor or ceiling, according to circumstances, and if large quantities of air are admitted the position is not so very important.

When they are near the ceiling the flow of air from them should be directed to the coldest side of the room. When they are near the floor it is perhaps best to have them at the coldest side of the room, and they should have large area, so the current of air will be slow as it passes them. The air soon finds the ceiling and afterwards follows the same motions as it would were it admitted at a higher level.

The outlets for the systems just mentioned should be at the floor on the inner sides of the rooms, and the vent flues are better in partition or at least warm walls.

In the designing of a building if it is inconvenient or impossible to get the heat flues near the outer sides of the rooms, they may be put in the inner walls. If the outer walls are hard finish, unfurred, and with much glass surface, then auxiliary coils are a great advantage for cold weather, as the two problems must always go together in cold climates—i. e., proper warming as well as proper ventilation.]

ONE-PIPE SYSTEMS.

WHY DO STEAM HEATING CONCERNS CONDEMN THE ONE-PIPE SYSTEM OF STEAM HEATING?

INQUIRER, Worcester, Mass., writes:

"Why is it that so many steam-heating concerns condemn the one-pipe system of house heating in every case, and will not acknowledge that it ever does, or can work satisfactorily, when it is well known that it works all right in many cases?

"Of course it is plain to everyone that two pipes are indispensable in any large building, when heated by a high-pressure steam-heating boiler.

"But there are many small or medium-size dwellings with about five to 15 radiators, heated with some of the various low-pressure apparatus, which have but one pipe, without even drip pipe at foot of risers, and work all right, without any noise or any trouble whatever.

"The writer of this is familiar with a number of such cases, and cannot see why it should be so utterly condemned, when it is simpler, and saves some expense, without any apparent disadvantage."

[We are not aware that the one-pipe system of conveying steam from boilers to radiators is unqualifiedly condemned. With certain radiators, and in the hands of careful fitters, very good results are obtained. With it in buildings that cover a good deal of ground it is often at a disadvantage, however, when contrasted with the two-pipe system. For instance, every coil or radiator will work with an inlet pipe for steam and an outlet pipe for water, and but very few coils will work with a single pipe for both purposes. Nearly all the modern radiators will work under one pipe if it is of large diameter; still there are some that will not, and these must have a return pipe.

One decided disadvantage the one-pipe and one-valve apparatus has is its great tendency to make noise when steam is let on a radiator, and its slowness to expel water when the radiator is once full. Let a radiator with a single pipe be shut off carelessly—that is, not tightly closed—or let there be a leaky valve to it so the radiator will condense itself full of water, then upon opening the valve there will be from 10 to 30 minutes of the most frightful racket experienced—technically known as water hammer—before the *statu quo* is established, and the affairs of that household go in anything like harmony again.

The system is however between 5 and 10 per cent. cheaper than the double-pipe system, and where it can be used we see no very great objection to it, provided the persons who are to use it are not led to believe it is the best.]

ONE-PIPE SYSTEM FOR HEATING TWO ROOMS BY STEAM.

J. S. N., Philadelphia, writes:

"While conceding fully the unquestioned superiority of the two-pipe, low-pressure heating system, yet for important reasons I am disposed to adopt the one-pipe simple plan as shown on sketch for heating two second-story rooms of a country dwelling-house, where the boiler pressure is ample at two pounds, unless you condemn it as objectionable. Would I be troubled with air-binding or hammering in this arrangement, and could I depend on the air being re-

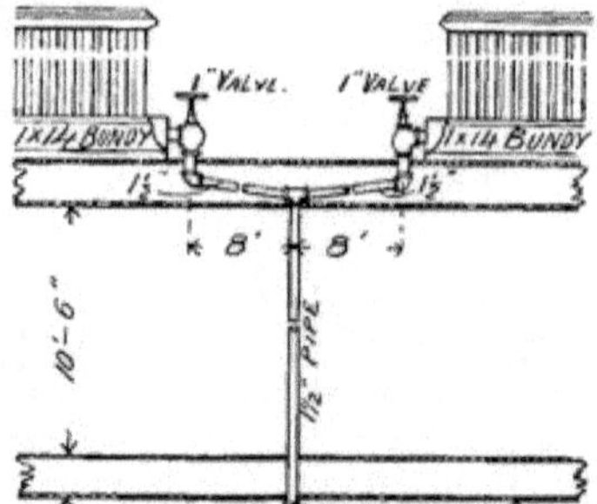

moved by a proper automatic air valve, and the water of condensation returning surely and easily to water level?"

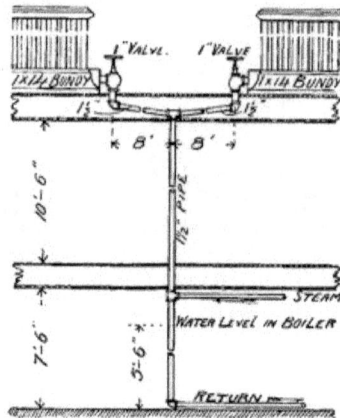

[With the pipes as you show them there is nothing to prevent your getting good results in warming, and even with the connections, from the tee at the head of the rising line, reduced to 1¼ inches each in diameter, the result will be satisfactory; provided you use the very best soft disk valves you can obtain and reliable air valves, with connecting pipe to sink or other suitable place of waste.

If the owner or user should happen to imperfectly close a valve, so a little steam will pass into the heater and condense there without being able to flow back, and to virtually fill or partly fill the base or pipes of the heater with water, instruct him to open the steam valve wide, and to wait until the steam has displaced the water, and to have no uneasiness if there is some noise, as there will be no danger. See that the ends of the radiators farthest from the valve are the highest.]

ONE-PIPE SYSTEM AND ITS RELIEF PIPES.

K. & L., Cortland, N. Y., writes:

"Will you be so kind as to give us your opinion on the *one-pipe system and its relief pipes* through the columns of your valuable paper and oblige a constant reader. We have just completed a job of steam heating in a private residence, wherein there are nine radiators, or 500 feet of radiating surface, and while doing the job we had several visitors watching the work, and among them a man who thinks *he knows it all*, and who has made some trouble for us.

We used the —— —— boiler, of Syracuse, and the job works good and heats the gentleman's house to 70 degrees in zero weather. This man *who knows it all* went and told the party that had the work done that his piping was all wrong, and that we should have carried our return pipes overhead. Some friend of the owner has so worked on him that he thinks of having it done, and we have told him that we could prove what we had done was right and would leave it to you to decide. He has no fault to find with the heat, and there is no noise or hammering, everything working quietly and the apparatus carrying only two pounds of steam in cold weather. He (this man who knows it all) says 'there is no use of taking a relief from every radiator pipe before we rise up to radiator.' We have nine radiators and nine relief pipes; the latter dropped to the cellar bottom and carried back to boiler on floor, and on the main lines on each end we have a relief pipe. All pipes pitch from the boiler. Relief pipe only ¾ pipe."

[We cannot decide on the merits of any particular apparatus on *ex-parte* representation, or at least without having a faithful diagram of the apparatus presented to us for publication. On a principle, however, we are free to give our views. The questions here involved, if we understand our correspondent rightly, are: (1) Whether a relief pipe from a *one*-pipe apparatus should be carried on the floor (and consequently below the water line) or overhead, and (2) whether a relief pipe from every end is a detriment or not. Our reply to these questions cannot be other than (1) relief pipes are best when dropped to the floor, (2) it matters not how many relief pipes are taken from mains if they all drop below the water line, provided there is at least one for every low end of steam pipe. A *third* question may be involved, which is, Whether the mains for a *one*-pipe system had better pitch *away* or *towards* the boiler? Answer: In short mains they *may* pitch towards the boiler, but in long ones they *should* pitch away from it.]

DEFECTIVE CIRCULATION IN A ONE-PIPE HEATING JOB.

F. R. Eshbach, Chicago, Ill., writes:

"Accompanying is a rough sketch showing the main and risers of a steam job I helped to do last winter. It is in a four-story double flat building. The boiler is a double 'Florida,' and it works all right except the place where it is marked *b*. Although there is a 1-inch drip in the main at that point, it fills up with water in that riser. Now it was changed a few weeks ago, and where it ran to the wall, then down and to the boiler, it now goes up and over into the header with a drip at the point of rising. Now the question is, Will it work? It has not been tried. The returns are all supplied with automatic air vents of the Marsh patent."

[This system is essentially a single-pipe arrangement, which consists of a continuous circuit that may be conventionally indicated by diagram, Fig. 2, where the steam is distributed through main S to vertical risers V V, etc., and returns, with all condensation water through same pipe and a branch R to the bottom of the heater at 7, the main being pitched continuously downward from 1, 2, 3, 4, 5, 6, to 7. If the pipes are large enough and properly arranged and connected, this system will work, but in the present case the arrangement is poor and lacks directness and traps are formed, as indicated by dotted lines at T T, Fig. 2, which would seal up the whole circulation if not drained by drip pipes D D, that discharge through pipe M to the return main R.

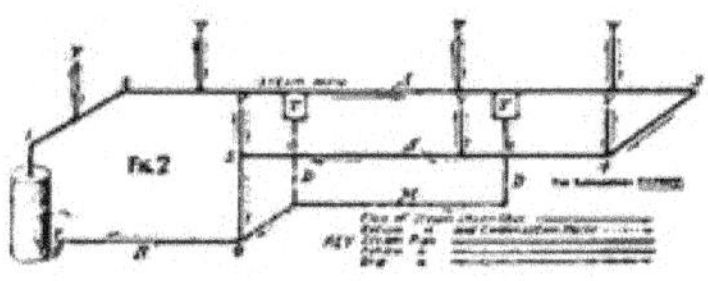

The reason for the water rising at *b* is undoubtedly the insufficient area of the horizontal mains of the apparatus for the work that it has to do, and the height of the mains above the water line of the boiler. At *a* the main has to drop under a girder, and here presumably the water trouble begins, the water standing too high in the drip, though it may not be apparent at that point. At *b* it becomes quite apparent, as the pressure is constantly becoming less, and the pitch of the pipe bringing it nearer the water line. The improvement you mention, and as shown by the dotted line *c*, is equivalent to enlarging the main, as it adds to its capacity, and connects direct from the boiler to the point of least pressure in the apparatus. We consider this an improvement, but whether it is sufficient to overcome the whole difficulty we are unable to say without more data. If not make the pipe C larger, say 2 or 2½ inches, and if this will not do, enlarge the mains throughout.]

THE COMPARATIVE MERITS OF THE ONE AND TWO-PIPE SYSTEMS OF STEAM HEATING.

STEAMFITTER, Brooklyn, writes.

"I notice in your issue of the 24th inst. an inquiry from a correspondent in reference to the one-pipe system of steam heating and your reply thereto, and would like to say a word in reference to the matter. You say, first, that the system is often at a disadvantage in buildings that cover a good deal of ground, when contrasted with the two-pipe system. This is not the case with the one-pipe system, as that term is now used; that is to say, where but one pipe is used for feed and for return for vertical pipes and radiator connections, and separate systems of horizontal feed and return pipes are used. The writer has seen a number of dwellings and similar buildings in which there was not a return or relief pipe of any description, the water of condensation being all returned through the steam mains, and of course discharged into the top of the boiler. These were all odd jobs, and the boilers had to be set very low in order to get the required inclination for the pipes. In regard to a radiator with but one pipe making more noise when steam is turned on than if it had two pipes, I must say that it is news to me, though I have seen steam turned on to a good many radiators with both styles of piping. That it takes longer to free the radiator from water with one pipe than with two I admit, but your remarks in regard to the results caused by a leaky valve, or one not propery closed, will apply equally well to both styles of piping, with this decided advantage, however, on the side of the one-pipe system, that there are only half as many valves to get out of order or be carelessly operated; for with the two-pipe system, as with the other, everything wants to be shut tight or else be wide open, though it is hard to make some people (and not a few either) believe that they haven't done their whole duty when they have shut the feed valve and left the return valve to take care of itself. Finally, I claim that in low-pressure heating, where the one-pipe system is admissible at all, it has some decided advantages which do render it the best: (1) It has but half as many valves to be operated and get out of order; (2) it has but half as many stuffing-boxes to leak; (3) it does not involve so much cutting of beams in running radiator connections; (4) where rising lines are exposed one pipe is certainly less objectionable, as

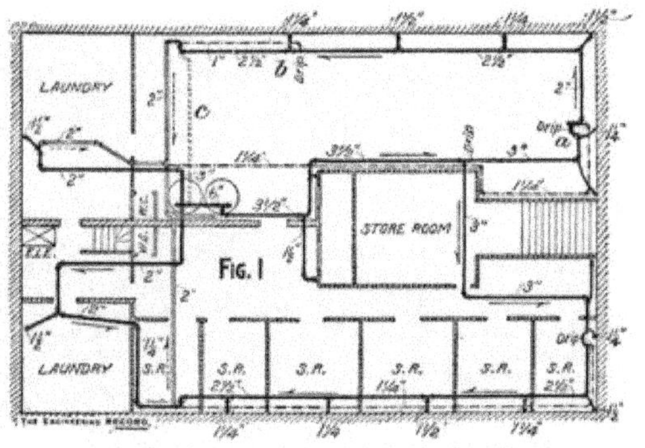

DEFECTIVE CIRCULATION IN A ONE-PIPE HEATING SYSTEM.

regards looks, than two, and any deviation from the path of rectitude and perpendicularity is less noticeable; (5) and, finally, as you yourself admit, it costs less. When properly put up one system will work as well as the other. When improperly put up neither one is fit to have in the house."

[There is very much less objection to a one-pipe system that has separate steam and return mains than there is to a one-pipe system with no separate return main. It is in fact a two-pipe system as far as the mains are concerned. The principal objection, however, to a radiator with a single pipe and valve for both the steam and the condensed water is, that when the radiator is full or partly full of condensed water there is a conflict between the steam to enter and the condensed water to run out. When steam is let into an empty radiator, as the specific heat of the iron is not great, and as it takes some time to expel the air, the influx of the steam as well as the condensation is comparatively slow, and thus the condensed water is enabled to run out along the bottom of the steam pipe in a contrary direction to the flow of steam without difficulty, provided the pipe is large enough. If, however, the radiator has been condensed full of water by being imperfectly closed the steam cannot flow in until the water flows out.

During the struggle for right of way portions of the steam are suddenly condensed and the vacuum thus formed sucks violently back the escaping water with the effect of causing that sharp, uncushioned blow called a water-hammer, so well understood by the experienced engineer and so disagreeably familiar to all users of defective steam-heating arrangements.

This water-hammer will occur also in the two-pipe system under like circumstances, but as in that case the steam is driving the water out before it, there is no struggle for right of way, and as the steam and water are less mingled the water hammer is not likely to be so severe and is certain to last for a much shorter time.

This is the principal objection to the one-pipe system and exists no matter how large the steam pipe or how well and carefully run, no matter what the pressure used, which is usually low—about one pound.

A two-pipe system, however, can be so arranged that it will not make a "water hammer," no matter how badly it is managed by the user, if it is run at very low pressure, and only one valve will be required to operate each radiator. When the return pipe of each radiator of a very low-pressure system is carried separately below the water line, the return valve can be omitted on the radiator, or, if put on, it can be allowed to remain open; then the radiator can be operated with all the ease and convenience of the one-pipe system, and as the radiator can never yet full of water, the noisy disturbance above described cannot occur.

This is plain to any one thoroughly versed in low-pressure heating, but for the information of our less experienced readers we will explain that when steam is shut off from a radiator whose return pipe runs separately to below the water line, unless the pressure carried exceeds one pound for every 28 inches that the radiator is above the water line the water will not back up into the radiator, and there will consequently never be any water in it, provided of course that the usual automatic air valves are used, which prevent the formation of a vacuum that might otherwise suck the radiator full of water when the steam was shut off.]

EXHAUST-STEAM HEATING.

HEAT OF EXHAUST STEAM.

STEAMFITTER, New York, writes:

"Your journal has been remarkably successful in giving satisfactory replies to the queries of practical men, who in their experience discover facts that seem to be at variance with established truths. I therefore submit to you the following questions for solution:

"(1) Why is it that one pound pressure of steam in a radiator supplied direct from a low-pressure boiler gives more heat than a radiator supplied with exhaust steam at one or two pounds pressure?

"(2) Is not exhaust steam at one pound pressure as hot as live steam at the same pressure?"

[Steam at one pound pressure at maximum density has a temperature of 215° Fahr.—omitting fractions—whether it is live or exhaust, and therefore should give the same heat in a radiator.

Exhaust steam, however, remains neither at a constant temperature nor density, whereas live steam ordinarily does. Practically, therefore, a live-steam radiator with a pressure of one pound gives a better result than an exhaust-steam radiator in which the maximum back pressure on the engine is one or two pounds.

If you will place a pressure gauge on the exhaust pipe of an engine—at some convenient point between the steam chest and the back-pressure valve—it will often be noticed that it jumps to two or three pounds at the moment that the exhaust valve opens and permits the steam to escape from the cylinder. It will also be noticed that for half or three-quarters of the time the gauge hand rests on the stop-pin. At these times—when the hand is on the stop-pin—the pressure of the exhaust steam has fallen below that of the atmosphere, but just how much we are unable to say, as it varies in different cases, depending on the style of engine used, the size of the exhaust pipe, the load on the back-pressure valves, the resistance of the coils, whether they are open to the atmosphere at their drip ends, and other causes. It is reasonable in any case to suppose that, owing to condensation in the coils, it falls more below the atmospheric line than it rises above it, and therefore if we assume a pressure of, say two pounds above for one-fourth of the time and four pounds below for the remaining three-fourths of the time our main effective temperature for heating will be but about 205 5 degrees—the temperature of two pounds above atmosphere being 219 degrees, and at five pounds below atmosphere being 201 degrees.

This would show that the exhaust steam, under circumstances that are probably more favorable than will occur in ordinary practice, though usually supposed to be above atmospheric pressure, really has a mean temperature of fully 10 degrees less than live steam at one pound pressure, and investigation may show that the difference may be twice as great in ordinary exhaust heating.

The diagram shows approximately the variations of pressure in the exhaust pipe.

The following arrangement has been used by the writer for some years to overcome this fluctuation of pressure: He places the back-pressure valve as far as possible from the engine, so that the resistance in the pipe will reduce the impact of the steam on the underside of the disk of the back-pressure valve, and in the same way reduces the shock to the spring of the back-pressure gauge by putting several turns in the gauge pipe and having the valve or cock under the gauge "choked" down, which, with the inertia of the water condensed in the gauge pipe, gives a steady motion to the index hand of the gauge. The stop-pin of the gauge is also removed, which, by allowing the hand to fall back, gives some indication of the lowest pressure in the pipe, which is nearly always below that of the atmosphere.

He then arranges a swinging check valve in the branch of the exhaust pipe that goes to the heating coils in such a manner as to receive as much as possible of the impact of the steam as it escapes from the engine. The check valve is thus forced open at the moment of exhaust, but instantly closes, and so maintains a considerably higher pressure in the coils than the average of that in the exhaust pipe.]

WHEN IS IT ECONOMICAL TO USE EXHAUST STEAM FOR HEATING?

ENGINEER, of New York, writes:

"A short time ago I saw in your columns a question substantially as the one above, together with your answer, in which you outlined the general conditions governing the economical use of exhaust steam.

"An actual example which recently came to my notice may prove of interest in connection with this.

"A certain engine, furnishing power for a manufacturing establishment, was tested with the view of obtaining a proper basis for charges for steam. Indicator cards were taken, and showed a heavy back pressure, something like 15 pounds, due to the exhaust being forced through a number of coils of 2-inch pipe into a tank which served the purpose of a feed-water heater. The question therefore at once presented itself whether heating the feed water in this way was economical, or whether better results could be obtained by pumping the feed water into the boiler at its normal temperature and allowing the engine to exhaust through a much larger pipe freely into the atmosphere; or in other words, whether it was more profitable to use the steam in the cylinder of the engine or in the heater.

"Calculations from the indicator cards showed that the amount of steam used per hour, with back pressure, was 595.63 pounds. The total heat in the steam exhausted per hour was 708,954 heat units, and this steam was found to raise the temperature of the feed water in the tank about 100 degrees. The amount of heat necessary to do this was, obviously, about 595 63 × 100, or 59,563 units. Of the total number of heat units (708,954) discharged from the engine there were therefore actually realized only the above noted 59,563, or about 8½ per cent.

"If this heating had been done in the boiler it would have required, theoretically, about four pounds of coal. When the feed-water heater was not in use, and the back pressure was reduced practically to zero, the hourly amount of steam used in the engine was only 444.77 pounds, showing a saving of 595 63 — 444.77 = 150.86 pounds. Assuming that the boiler evaporates, say, nine pounds of water per pound of coal, we find that $\frac{150.86}{9} = 16.76$ pounds of coal are required to produce the surplus steam which, in passing through the feed water, does work in heating which is equivalent to only four pounds of coal burned under the boiler.

"It was therefore a manifest disadvantage to employ the heater in this case and it was removed."

STEAM HEATING AT THE EDISON PHONOGRAPH WORKS, LLEWELLYN, N. J.

E. E. MAGOVERN sends the following particulars of the steam-heating work recently done at the Edison Phonograph Works, at Llewellyn, N. J.:

The "doll-shop" is shown in outline in Figs. 1 and 2. It is heated by a combination of the overhead and the ordinary floor systems. The building is of wood, weather-planked and lined inside with about one-half inch of yellow pine, tongued and grooved. The floor is of 1 inch white pine, raised 3 or 4 inches above the ground on brick pillars, the sides, front, and back being continued to the ground. The mean height of the room is taken as 16 feet. The windows measure 2'9½'x7', making for each window 17 25 square feet of glass surface. The surfaces and cubic contents are as follows:

Contents, Cubic Feet.	Square Feet Heating Surface.		
	Above.	Below.	Total.
A = 19,072	108	43	151
B = 10,624	84	37	121
C = 49,527	313	100	413
D = 30,800	210	63	273
E = 13,400	Heated by ordinary radiators.		
F = 8 800	68	26	

Ratios, Square Feet Heating Surface to Cubic Feet Contents.	No. of Windows.	Square Feet of Glass.
A—1 : 105	15	259
B—1 : 87	10	178
C—1 : 115	31	535
D—1 : 113	18	312
E—....	....	....
F—1 : 104	4	69

Total Exposed Surfaces.	Exposed Surface. Less Glass.	Assume Factor for White Pine = 80 1/1 .. = Equivalent Square Feet Glass.	Total Glass Surface.
A—1,203	888	72.5	315
B— 780	607	48.56	222
C—2,264	1,702	136.32	673
D—1,479	1,152	91.74	404
E—	....	....	....
F— 420	351	27.08	97

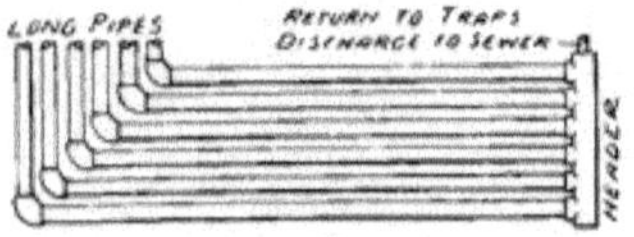

FIG. 3

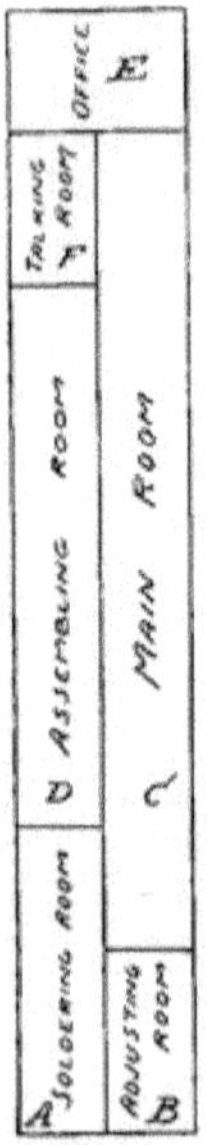

FIG. 1

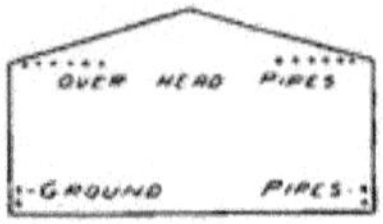

FIG. 2

STEAM HEATING AT EDISON'S PHONOGRAPH WORKS.

Baldwin's rule (see "Steam Heating for Buildings," by W. J. Baldwin, page 27) then gives for the number of square feet of heating surface: A, 167; B, 111; C, 337; D, 202; F, 42. These figures do not compare unfavorably with those above.

The spring due to expansion is taken up by the arrangement shown in Fig. 3. Exhaust steam only, at about atmospheric pressure, is used, and both the overhead and floor systems are fed from the center of the building.

METHOD OF USING EXHAUST STEAM TO WARM BUILDINGS.

STEAMFITTER, of Brooklyn, N. Y., writes:

"Will you for the benefit of your steamfitting readers generally and the undersigned in particular, give an illustration of a proper method of arranging for the use of the exhaust steam from an engine for heating purposes, and explain the reasons, etc., for the arrangement.

"My conditions are these: I have a pair of boilers, an engine, a feed-water heater, etc., to connect in a factory and I then desire to arrange the piping so I can send the exhaust steam of the engine to the heating pipes in winter and to the roof in summer, causing the least back pressure possible to the engine and securing the greatest that I can in the heating pipes. I also desire to know how the grease from the engine and exhaust steam is to be prevented from getting into the boilers with the return water.

"Any other pertinent information that you can add will be gratefully received."

[The subject of the arrangement of pipes, etc., for the use of exhaust steam from engines in the warming of buildings is an extensive one and there is more than one way to do it. There are general principles, however, that apply to all arrangements of this kind, an understanding of which can be obtained from the accompanying diagram, which illustrates the method used by William J. Baldwin in his factory practice and which he explains as follows:

Steam is taken from the boilers to the engine, as shown at the right of the diagram. In some engines the exhaust steam leaves the engine at the top of the steam chest, but in this case it is shown leaving it at the under side, which is preferable, as the water of condensation formed in the cylinder, etc., is at once taken away through drip pipe *d* to the sewer or elsewhere. From this point the exhaust pipe is carried under the feed-water heater, but with branches to it as shown. Often the exhaust pipe is so arranged as to force all of the exhaust steam through the heater. This, however, is not necessary, as the feed water requires only about one-fifth of all the exhaust steam to warm it to its hottest (212° Fahr.). It is therefore better to pass the main exhaust pipe under the heater, as shown in the diagram, with two branch pipes (1 and 2 as shown) connecting with the inlet and outlet of the heater. These pipes are furnished with valves, and a valve (No. 3 in the main exhaust pipe) can be introduced so as to force any desired amount of the exhaust steam through valve No. 1 into the heater; returning by valve No. 2 into the main exhaust pipe again. It has been found in practice that it is not absolutely necessary to use valve No. 3, as steam will circulate through the heater the same as it will through any ordinary radiator even if this valve is open. Many engineers, however, are not satisfied with this arrangement, and therefore valve No. 3 is introduced so that the engineer can adjust it to suit his own ideas of circulation. With the arrangement of valves and pipes, as shown underneath the feed-water heater, the resistance of the back pressure of the engine is less than it is when the exhaust steam is all forced through the heater, as is usually done.

Just beyond this point in the exhaust pipe it is well to use another drip *d* as shown, as much steam, about one-fifth the weight of all that leaves the engine, is condensed in the heater and should be got rid of at this point as quickly as possible, though many neglect to make this provision and keep the condensed water in the exhaust pipe, where it is forced to the top of the house or elsewhere, causing additional resistance and back presure on the engine.

To this point (where the main exhaust pipe rises), the arrangement of the exhaust pipe is the same whether the exhaust steam is to be used for heating purposes or not. Beyond this, to the left, commences the arrangement for utilizing the exhaust steam in warming. First, there is a check valve C V through which the exhaust steam is forced into the separating tank G T, sometimes called a "grease tank." The object of the check valve is to keep the pressure of steam constant within the grease tank and heating pipes, and equal to or a little above the atmospheric pressure, thus preventing the fluctuations of the pressure that must occur within the exhaust pipes, near the engine, from reaching the heating system.

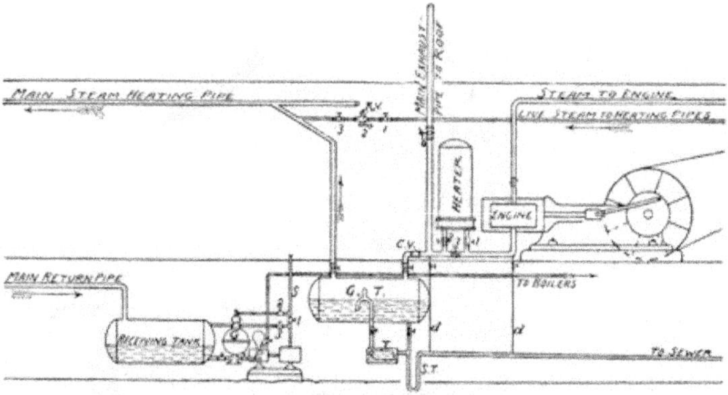

METHOD OF USING EXHAUST STEAM.

The grease tank has a twofold object—first to separate the grease from the exhaust steam, and second, to act as a reservoir. As the exhaust steam enters the tank it strikes the surface of the contained water, and by this forcible contact the greater part of the grease is caught and held by the water. The particles of grease that are not at once thus caught gradually separate from the steam as it passes slowly forward from one end of the grease tank to the other, and eventually also reach the water, where they remain. From the grease tank the steam ascends through the pipe at the left, and enters the heating system.

If the arrangement of exhaust pipes, check valve, etc., can be carried out as shown in the diagram, the water condensed in the heater can then be thrown directly into the grease tank, and the drip pipes *d d* will not be required. This can only be done, however, when there is a cellar or basement underneath the engine-room in which the grease tank can be located. It sometimes happens that the grease tank is above the horizontal part of the exhaust pipe, in which case the drip pipes *d d* are indispensably necessary.

In the present instance, as before stated, most of the condensed water from the heater passes into the grease tank; but in any case considerable water accumulates in the grease tank, and has to be removed as fast as it comes in. For this reason the bent pipe shown in the center of the grease tank is employed. It is simply a goose-neck made with fittings, and is usually about 2 inches in diameter. It is arranged so as to keep the tank about half-full of water, in which its inlet is submerged about halfway. The branch at the top of the bend is to prevent syphoning and secure a steady discharge of the water into the trap T. This water, it will be noticed, is neither drawn from the bottom nor from the surface, but at a point midway where it contains little or no grease. Since the light oils float on the surface of the water in the tank, and the heavy oils and earthy matters sink to the bottom, this prevents the grease and oil from being carried into the trap and thence into the sewers, and permits the oil to be drawn off and saved if desired.

The trap shown is an Aschcroft "open-bucket trap," which operates an ordinary plug cock. This trap will not discharge water hotter than 212 degrees. Should the water be hot enough to give off a vapor of above the atmospheric pressure, the vapor will raise the float and close the cock. Pulsations or variations of pressure within the grease tank will not materially affect the operation of this trap, as its discharge depends upon the temperature only. To the right of the trap, and joining the syphon trap S T, is a blow-off pipe from the grease tank. This pipe is used to draw the water from the tank when necessary, and if so desired the grease may be blown out in the same manner. The overflow from the trap T is connected with the sewer or some other suitable place of discharge, and the deep syphon trap S T prevents any back pressure on the trap from the main exhaust pipe, through the drip pipes *d d*.

The second object of the grease tank before alluded to, is that it forms a reservoir for the reception of the exhaust steam at the moment it leaves the engine. When the engine discharges directly into the pipes of a building, the resistance of the pipes is such that the pressure within them does not readily yield to the impulse from the engine. When exhausting into a large tank, however, the tank receives the whole, or nearly the whole, cylinder full of exhaust steam at the moment the engine lets go. It then has time to pass with slightly diminished pressure into the heating pipes during the interval before the engine exhausts again, and if properly arranged will keep the pressure in the pipes always above that of the atmosphere.

The water of condensation from an exhaust steam apparatus may be returned either to a receiving tank or pumped directly into the boiler, in which case the apparatus should have a pump governor as at G, by which pump P is automatically controlled so that its speed may correspond to the rate at which the return water come back. If the condensed water does not have to be raised to the boiler a Kieley trap will answer to return it.

The receiving tank can be omitted if desired, but it is safer to have one so as to be able to take care of the water of condensation for an hour or so should the pump or governor get out of order.

The usual method of connecting a pump governor is shown in the illustration. The steam pipe S leads directly from the boiler to the pump, and when valve 1 is open and valves 2 and 3 are closed the pump has to be controlled by hand. To control the pump by the pump governor G the valve 1 is closed and valves 2 and 3 are opened, and the steam to the pump then passes through the valve at the top of the governor; and this valve is controlled by a float within the body of the governor which opens it as the water rises, and *vice versa*.

It often becomes necessary to admit live steam into an exhaust-steam system. When there is sufficient exhaust steam for all heating purposes, of course this is not required except to supply heat before the engine starts or when it has stopped at noontime. It is usual to arrange a reducing valve for this purpose, as shown at R V, with a valve at each side of it for convenience of repairing or removing the reducing valve without either interrupting the exhaust-steam supply or shutting the live steam off at the boiler. A reducing valve should always be placed as close as possible to the system or apparatus which it is to regulate, and if placed near the boiler the pipe beyond it should be enlarged to correspond to the diminished pressure and increased volume of the steam. The resistance of a long pipe of small diameter between the reducing valve and the heating main is often many times greater than the pressure required in the heating system, and if the valve is adjusted to that pressure to begin with, and the pressure further reduced by the long, small pipe no satisfactory results can be obtained. Many a good reducing valve is made inoperative and the valve blamed for the ignorance of the man who put it in the wrong position.

Another serious case of trouble and waste of steam is when live steam for heating purposes is introduced into the exhaust pipe at a point between the check valve C V and the engine. This results frequently in a great loss of live steam through the exhaust pipe and back-pressure valve, and also increases the back pressure on the engine. A connection of this kind should never be made except as shown in the diagram.

The position of the back-pressure valve in the main exhaust pipe should also receive careful attention. It should be as near the point at which the exhaust steam passes through the check valve as it is possible to get it. When a long exhaust pipe, with a back-pressure valve on its upper end or near the roof, is employed, it forms a chamber into which the exhaust steam expands when released from the engine, instead of being forced directly through the check valve, and when thus arranged it is not possible to obtain as high a pressure in the heating system. The check valve for the exhaust steam should be of the swinging pattern. A poppet valve will do, but as a general thing they are very noisy and require more power to lift them.]

HEATING BY EXHAUST STEAM, ENGINE HORSE-POWER, SIZES OF FLUES AND REGISTERS.

JOHN GILLES, Milwaukee, Wis., writes:

"1. How do you estimate how much heating surface can be heated by the back pressure of an engine, supposing you had a 10 horse-power engine? 2. How do you measure the horse-power of an engine? 3. Do you use the same size registers for the same number of square feet of indirect heating surface for hot water as you do for steam? 4. How much larger must the register be than the flue?"

[1. You allow about 5 square feet of radiation to every pound weight of steam exhausted by the engine in an hour. If your engine is of the slide-valve type it is probably using from 45 to possibly 60 pounds of steam per horse-power per hour. Assuming that your 10 horse-power engine is using 45 pounds per horse-power per hour, you would have nearly 450 pounds exhausted, and this ought to heat $450 \times 5 = 2,250$ square feet of radiation.

2. The horse-power of an engine is expressed by the formula $H.\ P. = \frac{2\ P\ l\ a\ n}{33,000}$, in which P equals the mean effective pressure as calculated from an indicator card, l the length of stroke in feet, a the mean area of the piston, or one-half the area of the piston-rod subtracted from the area of the piston, and n the number of revolutions per minute.

3. In an indirect steam job the least flue area you should have should be 1 to 1¼ square inches to every square foot of heating surface, provided you have no

long horizontal reaches in your duct with little rise. Your register should have twice the area of the duct to allow for the fretwork.

4. For a hot-water job you need from 25 to 30 per cent. more heating surface and flue area than you do in one for low-pressure steam.

Mr. Gilles and other correspondents who desire prompt answers to their queries should not fail to give their full address, including street and number or post-office box. We will thus in some instances be enabled to mail a proof of the reply in advance of publication.]

WHEN IS IT ECONOMICAL TO USE EXHAUST STEAM FOR HEATING?

M. E., of Jersey City, N. J., writes:

"The inquiry in a recent number of your paper as to the economy of using exhaust steam for heating, and your reply to it, suggest a few thoughts on the subject in general.

"Where a manufacturing establishment is using steam for heating as well as for power, or where there are two establishments adjoining, one of which is using steam for power and the other for heating, and the amount of steam required by each is nearly the same, a plant can be erected to generate the steam for use in the two places at a small advance on the cost of a separate plant for each. The reason for this is, that while it is necessary to add about 1,147 heat units to every pound of water of a temperature of 32° Fahr. in order to change it into steam at atmospheric pressure, it requires the addition of only 33 heat units to give us the same weight of steam at a pressure of 75 pounds above the atmosphere. With this work can be performed in the engine, and then the steam can be delivered for heating at a reduced pressure, with but slight loss as compared with the gain resulting from its double use.

"In practice a good automatic cut-off non-condensing engine requires about 30 pounds of steam, of 60 pounds pressure, per hour for each horse-power. Of this about 22 pounds will be delivered as exhaust steam in the neighborhood of atmospheric pressure, and is available for heating purposes; or 71⅔ per cent. of the heat imparted to the steam in the boiler is actually available for heating after having passed through the engine. The difference between this available heat and the theoretically available quantity is lost in the engine by cylinder condensation, etc. In this manner it would be possible to obtain power and afterwards deliver the exhaust steam for heating. The proportionate cost of the coal would be about three-tenths for the power and seven-tenths for the heating.

"In other words, if an establishment had been paying $5,000 a year for the coal used in generating its steam for power, and it could subsequently deliver the exhaust steam for heating purposes to someone else, the quantity so delivered would have cost the one using it $3,500 a year for coal if generated by himself, leaving the cost of coal to the one using the steam for power $1,500 instead of $5,000 as before. This can be stated also in another way. If a party is paying $3,500 per year for coal for steam for heating purposes the addition of $1,500 worth of coal will add sufficient heat to enable this steam to be used to generate power, and then the same quantity of heat can be delivered for heating as before.

"Recently I was called upon by a large candle factory to determine what charge should be made for steam and power which was furnished to a lard refinery next door, where they were using about 4,150 pounds of live steam per hour for general purposes, besides from 15 to 20 horse power requiring about 64 pounds of steam per hour each, or altogether about 100 cubic feet of water per hour in the shape of steam. To generate this steam from water at a temperature of 122° Fahr. required about three tons of coal per day, or one pound of coal to 7½ pounds of water changed into steam of 60 pounds pressure. The exhaust steam from the engine was used in the candle factory to heat the feed water before entering the boilers, and also for heating a portion of the establishment, with such favorable results that it was estimated that an actual saving was effected in coal consumed of 80 tons per year for heating the feed water, and 30 tons for the heating done during one winter, being about 13 tons saved per month in winter-time by the exhaust steam generated by the consumption of 17 tons of coal per month. In this instance no back pressure in the engine was produced."

SYSTEMS OF PIPING.

RADIATOR CONNECTIONS.

William J. Wells, Monticello, Ill., writes:

"I have a job of hot-water heating with four radiators on the same floor with the boiler. Will it be well to make a connection at the top of the main by means of a nipple and elbow running over the wall and down to the radiators, or will I have to make a syphon and connect it with the expansion tank just before I drop to the radiator?"

[If you want to make a radiator work on the same floor as the boiler the only way to do so is by using very large pipes, say 1½ or 2-inch, to each radiator, being sure to tap the radiators as large as the pipes you use. Syphons do very little or no good, as the water cools as much in the upward as in the downward leg, and thus makes a balance. Be sure no air collects in the pipes at any point. If you go up and down to the radiators there will be high points where air can collect, at which you must place air valves.]

RETURNING WATER OF CONDENSATION TO A BOILER.

O. P., New York, writes:

"Is there any difficulty in heating by steam and returning the water of condensation from about 23 feet below level of boiler to a tank about 6 feet higher? We have a case of that class where a party wishes us to do their heating, and we have never taken a job of that kind."

[There is no difficulty in discharging water from heating apparatus 23 feet below the boiler to 6 feet above it if you will carry a pressure greater than, say 15 pounds per square inch. By such method you will be able to discharge the return water through a Nason trap into the tank, and thence pump it from the tank into the boiler; or, if you so desire, you may dispense with the tank and pump and substitute a Kieley or Blessing direct return trap and discharge the water directly into the boiler automatically.]

RETURNING WATER OF CONDENSATION TO A BOILER.

Harry B. Peacock, Hayes City, Kan., writes:

"I inclose a sketch of a high-pressure boiler to heat a mill. It is proposed to connect the return to the suction pipe that connects with the heater, and the heater is connected to a tank that feeds the heater, and from there to the pump. I told them that it would not work, but if they took the return direct to the boiler, as I have shown in sketch with dotted lines, it would work all right. I would like your opinion."

[If high-pressure steam is taken from the boiler and then through pipe A into coil B and through the return pipe C to the suction pipe D, it is very likely the high-pressure steam will blow into the heater H and thence into the water tank. If the pressure in the water tank is greater than the pressure of steam, then it appears to us that the pressure down from the tank will pass through the heater H, through the pipe D to the pipe C, and pass up into the coil. If there are check valves placed in the pipe D and the pipe C, all this, however, may be prevented; when the water from the coil will pass into the pump and be forced into the boiler along with the water from the heater. The plan you propose, as shown by the dotted lines, is the better of the two to follow, provided the pipes of the heating system are sufficiently large in diameter to get a circulation by gravitation; if not, some other method of forcing the water back will have to be followed.]

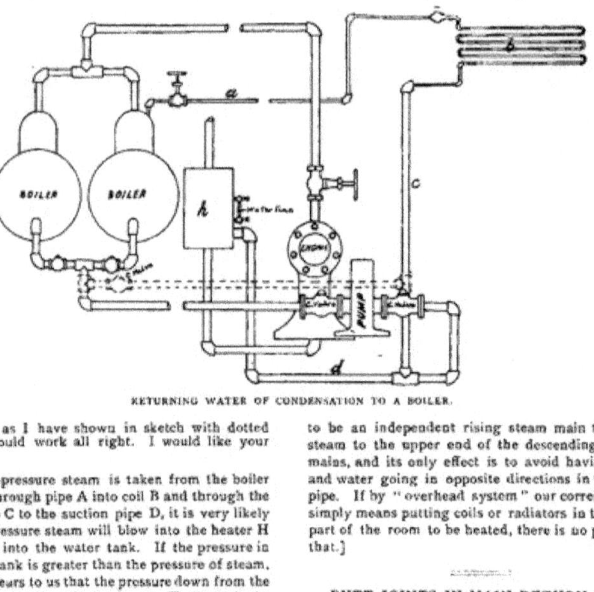

RETURNING WATER OF CONDENSATION TO A BOILER.

OVERHEAD STEAM HEATING.

M. B. & Co., Detroit, Mich., write:

"Will you kindly inform us in your next issue if there is a patent on the overhead system of steam heating?"

[There is no patent on simply taking the steam main to the top of a building and feeding downwards if you have separate steam and return connections on each radiator.

The "Mills Patent System," that many seem to think covers all methods of steam supply from overhead mains, really includes only the use of a descending main, answering at the same time both for steam supply and for return water, to which the radiators are connected by a single pipe each with no other connection. This is a single-pipe system as far as the radiators are concerned, but of course there has to be an independent rising steam main to supply steam to the upper end of the descending main or mains, and its only effect is to avoid having steam and water going in opposite directions in the same pipe. If by "overhead system" our correspondent simply means putting coils or radiators in the upper part of the room to be heated, there is no patent on that.]

BUTT JOINTS IN MAIN RETURN PIPE BELOW THE WATER LINE.

J. E. L., Goshen, N. Y., writes:

"A much-disputed point with steamfitters is the question: Does it make any difference in the working of a gravity steam-heating apparatus if the branch returns are brought into the main return pipe from directly opposite directions, connecting, for instance, the two main branches into the two runs of a 'bull-head' tee and the outlet of this tee into the boiler? I inclose rough sketch to illustrate the point. In this case two-thirds of the radiating surface is located in one end of the building and one-third in the opposite end. Some workmen contend that if connected up butt-joint, as in A, there will be trouble about the return in the smaller end. They claim that it will be held back, to a certain extent, by the greater weight or volume of water returning from the larger amount of condensation, and should be connected as at B. Others say it makes no difference whatever, so long as the connections are below the water line, and all the pipes, including the steam supply, are large enough and are properly proportioned."

[With properly proportioned return pipes for the service required, and a free steamway to the radiators, so that an equal steam pressure is maintained on top of the water in the vertical return legs, there will be no practical difference in the flow of water from either section or through either form of connection. Under those conditions that section which condenses the most steam, and the resultant water of which cools the faster, will pass the most water to the boiler. The water which retains its heat is lighter in specific gravity than the colder water, and in that proportion the colder and heavier water will fall, and so force the water in the lower pipes towards the boiler. Should one section have a free steam head, and the steam valves on the other section be closed so as not to deliver the volume of steam which those radiators should condense, the result is a partial vacuum in those radiators and the holding back of their return waters. Under proper conditions the

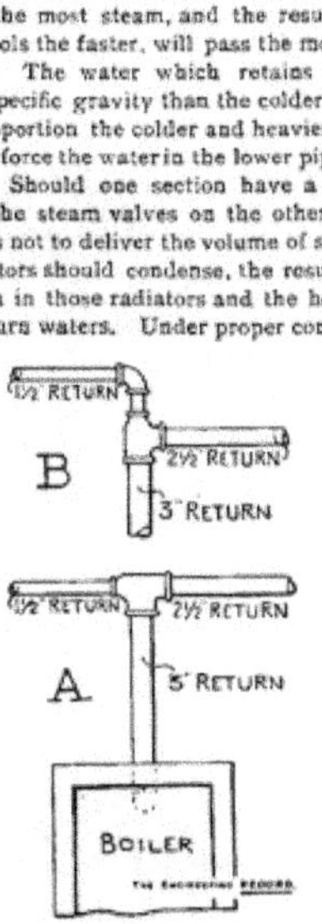

other section would then cut off the flow from the opposite section, even though it be much colder and heavier. If the returns from one section are of proper size and from the other insufficient, all other conditions being equal, the water from the former will take precedence in this same proportion.]

RADIATOR AND COIL CONNECTIONS UNDER THE MILLS SYSTEM.

EDWARD E. MAGOVERN writes:

"In reviewing a heating apparatus of the Mills overhead single-pipe class, erected in a large office building in New York City, the writer was impressed with the method of connecting radiators and coils, which, though customary, is to his mind unscientific and incorrect, invariably leading to defective circulation, and consequently to lack of heat in the building. It is his purpose, in this communication, to show wherein the connections were improper and to suggest the remedy to be applied to insure the efficiency of the apparatus.

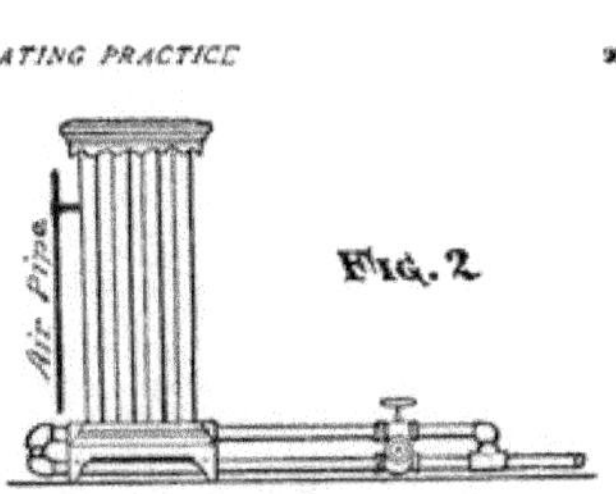

FIG. 2

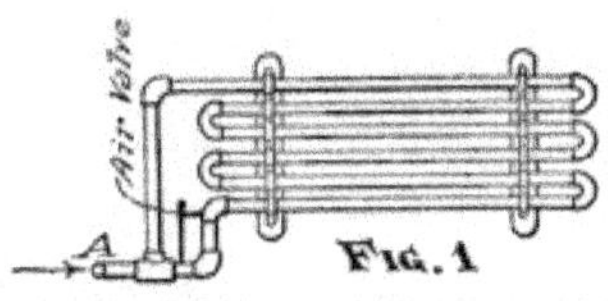

FIG. 1

"Referring to Fig. 1, the pipe, 1¼-inch, leading from the riser, is shown at A. It terminates in a tee. From the outlet of the latter a 1¼-inch pipe leads to the top of the pipe coil. The run of the tee is made 1¼-inch to ¾-inch, and from the latter a ¾-inch pipe is run, terminating in a reducer connected with the bottom pipe of the coil. On the ¾-inch pipe an air valve, connected by a ⅜-inch air and drip pipe to the main ½-inch drip pipe running parallel with the riser, is placed. The pipe A, together with the ¾-inch nipple, is given pitch toward the riser, to facilitate the discharge of the condensed steam. The theory of this method of connection is that the steam from the riser, *via* pipe A, rises, driving the air and water of condensation ahead of it, through the pipes and return bends, until the air valve is reached, at which point the air is discharged, the water following the ¾-inch pipe, and by way of A reaching the riser. What really happens is this: The steam enters the coil by both pipes, the 1¼-inch and the ¾-inch, the entry through the latter being facilitated by the air valve, and the flow in both directions imprisons air in the center pipe or pipes, thus reducing the heating surface. In the case of the radiator, Fig. 2, it will be found that the center pipes are cold for the same reason. The remedy in the case of the coil is to provide but one connection, say of 1¼-inch pipe, to be attached to the coil in place of the present ¾-inch pipe. Stop off, by plug or cap, the present connection to the top pipe, and attach the air valve thereto. With the radiator, removing the ¾-inch pipe and plugging the outlet will have the desired effect."

A PUMP-GOVERNOR HEATING SYSTEM.

C. A. F., of Idaho Falls, Idaho, writes:

"A new hotel now building in this place requires a system of heating and I have been asked for information. My experience in this line is very limited, therefore apply to you for general information.

"The temperature in this locality during winter months is variable, falling sometimes down to 30 degrees below zero at night; rising during the day to 20 degrees or more above. The prevailing winds occur in summer, with light breeze only in winter.

The hotel will have three stories and a basement to be heated, and contains about 141,000 cubic feet of air to be tempered. Walls brick and stone.

"Which system would you recommend; low pressure steam or hot water? If steam, direct or or indirect radiation? If hot water, direct or indirect radiators? What plant can you recommend? I have no practical knowledge of any."

[Our advice would be to use steam for warming a hotel, ordinary direct radiation, with the radiators in the rooms, the simplest and best method. The pressure can be anything from two pounds to 40 pounds, or even higher. In hotels the cooking is often done by steam, and then you may want a laundry engine or electric light, for which you will require steam. Let the apparatus be a gravity return one if possible, and if not, make it a pump-governor system; that is, catch the returns (condensation) in a tank and pump them into the boiler by an automatically controlled pump, as shown in the annexed cut.

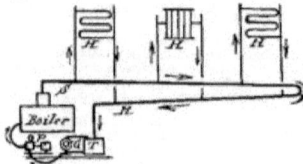

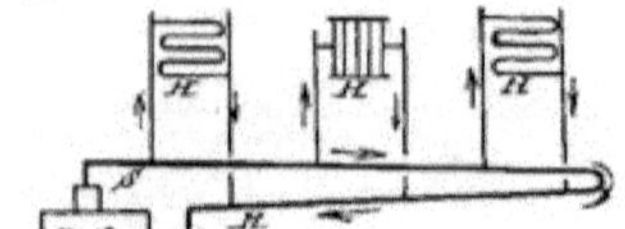

PUMP-GOVERNOR SYSTEM.

Perhaps this system is the best for those who are not experts to have to construct. There is less chance of failure. In the cut the following notation is used:

S—Steam pipe (main).

R—Return pipe (main).

H—Heaters (any kind).

T—Tank to condensation.

G—Governor (float to operate pump).

P—Pump.

Any good horizontal boiler will do for steam, and there are also many types of water-tube and cast-iron boilers now used for heating.

One square foot of heating surface to each 30 cubic feet of a span will do for your climate, with a pressure of about 10 pounds per square inch. The elevation will affect the temperature of the steam, but not enough to be appreciable, as long as you have a range, say, from 10 to 40 pounds. In cold weather, or at night, run the pressure up. If this is not ample enough, write again.]

A BY-PASS AROUND A STEAM METER.

E. E. Magovern, M. E., consulting-engineer, formerly in the employ of the New York Steam Company, sends the accompanying sketch of a by-pass around a steam meter, and says:

"This 'curiosity of crime' was recalled by seeing in a recent issue of The Engineering RECORD the by-pass through a steam trap. It really happened in a store in New York City. There was a small engine, about five horse-power, doing very little work; about one horse-power turned a coffee-mill, and there was a small return-bend coil, both con-

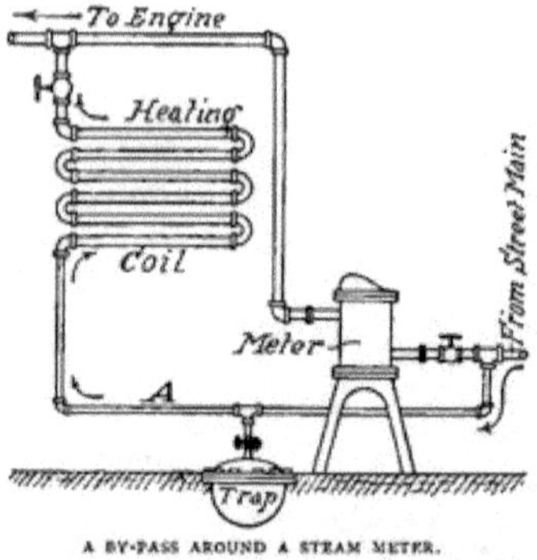

A BY-PASS AROUND A STEAM METER.

nected to the same high-pressure pipe. The drip from this coil and the drip on the inlet of the meter were both connected without check valves to the same trap. As it required a pressure of two pounds per square inch to raise the piston of the meter, the result was that the steam took the easiest course—that is, through the drip pipes, up the return of the coil, through the coil to the engine, and the meter of course failed to register, though a counter on the engine showed that the engine was being used."

[We are obliged to Mr. Magovern for his "curio." It is of course understood that the by-pass was a blunder of the steamfitter who made the connection for the Steam Company, and was not due to any fraudulent purpose on the part of the customer. We do not know how the defect was remedied, but a check valve at A would have been a simple and obvious remedy, provided the coil was high enough above the check, as the difference of pressures due to the resistance of the meter would keep nearly 5 feet of water always above the valve, if one were used.]

GRAVITY VERSUS RETURN TRAP SYSTEMS OF HEATING BY STEAM.

B. J. H., of Elizabeth, N. J., writes:

"I have a copy of 'Steam Heating for Buildings,' and would like the following information, which it does not contain:

"(1) What would be the effect of having the steam mains of a low-pressure gravity apparatus about 2 feet above the water line?

"(2) How much advantage in coal consumption or better working of apparatus would be gained by making the difference 5 feet by putting in a return trap?

"(3) Are the upright multitubular drop-tube boilers manufactured for the trade? If so, where can they be obtained?"

[(1) Unless they are large and well run and every detail of the apparatus carefully carried out, water will rise into the mains and "pounding" will be the result.

(2) If the apparatus works well as a gravity apparatus there is no advantage in putting in a return trap. There is an advantage by making the

difference of level 5 feet, if the apparatus does not work properly at 2 feet; not otherwise.

(3) We know of no one who makes the boilers you refer to. The author of "Steam Heating for Buildings" designed them and used them. Any person, however, is at liberty to make them.]

WHERE TO PLACE A REDUCING VALVE.

Stationary Engineer, of New York City, writes:

"On page 247 of your issue of April 6, 1889, in the reply to 'Steamfitter' you say, 'Many a good reducing valve is made inoperative, and the valve blamed for the ignorance of the man who put it in the wrong position.' Your reply reminds me that I have a reducing pressure valve (1½ inches) close to a boiler *a*, and the place one is very likely to place it, and that from this valve I carry steam for about 150 feet through a 1½-inch pipe to the point where it enters the steam-heating main. The pressure in the boiler is 50 pounds, and the back pressure on the engine is not much over one pound, still I find I have to load my back-pressure valve until the gauge just outside of it shows from 10 to 15 pounds, and with less pressure in the pipe on the low pressure side of the reducing valve I have been unable to get the live steam into the exhaust-pipe system.

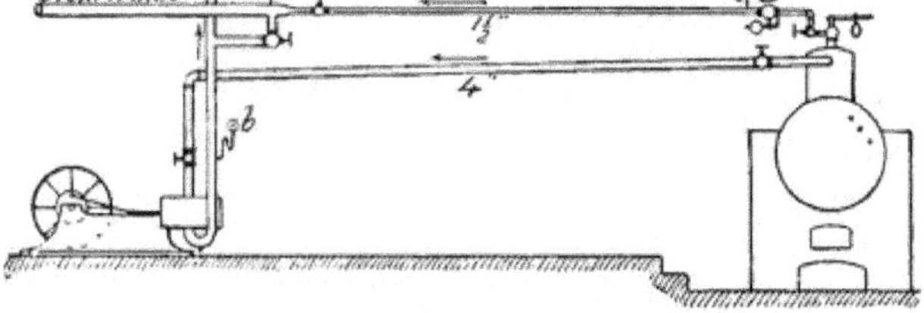

WHERE TO PLACE A REDUCING VALVE.

"Your reply, however, to 'Steamfitter' makes the reason for the higher pressure somewhat clear to me. I have great trouble, however, to regulate my pressure. At times I am forced to set the regulator to 15 pounds to get the desired amount of extra live steam I require in the heating apparatus, but in an hour later I may find that the back pressure on the engine has increased to five or six pounds, and that I have to regulate the reducing valve over again, and often I find I have to come down to 10 and even five pounds before I reduce the back pressure to its proper condition.

"Again, I find just the reverse. Everything will work properly for an hour or so at eight or 10 pounds at the low-pressure side of the regulating valve, and then I am called on for more steam and I find I have to increase the pressure to 15 pounds again.

"Now, sir, will you kindly explain the cause of the fluctuations, and also inform me how to connect my regulating valve to remedy the trouble? The accompanying sketch will give you an idea of my apparatus."

[The fluctuation of pressure near your regulating valve at *a* is due to the fact that the use of live steam in your heating pipes is not uniform. If the load on your engine is variable the amount of exhaust steam will vary also, and even with a constant load the heating apparatus will use more steam on a cold morning than it will at noon, especially if the day turns out fine.

When the engine is not doing much work and therefore not furnishing much exhaust steam, there is a heavy draft on the regulating valve to make up the deficiency, and consequently the frictional resistance in the long 1½-inch pipe is increased; and to get the necessary amount of steam through it the pressure must also be increased; hence the 15 pounds.

On the other hand, if the load on the engine increases the exhaust steam is proportionally increased and there is less draft on the live steam, but the reducing valve being set for 15 pounds will not close until that pressure is reached at *a*; consequently, an excess of live steam passes into the heating pipes and the excessive back pressure becomes apparent on the gauge at *b*.

The remedy is to move the regulating valve from *a* to *c*, taking care at the same time that the 1½-inch pipe has a pitch backward so as to drain into the boiler, or is provided at *c* with some method of getting rid of the water of condensation if the pitch of the pipe is towards *c*.

To sum up briefly, the trouble with your arrangement lies in the fact that the amount of pressure lost by the steam in passing through the pipe *a c* is a variable quantity, depending on the rate at which the steam flows and that you cannot by keeping a uniform pressure at *a* obtain the desired uniform pressure at *c* with this variable resistance in between.]

STEAM RETURNS NEAR THE WATER LEVEL.

Edward E. Magovern, of New York, writes:

"In the design of heating systems the engineer, for obvious reasons, gives a good margin between the level of the radiators and that of the water in the boiler. Then, by utilizing this margin, giving a good pitch to the returns, there is insured a good circulation in the system. It will happen frequently, however, that the margin spoken of is small, and though no theoretical difficulties present themselves, the designer fears, on close work, that incompetent or careless workmen may cause a failure in circulation.

"A case recently came to light in the writer's practice that possibly possesses sufficient novelty to warrant publication. At the building in question, owing to lack of head room and the impossibility of lowering the boiler for fear of reaching tide water, it was necessary to run the steam and return of a Mills overhead system of heating at a very slight difference in level. The figure shows the system, the main pipe being 3 and the return 2-inch. The two radiators, whose connections are shown at R R, were originally fed from a continuation of the 3-inch main A. As the portion of the building heated by R R was to be used separately, and during longer hours than the remainder of the premises, there was, after the job had been originally completed, a separate connection C, of 1½-inch pipe, dripped at D with 1¼-inch; both connections run outside the valves E F. The remainder of the system will readily be understood from the figure; the ends of the downfalls being collected at the second story in a 2-inch pipe, and dropped by G into the main return in the cellar H, which was also 2-inch pipe. The object of collecting the returns at the second story was to do away with cutting into the store floor as far as possible. To the 2-inch return was also connected the drip D previously spoken of, and also the drip I on the main riser J. Prior to making the independent connection C, the job was tested and found to circulate perfectly. At the request of the owner a fitter was detailed to make the independent connections and provided, prior to commencing, with a rough sketch of the job. On receipt of information that this work was completed, the pipes were covered and steam raised. It was then found that the radiators R R circulated well, but those in the remainder of the building were scarcely heated. A petcock placed on the top of the riser J showed a mere vapor, though the gauge on the boiler read three pounds. The gate valve E was examined, to see if by chance the gates had become detached from the steam, and a flange union on A disconnected with the expectation of finding an unperforated gasket, but no such solution of the difficulty presented itself. On personally visiting the premises the writer had the felting stripped off the pipes at the point of the more recent connections, and discovered a 2-inch swing check valve introduced at K by the fitter for some unknown reason. The difference in level between the top of A and that of the

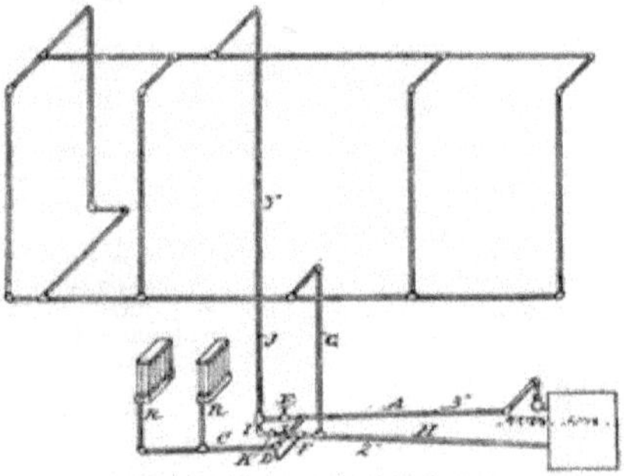

STEAM RETURNS NEAR THE WATER LEVEL.

water in the boiler was but about 6 inches, giving less than a quarter of a pound effective pressure, in the event of the 3-inch pipe being water-sealed, to move the check. Upon removing the check circulation was perfectly established, showing that the system had at some time been flooded by the water-feeder, and the resistance of the swing check was sufficient to hold the water back, thus sealing the 3-inch pipe."

HEATING COILS IN A STEAM BOILER FIREBOX.

MARSHALL, Pike County, Pa., writes:

"In a recent issue your answer to 'Selim' notes conditions and requirements which nearly correspond with our own. Some time ago our business required hot water upon several floors. We use water power, but heat by low-pressure steam. There is no plumbing shop in our immediate vicinity, but as the man in charge of our heating and the gas machine for lighting is an excellent mechanic, and has served his time at the plumbing trade, the job was turned over to him. He set up a galvanized-iron tank, connecting it up in substantially the manner you suggested in the article referred to. We had an abundance of hot water during the heating season and without extra cost. He now proposes to arrange so that we may have our water heated during the summer months by burning gas from our gas machine. Upon the chance that it may be interesting to others similarly situated, and in the belief that it has merit, I send you a sketch made by the designer of our heating coil under the steam boiler. Can you give us any pointers on this job?"

[We are able to take advantage of our correspondent's sketch to illustrate also the answer to an inquiry

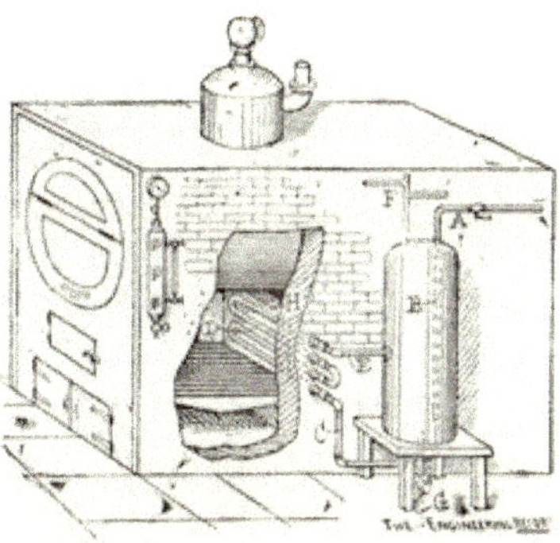

from Muncie, Ind., upon the same subject. In the accompanying drawing A is a cold-water pipe from the roof tank with a shut-off cock for use in case of leakage or repairs; B, the conduction tube inside the tank; C, a cold-water pipe from the tank to the heating coil; D, the heating coils against the fire wall H under the boiler; E, hot-water pipe from heating coil to tank; F, hot-water distribution to the required points; G, sediment cock and pipe for blowing off tank and coils. As no figures are given we advise the use of a large tank with heating coils not less than 1½ inches inside diameter, as the sizes ordinarily used in connections for domestic use might not stay primed when brought into contact with the greater heat of a boiler firebox. The larger tank supply acts as an absorbent of heat and tends to prevent foaming or "kicking back" in the heating coils. Care should be taken to lay C as far below the entrance to the heating coils as possible. The lower heating pipe should enter the firebox and continue at a good incline. All pipes should be inclined so that the current of water will be continuously upward until it enters the tank. If this is not observed trouble will ensue. The end of each heating coil should be connected up on the outside of the boiler brickwork by a right and left coupling or flange joint for convenience in repairing or the removal of heating coils. Unions should in no case be used for this work. Heavy malleable fittings should be used on inside work.]

PUMP RETURN SYSTEM OF STEAM HEATING.

ARCHITECT, of Providence, R. I., writes:

"The accompanying plan is meant to show the arrangement of buildings to be heated by the low-pressure gravity system of steam heating. The pipes are all supposed to have proper pitch for drainage and to be of proper sizes, and the water of condensation will run back to tank, from which it will be pumped into the boiler. If the boiler-house had been below the lowest building, of course there would be no occasion for a pump.

"(1) Will both buildings receive their supply of steam with the same facility and of equal dryness? (2) Will it be necessary to have any check or other valves except such as would be necessary to shut off steam, and will any traps be required? (3) Will it be possible, under any circumstances, to produce a vacuum by which the water can be drawn out of the boiler into the tank? (4) The radiators in the building on the low level will be about 10 feet below the water line in the boilers."

[The accompanying sketch explains the arrangement of the buildings in question. A is the boiler-house; B, building with radiators below boilers; C, building above grade of boiler-house; T, tank low enough for water to return by gravity from lowest radiators; from this the water is to be pumped back to the boilers; S and S' are steam and return mains respectively.

(1) If the steam and return pipes are sufficiently large and properly run there will be no difficulty in having equal dryness and nearly equal pressures at all parts of the apparatus. Absolutely equal pressures cannot be obtained through any system of piping, but differences below the limit of one pound can be secured.

(2) If your tank is a closed one, and your pipes are large enough, no checks or traps will be necessary, and a pump and pump governor will take water from the tank to the boilers. If the tank is an open one traps will be required on each main return.

(3) If the pump is properly connected with the boilers no vacuum in the tank can draw the water from the boilers, any more than the pressure within

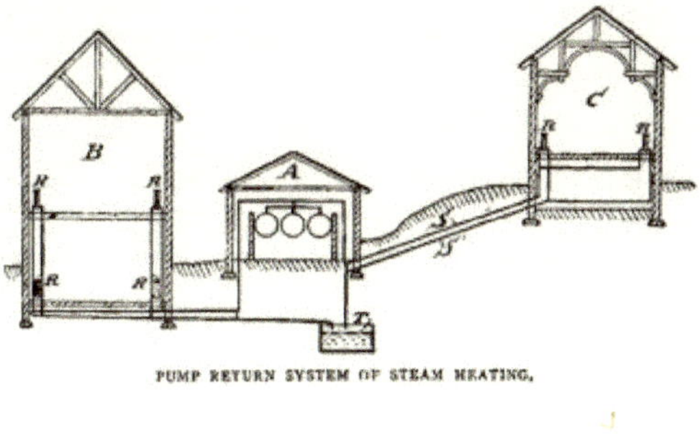

PUMP RETURN SYSTEM OF STEAM HEATING.

the boiler can drive the water out through the check valves and pump.

(4) It matters not how much the radiators are below the water line of boilers in a system such as you propose, so long as they are sufficiently above the tank to allow the water to run to the same by gravitation.]

CONNECTING STEAM AND RETURN RISERS.

STEAM HEATER, of New York City, writes:

"Do you consider the system of piping shown in the inclosed diagram a suitable one for a gravity system, or in fact for any system of steam heating?

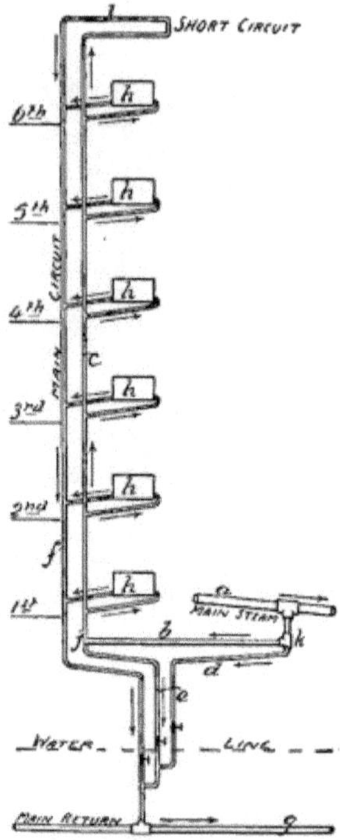

ABOUT CONNECTING STEAM AND RETURN RISERS.

"It is in use in a large building in this city, and though it works reasonably well, I claim it would work better if the connections *d* and *i* were omitted, especially the top connection *i*."

[We do not see the matter in the same light that you evidently do. The connection *d* can in no way interfere with the circulation of the steam through the radiators, especially since it connects with the return below the water line, and if the pipe *b* is of considerable length, or is higher at *j* than at *k*, the relief *d* is an absolute necessity. It is customary also to use the relief *e* to take care of the water that falls down the riser, and if *k b j* were not too long or had a sufficient pitch toward *j*, it would answer alone without relief *d*.

Had the connection *k b j* been above the main instead of below it, the drip *d* would of course have been unnecessary, as the water in the connection could then either fall back into the main or flow on to the drip *e*.

With regard to the loop at the top of the rising lines at *i* joining the steam and return pipes, there can be no more objection to it than there could be to putting another radiator at that point, even if it were kept always open. You have marked it a "short circuit," but we desire to point out the fact that it is no more a short circuit than any radiator on the line. Suppose you removed it and had the radiator on the sixth floor open, would not that radiator bear the same relation to the radiator on the floor below it that the loop now bears to the radiator on the sixth floor?

The loop is a proper connection and keeps up a circulation in the line when every radiator is shut off, thus keeping the pipes warm and ready to supply steam promptly when needed. It also helps to prevent the "water hammer," by keeping the water down in the return riser, to near the water line, whether the radiators are shut off or not.

Of course we advise taking the riser connection *b* from the top of the main steam pipe when possible. Still, the apparatus, as shown, indicates a thorough appreciation of the subject by the fitter who put it up, and in a case where the tee in the main "looks down" with any considerable length of pipe between the main and the riser, no better arrangement could be used.

We think you may have been a little confused by trying to apply some hot-water ideas to a steam system, for it should never be forgotten that the principles governing the circulation of steam and hot water are radically different, and that any attempt to apply the rules of one to the practice of the other will only result in confusion and trouble. With hot water, a fluid of practically unvariable volume has to be moved bodily through the pipes by difference of pressure, and it will take advantage of every opportunity to "short circuit" and escape the resistance of longer travel; with steam, however, when used for heating purposes, there is practically no such thing as a short circuit. When steam is once admitted to the pipes of a heating system it, like the water, is impelled by difference of pressure, but its condensation on a cold surface practically destroys the pressure, or a portion of it, in that direction, so that a continual flow of steam from all directions to the condensing surface will commence and continue as long as the surface continues to condense it. It is only necessary to provide for the escape of air and the removal of condensed water, and the circulation will take care of itself so long as the steam can reach the cold surface, which, in a heating system, is the surface of the radiator that, though warmer than the outside air, is colder than the steam within.]

EXPANSION OF PIPING.

PIPE SUPPORTS AND CONNECTION FOR A BOILER.

John C. Carlton, New York writes:

"In conversation with a brother steamfitter a few days ago he told me he had seen an account of some improperly supported pipes, and he sketched them out as shown.

"He did not say that the connection itself was dangerous, but simply the manner of supporting the pipe, claiming that when the boiler expanded it raised the main off its bearings and that all the weight came on the safety valve connection, and that in some cases the pipe was broken off. If this is the

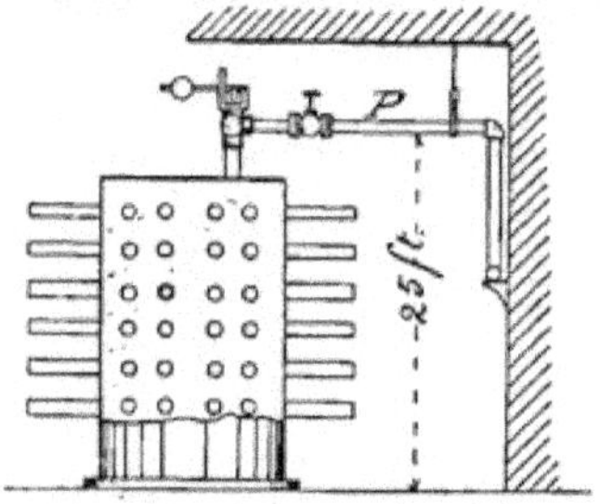

PIPE SUPPORTS AND CONNECTION FOR BOILER.

case will you kindly show me how they should be supported? The boiler is what is known to steamfitters as the porcupine boiler."

[While we cannot recall any case where breakage of a steam pipe or connection occurred in the manner stated, it is possible for an accident of this kind to occur, where the pipe P is short and of large size, making a comparatively rigid connection. The vertical expansion would be about three-fourths inch, and with even a moderately long pipe there would be sufficient spring to avoid trouble. There are cases where a spring bend might be put in the pipe P to good advantage. This would allow for expansion in both vertical and horizontal directions, and thus avoid also any trouble which might be experienced from the closeness of the wall at the right to the downward bend of the pipe.]

EXPANSION OF STEAM PIPES.

M. B., Providence, R. I., writes:

"Will you inform me how much allowance to make for the expansion of pipes under steam pressure?"

[W. J. Baldwin in his book on "Steam Heating for Buildings" says that wrought-iron pipe expands $\frac{1}{150000}$ of its length for each degree Fahrenheit to which it may be subjected within the limits used by the steamfitter. The length of pipe in inches, therefore, multiplied by the number of degrees to which it is heated and divided by 150,000 will give the expansion, for that difference in temperature, in inches, or fractions of an inch. For example: Find what the length of a 100-foot line of pipe will be when heated to the temperature of 100 pounds steam pressure, its initial temperature being zero. Applying the above rule we have $\frac{100 \times 12 \times 338^\circ}{150{,}000} = 2.7$ inches, 338 degrees being the temperature corresponding to 100 pounds pressure.

Cast iron expands $\frac{1}{162000}$ of its length for each degree Fahrenheit within ordinary limits.]

K. G., New York, asks why steamfitters do not use more expansion joints in their work than they do. He says he finds large pipes, of 100 feet in length or more, put up without any provision for the expansion, which he thinks is "considerable."

[The expansion of 100 feet of wrought-iron pipe from a temperature of 32° Fahr. to the temperature of steam at one pound pressure is about 1.47 inches; at 25 pounds pressure, 1.78 inches; at 50 pounds, 2.12; and at 100 pounds, 2.45. If a pipe under such circumstances has a single right angle turn in it or at its end, the movement towards the corner (where it should not be fastened) compensates and answers for an expansion joint. With pipe of 1 inch in diameter a turn of 5 feet is ample to provide for the movement without danger of breaking. As the pipe grows larger in diameter it will require a longer spring piece; so that a 6-inch pipe would require a spring piece of about 20 feet.]

TROUBLE WITH APPARATUS.

DEFECTIVE CIRCULATION IN A STEAM-HEATING JOB.

H. B. Peacock, Easton, Pa., writes:

"I send a draught of a steam job that will not return in the boilers, and I made a draught of a box with a float in the return in the box. The float rises and the pump draws the condensed water from it, and when it is very near empty it shuts it off till it fills again. Don't you think this way will work, or what would you advise me to do?"

[As regards the job about which our correspondent asks, its whole trouble lies in the descending loop of the steam pipe, into which all the water that is carried over from the boiler, or that may condense between the boiler and the coil, will inevitably fall, and if not drawn off as fast as it comes will, in a very short time, stop the flow of the steam.

Our correspondent's sketch indicates much greater difference in size between the steam and return pipes than is shown in our illustration, and there may be other difficulties not hinted at, but as far as can be judged from the sketch, all that is necessary to secure proper circulation is to have the steam pipe kept free from water. This is apparently intended to be provided for by the "cock for drain." If this is left open all the time, and wide enough to make sure of carrying off all the condensed water, the job would probably work, but the waste of steam would be great. If only opened at intervals the job will circulate but a few minutes after it is closed. If left open, and a trap put on to take care of the condensed water and keep the steam from blowing through, there is no reason why the job should not work, but some hot water would be wasted. Our correspondent has shown the drain cock and return pipe in such a position that it is not certain that there is no connection between them. Such a connection, if it exists, would of course be worse than useless, as the water from the return would flood the steam pipe and stop circulation at once

The proposed remedy of discharging the return into a tank, and then pumping it back into the boiler is not necessary, nor would it meet the difficulty if low pressure is used, and even if the pressure were high enough to make it work it would be very noisy from the water hammer in the pipes. If the steam can get to the coil the condensed water will have no difficulty in getting back to the boiler, provided the coil is high enough and the return pipe of sufficient size and unobstructed. We should prefer a stop valve instead of the check near the boiler, as being less of an obstruction and more reliable. Even if pumping from a tank were likely to remove the difficulty, we should not think of controlling the suction to the pump by means of a ball cock, for when it closed the pump would lose its suction, and pound away until steam was shut off. If such an arrangement were used at all the float should control the throttle valve on the steam pipe supplying the pump. If the tank were made tight so that the pump could get a suction on the return pipe, it might be able to pull steam, water and all around through the coil if the latter was not too high and the boiler pressure was sufficient, but, as in the previous case, the water hammer would give much trouble.

The only thing we can advise, therefore, is to trap the water out of the lowest part of the steam pipe and waste it into the drain, and if the boiler does not foam and the steam pipe is properly protected the loss will not be very great.]

IMPROPER ARRANGEMENT OF DRIP PIPES IN A HEATING AND POWER SYSTEM.

Observer, New York, writes:

"It is of course known to many of our readers that all steam pipes running long distances, especially where pockets or drips occur, should be fitted with drip pipes. Whether to lead these several drips to one trap, or to give each a separate trap is supposed

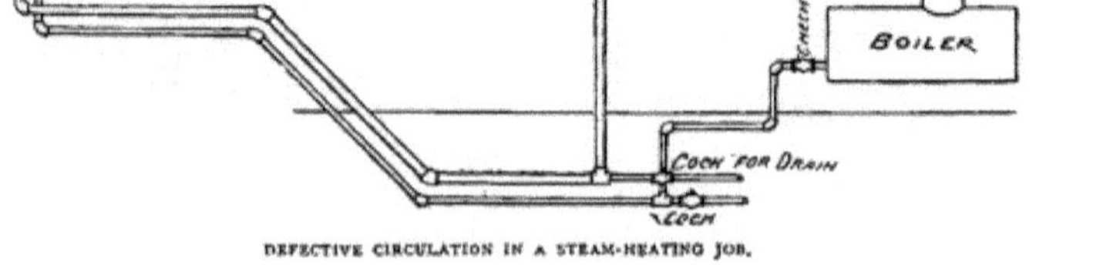

DEFECTIVE CIRCULATION IN A STEAM-HEATING JOB.

by some to be entirely dependent on the position of such drips.

"I have seen drips arranged in the manner shown in the accompanying sketch, in which the faultiness of the plan is at one apparent. In this sketch D represents the discharge pipe from the trap T; H is a drip from a low-pressure steam-heating system; and S is a drip from a high-pressure power system. It is readily seen that steam will circulate from the high-pressure or power system into the heating system by this method, and that when either system is shut off, it is fed by the other. Serious accidents may occur from this arrangement of drips. It may be thought that the introduction of check valves in both drip pipes would be a remedy, but it is well to discourage their use, since check valves are not always, in fact are rarely, tight. It is generally best to locate the trap close to the drip and to allow one trap for each drip when there are both heating and power systems in a building. If there be only one system, two or more drips may be connected to the same trap.

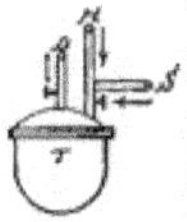

"The practice illustrated in the sketch has frequently given rise to complaints of inefficiency of regulating valves, when, in reality, the latter are trying to do their proper work, but are foiled by having a by-passed steam supply through the drips."

[The above of course shows a faulty arrangement, as it is evident the low-pressure system cannot open its check valve against the greater pressure in the other pipe. It is often advantageous to allow the trap from a high-pressure system to discharge into a system of a much lower pressure. In this case some of the water of condensation of the high-pressure system flies into steam again, simply by the reduction of pressure, and becomes utilized in the low system, the water again running off through the low-pressure trap. In the same way a high-pressure system trap may discharge into a pump-governor system and the escape steam and water be thus saved.]

AN ELEVATED RETURN AND WATER LEVEL.

S. F. C. writes:

"Last fall I had occasion to change a steam-heating job. The returns were all against the outer walls and it was desired to assemble them in a passageway on the first floor and to keep them overhead. The boiler was in the basement. It was considered advisable to keep the returns wet as the job was an old and noisy one. This was accomplished by running the main return A as shown and bringing the lateral returns B into it. At the end next the boiler, the syphon D was formed, giving the water level C as shown by the dotted lines. To prevent the longer leg of this syphon from drawing the water from the upper level and so breaking the seals of B B, and to prevent the accumulation of air or steam at the top of D, the pipe E was connected to the main steam pipe F to act as an equalizer or escape pipe. Arrangement was provided at G for emptying the return to the sewer. The steam pressure at times reached 10 pounds. The distance from F to the lowest point of A was 32 inches. The job has worked smoothly the entire winter."

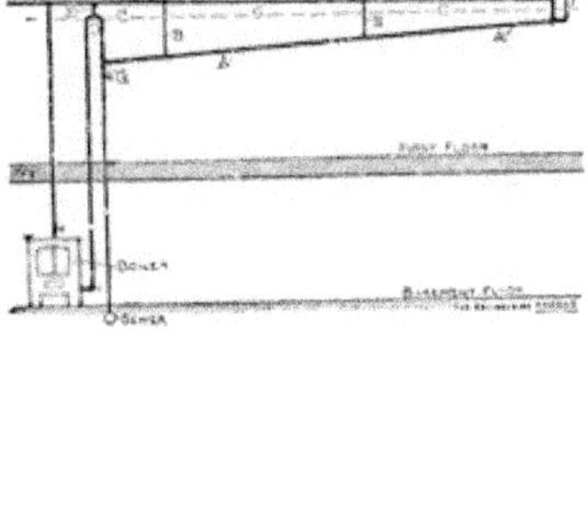

TROUBLE WITH A STEAM-HEATING PLANT.

Guy Tilden, architect, of Canton, O., writes:

"I inclose you herewith a blue-print showing the arrangement of the artificial water line of a steam-heating job in this city, which is giving some trouble. The job contains about 2,000 feet of direct radiation, is a two-pipe job, with separate return from each radiator which connects with the main return. There are straightway stop valves in the main supply and main return, but no check valve in the return.

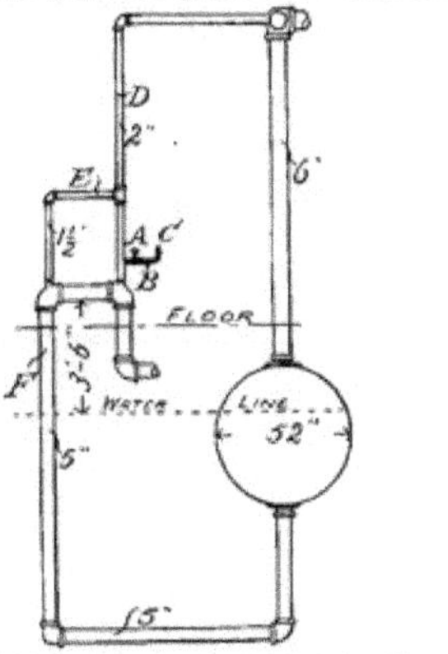

"The trouble is that the water syphons out of the returns with considerable force, which takes about half a minute, and then everything is quiet for 15 or 20 minutes, then it will syphon again, finishing with snapping, cracking, and a little pounding, then all is quiet again.

"The 2-inch pipe D we had on first, and, thinking that this choked when syphoning began, I had the 1½ inch pipe E put on. This helped it some, but only a little. Then I had the pipe E disconnected from D, and increased to 2 inches and run direct to the boiler. This helped still more. I also had a 1½-inch pipe connected with the return at F and run direct to the boiler, but this did no good at all. I also had a ½-inch pipe and stop valve connected at A, a compression gauge cock at B, and a check valve at C, and found that when syphonage occurred with eight pounds pressure on the boiler the check would momentarily open, and if we opened the gauge cock B water and steam, principally water, would flow through at a great rate."

[We have known of trouble similar to yours in an apparatus arranged with an artificial water line. The cause of it was never satisfactorily explained, but it was remedied by using a swing check valve and stop valve in the pipe D as close to the top of the syphon as it was possible to place it. The check valve, of course, should open from the main steam pipe toward the return pipe. The stop valve was used simply to vary the size of the passage through the pipe D, and is probably not essential. It, however, may save further trouble to put it in when the alteration is made for the check valve. We believe that the engineer in charge found the contrivance to work best when nearly closed. It will be necessary to alter the pipe D above the branch E so as to use a horizontal check valve. This, however, is a matter of detail which any fitter should understand.]

FAULTY ARRANGEMENT OF CYLINDER DRIPS.

HOUSE HEATER, of Brooklyn, writes again:

"I see you printed what I sent you a while ago about a by-pass through a steam trap, so I send you a sketch of another almost or quite as bad, which I used to see for a long time in the window of a tea store on Fulton Street, where they had a small steam

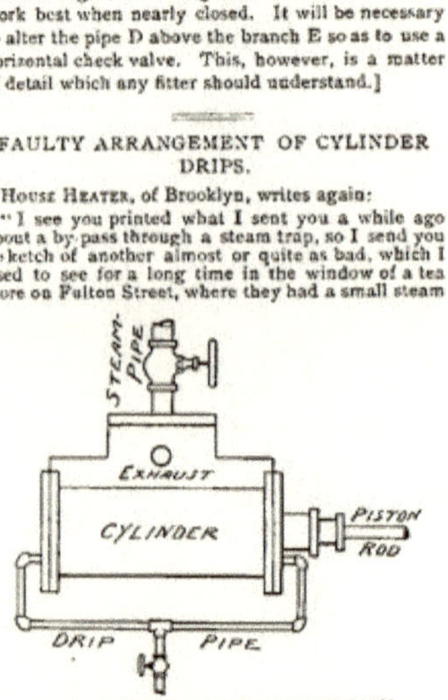

FAULTY ARRANGEMENT OF CYLINDER DRIPS.

engine for grinding coffee. I suppose the man that put it up thought he would be smart and save the price of one valve on the job; but it would be interesting to know how much that piece of economy cost first and last for extra steam or coal. If the engine ran all the time I suppose the extra valve would have saved its cost every month."

DECREASED HEATING POWER OF COILS.

C. F. A. writes:

"Our building is heated by steam. The piping has been in about 10 years, and the results are not so good as in the earlier years of the use of the system. Indeed, during the past winter the arrangement might be classed as a failure. The heating is mainly by coils, placed about the walls, and originally intended to heat with exhaust steam, which entered the coils directly after passing through a small heater. Last winter we used live steam, and the results were not as good as formerly secured with the exhaust steam. In repairing a coil I noticed a pasty deposit on the inside of the pipes. In some places it was quite thick and hard. Can this have anything to do with the loss of heating power in the pipes? How does it get into the pipes?"

[As you do not state that any constructive changes have been made in your heating system, it is fair to assume that you have located the trouble in the internal condition of the pipes. If you had no oil extractor in service while using exhaust steam, the oil used for lubricating your engine cylinder was undoubtedly passed in an atomized state with the exhaust steam, and was precipitated on the internal surfaces of the heating pipes. Your boiler may have foamed, the scum passing off with the live steam through the cylinder on the principle of the surface blow-off. The deposit you mention could have been formed from either of these causes, and this precipitation continued from year to year would account for the decreased heating power of the coil.]

FAILURE OF BOILER TO HEAT WATER.

HUTCHENS & MONTGOMERY, Huntsville, Ala., write:

"Will you please publish answer to the following question, if you can? About seven years ago, Mr. Montgomery, my partner, put up a 30-gallon iron boiler, connected with kitchen stove, to supply hot water. It has been in use, giving perfect satisfaction, for said period until about 10 days ago, when it ceased to heat the water. We disconnected the boiler, found the pipes all clear, put it back; it still refused to work. We then took the boiler down; found the vent in the pipe in the boiler rusted up. We then put in new pipe; it still refused to work. Another firm in the city went down and told the party the work was not exactly all OK; that the pipes were trapped or sacked; said they could make it work. Party told them to go ahead, with the provision that when it worked he would pay. They disconnected the boiler put in new pipes, new coil, and still it refused to obey. They then said it was in the boiler. Party told them to put in new boiler, which they proceeded to do. After everything was made new and fire built in stove until it was as hot as hades, the water in the boiler still refused to get warm. They even shut the supply to the boiler off, and then it did no better. What is the matter? The circulation pipes are all open and free, being new pipe, but what keeps the water from heating has puzzled us all."

[For seven years a domestic hot-water plant has worked satisfactorily and then failed to warm the water. In the endeavor to account for the lack of heat everything about the apparatus seems to have been examined, and much of it renewed, except the water-back in the stove. We would now advise a thorough examination of the water-back, and if it is found to be nearly filled with lime or other deposit, we would put in a new water-back. Had we been called in in the first instance, we would have commenced by examining the water-back before we did anything else. It is evident that should the heat of the fire be absorbed by the water in the back, it must either get vent by circulation or manifest itself by noise or an explosion. The trouble must be with the water-back.

A disarranged grate also often affects the heating of the water.]

TROUBLE WITH A STEAM-HEATING AND POWER PLANT.

ALBERT SPIES, New York City, writes:

"In a steam-heating and power plant it was found, at the beginning of cold weather, that not enough steam could be obtained to keep the engine up to speed, and at the same time heat the building. The conditions are shown in the annexed sketch, in which K is a 12x14-inch high-speed automatic engine driving an Edison dynamo, two elevators, and three cooling fans in the engine-room; A is a 4-inch exhaust pipe leading to the roof; from it branches a 3-inch pipe B, connecting with the heating system through the rising main F, which, further on, runs horizontally, and is designed to supply the system with exhaust steam from the engine; E is a live-steam supply pipe

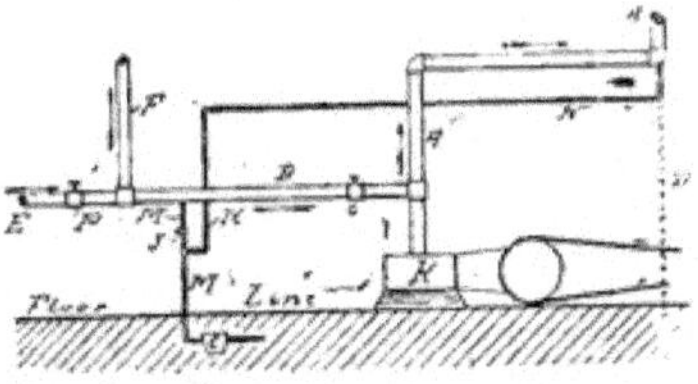

to the heating system, the intention being that, if desired, live steam could be used for heating, either alone or in conjunction with the engine exhaust. If the live steam alone is to be used for heating, the stop valve P is opened and the valve G is closed, and the weight is taken off the back-pressure valve (not shown) in the main exhaust pipe A. If exhaust steam only is to furnish heat, the valve P is closed, the back-pressure valve is weighted, and valve G is opened; M and N are drip pipes from the pipes B and A respectively, connecting with a Nason trap C. The latter discharges into the sewer. There are several other traps connected to the heating system returns. J is a stop valve. From the 3 inch main steam supply pipe to the engine the 2-inch branch E is taken, without the intervention of a pressure-reducing valve. The live steam admitted to the heating system is simply throttled down by means of a stop valve.

"With no steam, either live or exhaust, in the heating system, the engine ran satisfactorily. With the engine exhausting into the heating system, the electric lights fell below their normal brilliancy, and the engine gave signs of laboring. With only live steam turned on for heating, the engine slowed down and entirely failed to do its work, and no heat could be obtained in the building.

"Indicator cards taken from the engine showed, with the exhaust steam going into the heating system, a back pressure of something like 10 or 12 pounds per square inch.

"The indicator cards showed conclusively that the heating system was not originally intended for exhaust steam use, and could not be expected to give good results with it. Indicator cards taken from the engine when live steam was turned into the heating system showed a reduction in the initial pressure in the engine cylinder of fully 30 pounds. Since the engine did not get its proper steam supply and since the heating system also could not be warmed up, it was concluded that there was an unsuspected outlet for the steam somewhere which, if stopped, would remedy matters. The correctness of this was proved by the existence of the connections to the trap C, the drips M and N coming from two different systems, represented by the pipes B and A, in one of which (B) the steam pressure was quite high, while in the other (A) it was only about three pounds per square inch. With the valve J open, as it was, the live steam from pipe E therefore took a short cut (along the line of least resistance) down the drip pipe M, through valve J and then up through pipe N into the main exhaust pipe A. Closing the valve J remedied the difficulty, but left the pipes E and B without an operative drip connection, thus allowing water of condensation to accumulate in them. As a temporary expedient, therefore, the connections were left undisturbed, and the valve J was opened several times a day for short periods, enabling the collected water to be blown out. These periods of opening were never long enough to perceptibly affect the running of the engine. This makeshift was permitted only because of the intention to completely overhaul the heating system in the coming spring, and the consequent desire to avoid expense at the present time for patching up an unsatisfactory job. The proper thing would have been to provide a separate trap for each drip, M and N, leaving the trap C as at present, for the drip M, and carrying the drip N down, as shown by the dotted line D, and connecting it with an independent trap.

"Such faulty trap connection, as is here shown, with the intention of making one trap serve two systems under different pressures, is found every now and then, and will always lead to trouble. It is claimed by some that the introduction of check valves in both drip pipes would be a remedy, but it is well to discourage such practice, as it will rarely give full satisfaction."

[We should indorse the proposition to trap the pipes M and N which will help to operate the plant through the winter. The scheme to put check valves in those pipes is not well-advised. Such valves in pipes arranged as shown in the sketch would always be open or shut when steam pressure was turned on.]

NOISE CAUSED IN THE MAINS OF A STEAM-HEATING APPARATUS BY AN IMPROPERLY-ARRANGED RELIEF PIPE.

H. B. S., Grand Forks, Dak., writes:

"We have lately warmed a separate building 200 feet distant from our main building. The pipes in the new building are exceedingly noisy at times, snapping and pounding, and all our efforts to prevent it have thus far been only partially successful. We therefore apply for advice through your journal, sending sketch made by our janitor, and hope you will be able to suggest a remedy.

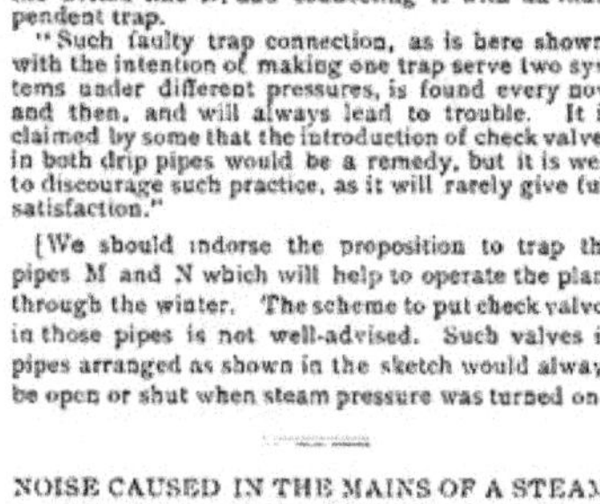

FIG. 2

"A shows connecting pipe, 1¼ inches in diameter, between the live-steam pipe B and the return pipe C. This connecting pipe, which is only 2 or 3 inches long, is open, and allows the live steam to meet the returning water. At least I so understand it, but I am not engineer enough to know the philosophy of it."

[From the data sent we are able to give you very little advice in the matter of the noisy pipes. In other words, we are unable to determine whether the trouble is a defect of principle or of detail. We are of the opinion, however, that should you alter the conection A in such a manner that you can put a check valve in it the noise may stop. If the tee B is

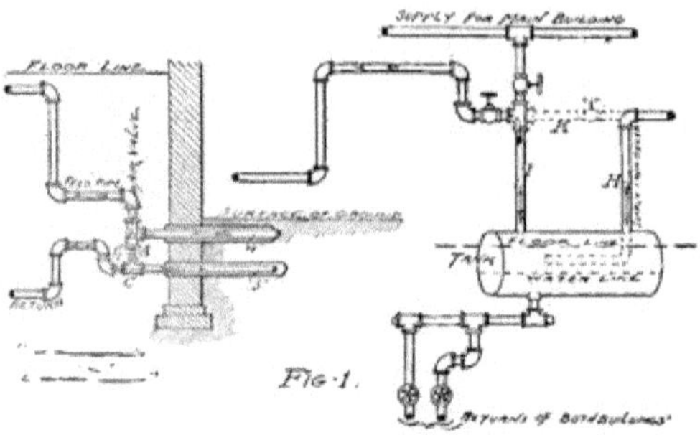

Fig. 1.

too close to the tee C to put in a connection with a check valve, connect B to some other part of the return pipe, say as shown by the extra lines in Fig. 2. Use a swinging check valve the full size of the pipe (1¼-inch), and have it open from the main into the return pipe.

We are unable to see the good accomplished by the pipe H—the main steam supply from boilers—entering the receiving tank as shown, and the steam thence passing from the tank through the pipe I to the coils, etc. We rather consider it a disadvantage by causing unnecessary friction and condensation. If after trying the check valve in the relief main the water still rises or remains in the main at B, connect the pipe H to I by the direct pipe K (dotted lines), and either remove the pipes H and I or put valves in them so they may be closed. If you do the latter, put a valve also in K as shown.]

FAILURE IN STEAM HEATING FROM CARELESS MANAGEMENT.

In a steam-heating system put into a building in New York to heat a large store on the ground floor, much annoyance was caused by the failure of the radiators to heat up, notwithstanding the fact that the steam supply through the mains was abundant.

An examination showed the pipe arrangement to be substantially as shown in the accompanying illustration. Steam was taken from the street main of the New York Steam Company, and after passing through a meter was carried along the whole length of the store in a pipe S, from which supply branches led off to the radiators R. The return main P led to a trap designed to discharge into the street return. A cross-connection between the supply and return mains S and P, forming practically a main circuit, was fitted with a stop valve, as indicated. The object of this cross-connection is not quite clear, except that it probably served some purpose in enabling the system to more readily clear itself of water when turning on steam in the morning, the stop valve being open. Through ignorance, however, this valve was allowed to constantly remain wide open, and, with sticking of the trap, which permitted the steam to blow through directly to the street return pipe, was responsible for all the trouble. The steam naturally took the easiest course through the main circuit, and the radiators received little or no supply.

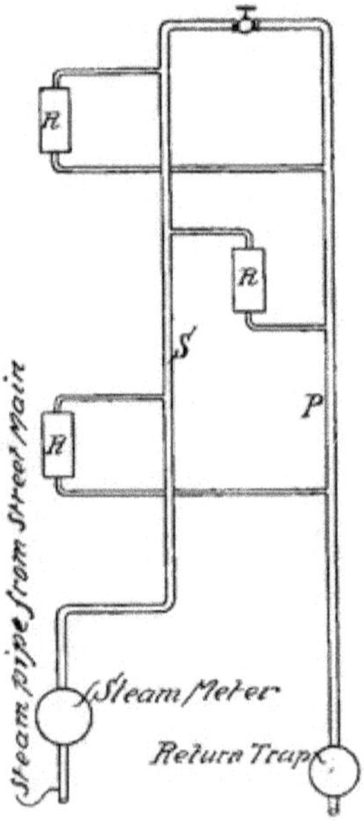

TROUBLE WITH A STEAM-HEATING SYSTEM.

Adjustment of the trap and an injunction to keep the stop valve in the cross-connection closed proved an effectual remedy.

PIPE SIZES.

A STEAM-HEATING PROBLEM.

N. K. Howard, Lincoln, Neb., writes:

"I send you a sketch of steam main as made by an architect. The cut shows a one-pipe system composed of four mains branching off from one large main, mains shown by solid lines to have a gradual fall and return into one main return pipe, shown by lines broken by a single dot, or another way is to have a water seal on each of the return pipes as marked in dotted lines near the main return. I never saw a job of that kind, and I do not see how it could work. If that will work I might have saved a great deal of pipe on some jobs I have worked on. This architect says he has had 20 years' experience in heating and plumbing, and in New York at that, so I presume he knows what he is talking about. All

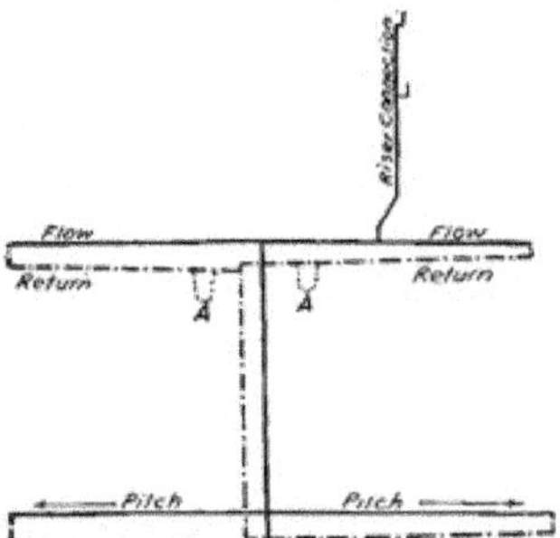

of these mains were to be above the water line in the boiler, thereby making it a dry return. I would like to know if it would work."

[Such an apparatus will work if the pipes are sufficiently large in diameter. The "seals," or "traps," or "dips" A A, etc., are simply to prevent a current of steam backward through the return pipe. The apparatus will work, even without them, but it is found in practice that the air is better taken care of when these traps are in the return. They offer very little resistance. Still, they offer enough to produce "circulation," as it is called.

We wish to lay stress on the question of the diameter of pipes with such an apparatus; they must be larger in diameter than for work that runs below the water line. The balance of pressure in a water-line apparatus is regulated by the column of water in the return pipe, and of course the diameters of all pipes must be so great that any difference of pressure will not be great enough to raise water into the steam mains.

With an apparatus as shown here, the diameter of the pipes must be great enough to prevent water from rising into the horizontal return pipes. If water fills or partly fills horizontal returns (above the water line) the work will be noisy. The returns must be either under water or far above the water.]

QUESTIONS ABOUT STEAM HEATING.

Anxious Inquirer, of Leicester, England, writes:

"Would you kindly put me right as to the meaning of the following, which occurs in tables 7, 8, 9, and 10 of Robert Briggs' 'Steam-Heating.' At the head of each table it says, 'Internal diameters of steam mains, with total resistance equal to' so many 'inches of water column.' I note, also, as the 'water column' increases the size of the pipes diminish.

"This is hardly clear to me, especially the 'water column' and 'head of steam,' as mentioned in page 84.

"Would you also inform me what sized pipes you would recommend for the 'trap' system of heating? Baldwin recommends 1-inch pipe per 100 square feet of heating surface for a low-pressure gravity apparatus, but I think this too large, where the return water is simply flowing back to a hot well under atmospheric pressure. Is there any rule so that the sizes may be got from?"

[Mr. Briggs intends to show by his tables that with a certain diameter and a given length of pipe a radiator of a stated superficial area could be supplied with steam at various pressures with the loss of pressure mentioned at the head of each table. The loss of pressure and the "inches of water column" are the equivalent of each other. Referring to Table IX., page 92, he says the "total resistance (loss of pressure) is equal to 12 inches of column," and by the same table he shows that a 100 square foot radiator, 10 feet from the source of supply, would require a pipe whose diameter was .52 of an inch when the initial pressure was 10 pounds, and that when the same radiator was 60 feet off the diameter of the pipe would have to be .75 of an inch, and that the water in the return pipe would be 12 inches higher than in the boiler, the loss of pressure, after supplying the radiator, being somewhat less than a half-pound.

Baldwin's sizes are larger than those given in the Briggs' table and are given for a general condition only—namely, two to five pounds pressure with a loss of about half a pound between boiler and water line in the return pipes. He assumes that when pipes are large enough for the above conditions they will be large enough for all others, which is true, as the higher the pressure the smaller the pipe may be. In designing an apparatus, however, it must be remembered that in getting up steam you will have all ranges of pressure from 0 to 50 or more, and that to avoid pounding the pipes must be large enough for the lowest pressures.]

AIR VALVES.

CIRCULATION IN A CHURCH STEAM-HEATING SYSTEM.

ARCHITECT, St. Catherine, Ont., writes:

"I take the liberty, as an old subscriber, of inclosing herewith a sketch plan of part piping and coils of a heating job just completed in a church buildnig. In addition to coils shown under seats of pews, there are 500 feet of surface in indirect pin radiators, and 800 feet of surface in direct cast-iron radiators. The steam pressure is from one to five pounds per square inch. Everything is working satisfactorily except the lower piping in coils under pews. It takes from one to two hours for the air to work out of the mentioned part of coil, even with the petcock full open at termination of ½-inch air-vent pipe, consequently about one-third of coil is non-effective as heating surface for a length of time. There are 17 coils, all connected with the 3-inch supply as shown. The ½-inch pipe for venting coil is tapped with the upper T at *a*, and connects with ½-inch main air-vent pipe at *b*.

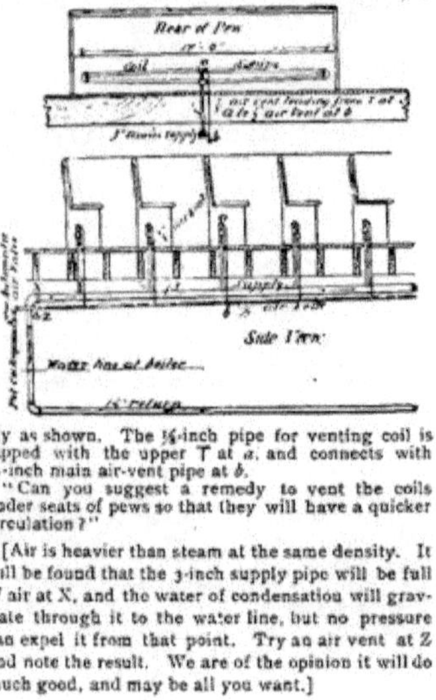

"Can you suggest a remedy to vent the coils under seats of pews so that they will have a quicker circulation?"

[Air is heavier than steam at the same density. It will be found that the 3-inch supply pipe will be full of air at X, and the water of condensation will gravitate through it to the water line, but no pressure can expel it from that point. Try an air vent at Z and note the result. We are of the opinion it will do much good, and may be all you want.]

AIR VALVES FOR STEAM COILS.

A CANADIAN steamfitter writes:

"I have just read Mr. Baldwin's book on 'Steam Heating for Buildings,' from which I have gained much valuable information.

"There is one point, however, which I cannot quite understand. It is where he speaks of the proper position for air valves. If it would not trouble you very much, I would be much obliged if you would advise me where you would recommend placing the air valves on coils Nos. 13 and 14, as shown in Plate I. of his book (flat coils and header coils)."

[The air valve is usually placed on the upper end of the return header on coil 14; on coil 13 it is usually placed in the lower bend near the return pipe. Usually those positions are satisfactory. It happens,

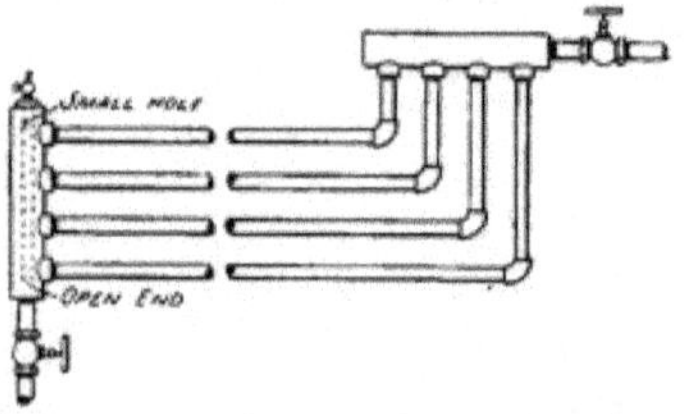

however, at times that even in these positions the air does not go off satisfactorily. It generally can be traced then to some fault of construction in the apparatus. We have remedied it in long coils, such as No. 14, in the manner shown in the annexed sketch.

Drill the plug in the top of the header, and carry a small pipe, one-fourth or one-eighth, down inside the header to about the level of the lowest pipe of the coil, leaving the lower end open. Then make a small hole in the pipe near the upper pipe of the coil, turning the hole away from the top pipe as shown. This plan draws the air off from the lowest part of the coil, while it also prevents the water from running from the air cock, should water stand in the lower pipes of the coil, as is often the case.]

CAN AN AIR VALVE ON A RADIATOR SYPHON WATER FROM A BOILER?

AIR VALVE writes:

"I have heard it stated that the use of automatic air valves on radiators will cause (under certain conditions) the syphonage of water out of a boiler. Is this true, and what are the conditions?"

[The only conditions to which your query could apply, as far as we can understand it, would be those of a radiator above the water line of a low-pressure boiler and provided with an automatic air valve, and of which the steam valve had been shut off, while the return valve was left open. Under these conditions the steam would soon condense in the radiator, forming a vacuum, which would be filled by drawing up water from the boiler, provided the steam pressure was sufficient and the height of the radiator not too great. As soon, however, as the radiator was sufficiently cool, the air valve would automatically open, and by admitting air to the radiator would permit the water to flow back into the boiler, or as much of it as the steam pressure would allow. In this case you will see that the automatic air valve really prevents the continuance of the so-called "syphonage," a term that we do not think should be used except when referring to the action of a regular syphon.]

MISCELLANEOUS.

TO PREVENT RUST IN HEATING BOILERS DURING THE SUMMER.

M. C. Meigs, Washington, D. C., writes:

"I notice in some of the trade circulars of firms engaged in steam and hot-water heating directions for preservation of the pipes and radiators, to, at close of winter's work, empty all the apparatus and then to fill again with fresh water.

"Is not this a mistake? Water itself, in absence of air, does not corrode iron. But living water, fresh from streams or springs, delivered through city pipes, has always a certain quantity of air, richer than atmospheric air in oxygen, absorbed by the water. It is this free oxygen which supports the life of fishes, and which supplies the oxygen to rust iron in contact with it. Boiling expels this air and oxygen, and the water which has remained unchanged for weeks in a steam or hot air heater should be thoroughly free from air and free oxygen and the best fitted to prevent oxidation or rusting or corrosion of the inside of iron boilers and pipes.

"Does experience contradict this reasoning, which seems at least to be justified by the laws of chemistry?

"Is it not safer to preserve in the apparatus the water long used and freed from air and free oxygen by repeated boiling?"

[It is a mistake to draw off the water from a heating apparatus in the spring and fill it with fresh water. The condensed water that remains over the winter with more or less oil, etc., in it should be let remain. It is more desirable, however, to fill the boiler to the stop valve with water than to draw it off and leave it empty, and therefore when there is no condensed water to fill it with, fresh water has to be added. Draw all this water off in the fall, clean the boiler, and fill with fresh, clean water for the winter's use.]

CIRCULATION IN HEATING TANKS.

Ignoramus, San Francisco, writes:

"1. Will you kindly give me your experience and opinion upon the following methods of heating water in closed tanks? One is provided with an injector device for heating the water with live steam at 90 pounds pressure. This device is introduced in the middle of a pipe, the ends of which are connected to the ends of the tank by return bends. The other method is to heat the water by the usual pipe coil, which is cross-connected to the exhaust and live-steam pipes, the live steam being reduced to five pounds pressure before entering the heating coil. Which tank will give the most hot water at a temperature of 150° Fahr., the steam coil to condense all the steam that will flow through a ½-inch steam pipe at 90 pounds pressure?

"2. If there is a difference, why is it and why does the water circulate better with the coil than with the jet?

"3. What is the best arrangement for heating salt water, an iron tank or coil or composition coil with iron or steel tank?

"4. Why can a building be heated with less coal by expanding the steam through an engine or pump than by turning the steam directly into the heating system from the boilers through a pressure-reducing valve or by throttling?"

[1 and 2. We know of the injector device you mention, but do not know how successfully it has been used. It would be impossible to say which method would give the best results, as there is no data at hand to work from. The ½-inch pipe you mention, if of short length, would probably supply a coil of 300 square feet of surface with steam at 90 pounds pressure. The injector device would be limited by the frequency with which the water in the tank could be changed, for if it did not change it would soon be heated up to the temperature of the steam supplied to the so-called injector, and the condensation of steam would cease, and hence the circulation of the water would stop also. In other words, the steam condensed in the injector would depend entirely upon the rapidity with which the hot water was drawn from the tank.

3. A cast-iron tank with a copper coil will probably give you as good a result as anything for salt water. There would probably be little difference between an iron and steel tank, but it is generally conceded the nearer a metal is in its composition to its ore the better it will resist oxidation.

4. A building cannot be heated with less steam by first expanding it through an engine or pump than by passing through a reducing valve. It is entirely a question of the amount of heat in a unit weight of steam, and if the pressure be the same in both instances the heat in it will be the same also. There is this point, though, when you heat with exhaust steam you warm your building with heat that would have otherwise been wasted by going out of the exhaust pipe. By the other method you draw your steam directly from the boilers, and it takes coal to make it. By heating with exhaust steam you would not affect your coal pile.]

RESPONSIBILITY FOR FREEZING OF STEAM COILS.

G., New York, writes:

"A disputed point often arises between the steam-heating contractor and the owner in unfinished houses where the apparatus is operated by the owner for his own convenience before the apparatus is accepted The coils used for indirect heating are often allowed to freeze. What is the best way to prevent freezing?"

[When an owner insists on the use of an apparatus before its completion he should be responsible for any damage that may occur, and the contractor would do well to protest in all such cases and refuse to allow the apparatus to be used unless the owner accepted the responsibility. It is the habit with some contractors to furnish a man of their own selection to run the apparatus for the owner's benefit on the payment of a stipulated amount per day. In such cases the contractor is the responsible party. No complete and well-constructed apparatus will freeze unless it is neglected. When steam is in the pipes they *cannot* freeze, as the heat will prevent it; and when steam is down, the pipes and coils should be empty as low down as the water line, and consequently cannot freeze, as there is nothing within them to freeze. If the return pipes are allowed to remain filled with water in a cold and open building they will of course freeze, and under such circumstances they should be drawn off.

A poor fire will cause freezing in a steam apparatus by allowing a vapor to go over into the pipes where it will condense and freeze, but when sufficient steam is sent over this cannot happen. Where valves are used on indirect coils, and either one of them becomes closed by accident or design, the coil will freeze if the air passing over it is sufficiently cold. When the upper or steam valve is closed the supply of steam to the coil is interrupted and what remains within it condenses, forming a vacuum that draws the water from the return pipe into the coil and allows it to freeze. When the lower or return valve is closed the water of condensation accumulates within the coil and it freezes. Should either or both valves become sufficiently closed to lessen the pressure on the coil a pound or two, according to their height above the water line, they (the coils) will freeze, or should the valves leak when closed from a defect, or by being carelessly closed, so as to allow the return water to rise slowly into the coil or a leakage of steam into the coil, they will freeze. A good way for private house work is not to use valves in the coil for indirect heating. Then the chances of freezing are reduced to mismanagement at the boiler only. Some will say this will result in an unnecessary waste of fuel. It will result in some waste, but if the air ducts and registers are closed tightly this waste will not be great. The vapor from a banked fire will go over into a steam apparatus, and should it blow suddenly cold during the night the apparatus will freeze. All causes of freezing, except those caused by leaky valves, are those of management, in a properly-constructed apparatus. If an apparatus is so made that the water in the indirect coils will not fall below the bottom of the coil under proper manipulation, then freezing may follow in very cold weather.]

OBJECTION TO THREE LUGS ON A BOILER.

STEAMFITTER, Pittsburg, asks if there is any good objection to putting three lugs on the side of a horizontal boiler that is to be set in brickwork.

[Our answer is, there is a good and valid reason why not more than two lugs should be used on the side of a boiler of ordinary length, which is, when *three* or more lugs are used on a side the middle pair of lugs will have to carry the whole weight of the boiler should the foundations or the walls at the ends of the boiler settle.

Where three lugs are used, the danger of breaking off a lug or of pulling a piece out of the side of the boiler is increased. Indeed the initial rupture in a boiler shell is often traced and found to start at the rivet holes under the lugs.]

MEASURING PIPE IN FORTY-FIVE DEGREE FITTING.

A CORRESPONDENT writes:

"Having frequently observed the awkward and slow way in which many steamfitters lay out and measure work when using 45-degree ells, I would direct attention to a very simple method:

"Let A be a line of pipe to be joined to another line B, by the 45-degree connection C. It is assumed to be found by measurement that the perpendicular

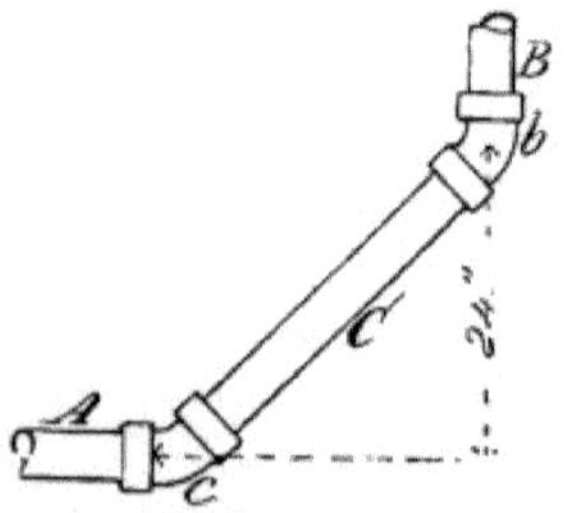

FORTY-FIVE DEGREE PIPE FITTING.

distance from the center of the 45-degree ell *b* to the central axis of the pipe A is 24 inches. Add to 24 inches thirteen thirty-seconds of 24 inches and it will give with all requisite accuracy the disance from the center of the ell *b* to the center of the ell *c*, thus:

$$\frac{13}{32} \times 24 = \frac{39}{4} = 9\tfrac{3}{4}$$

$$24 + 9\tfrac{3}{4} = 33\tfrac{3}{4} \text{ inches.}$$

"The reason for this is perfectly simple. The ells *b* and *c* being 45 degrees, the connection C is the diagonal of a square, a side of which is 24 inches, and the ratio of the diagonal of a square to one of its sides is $1\frac{13}{32}$ very nearly, or if decimals be preferred, 1.41."

CONTINUOUS USE OF WATER IN A STEAM-HEATING BOILER.

N. I. B., Peoria, Ill., writes:

"Last fall I put a steam-heating apparatus into a residence now approaching completion. There are 1,050 square feet of heating surface in it. The water main leading to the house is now frozen and there is no other water convenient. The hardwood men, decorators, and other mechanics are working in the house. We have three gauges of water in the boiler now. We do not want to let down steam, as that affects the hardwood work. I have had no such experience before and should like to know how long the present supply of water will probably last."

[Take precautions against the common practice of drawing hot water from the boiler, or steam for making glue, paste, etc. Remove all valve handles which might be tampered with, and see that air only is let out of the air valves. Then if you have no leaks, the amount of water you have on hand may serve you until the main is thawed out. In any event the loss will be so small and gradual that you can by attention guard against damage.]

O. L., St. Paul, Minn., writes to know the expansion of air and the true proportion an inlet duct to a radiator should bear to the flue in the wall.

[The expansion of air is about $\frac{1}{491}$ of its bulk for each degree Fahr. it is warmed. A good common rule among heating engineers is to have the heat flue one-quarter larger than the inlet, or, perhaps more properly speaking, to have the inlet not less than four-fifths of the hot-air flue.]

ABOUT A STOP VALVE ON A HEATING MAIN.

A. D. R., of Bristol, Tenn., asks:

"Do you think it advisable to put a steam stop valve on the main leading from a low-pressure steam-heating boiler? Is such a valve detrimental to such a job? If it is or is not please give your reasons."

[In low-pressure work, such as house-heating, it is safer to dispense with a stop valve in the main than to use one. In fact, if one is put into the pipe it is practically impossible to use it to any advantage.

Should there be a strong fire in the boiler the stop valve cannot be closed without a dangerous increase of pressure in the low-pressure boiler unless the safety valve relieves it, but even in that case it is not desirable to blow off into the cellar of the house, and an escape pipe from the safety valve of a house apparatus is not advisable. Of course an engineer is liable to reply to this and say, "The door of the furnace should be thrown open and the fire checked," which is very true, but it is also true that throwing the furnace door open on a low-pressure apparatus—carrying, say one or two pounds of steam—will, as a general thing, check the formation of steam so rapidly that a partial vacuum will exist in the pipes in two or three minutes, and this being the case there is no necessity for a stop valve on the main, as simply opening the firedoor answers the purpose of interrupting the steam better than closing a valve when pipes are to be repaired or radiators disconnected.

With a good fire and hot water in the boiler, 10 minutes after the door is closed again steam will be formed and flow into the pipes and the apparatus will be in operation again. There is danger also in the use of a main stop valve where no check valve is used in the return pipe. Should the main valve be closed and the return valve neglected, the water will "back" out of the boiler and it will be "burned" for want of water.

To sum up, no purpose can be accomplished by using the valve that cannot be equally well attained by opening or closing the furnace door, and closing the valve, without such care of the fire as an ignorant servant is not likely to take, may result in the destruction or explosion of the boiler.]

COMBUSTIBLE GAS FROM A HOT-WATER HEATER.

H. H. Hill, Brandon, Vt., writes:

"I have a hot-water apparatus in my house. I was told to open the air cocks and let the air out. I did so as often as two or three times a week, and one evening, with a lighted lamp in my hand, I opened one and it took fire, burst the chimney and came very near setting the house on fire. After two days any of them will burn like a gas jet for about one-fourth to one-half a minute. This appears to be a little dangerous. What is the matter? It did it all last winter and I thought it would get over it this summer, but it hasn't. What is to be done? Has any steamfitter made a mistake and sent me a gas generator instead of a hot-water apparatus?"

[The Engineering Record has frequently noted this occurrence. There is nothing whatever dangerous about it, as it is presumably only a hydrocarbon gas. Whether it is formed by the decomposition of the water direct, or by some impurity in the water, we do not know. It is well known, however, that particles such as oil, etc., will decompose on the inside of a hot surface, and that the gas thus formed is insoluble and will separate from the water. There is no danger from it, however, as it cannot ignite until mixed with air, and the quantity is inconsiderable.]

TROUBLE FROM PRIMING.

Greenwich, New York City, writes:

"I have a vertical boiler which is used partly for furnishing steam for power, and partly for heating a building, which is giving trouble from priming. You may be able to suggest some remedy for the difficulty. The boiler is 6 feet in diameter and 9 feet in height. It contains 420 2-inch tubes, 6 feet 6 inches long, and the diameter of the firebox is about 5 feet 6 inches. The tubes are laid out within a circle having a diameter of about 5 feet. It has been suggested to me that the difficulty is produced by the great number of tubes and the deficiency of room for circulation. If this is the case would it be well to take out some of the tubes, and if so, what number and in what locality? The steam is discharged from the boiler through a nozzle attached to the side of the shell. In the interior a plate covers the mouth of the nozzle, standing off a distance of 3 inches and

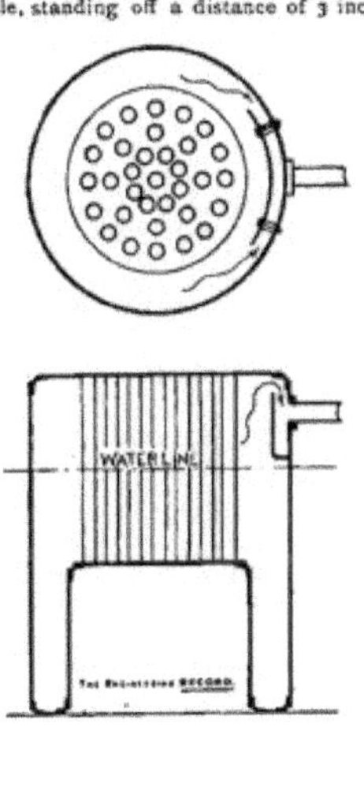

preventing access to the same except through the opening which is left at the extreme top, close to the upper head."

[We have had a somewhat similar experience in our own practice. We lowered the water line as much as we consider safe and found an improvement. We also tried several methods of using a baffling-plate over the inside mouth of the steam nozzle and found that a large plate set out sufficiently far from the outlet so as not to appreciably contract it and open at the top and sides, but closed at the bottom, which bottom was just about the water line, overcame the trouble. Our theory was the agitation attending on ebullition threw the water against the steam pipe and it was carried out with the current of steam, which was rapid at the mouth of the pipe. When the plate, which was about 2 feet long, for a 4-inch pipe and bent to about the circle of the boiler was applied the velocity of the escaping steam at any part of the edge of the plate was so slow it permitted the water to fall back into the boiler and none but fairly dry steam to escape. The diagram shows how this was arranged. The plate should not be set more than the diameter of the pipe from the shell of the boiler and as much closer as possible without withdrawing the steam. The plate can be held by four sockets and bolts passed through shell of boiler and plate. If the manhole or handhole through which the plate has to be passed is not large enough to admit the plate whole make the plate in parts and bolt it together within the boiler.]

RADIATING SURFACE AND REDUCED STEAM PRESSURES.

PRESSURE, Brooklyn, N. Y., writes:

"The undersigned, who is a working steamfitter, has had the question asked him several times whether you cannot use a smaller pipe and a smaller amount of radiating surface to give the same result on a steam-heating job where a reducing valve is used than you can in an ordinary low-pressure gravity system of, say five pounds, five pounds also being the pressure to which the steam is reduced in the other system."

[There should be no difference between the heating surface of the two systems. The heat given off per square foot of surface depends upon the difference in temperature between the steam in the radiators and the air in the room to be warmed. In each of the two cases you mentioned, if the steam is at five pounds pressure the number of heat units transmitted by the two systems will be equal, provided that the amount of heating surface and the other conditions are the same. The fact that the steam in one system passes through a reducing valve in no way affects the efficiency or value of the surface.]

STEAMFITTERS' KNOCK-DOWN BENCH.

A VERY convenient bench for steam and hot-water pipe has been devised by George Andrews, of Minneapolis. It is strong, simple, easily made, and very rigid and can be instantly taken down and folded up into a thin flat shape easily packed and carried, and of a size that will enter any ordinary door or window. The table is bolted to side and end pieces E E, and has the legs attached by bolts through the iron straps S S so as to form hinges about which they may be revolved up to the table in the direction of the dotted arrows. Each of the four braces D D has three hinges, A a 10-inch tee and B and F 16-inch straps, which enable them to fold with the legs. The legs are re-enforced by an iron face strap H H, which at the top forms one part of the hinge, and at the bottom turns out to form a plate through which lag screws are used to secure it to the floor.

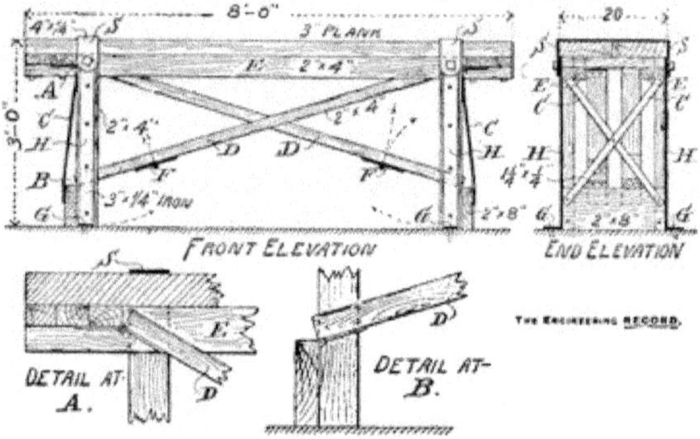

STEAMFITTERS' KNOCK-DOWN BENCH.

HEAT-CONDUCTING PROPERTIES OF BUILDING MATERIALS.

WILLIAM ATKINSON, Boston, Mass., writes:

"Can you refer me to any experiments on the relative non heat-conducting qualities of brick, especially the modern kinds of light-colored and speckled brick, and of brick as compared with stone in this respect; also to any discussion of the merits of painting brick walls to make them impermeable to moisture? I am familiar with what little there is in Baker on these subjects."

[Thomas Box, in his "Practical Treatise on Heat," page 211, gives a table of the conducting powers of materials prepared from the experiments of Péclet. It gives the quantity of heat in units transmitted per square foot per hour by a plate 1 inch in thickness, the two surfaces differing in temperature 1 degree:

Fine-grained gray marble	28.00
Coarse-grained white marble	22.4
Stone, calcareous, fine	16.7
Stone, calcareous, ordinary	13.68
Baked clay, brickwork	4.83
Brick dust, sifted	1.33

Hood, in his "Warming and Ventilating of Buildings," page 249, gives the results of M. Depretz, which, placing the conducting power of marble at 1, give .483 as the value for firebrick. For further information on M. Depretz's experiments Hood refers to "Traité de Physique," page 201. We know of no experiments on different kinds of bricks.

As to your last question, we believe that there are several patent processes for painting walls to make them impermeable to moisture which have proved successful.]

THE SMEAD SYSTEM FOR SCHOOLS.

N. K. Howard, Lincoln, Neb., writes:

"The School Board here is considering the Smead system of heating. I would like information on the system, and to have your opinion on the subject. I think there are enough other steam and hot-water systems that would give much better results."

[Just what is proposed by the so-called Smead system for Lincoln, Neb., does not here appear. There has been considerable controversy regarding the merits of the system after its adoption for certain school buildings in Detroit, Mich., and our correspondent is referred to the health officer of Detroit for information. If it is proposed in Lincoln to evaporate and dry the excrement and urine in the building, the following query and the reply thereto, reprinted from The Engineering RECORD of August 24, 1889, gives our opinion at this time also:

"William Moran, of Titusville, Pa., asks the following questions:

"1. In a school building, is it good sanitary practice to connect the ventilating system of the rooms with the vaults and urinals located in the basement?

"2. Is it good sanitary practice to evaporate the excrement and urine of 400 persons, conveying the contaminated air through brick and mortar ducts, with which all the rooms have direct connection, by underfloor spaces and open registers?

"3. In a school building, heated by warm-air furnaces, is it possible to evaporate, from day to day, all the solid and liquid excrement of 400 persons, deposited in water-tight brick-lined vaults, located in the basement, by means of such currents of air, drawn from the rooms above, as may be induced to pass through the vaults on their way to base of the ventilating shaft, without using other means than the draft of a warm flue to create such current?

"4. In the arrangement mentioned in question 3, will not the currents of impure air sometimes be reversed, to the great peril of the occupants of the building?

"5. Would not the foul air leaving top of shaft, under certain conditions of the atmosphere, settle down to the earth, and be an unbearable nuisance to all the neighborhood immediately surrounding the building?

"It is not good sanitary practice to place the vaults and urinals of a school building in the basement of that building. It is not good sanitary practice to attempt to dispose of the bulk of the excrement and urine of several hundred children by evaporation by means of a current of air drawn over such material and sent up a shaft. For a time such a system may be made to work without much danger of creating a nuisance in the form of offensive smells, but ultimately the brickwork of the flues and walls through which the offensive air is drawn is likely to become saturated and give off an unpleasant odor; the foul vapor escaping from the top of the shaft may be precipitated by rain, or in dry weather dangerous particles may fall in the vicinity in the form of dust. For these reasons the system should not be used in a schoolhouse in a thickly-populated locality; it should not be used where it is possible to connect the water-closets of the schoolhouse with sewers, and in thinly-settled localities, where there is abundant room for separate buildings for water-closets, they should never be placed in the main building. With regard to questions 3 and 4, it is theoretically possible to maintain a constant current of air in one direction which will produce the evaporation referred to, but to do this will require much more constant care and watchfulness on the part of the janitor than it is reasonable to expect from such an official."]

COLD AIR FROM A STEAM-HEATING RADIATOR.

F. W. S., Boston, writes:

"We have a steam-heating radiator in our office. At times I want more heat, and on opening the air valve on the radiator a rush of air comes from it, cold enough to be from a refrigerating machine instead of a steam heater. At first it is cold enough to chill the hand, but it gradually grows warmer. Where does this cold air come from?"

[It is not the low temperature of the air which comes from the radiator that chills your hand. It is the sudden expansion of the volume of air expelled so rapidly from the radiator by the steam pressure behind it, which air absorbs heat from the surrounding air, and your hand being well within this influence and warmer than the air makes this action the more apparent. The dryer and colder the air in the radiator and the warmer and damper your hand the more perceptible this sensation will be. Compressed air injected into damp summer air will create snow, or then an icicle, if the conditions are favorable.]

METHOD OF REGULATING DRAFT BY EXPANSION TANK.

A READER, Billings, Mont., writes:

"In answer to W. P. Powers' inquiry, in your last issue, in regard to an automatic device to govern dampers on hot-water apparatus, I respectfully send the inclosed cut that I think will explain itself to any good workman. I have used it with good success in

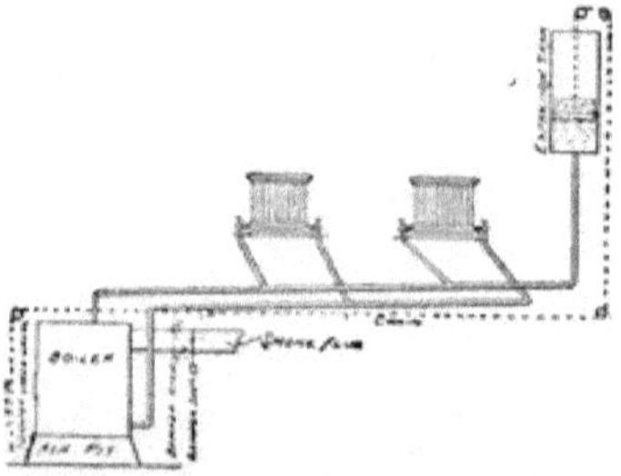

my practice. The device, of course, is intended for an open-tank system, and can be regulated for any temperature of water by adjusting chain. The float and damper should be counterbalanced. On the end of lever, attached to smoke flue damper, there should be an iron or lead weight to close damper when float rises."

[Our correspondent's method is not unknown to many in the hot-water trade. For some reason, however, it has not received the recognition it would appear to command. Like almost everything else it has a weak point, which is the trouble to keep the desired level in the expansion tank. It cannot be used when a ball cock is used, as the ball cock keeps a constant level unless it is submerged to operate only at a very low point; and when there is no automatic supply, unless the apparatus is absolutely tight, the variation in the tank destroys the regulation. In the hands of one who will look after the careful regulation of the height of the water it should give good results.]

LETTING COLD WATER INTO A HEATING BOILER.

A. H. I., Rutland, Vt., writes:

"The automatic water regulator on my steam-heating boiler is out of order. Please let me know how often I should put water in it, and when, or how to manage it."

[If your regulator has done good work, have it repaired by a competent man, or perform the functions of the regulator yourself as nearly as you can, by letting in a small stream of water when the water level has dropped to the lowest line decided on, which should be shown by a mark on the boiler water-gauge glass. A good time to let water in is in the morning when the steam and fire are low, or preferably, at any other time when those conditions exist, as the chilling of the water in the morning will delay the getting up of steam. A safe practice is never to turn on a full head of cold water into a boiler when it is warm. A small stream, though requiring a few minutes longer to fill, may save a large repair bill.]

HOT-WATER HEATING—NOTES AND QUERIES.

GREENHOUSES.

HEATING A GREENHOUSE.

FITTER writes:

"We submit a ground plan and section of greenhouse, with dimensions, and which we would be pleased to have your ideas upon, and instructions with regard to how to make a neat job, and one that will heat properly the space, say, when the weather is down to 20 degrees below zero. It is in the city and we want to make a neat job of it. We want you to render us advice as to location of stand-pipes or tanks, air cocks if you advise any, or other steps if used; the best means to obtain a good circulation and proper heat with the least amount of 4-inch pipe; the best compound for rust joint, etc. The boiler has two outlets as shown. Now, will you draw the runs and the number for the side and middle beds and returns (the return outlets are on bottom of boiler, not shown), and mark sections of the pipes on the section plan? Boiler is located in the pit where stairs go down.

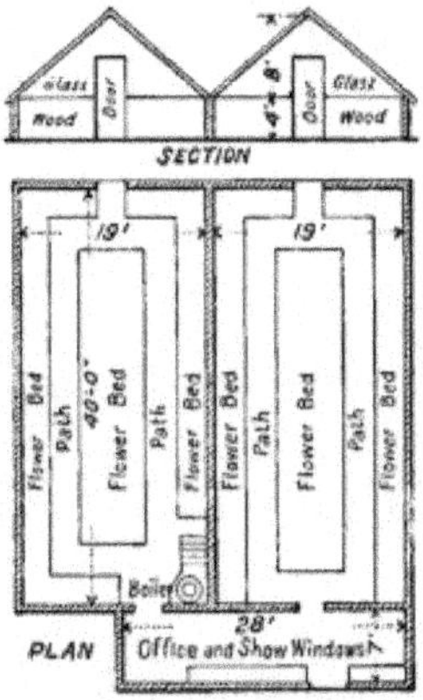

"Mark location of stand-pipes or tanks and give us the proper grade you would run the pipe.

"Any pipes run near or on path at the office will be boxed with steps, but we do not want to cross the rear door with the pipes if it can be avoided.

"Size of boiler used, 30x48 inches. Approximate heating power of 4-inch pipe, 1,250 feet. Heating surface, 4,082 square inches."

[The glass surface of your building is about 2,400 square feet. A good common rule to follow in so cold a climate, in determining heating surfaces for greenhouses or nurseries, is to allow 1 foot of 4-inch pipe to each 2 square feet of glass in a perpendicular form. This would call for 1,200 feet of 4-inch pipe.

In your case, however, the usual slanting roofs prevail, and in this case 1 foot of pipe to 3 of glass is considered enough. This would call for 800 feet of 4-inch pipe. It will be noticed in your plan that the total length of your outside "benches" or beds is about 200 feet. If you use four pipes the length of these beds it takes about the length of pipe you require. You can, in your judgment, use all this pipe under the outer branches, or you can divide it and use some under the inner ones, though the latter we do not advise, as we consider it unnecessary.

From your plan we assume the boiler to be on the level with the floor. In such a case you will have to use your expansion tank near the boiler, so you can pass over the doorways. Place the tank high up, and if you use four coils, which presumably you will for the four outside beds, take four separate flow pipes to the four coils above all doorways, etc. This may compel you to take the flow pipes for the right-hand side out-of-doors. In such case use two flow pipes from the boiler to two expansion tanks, placing one in each house high up, then flow from the tanks to the coils. In returning you can carry the return pipes with a little dip to get past the doorways. Let the pipes have a slight downward inclination and you will require no air cocks.]

HEATING WATER FOR WATERING GREENHOUSES.

JOSEPH HEACOCK, Jenkintown, Pa., writes:

"I take the liberty of inquiring of you for the best arrangement to heat the water that we use to water our greenhouses with. The water is taken directly from the street main, and in cold weather, in winter, it is only about 40 degrees, and we want to raise the temperature to about 70 degrees when we apply it with the hose.

"We do our heating with two 50 H. P. return-tubular boilers, and have an abundance of steam that can be used for heating the water. I have thought that a circulating boiler, filled with tubes for the stream to pass through, might answer, providing there is any arrangement by which a valve at the top, regulating the steam, could be made to work automatically. We use the gravity system of heating, all condensed steam returning to the boiler, carrying two to three pounds of steam. The water is taken through a 2-inch Worthington meter, under 80 pounds pressure, and is used in each house through ¾-inch hose.

"We find in practice that *some* varieties of roses will not stand being watered with water at 40 degrees, when the temperature of the house stands at

70 degrees, hence the great importance to us of some arrangement that would automatically keep the water at the desired temperature.

"I would be exceedingly obliged to you for any information or suggestions that you might be kind enough to give me."

[We have referred this matter to Mr. William J. Baldwin, heating engineer, of New York, who sends us the following:

An ordinary feed-water heater or any other tank with the steam coils within it, may be used for warm-

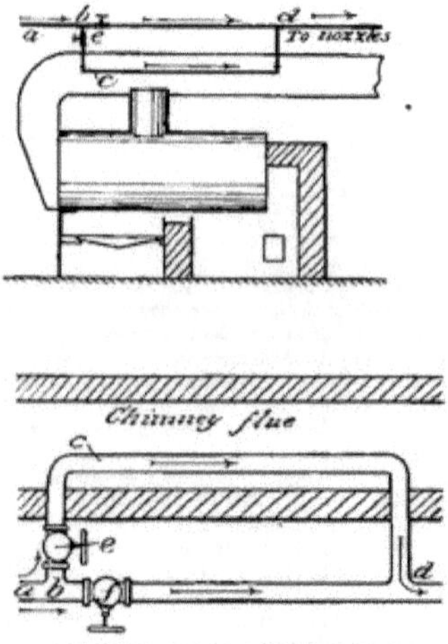

PIPE ARRANGEMENT FOR HEATING WATER.

ing the water from 40 to 70 degrees, just as well as from 40 to a much higher temperature, provided the inlet and outlet steam supply is controlled by an automatic or thermostatic valve, and there are several electrical controlled valves that will answer this purpose nicely. The water of condensation, however, from such an apparatus cannot be returned by gravity into the boiler, and must either be wasted or returned by pump and governor, or their equivalent, all of which makes a large and more expensive apparatus than would be necessary for all the sprinkling water required. A cheap method of accomplishing the same, and one that will utilize waste heat, would be to take the cold-water supply pipe *a* and branch it at *b* into the chimney or smoke flue, returning into the pipe *a* again at *d*. At *b* two valves are used, *e* and *f*. When the valve *f* is closed and the valve *e* open, all the water is made to travel through pipe *c* to the points of delivery, and thus can be delivered comparatively warm at the nozzles. To regulate the temperature at the point of delivery, the valve *f* may be opened partially and *e* partially closed, so that the current of water is divided at *d*, part going through the direct pipe and part through the heater in the chimney, thus giving almost any regulation of temperature by the manipulation of the valves *e* and *f*. The length of the pipe *c*, of course, will largely depend on the quantity of water drawn. The quantity for sprinkling flowers, however, not being great, from 10 to 12 feet of 2-inch pipe is probably all that would be required in the heat of an ordinary chimney. An arrangement like that shown can be introduced into the dome of a boiler, or into a large steam pipe, just as well as into a chimney, in which case of course the length of the pipe *c* can be greatly reduced.]

HOT-WATER HEATING OF A GREENHOUSE.

W. F., Binghamton, N. Y., writes:

"I have been considerably puzzled by hot-water heating work. In the heating of a greenhouse shown by accompanying sketch, where the heater is 10 inches below the coils of radiating pipes, there seems to be very poor circulation. Can you suggest a means of improvement?"

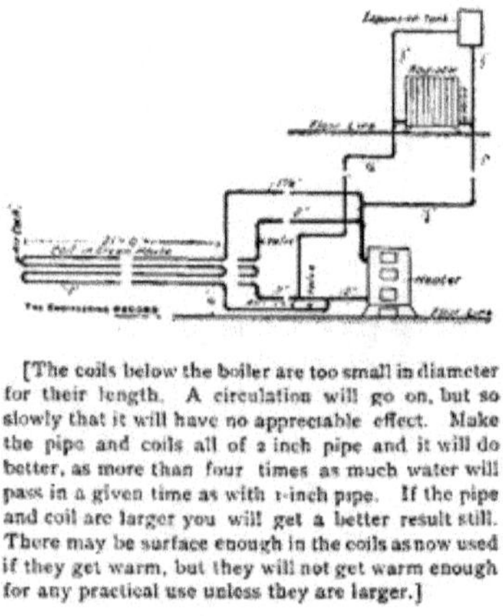

[The coils below the boiler are too small in diameter for their length. A circulation will go on, but so slowly that it will have no appreciable effect. Make the pipe and coils all of 2 inch pipe and it will do better, as more than four times as much water will pass in a given time as with 1-inch pipe. If the pipe and coil are larger you will get a better result still. There may be surface enough in the coils as now used if they get warm, but they will not get warm enough for any practical use unless they are larger.]

TROUBLE WITH APPARATUS.

TRAP IN A HOT-WATER HEATING RETURN PIPE.

C. V. Z., Providence, R. I., writes:

"My house is heated by the indirect hot-water system, all of the radiators being hung from the basement ceiling as shown by the accompanying sketch. The cast-iron heating radiators A are encased, outside air being carried to them through air ducts B. The warmed air passes into the rooms on the first floor through a short tin neck C C, and register set in the floor, and to the second floor through tin flues D D set in the partitions with registers in the walls. The hot water passes from the boiler to a large pipe or drum E which connects together its several sections. From this drum the supply pipes F lead to the several heating stacks. The return pipes G G are gathered at convenient points, one-half of them entering each side of the boiler. On one side the return pipe is exposed, on the other we had to run it under a stone wall about 18 inches deep, as it would have been a hard job to cut through it. Now the heaters on the left-hand side of the house worked satisfactorily, but those on the right side were a failure. We even increased their size, but their service is not what it should be. Can you suggest the cause of the trouble?"

[Whatever else may be wrong with your apparatus, your sketch shows a very serious defect in the diving return pipe carried under the wall. We assume that this is the right-hand side of your building and the side which has given you trouble. It requires but a small impediment to direct the flow of hot water in the opposite direction to that laid down for it. In this case the impediment is furnished by the colder and heavier water in the trap under the wall. The conditions on the other side of the house are proper and favorable for an unobstructed flow. Disconnect from this trap and connect the return pipe as shown by the dotted lines II, and if all other conditions are proper you will have no farther trouble of this character.]

IMPAIRED CIRCULATION OF A HOT-WATER HEATING SYSTEM.

F. T. M., New York, writes:

"I inclose sketch of a part of a hot-water heating apparatus with which I have had a little trouble. The sketch shows the sizes of pipes and radiator. I found that the radiators marked A and C worked all

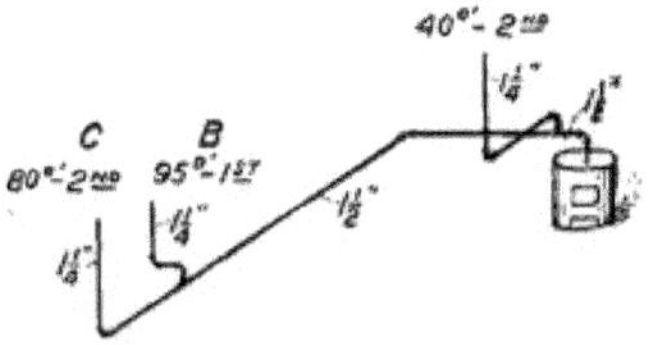

the time, while the circulation in B was always very feeble. Can you tell me what should be done to remedy this, and why should the radiator C get plenty of hot water while B got little, if any?"

[We believe that a 1½-inch pipe is too small to supply the radiators shown. We would advise your cutting out the radiator A from the branch shown and running separate connections to the boiler. The trouble is probably due to too small pipes, and possibly to a fault in the alignment at some point. The radiator C is favored by the tendency of the water to flow past the tee through which B draws its supply, and also because C is on the second floor, this giving a greater difference in pressure between lower ends of the return and flow risers. The latter probably more than compensates for the increased friction due to the longer length of pipe.]

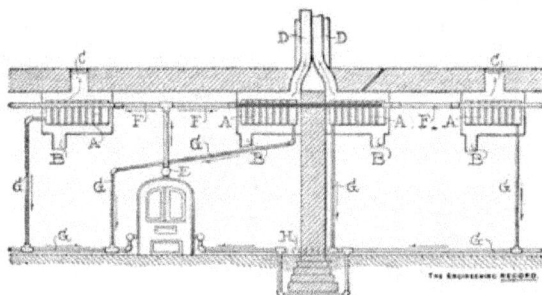

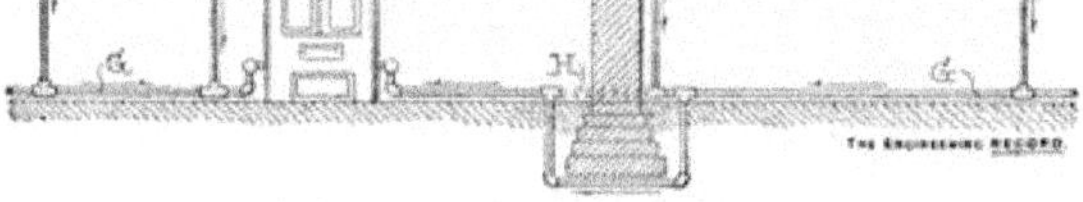

TRAP IN A HOT-WATER HEATING RETURN PIPE.

TROUBLE WITH A HOT-WATER HEATING SYSTEM.

LOCAL HEATER, Pittsburg, Pa., writes:

"We have just finished and tested a hot-water heating system for a hospital where the lines are run about as shown in the accompanying diagram. This is not drawn to scale, but shows the arrangement of pipes and radiators. All the radiators are on the first floor except those marked S, which are in the second story. The heater is of a satisfactory pattern and of sufficient size, and the pipes are all nicely pitched 1 inch every 10 feet. Their sizes are increased above the first design, and the change was approved by an experienced designer. All the lines marked A work to perfection but the 5-inch line does not work satisfactorily near the end. The line B seems to feed through the return pipe. Tank C is on the return pipe, and there is a small radiator on its line. From tank C to line B both the feed and return pipes are hot on top and almost cold on the under side. The radiators D all get cold when the temperature of the water falls below 140° Fahr. When we close the hot-radiator valves on the 5-inch line, the radiators D all get hot, but cool off as soon as the others are turned on again.

"Has the return pipe too much fall or are the pipes too large? Heretofore we have never found pipes too large, but rather too small. This is the first trouble we have had in 37 jobs, and we are anxious to have your opinion of the defect so that we can correct it immediately and make the entire plant work properly."

[The system of flow pipes shown in the accompanying diagram appears to be ample, and is very evenly proportioned. The lengths of course are not given, but we assume the runs are not excessively long. The arrangement of the 6-inch pipe into the 6"x4"x5" tee favors the 4-inch run of pipe by the direction of flow being past the 5-inch branch. In like manner, if your return pipes are exact counterpart of the flow pipe the 4-inch return will also be favored. This may make sufficient hindrance to the 5-inch circuit as to cause the dead ends D and D. Assuming therefore that the alignment is perfect in all parts, and that "air traps" do not exist, we see but little the matter with the apparatus. But as trouble does exist we would favor the 5-inch return by cutting it from the 6-inch pipe and taking it into the boiler separately. In like manner we would favor the 5-inch flow, either by taking it direct from the boiler, or by changing the pipes so as to give it the direct flow or at least putting the 6-inch pipe into a "bull-head" tee. Were it not that the ends of two branches give trouble and two more do not we would advise you to look for a partial stoppage in the 5-inch pipe. We know of a case where a piece of scantling was taken from a piece of 6-inch pipe not long ago.]

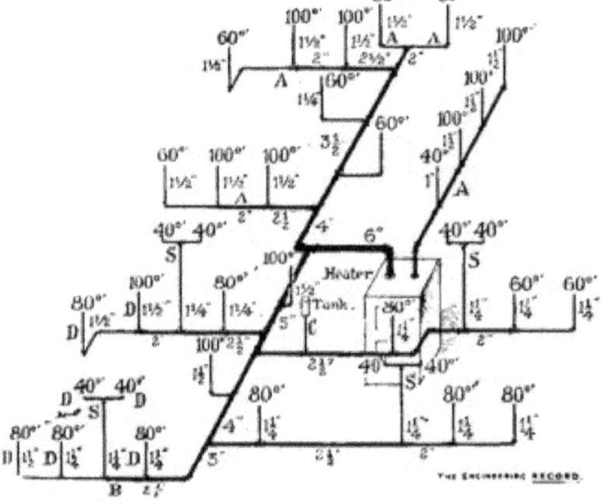

TROUBLE WITH A HOT-WATER HEATING SYSTEM.

ONE-PIPE HOT-WATER JOBS.

ONE-PIPE HOT-WATER JOBS.

T. F. E., Gloversville, N. Y., writes:

"Will a hot-water job constructed on the one-pipe system work? By a one-pipe system I mean, to have a flow pipe carried around the cellar and back to the boiler, and to take hot water from the top of the pipe to supply radiators and connect into the side of the same line. I have never tried this, and as I am thinking something about doing it I thought I would write to you first for your opinion."

[A one-pipe job as you show it, which means a circulation or large pipe around the basement from which you take a flow pipe on top and a return running into the side, will work. Hood, in his work on the "Warming and Ventilation of Buildings," describes a one-pipe system of this class. Of course it is not a strictly one-pipe system, as there is both a flow and return to every radiator, and when the circulation through the basement is of large diameter the circulation becomes practically a part of the boiler, and there is no reason why a good circulation should not go on in the radiators. The mistake made with these so-called one-pipe systems is that the circulation around the basement is pipe of an ordinary diameter, so that the quantity of water flowing through the circulation in a given time is materially reduced in temperature, in which case each successive radiator, starting from the boiler, must have a mean temperature lower than the one just preceding it. With small jobs and few radiators this difference may not be appreciable, but in work of any considerable magnitude the radiators furthest from the boiler are very much too cool, and hence you would have an impaired circulation.]

FUEL CONSUMPTION.

EXCESSIVE FUEL CONSUMPTION IN A HOT-WATER HEATER.

ARCHITECT, BUFFALO, N. Y., writes:

"One of my clients has the novel experience of burning a ton and a half of hard coal a week in a hot-water heater rated to carry 1,300 feet of radiating surface. There are less than 900 feet in radiators, and including piping about 1,100 feet. The thermometer registers from 120 degrees to 130 degrees. Circulation and radiation seem to be perfect, but the heat seems to go up the chimney. The smoke pipe is connected with two flues to increase draft, 9x13-inch and 9x9-inch flues, yet the draft seems sluggish, so that the ashpit door is open most of the time. Do you think draft is to blame, or is the boiler? With good or bad draft, is it possible to burn a ton and a half of coal a week with no better results if the boiler is not at fault?"

[Your client is burning pretty nearly double the amount of coal that he should in the apparatus mentioned. The combustion in the heater is not complete, as destructive distillation is probably going on in your furnace causing a large amount of carbonic oxide to pass up the chimney unconsumed. We do not think that he should use the two flues mentioned, as the results thus obtained are seldom satisfactory. If one is not sufficient, use the other, but knock out the partition between them.

As to the quantity of coal burned, you say that he uses 1½ tons, or 3,000 pounds, a week. With a well-proportioned hot-water apparatus he ought to realize 12,000 heat units from each pound of coal burned in a hot-water apparatus, or 36,000,000 heat units from 3,000 pounds. We will now compare this with the heat ordinarily radiated. Baldwin's "Hot-Water Heating and Steam Fitting" says that 2 heat units will be radiated from a square foot of surface for every degree difference between the temperature of the water and the temperature of the room. You have 1,100 square feet of surface, and the difference between the water (125 degrees) and the room (70 degrees) will be 55 degrees. Hence in one hour you will radiate $1,100 \times 2 \times 55 = 121,000$ heat units, and in a week 20,328,000 heat units. But at the rate of combustion stated (1½ tons per week) your heater ought to furnish 36,000,000 heat units in a week, but as only 20,328,000 are needed to heat the building, the difference, about 43 per cent., is evidently going to waste up the chimney in the form of unconsumed gases.]

HEATING BELOW THE BOILER LEVEL.

HOT-WATER HEATING ON THREE FLOORS.

ROBERT E. MORRIS, 2138 North Thirty-second Street, Philadelphia, Pa., writes:

"I have noticed the articles in THE ENGINEERING RECORD on hot-water heating at the boiler level, and would like your views regarding a job on three floors, with radiators on each floor, which seems to me to be a more difficult piece of work and one in which failure of circulation would be liable to occur. The ratio is: Basement, 1 in 30; first floor, 1 in 37; second floor, 1 in 48; total amount of radiation, 468 square feet."

[We never advise putting radiators below or even on the same floor with a hot-water heating apparatus. If, however, it must be done, it can only be accomplished, and then with only medium satisfaction, by exceedingly large piping.

If the problem is given us to run the piping for the radiators as shown in the sketch, we would start with a 4-inch circuit from the top of the boiler and run to the extreme end of the lower story, drop to the floor, and return to the boiler, all with 4-inch pipe. This will make a 4-inch circuit, through which the water will flow with more or less velocity. From this 4-inch main we would rise with 1¼-inch as far as the first radiator and 1-inch pipe from there to the second radiators, the return pipes being similar but carried down to the main return on the lower floor or under it. In the case of three radiators on the lower floor we would supply each with a 2-inch pipe, going into the top of the radiator with a 2-inch return pipe

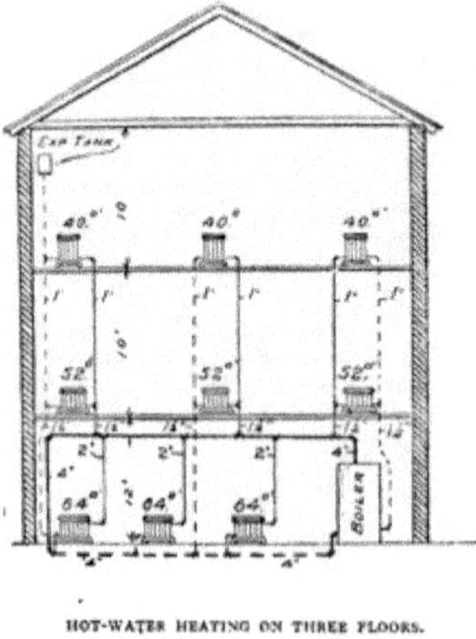

HOT-WATER HEATING ON THREE FLOORS.

at the other end of the radiator, connecting with the 4-inch circuit at or under the floor. For details see sketch.]

HOT-WATER HEATING AT THE BOILER LEVEL.

J. A. F., Boston, Mass., writes:

"In THE ENGINEERING RECORD of November 25, 1893, I noticed the remarks of J. W. H., of Montreal, in regard to connecting hot-water radiators on the same level as the boiler. I beg to disagree with J. W. H. as to his methods of connecting radiators, and would suggest to any fitters who are practicing

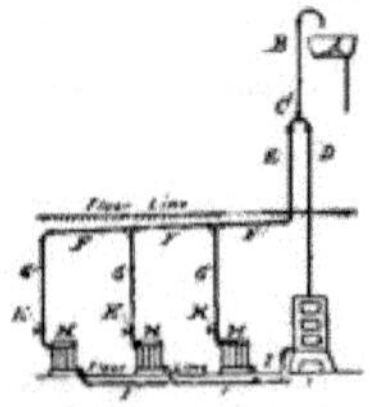

heating by hot water, that when they have any work to do where the radiators are to be set upon the same level as the boiler, that they connect the pipes that drop from the overhead main pipe, and marked on plan shown, G G G, connecting them into the top of the radiators instead of the bottom.

"J. W. H. is aware that in Montreal a box coil for hot water would be connected at the top instead of the bottom. Why not a radiator? I know the habit has become general among manufacturers of hot-water radiators to tap them at the bottom, but they are usually to be used for connection above the boiler; but as above stated, when a job of overhead piping is to be done, and radiation is to be placed on the same level as the boiler, if the radiators are connected at the top they will be found to have a more positive circulation, and very much quicker action, than when connected at the bottom. It stands to reason that radiators connected at the bottom stand full of water of a higher specific gravity, consequently the water in the radiator heats only by diffusion and not positive circulation. By connecting the radiators at the top the hot water is allowed to enter and the cooler water falls into the return pipe, and by its own specific gravity finds its way back to the heater.

"I think J. W. H. will agree with me that in connecting hot-water radiators at the top it is the right principle, especially in this case, and as we are trying to educate the fitting fraternity to do hot-water heating properly and get the best results, I trust my suggestion will be in order. I inclose you sketch which I would like to have you print with my letter, showing my way of connecting the radiators. In it A is the tank; B, the air pipe; C, the return bend; D, the flow pipe; E the return; F, the main; G G G, branches; H H H, radiators; I, return main; J, emptying cock; and K K K, valves.

"I agree with J. W. H. that there is no difficulty in having hot-water radiators circulate on the same level as the boiler, if connected as shown in my sketch and with ordinary size mains. I presume he meant a 2-inch main would carry radiation up to 400 feet, and so on in proportion. I have had as high as 400 feet of radiation on a continuous 2-inch main overhead pipe, and found it to give the very best of satisfaction."

HOT-WATER RADIATORS ON A LEVEL WITH BOILER.

ARCHITECT, of Cairo, Ill., writes:

"Can a hot-water heater be set up as well on a level with the ground floor—that is, in the first story of a house, as in the basement or cellar? Can it work to as good advantage in heating the first and second stories? In this town there are no cellars."

[In THE ENGINEERING RECORD of July 12, 1890, we show an arrangement of securing circulation through hot-water radiators placed below the level of the boiler. The flow pipe from the boiler is first carried up some distance, in this case to the top of the first story, and is then brought down to supply the radiators.

The plan has given entire satisfaction. The same method is frequently applied where indirect hot-water heating is employed, and where the indirect radiators are only slightly above the level of the top of the boiler. The working conditions of such radiators correspond very nearly with those on "Architect's" first floor, the boiler and radiators there being on the same level. If our correspondent, therefore, will apply the principle illustrated in our former issue he will have no trouble in getting proper circulation in his first-floor radiators. The second-story radiators call for no special considerations, and ought to work well when connected in the customary manner. We would suggest, however, that independent circuits be used for each story. With this arrangement each set of radiators will have its own main flow and return pipe, and there will be no opportunity for the upper radiators to circulate at the expense of the lower ones.]

HOT-WATER RADIATORS ON A LEVEL WITH THE BOILER.

WILLIAM J. BALDWIN, New York, writes:

"In your issue of August 9, 1890, 'Architect' asks: 'Can a hot-water heater be set up as well on a level with the ground floor—that is, in the first story of a house as in the basement or cellar? Can it work to as good advantage in heating the first and second stories? In this town there are no cellars.'

"In reply, you say in your issue of July 12, 1890: 'We show an arrangement of securing circulation through hot-water radiators placed below the level of the boiler. The flow pipe from the boiler is first carried up some distance, in this case to the top of the first story, and is then brought down to supply the radiators.

"'The plan has given entire satisfaction. * * * The working conditions of such radiators correspond very nearly with those on 'Architect's' first floor, the boiler and radiators there being on the same level. If our correspondent, therefore, will apply the principle illustrated in our former issue he will have no trouble in getting proper circulation in his first-floor radiators.'

"You evidently do not intend to convey the idea that the plan will give as good results as a *circulation above the boiler*. Still this is the construction many of your readers will be apt to apply to it; hence my letter to you on the subject.

"When you tell your correspondent to apply the principles shown in your issue of July 12, and tell him he will have no trouble in getting proper circulation, of course you mean that he will get as good circulation as can be expected under such conditions, always assuming, of course, that his pipes are properly run and sufficiently large. Circulation below the boiler is seldom satisfactory. It is always sluggish and requires very much more heating surface to accomplish a given result than would be required for similar conditions when the boiler can be below the heaters. My advice to 'Architect' is *not to do it* if there is any other way out of the difficulty.

"Would it not be better to put coils near the ceilings?"

[From the nature of "Architect's" inquiry, we took it for granted that he contemplated a job in which the lowest radiators had to be used either on a level with the boiler or not at all. In our reply we therefore attempted to show how the problem could be solved under such unfavorable conditions. As Mr. Baldwin points out, they are to be avoided if in any way possible.]

PIPING FOR HOT WATER RADIATORS ON BOILER LEVEL.

A. T. ROGERS, New York, writes:

"This question has been asked me two or three times within a few weeks, as a test question:

"Supposing a building to be heated by hot water, in which it is necessary to place one or more radiators on the same floor as the heater and others on upper floors, and in which doors, etc., prevent the running of the return pipes along the wall above the floor. If the return pipes drop below the floor, there is produced a pocket of colder water below the bottom of the heater, which will tend to lie stagnant. How will you overcome it?

"The solution proposed by the questioner was this: From the point where the flow pipe drops to the radiators on the lower floor run a single 1-inch pipe up to the level of the expansion tank; this will give a head which will overcome the stagnant pocket before mentioned.

"Now, to speak plainly, I don't see it. I think that the single pipe running upwards is a waste of material, for no head is produced by it that was not already there, due to the height of the expansion tank. But if the main for the lower floor be run up full size to just below the level of the tank, with an air cock or air pipe at its highest point, and then run down again to the radiators, we have a high column of cooler water overbalancing a column of warmer water, and which would tend to overcome the cold pocket below the heater. Upon my suggesting the double pipe, as above, my questioner rejected it because of the expense, substituting the single 1-inch pipe. I have never heard of a job done as he suggested, but have seen a job near New York where this condition exists, and where the connection is like this—upward from the horizontal main at 45 degrees, then horizontal, then vertically downwards to radiator. I have not seen this job work, but am told by the fitter who did the work that the radiator on the lower floor worked as quickly as any on upper floors.

"I have seen jobs where the return has risen from the radiator to pass over doors, etc., dropping again to the heater, and when the descending leg of the syphon was the longer the circulation was generally good. In all such cases as those I have mentioned care is needed to favor difficult branches as against those having straight upward runs.

"I should like to hear the experience of others, as these are somewhat frequent problems."

[A dip, or a "trap," below a doorway, or in crossing a hallway or other passage, is a very common occurrence in hot-water apparatus. They should be avoided if possible, but when unavoidable they may be used, and, unless an air pocket is formed that will become air-bound, they offer very little obstruction to the flow of the water. A very feeble circulation may be stopped or further impaired by it, but a good circulation, with ample diameter of piping, will go on unless the air stops it.

A 1-inch pipe, run from the stagnant pocket to the expansion tank, cannot increase the head and will not help the circulation directly. It may act as an air pipe, however, in some particular case and the person finding the circulation improved attribute the result to the wrong cause. A pipe carried up and down again as you propose will add something to the force of circulation. Its value, however, is overrated, as the "down" side must have nearly the same temperature as the "up" leg if the circulation is rapid. In special cases it helps, but a well-designed job will not be benefited by it. Carrying the return pipe over obstacles is not good. It can be done, however, but care must be taken to take off the air at every point.]

EXPANSION TANKS.

DANGER FROM CLOSED HOT-WATER APPARATUS.

No Safety Valve, Boston, Mass., writes:

"I noticed an article in your valuable paper of December 24, 1887, in which you show a diagram of an expansion tank for a hot-water heating apparatus, explaining the danger of running under pressure, or with a closed tank having a safety valve attached. I have followed the vocation of steam and hot-water heating engineer for years, and in that time have had a large practical experience in erecting low-pressure hot-water heating apparatus. As a great many

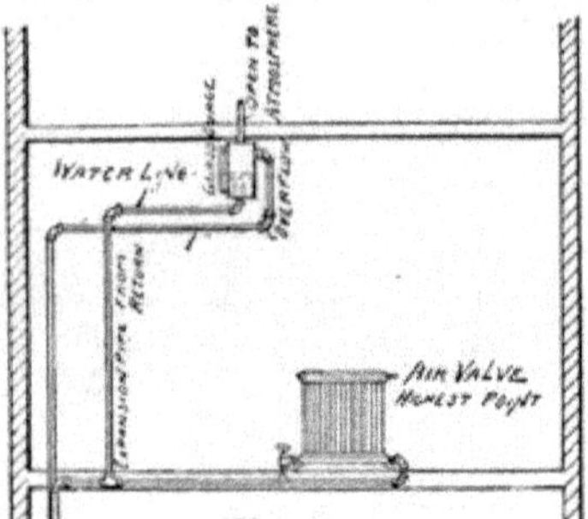

Fig. 1

fitters have an erroneous idea that it is necessary to run a hot-water heating apparatus under pressure, and as I have known many accidents to occur by closing the expansion tank from the atmosphere, such as the breaking of the radiators and boilers (all of good make), and as fitters, generally speaking, are to a certain extent ignorant of the principles involved in hot-water circulation, I take the opportunity of writing, with the hope that it may be read by those who are doing hot-water fitting; with the assurance that if their work is done upon the low-pressure system (by having the tank open to the atmosphere, as shown in Fig. 1) they will run no risk of danger.

"By way of example I inclose you a sketch (Fig 2) showing the connection of an expansion tank to a hot-water apparatus, in which a radiator and heater were broken; the fitter who did the work was evidently not aware of the danger of confining the water by using a safety valve to take care of the increment of pressure, and to show the danger of doing work in this manner I would state, that supposing the expansion tank to be set up in the manner herein shown (Fig. 2), and the safety valve on the tank loaded to 10 pounds pressure, it is equivalent to adding 23 feet to the height of the apparatus so far as pressure is concerned, thus making the pressure in a three story house 30 feet high, 25 pounds to the square inch in round numbers.

"Having conducted some experiments the past winter as to the amount of pressure obtainable in a closed apparatus, I find in a tank 3 feet high and 12 inches in diameter, placed at the height of 30 feet above the boiler and closed to the atmosphere, that with the water line in the tank 1 foot from the bottom at (A, Fig. 3) the pressure was 15 pounds, and at (B) 30 pounds; at (C) 45 pounds; at (D) 105 pounds; at (F) 225 pounds; all extra pressure, and as yet the water had not reached a temperature of 212° Fahr., so you see what an enormous pressure can be put on a hot-water heating apparatus by using a closed tank.

"The pressure is from the compression of air in the top of the tank, caused by the expansion of the water, and as the expansive power of water is irresistible, it goes without saying that 'a word to the wise is sufficient.' Some fitters have an idea that a closed apparatus will circulate better than an open one, which is erroneous, as practically all that can be gained by running under pressure is a little more heat at the radiators, and of course this is not obtained without heavy firing and necessary increase in the consumption of fuel. A closed apparatus under pressure is therefore of no practical service unless you desire to raise the temperature of the water over 212° Fahr., and makes an apparatus, which is otherwise safe in every respect, absolutely dangerous by confining the water. If fitters will take advice, when doing low-pressure hot-water heating, and leave their expansion tank open to the atmosphere, they would save themselves a large amount of annoyance and anxiety.

"By low-pressure hot-water heating is meant that which is operated under the pressure due to the column of water only, and usually heated to any point

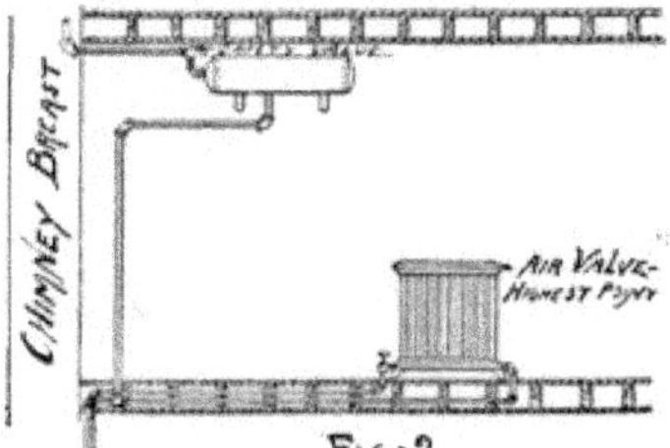

FIG·2

less than 212 degrees, and the ordinary working temperature is from 180° to 200° Fahr."

[So little can be gained in the temperature of a closed apparatus without approaching a dangerous pressure that it should not be resorted to in private house heating.

For drying purposes, or where a high temperature is actually necessary, then it must be resorted to if water is the heating medium.

In THE ENGINEERING RECORD of December 24, 1887, we discussed this subject very thoroughly and gave a diagram of pressures for different temperatures where the tank was one-twentieth the capacity of the pipes and boilers.

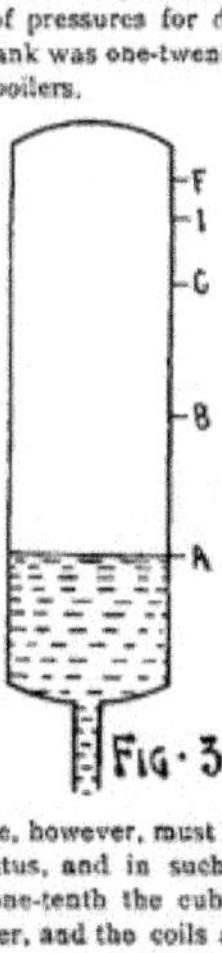

FIG·3

The safety valve, however, must be resorted to on all closed apparatus, and in such cases the tank should be about one-tenth the cubic capacity of all the pipes and boiler, and the coils and boiler should be special so as to bear the high pressure.

In the diagram, Fig. 3, the increase of pressure does not seem to agree with the law of the compression of air at constant temperature. It probably so happened in practice, however, that the temperature did not remain constant and that the gauge was near the boiler, so that it had a constant head of water of about 15 pounds at the start to be deducted from the gauge reading to show the pressure of the air in the tank.]

EXPANSION-TANK CONNECTION.

GREENHORN, York, Pa., writes:

"The question for proper connection for expansion tanks in hot-water heating apparatus is one requiring considerable attention and study in order to secure the best results. We notice in your paper several inquiries and answers in regard to this matter. We have found in experience that the greatest difficulty to be overcome is to prevent the water in the boiler and radiators from being driven out by the formation of steam. Even when the pipe connection from the heater to the tank is taken directly from the top of

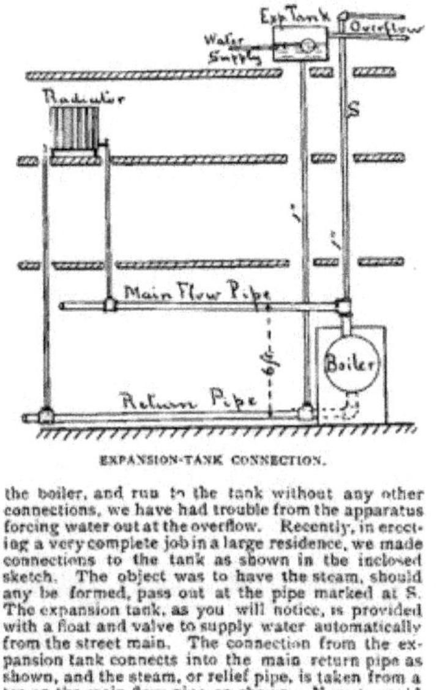

EXPANSION-TANK CONNECTION.

the boiler, and run to the tank without any other connections, we have had trouble from the apparatus forcing water out at the overflow. Recently, in erecting a very complete job in a large residence, we made connections to the tank as shown in the inclosed sketch. The object was to have the steam, should any be formed, pass out at the pipe marked at S. The expansion tank, as you will notice, is provided with a float and valve to supply water automatically from the street main. The connection from the expansion tank connects into the main return pipe as shown, and the steam, or relief pipe, is taken from a tee on the main flow pipe as shown. Now it would appear reasonable to suppose that the steam when formed would escape through this relief pipe. In firing up the apparatus, however, and running the temperature up to the boiling point, we were surprised to find the water forced into the expansion tank and out at the overflow in place of passing out at the relief pipe. The overflow in the expansion tank is about 12 inches lower than the top of the relief pipe, which might possibly be the difficulty. We might add, that when the water was being forced out of the overflow pipe from the expansion tank that the radiators on the second floor worked dry steam, and no steam appeared to be coming through the overflow pipe, and the relief pipe showed neither steam nor water at the outer end. The relief pipe having such a direct connection from the tee on top of the heater should, in our opinion, give steam before the water would be forced up into the expansion tank through the return pipe, which is 6 feet lower than the flow pipe.

"Please give us your views on this question."

[The difference in level between the overflow in the expansion tank and the top of the relief pipe S is, in a measure, objectionable, but cannot be held accountable for the difficulty. We are inclined to think that the pipe S is not run up directly as shown in the sketch, but probably pursues a more roundabout course. If this be the case it is not unlikely that an air trap is formed somewhere in the line of the pipe. This would tend to entirely defeat the purpose of the pipe, its small diameter also working against it. We would advise a careful examination of this pipe S with a view of detecting any possible air trap. Where an obstruction is to be run around, or where, for any reason, the pipe departs from a vertical line, it requires but little miscalculation in the pitch to create trouble. A downward grade at any point should be carefully avoided. We would suggest, also, that some obstruction may have entered the pipe S and blocked the passage. In any event we do not see why the pipe S was not let directly to the expansion tank instead of using it as a relief and connecting the tank with the return main by another pipe. This plan would certainly have been simpler. There may have been some objection to this which is not apparent from the sketch and particulars. The boiler, also, is very probably too large for its work, imparting to the water an unnecessarily high temperature and creating a constant tendency to form steam. With the pipe S unobstructed, and run in a direct line as shown in the sketch, we can see no reason why it should not efficiently serve its purpose as a relief.]

POSITION OF EXPANSION TANK IN HOT-WATER HEATING APPARATUS.

John C. Fesigen, Jr., of the New Orleans, La., Railway and Mill Supply Company, writes:

"I inclose herewith a sketch of a heating apparatus, and would like to hear from your correspondents in regard to the position of the expansion tank connected with this apparatus. Baldwin, in his 'Hot-Water Heating,' treats extensively of expansion tanks, but only with boilers using one flow pipe. The boiler used in the apparatus referred to is the 'Plaxton' boiler, from which there are five flow pipes. This boiler is capable of carrying about 200 feet more radiation surface than is now attached to it, and with excessive firing the boiler is liable to make steam. I would like to hear some comments on the position of the tank as shown, and to hear from some of your correspondents what they would do in this case to prevent the formation of steam in such quantities as to drive the water out of the radiators and pipes. My idea is that this expansion tank should be connected with the boiler through a manifold connection from each one of the 45-degree elbows connected to flow pipes. With such an arrangement the steam forming in the boiler could be readily carried off through the expansion tank, whether closed or open, provided the closed system had a low-pressure safety valve attached to same."

[Connection of the boiler with the expansion tank by means of the pipe E in the manner shown in the

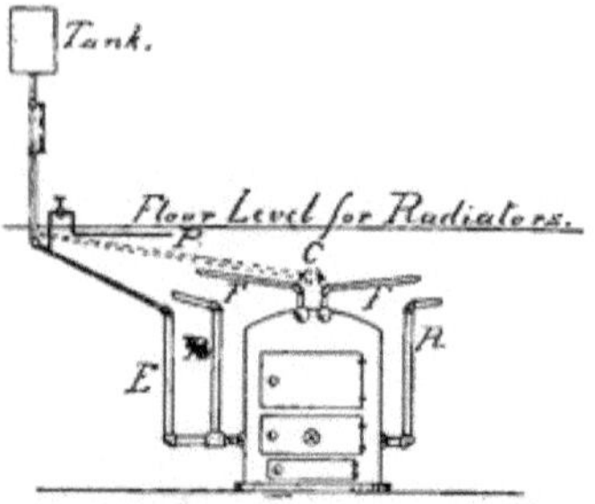

sketch is not to be commended, since with this arrangement it becomes possible to drive all the water out of the boiler if steam be formed in sufficient quantity. In Mr. Baldwin's book on "Hot-Water Heating" attention is directed to this circumstance on page 307. The expansion tank pipe should be taken from the highest possible point of the boiler, and while the arrangement suggested by our correspondent, and shown in dotted lines, would answer, it seems to us a needlessly complicated one. We would advise taking a branch from one of the flow pipes, F, to the expansion tank, or, if all the flow pipes from the boiler are required for hot-water distribution, to use one of them directly as a tank pipe. We have no doubt that all the trouble from steam formation will be overcome in this way.]

CONNECTION TO AN EXPANSION TANK.

C. B. B., of New York writes:

"In reply to C. M., of Hamilton, Ont., I do not think you have made your answer full enough to be

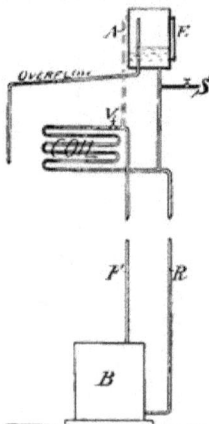

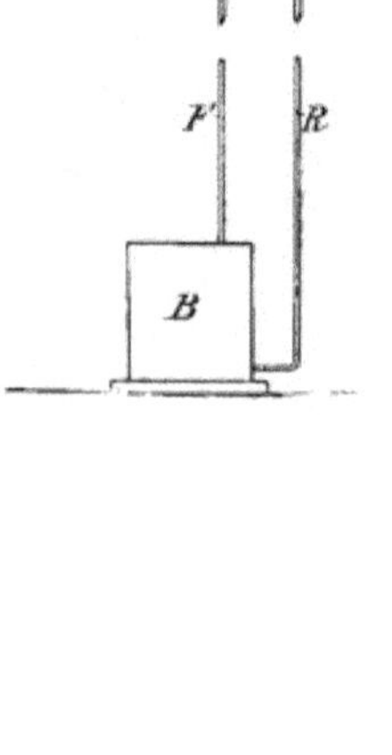

clearly understood by some men who may be called upon to fit up hot-water heating apparatus, and who do not know as much about the matter as you do.

"I inclose the sketch which is the subject of this controversy and have applied the proper connections to insure successful circulation. I have also placed a free open way radiator valve on the coil in its proper position if a valve is required."

[The pipe A and the valve V are those supplied by our correspondent. The pipe A is an air vent and also a steam-escape pipe, should steam form through overfiring or by a stoppage of the circulation with an ordinary fire, as when the valve V would be closed. This is substantially what we advised; excepting the valve, which of course is nearly always applied and used either to retard the circulation through the coil or to stop it altogether. It is understood, of course, that there are many coils in the work, the one given only showing the principle involved.]

METHODS OF PIPING.

ON WARMING THE WATER SUPPLY BY STEAM.

George E. Stauffer, of East Stroudsburg, Pa., writes:

"In a short time I expect to heat a summer hotel (for late and early season) with steam, and also expect to put in a water-heating apparatus connected with the boiler similar to the plan you give on page 125 of your 'Steam-Heating Problems,' but I expect to connect the supply direct to hydrant instead of a tank overhead; the pressure is about 40 pounds. I would ask why it would not work just as well as from a tank? Another question I would like you to answer is: Can I carry hot water, say 100 feet horizontally with a fall of 3 feet, and have it circulate so as to keep the spigot hot while not in use. I carry the water from one building to another through the yard, consequently cannot have more fall, as I must keep above the water line in boiler, as I want to drip the condensed steam in boiler. I inclose a rough sketch. I expect to connect hydrant and circulating pipe at same place in hot-water boiler; will put a check valve in circulating pipe so as to prevent drawing cold water direct from hydrant, should they open several spigots at same time, as with a heavy draft at spigots it would flow both ways."

[There is no reason why a connection from the hydrant, such as you describe, is not just as good for your purpose as one from the tank, except during interruptions of the street supply. Do not put a check valve in the supply pipe leading from the hydrant or it may result in the bursting of your tank or piping. When a tank is used it is in the top of the house, and no one is likely to think of putting a check valve on the pipe between it and the heater, but with a street supply many think a check valve is a good thing to prevent the syphoning of the water back into the street main should the water be drawn off in the latter. Such a valve prevents the increment of expansion of the water when heated from flowing back into the street main and the result may be a rupture. In other words, if there is no check valve on the supply pipe, then, unless the house connection is shut off, the pressure in the house pipes can never exceed that in the street, while if a check valve is used the possible pressure is only limited by the strength of the house pipes, etc., and the temperature to which the water may be heated.

Do not use the check valve you have shown and refer to on the circulating pipe, as the probability is that it will stop your circulation, unless it is something special and very light. Connect the hydrant pipe with the hot-water boiler at the further end, as shown at *a*, by the dotted lines, and you will probably not be troubled by drawing cold water at the faucet. In other respects the apparatus, as shown, will work, if you use pipes of ample diameter. Your sketch shows a *rise* of about 3 feet (including the height of the hot-water boiler), and it is to that we presume you refer when you speak of "*fall*."

When the return circulation pipe is smaller than the flow pipe, it helps to prevent the "back flow" of cold water through the former. Therefore if you use a 1½-inch flow pipe use a 1-inch return, or in about that proportion. A stop valve in the circulating pipe in place of the proposed check valve may be

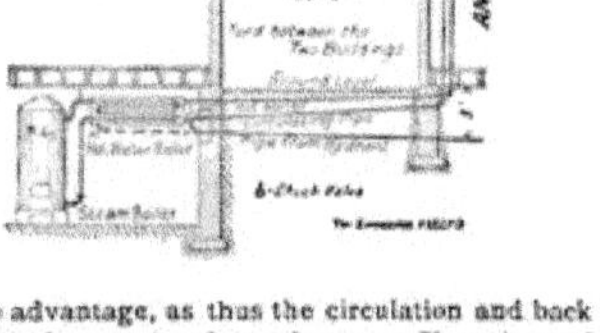

used to advantage, as thus the circulation and back flow may be regulated at pleasure. Use pipes of large diameter in the steam coil and its connections, say not less than 2 inches for supply and coil and 1¼ for the return, or in about that proportion if you use larger pipes.

You will do well to study carefully the "Problem in the Circulation of the Hot-Water Supply," in The Engineering RECORD of April 6, 1889, from which you will see that the higher you can run the circulating loop in the annex the more efficient the circulation is likely to be, and should it not be satisfactory and the extension of the pipes upward be inconvenient, it would be well to run the pipe horizontally a ways before descending, even if you bring it back to the same place, as the sole cause of circulation is the difference of temperature between the water going up and that going down, and hence the more you can cool your water before it descends the better.]

WARMING A JAIL BY HOT WATER.

THE following description and illustrations of the method of warming the Schenectady Jail were sent to us by J. V. Vrooman & Sons, of Schenectady, N. Y., who did the work, and as it differs from most hot-water work by having the return pipes start at the first coil and flow in the same general way around the building as the flow pipes, growing longer as the flow pipes decrease in size, we give it in full in their own words:

"Our jail was built of stone, without cellar or pit for the heating apparatus, and it was at first proposed to heat it with steam. There being no cellar made it a difficult matter to put radiators on the main prison floor or in the lower tier of cells.

"The prison part of the building is lined with ¼-inch steel, the corner being made by riveting to angle irons. The main floor is of 6-inch Tribes Hill stone, laid in large blocks; the gallery and floors of the second tier of cells are of ¼-inch steel.

"The system of heating used is hot water with a No. 28 'Gurney' boiler. The pipes are run as shown in the diagrams; two 2-inch pipes leave the boiler in the entry and pass up to near the ceiling of the first story, ascending to A, Fig. 2, where the air pipes are taken out and connected with one ¾-inch pipe, which is carried, still ascending, over the ceiling timbers of the second story until it reaches a point over the expansion tank, near the second-story ceiling; there it ends in a tee, from which two pipes extend, one to the top of the expansion tank, and the other up about 18 inches, so that the air can escape from the pipes and coils, and so that if the water should reach the boiling point it can overflow into the tank.

"The vent pipe and expansion tank are shown in Fig. 5.

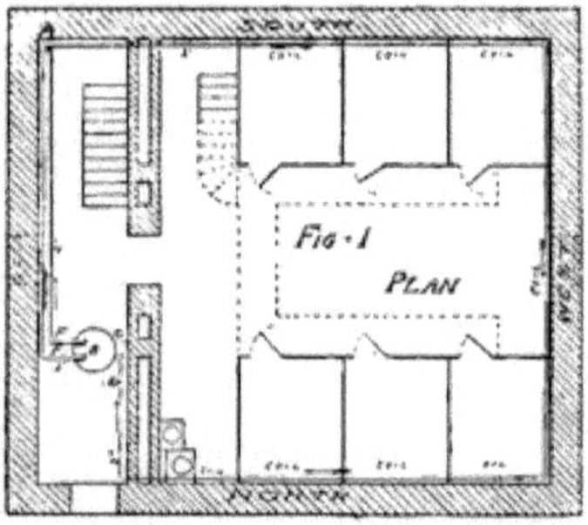

"After leaving the vent pipe at A, the two flow pipes F F go through the brick wall and steel lining to the prison proper, when they separate, one running through the upper tier of cells, the other through the lower tier, both near the ceiling, as far as the last cells, where the one in the upper tier is capped on the outside of the steel plate and the one in the lower tier ends in a miter coil in the closet (see Fig. 4). From each of the main flow pipes a 1-inch pipe is taken in each cell and carried down to a return bend radiator of three or four pipes as required, and then into the return pipes G G, that extend through the cells near the floor and on through the wall to the boiler. Under the window is placed a coil of 13 1-inch pipes connected with the first-tier pipes as shown in Fig. 3.

"During the last winter, with its severe weather, the apparatus was used for heating the building, so that the masons and carpenters could finish their work. With windows partly open and a large skylight over the main prison only covered with matting,

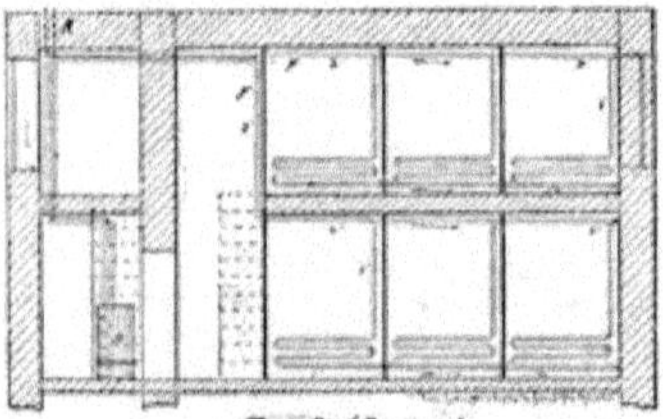

Fig·2 (South)

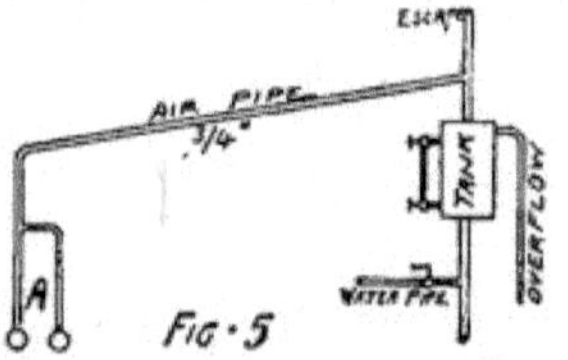

Fig·5

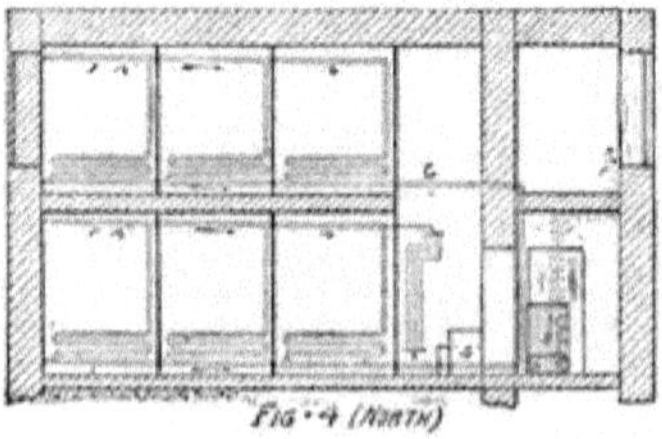

Fig·4 (North)

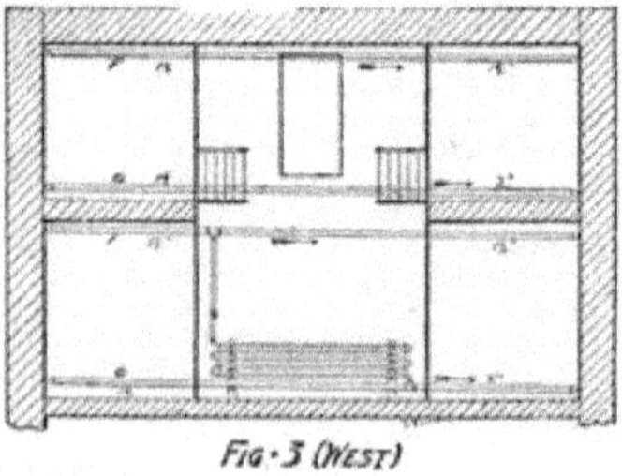

Fig·3 (West)

WARMING A JAIL BY HOT WATER.

there was no trouble from freezing or bursting of pipe, and the quantity of fuel consumed was no more than one stove would use, though a stove used before the heating apparatus was finished did not begin to keep the building warm. A large room in the second story was then open to the main floor, but has since been closed. We have received very flattering testimonials as to the efficiency of the apparatus."

A HOT-WATER CIRCULATION QUESTION.

FRANK J. GRODAVENT, of Denver, Colo., writes:

"I have a question relative to hot-water heating which I would be very glad to have you answer. A party here has contracted to heat a dwelling by hot water, the total amount of radiation being about 900 feet, the greater portion of the first floor to be heated by indirect radiation. The boiler to be used will be a Richardson & Boynton 'Perfect.'

"It was originally intended to have the boiler located in the cellar of the building, but lately the owner has decided to place the boiler in the basement of the stable, which is located at the rear of the house and just forward of the line, leaving a 4-foot passage or walk between the two. The boiler can be placed at as great a depth as may be desired to get the proper rise to flow and proper fall to return pipes.

"The contractor doing the work claims that he can get just as good results by having one flow and return pipe leading to the most central point and taking branches as desired.

"I claim that he will get better results if he runs separate flow and returns for the first and second stories, and that he can supply the small amount of radiation in the attic from the second-story risers. I claim that with the single pipe there will be more danger of the upper radiators getting the heated water at the loss of the indirects which will be in the basement or cellar of the house.

"Another point I claim, that he will require air pipes carried from the highest point of the flow and return mains to these indirect radiators, as the mains rise, run level and run down to radiators and will thus form air pockets, which will stop the flow of water. I am aware that at times air valves are used at these points, but believe the air pipe better.

"I do not claim to be a hot-water expert and have had only a small amount of the practical part to contend with, yet I have had a deep interest in this system of warming and will feel very grateful if you will kindly give the information desired."

[The results will be just as good if one large flow and a similar large return pipe are run to a central point and the branches taken therefrom accordingly. Of course we do not say that any one of the proposed methods cannot be botched. We assume, however, that the work is in the hands of a capable person. He should avoid air traps in his mains, and if he cannot do so the method you propose is the one we consider the best. The air pipes should rise to the expansion tank.]

THE PITCH OF HOT-WATER HEATING PIPES.

W. W. STRONG, C. E., Northampton, Mass., writes:

"In hot-water heating, should the supply and return pipes be run level or on a grade? If on a grade, at what inclination? Which pipe, if either, should have the most grade? Will it cost more to do the same heating if pipes are level than if on a grade?"

[Hot-water pipes should have a pitch, upward as they flow away from the boiler and downward as they return to it, which makes substantially two pipes side by side, with an upward grade as they go from the boiler, the one the flow and the other the return pipe. The object of this pitch is to get rid of the air. By getting rid of the air you secure circulation; therefore, we may say the grade is the cheaper, as on the level, except in very large pipes, air binding will follow. With good alignment and straight tubes a very small grade will do, say one-fourth of an inch to 10 feet. With ordinary small pipes, up to 1½-inch, a pitch of one half to three-fourths of an inch to 10 feet may be required. Flow and return pipes should be of the same diameter and of the same pitch or grade.]

AN INCREASED HOT-WATER SUPPLY WANTED.

JOSSELYN & TAYLOR, architects, Cedar Rapids, Iowa, writes:

"In a Y. M. C. A. bathroom connected with four shower and two bath tubs is an 82-gallon iron boiler, placed horizontally and containing 20 feet of 1-inch steam pipe connected with the boiler that heats the building. The pressure of city water is about 40 pounds and of steam seven pounds.

"This supplies 40 baths, but when, say 60 or more are needed in an evening continuously, hot water fails. The membership is increasing and it is desired to learn, if possible, how to do more than guess at a way to make adequate provision.

"Another boiler of the same size can be placed by the side of the present one, but no larger, nor are there other convenient places. This is directly over the shower and 75 feet from steam-heater boiler.

"The waste steam goes through a Hawes trap, as it cannot return to the boiler. The following questions occur to us:

"1. What is the best size, amount, and kind of steam pipe to put in the boilers?

"2. How should the pipe be run—lengthwise or around the boiler spirally?

"3. If two boilers are used, what is the best way to connect them and the supply, etc.?

"It is to be remembered that the users of the baths at certain times come thick and fast, and are not careful about economical use of the water.

"A minimum expenditure of money in the additional apparatus is desired.

"Any information and suggestions will be thankfully received."

[1. Brass pipe not smaller than 1 inch in diameter.

2. Lengthwise. Spiral coils are not good for this purpose.

3. Connect them as one boiler with leveling pipes top and bottom.

You require about 1 square foot of pipe surface to each bath, if the demand is sudden and wasteful. The larger your cylinders or tanks are outside the coils, the better the result. When there is a sudden draft on small tanks, the water that warmed between times is drawn off first, and the last drawn, of course, is colder. If there is a large body of water, it acts as a reservoir of heat by absorbing it from the coils continuously, whereas a small body of water is made very hot in a few moments, and as soon as it is as hot, or nearly as hot, as the steam, condensation ceases, and the supply of hot water is intermittent.]

LARGE VS. SMALL DIAMETERS FOR HOT-WATER HEATING PIPES.

C. S., of St. Louis, Mo., writes:

"Will you oblige a subscriber by answering the following questions? It occurs to your correspondent that by using smaller mains and risers, everything properly proportioned, the water in a hot-water heating system would be kept at a higher temperature and would be delivered hotter to the radiators, though in smaller quantities, than if larger pipe is used.

"This does not agree with the tables, etc., in Baldwin's book on 'Hot-Water Heating and Fitting,' which I am not disposed to contradict; but I do not see why, because a pipe is smaller the temperature of the radiators should not be as high as with large pipe. Is it not true, that what is lost in quantity is made up for by a higher temperature? There is less water to heat while the same amount of firing is done as with large pipe; there is less water present and it must get hotter.

"A pint of water in a tin cup held over a candle-light might be brought to boiling while a gallon of water, placed similarly, could not, even in double the time, be brought near the boiling point.

"There is no doubt more friction in small piping, which of course is objectionable. Please favor me with an answer."

[A reduction in size of pipes, meaning their diameters (but all other things, such as position and method of running pipes, remaining the same), will result in cooler water at the radiators instead of hotter, as you suppose. We must assume, for comparison, that the water leaves the boilers at the same temperature in two separate cases; one with large pipes and one with small pipes. In the case of the large-pipe apparatus two or three times as much water will flow out in a unit of time, say one minute, for the simple reason that the pipe is bigger, the velocities being about the same in both cases (but slowest, if there be any appreciable difference, with the small pipes). It is evident that when two measures of hot water are carried through a radiator in a unit of time the loss of temperature in one radiator will be only half what the loss would be in the one that received only one measure. Starting even they will both begin to cool, and the one through which the greatest flow of water goes will show the least loss of temperature in a given time, and consequently do the most work.

You cannot quicken the velocities to compensate for the loss of diameter; they will be slower instead of quicker with wa er at the same temperature in both, and even with the greater temperature in the small pipe at the sta-t the increase of velocity can never compensate for the decrease in quantity under any range of temperature at which heating apparatus works. It is true that you can warm a pint of water in much less time than you can warm a gallon, and it is equally true that it requires only one-eighth of the heat to warm a pint than it does to warm a gallon, and that in cooling a pint of water will only give out one-eighth of the heat that a gallon will, and consequently do only one-eighth the work. It is well to bear in mind the maxim that you cannot get something from nothing.]

C. S., of St. Louis, Mo., writes:

"I am much obliged to you for your answer to my inquiry on a hot-water heating question which has puzzled me very much. Please pardon me for being still inquisitive, because your explanation opens up another question for me which I cannot answer. If, as you say, the temperature of the water in the small pipe system would be, perhaps, less than that in a large one, what becomes of the heat in the small pipe system which, because by using small piping and radiators with very small passages will carry, say, perhaps, 25 per cent. less water than the large pipe system, having the same size boiler as the other one? Does it not seem reasonable to suppose that the system with the least water must run the water hotter, because the same amount of heat is applied to one as to the other? If the water in the one system which carries, say, perhaps, 300 gallons of water, does not run hotter than the water in the other system, which carries, perhaps, 400 gallons of water, kindly state what becomes of the heat? Is it all lost by friction on account of the smallness of the pipes, and small passages of radiators, or does it go up the flues? It is perhaps wrong for me to take such liberty and expect you to answer all these questions, still, after reading Baldwin's book, which I have found very profitable to me in my business, the understanding of same would not be complete unless some light was turned on these questions"

[To your question, "What becomes of the heat in the small pipe system?" the reply is, "It was not there." We do not mean to say that the water was not as hot at the start as in a large system, as the temperatures may have been the very same, but the total heat contained in the pipe was less, and consequently the temperature in the small pipe fell the most rapidly. You must not imagine in your reasoning that you can use the water generally hotter in a small pipe than in a big pipe system, as the large pipe system has equal advantages in this respect. The amount of water in the system, whether 300 gallons or 400 gallons, has little or nothing to do with the heating capacity of an apparatus. It is simply the number of gallons that will flow past a certain point in a given time, and that entirely depends on the size of the flow pipes. There may be an advantage in having a small quantity of water in the boiler, and radiators that will contain a comparatively small quantity of water, but this does not affect the flow in the pipes, unless the water passages in the boiler or in the radiators are stunted, when of course it will affect the circulation detrimentally.

Take two apparatus, for examples: In one there is a boiler holding 100 gallons, with the pipes holding 100 gallons, and the radiators 100 gallons more, making 300 gallons in all. In the other apparatus the boiler holds 200 gallons, but in every other respect the apparatus is the same as the first, except in holding 100 gallons more water. These two apparatus will do exactly the same work in the same time with the same fire, assuming the boilers to be the same except in their holding capacity. Now take the smaller boiler (or, if you please, the larger one; it matters nothing to us which you take), and reduce the flow pipes in diameter. The water cannot then get away from the boiler fast as before, and the result will be colder radiators, and probably a very hot boiler. Extra firing will not force the circulation; it will simply result in making steam, and in making your effort abortive. Another way of looking at this whole question is: Suppose you have one large boiler that is fired and kept just below the point of making steam; then if you connect a small pipe system to one end of the boiler, and a large pipe system to the other end of the boiler (the radiators and pipes in both cases being exactly the same), which will do the most work? Our answer, and we have no doubt your own answer, is "Why, the apparatus with the larger pipes," as it is quite apparent to any one who has considered this subject, that it is only necessary to keep on reducing the size of the pipes in the small pipe apparatus, until the apparatus will cease to work entirely.

A HOT-WATER RADIATOR CONNECTION TO A STEAM-HEATING BOILER.

AN Arlington, N. J., correspondent writes:

"My house is heated with an ordinary portable house-heating steam boiler by the direct closed gravity return system. I have been thinking of trying to heat an isolated room by hot water from the same boiler, and propose to tap my steam boiler a few inches below the water line, as at A on the inclosed sketch, and take my hot-water supply from that point to the room to be heated. In this I will place a regular hot-water radiator, and use a separate return for the hot-water system, as well as supply. Will this operate successfully, and can I heat this room by hot water at the same time that I am heating the rest of the house by steam? Would a 1-inch pipe be large enough for supply and return? The room is a small one, only 8'x9'x8½' high."

[Your proposed method of warming a room by a hot-water circulation, taken from a steam boiler, will work and be satisfactory just so long as the pressure of steam in the boiler will be great enough to maintain the water in the pipes and keep them constantly full. Suppose the top of your coil is 20 feet above the water line in the boiler. Then as long as you carry, say 10 pounds of steam, or higher, the water coil will circulate. On the first floor above the boiler four or five pounds of steam will do. The pressure is necessary to keep the pipes full. One-inch steam pipe is large for such a room.]

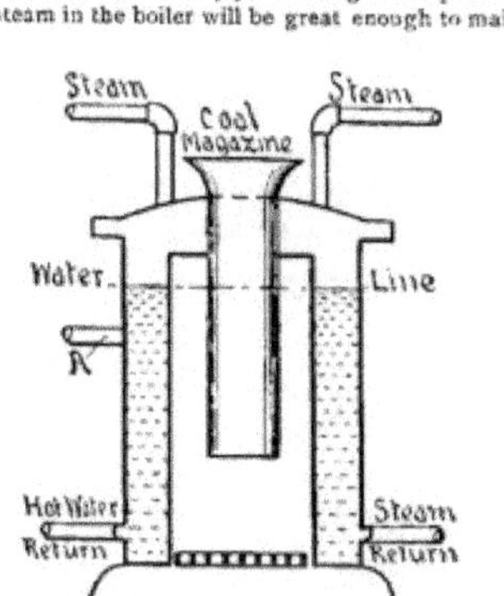

HOT-WATER RADIATOR CONNECTION TO A STEAM-HEATING BOILER.

A. B. BAKR, Chief Engineer of the E. H. Cook Company, of Rochester, N. Y., writes:

"On page 392 of THE ENGINEERING RECORD of November 14, 1891, I notice an inquiry on how to heat a room by hot water from a steam boiler. The writer had occasion to do a small hot-water job from a steam boiler which was used to run a steam engine, and accomplished it in a very satisfactory manner

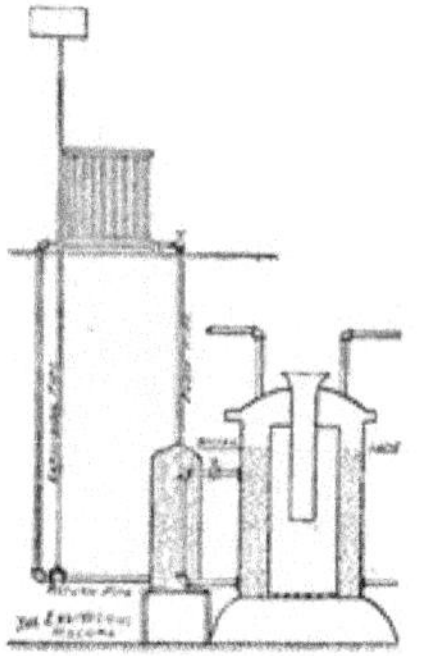

by having a small range boiler made with a spiral coil of brass pipe placed inside and attaching the brass coil to the boiler below the water line, and then using the range boiler in same manner as any ordinary hot-water boiler. A straight piece of, say 1½-inch pipe, 3 or 4 feet long, would be sufficient for a small radiator. A valve should be placed in pipe between the boilers to cut off or regulate the heat. If the apparatus was arranged in this way the hot-water radiator would always have some heat so long as the water in steam boiler remained warm, even after the steam radiators throughout the house were all cold. I inclose a sketch which explains itself."

HOT WATER FROM THE RETURN PIPES.

SELIM, Piscataquis, N. H., writes:

"My opinion was asked as to a proposed plan for bringing hot water to washbowls and sinks. The engineer considered it a 'happy thought.' His plan was this: He has a low-pressure or gravity system heating the building by steam, and was to make return pipes (one or more) supply the hot water to the bowls, etc. Would a man having any clear idea of the principle of steam heating attempt such things? I gave him my opinion in very plain English. I then asked him how high above 'water line' his bowls and sinks were located; how much pressure he proposed to carry on his boiler; if he was to have a fireman in constant attendance, or to control by automatic damper regulator; where his hot water was to come from? I asked if he had a feed-water heater and pump, or injector, and if so, why he called it a gravity job; and finally, why he did not put in a small hot-water boiler or tank, with brass coil connecting with his steam and return, and thus safely supply hot water to his bowls and sinks? He has a horizontal tubular boiler of 40 horse-power. With the hot water at several sinks, running—left running, thoughtlessly, as they are very likely to be—what would be the very probable result?"

[This proposed plan of hot-water supply is too ridiculous to be entertained, and but for the fact that just such men as would plan a job of this sort often, by their unskillfulness and ignorance, cause great inconvenience and injury to others, even placing human life in jeopardy, we would not feel justified in going into details, in answering the query of our correspondent. No person properly trained as a heating engineer would lay out such a job, and employers should consult their own interests by not entrusting work to such impracticable and dangerous men. Assuming that the job was installed upon the plan indicated, only steam could be drawn upon the top floors, steam and water from the cocks near the water line, and water from those below the water line. Water drawn from such a system would not be fit for domestic use. It would be full of rust, and at times would emit a disagreeable odor, such as is often detected where air is drawn from gravity coils.

One of the first laws of steam heating which a fitter should learn is to allow no water to be taken from the returns. Experience has taught that this practice has caused the "burning" of more boilers than all other causes combined. Many heating contractors, in recognition of this danger, will not connect a "blow-off" directly to a sewer. This restriction we heartily indorse for small jobs or places where an engineer is not employed

Your plan of a hot-water tank with brass heating pipes through which the steam and return pipes would connect, is very proper and is the best that can be done under some conditions. We would suggest in this case, using a hot-water circulating boiler of sufficient size and of the character used in the plumbing of dwellings. If there is sufficient pressure in the main service pipe it will force the hot water from the boiler to the several points for use; or if not, a tank should be placed sufficiently high and so connected that when in service it would act as a head, giving the desired pressure. The water in this boiler or tank may be heated by connecting flow and return pipes into the firebox of the steam boiler, on the same general plan as is used in connecting a kitchen range and tank. The pipes can be laid against the bridge wall. The hotter the place the better, if much hot water is required, but great care must be taken to have the connecting pipes properly run, otherwise there will be endless noises and repairs. Any good plumber should know how to arrange the job. You ask what would be the probable result of drawing hot-water service from the returns of a gravity system. It might be annoyance, stench, dirty water, scalding by steam with chances favoring a burned or cracked boiler with a heavy boiler-maker's bill, or an exploded boiler with attendant damage to property and peril to life, and the incidental inquiry—after the event—"How did it happen? Who is to blame?"]

MISCELLANEOUS QUERIES.

GAS IN HOT-WATER RADIATORS.

A. E. Kenrick, Brookline, Mass., writes:

"I read, with a great deal of curiosity, the inquiry of Messrs. Mooney & Baine, in regard to gas in hot-water radiators. Strange as it may seem, on the same day that I saw the inquiry a customer, for whom I erected a hot-water heating apparatus last year, came into my office greatly excited and asked me to explain the reasons for what he had just discovered in his apparatus. He said that he found that one of his radiators was not heating properly and needed to be vented; he therefore went into the basement, got an old tomato can and commenced to vent the radiator. As it was dark where the radiator was located he lit a match so that he could see when the water came. To his amazement the air ignited and burned for a second or so with a blue flame, and he burnt the paper on the side of the can with it. He came to me for information; he said that he was not aware that you could light as well as heat houses with hot water. Since then I have tried radiators in my own house, and have been able to obtain the same results.

"My theory is that the hydrogen in the water became separated from the water by heating, and when liberated burns with a blue flame if ignited. As for the milky color, that may be seen almost any time that the water is drawn off and the radiator refilled without venting, especially if filled from a street main with considerable pressure. It is due simply to air in the water. If a glassful of the water be allowed to stand a few minutes the air will escape and the water will become perfectly clear."

[Mr. Kenrick's communication would seem to confirm the statement that inflammable gas is found in radiators, although it has been received incredulously in some quarters. We cannot, however, agree to the theory that the gas is hydrogen arising from the decomposition of the water, since the temperature of dissociation of water, or in other words, the temperature at which it is separated into its component parts, oxygen and hydrogen, is somewhere in the neighborhood of 2,000° Fahr. It is almost unnecessary to say that this temperature is never attained in a hot-water radiator. The explanation which we offered in connection with the letter from Messrs. Mooney & Baine, that the gas is due to the distillation of some light mineral oil used in making up radiator joints, or perhaps to decomposition of vegetable matter in the water, is probably more nearly correct.]

THE HEATING AND VENTILATION OF A CHURCH.

A. M. P., New Orleans, La., writes:

"I venture to ask your opinion upon a subject which has been under discussion here. St. Paul's Episcopal Church has contracted to have a hot-water heating apparatus put in. The contractor proposes to put radiators at points marked A upon the plan as shown. He runs a 4-inch feed pipe from the boiler on each side of the church to supply both the direct and indirect radiators. In the side aisles he places indirect radiation, the register being about 18x24 inches. In the center aisle he places six registers, connected with a vitrified pipe. The pipe under the floor, 15 or 16 inches in diameter, stands at C'; at C" the pipe is increased in size, and is further increased at each of the six registers. The pipe is last connected to a flue about 50'x50'x50' high. There is a register and indirect radiator in the floor of the chancel at B". This register and indirect radiator connect with the vitrified pipe, and also to a cold-air flue.

"There are a great many feet of pipe in the foul-air shaft. The contractor proposes to heat the church as follows: Before the congregation assembles the damper in the cold-air duct supplying the register in the chancel is closed and also the damper in the foul-air shaft. He holds that the cold air will settle on the floor, be drawn down through the registers in the center aisle to the indirect radiator under the chancel and will then pass up through register B' into the body of the church. This circulation will continue until the congregation assembles. Then the damper in the cold-air duct supplying indirect stack B' will be opened, the damper in the foul-air flue opened, and the damper in the duct connecting the vitrified pipe and indirect stack B' will be closed. The foul air will then be drawn out of the church through the flue. In the base of the flue he has put several hundred feet of pipe through which a current of warm or hot water will flow. In the cellar under the church he has put a Bolton heater. The contractor holds that this plan will quickly and economically warm the church, and will remove the foul air which he says largely settles on the floor. We know little or nothing about heating and ventilation down here, and will be very much obliged for an expression of your opinion as to the efficacy of this plan in the next issue of your widely read and appreciated journal."

[Heating apparatus on the general principle indicated in the above letter—i. e., causing the air to circulate round and round, passing repeatedly over the heating surfaces while the room is unoccupied, and then changing to a continuous system of fresh-air supply when the room is filled with people, has been tried in several cases and with good success.

We presume that the lowest outside temperature calculated for, in fixing the amount of heating surface for this church, is 32° Fahr., and that the number of persons in the auditorium to be supplied with fresh air is about 1,000. The foul-air aspirating shaft is of dimensions sufficient to carry off 1,100 cubic feet of air per second, but there seem to be only 25 square feet of fresh-air register surface, and no doubt part of this is obstructed by the iron-work of the registers, so that not more than 15 square feet of clear fresh-air inlet are provided for the auditorium. Even if a

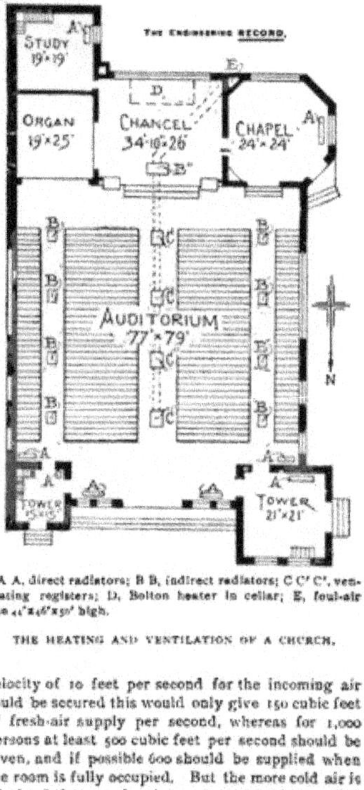

A A, direct radiators; B B, indirect radiators; C C' C", ventilating registers; D, Bolton heater in cellar; E, foul-air flue 4'6"x4'6"x50' high.

THE HEATING AND VENTILATION OF A CHURCH.

velocity of 10 feet per second for the incoming air could be secured this would only give 150 cubic feet of fresh-air supply per second, whereas for 1,000 persons at least 500 cubic feet per second should be given, and if possible 600 should be supplied when the room is fully occupied. But the more cold air is admitted the more heating surface must be supplied to secure comfort, and if the heating surface has been proportioned to 150 cubic feet of fresh-air supply, it must be increased if this supply is tripled or quadrupled, as it should be.

The tendency of engineers in the North, for buildings of this kind, is to use mechanical means of moving the air in the shape of a fan or blower in preference to using an aspirating shaft.]

HEATING A CARVING TABLE.

NICHOL & RYAN, Appleton, Wis., write:

"We send you sketch of job we have just done, and the job works all right except the carving table. We would like to get your opinion in regard to the best system of heating the table successfully. The fixtures on the first floor were already in, but we have added the boiler and heater in the basement."

[We cannot advise a much better method of warming a carving table than to use a hot-water circulation. Steam and gas may be substituted for the hot water, but a well-arranged hot-water apparatus is the simplest. There appear two reasons why the carving table does not work to your satisfaction. In the first place it is very probable that the table remains air-bound or partially so. It is the highest section of the

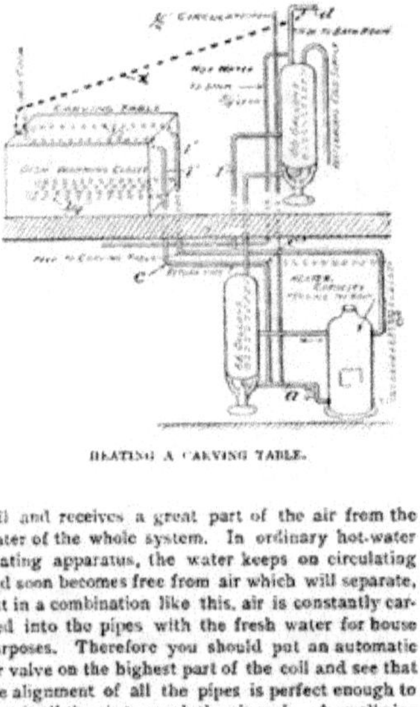

HEATING A CARVING TABLE.

coil and receives a great part of the air from the water of the whole system. In ordinary hot-water heating apparatus, the water keeps on circulating and soon becomes free from air which will separate, but in a combination like this, air is constantly carried into the pipes with the fresh water for house purposes. Therefore you should put an automatic air valve on the highest part of the coil and see that the alignment of all the pipes is perfect enough to permit all the air to reach the air cock. A small pipe from the coil to the branch to bathtub, as shown by dotted line *x*, will be a good substitute for the air cock, if it is convenient to put it in properly. This little pipe must rise gradually to the point of connection with the pipe *d*, if you use it.

The second reason why the apparatus does not work satisfactorily is that you are trying to return all the water to the heater through the pipe *a*. Cut off the pipe *c*, at the elbow *c'*, and removing the portion between that point and pipe *a*, extend it, as indicated by dotted line *b*, running it into the heater as shown on the opposite side. This will improve the circulation.]

TEMPERATURE OBSERVATIONS OF HOT-WATER PIPES.

M. C. F., St. Louis, writes:

"Do you know of any way in which I can find out the temperature of the water in the pipes of a hot-water job without breaking the pipe line to put in a thermometer cup?"

[Place the bulb of a thermometer against the pipe and put a lump of putty over the bulb so as to press the bulb against the pipe. You might further prevent radiation from the bulb by putting cotton waste outside of the putty. String can then be wrapped about the whole so as to hold it in position. If this is carefully done the thermometer will register within 1 degree of the temperature of the water in the pipe.]

FRICTION OF ELBOWS IN HOT-WATER PIPE.

C. B., San Francisco, Cal., writes:

"Mr. Baldwin's book on hot-water heating is authority for the statement that the friction of two 45-degree elbows is equal to that of one 90-degree elbow. I have generally understood that the friction of a 45-degree elbow was scarcely perceptible and of no consequence in steamfitting. I have seen two of them used in hot-water work in place of a 90 degree elbow. Suppose now I am running a line of pipes through a cellar and wanted to take outside branches and turn on 90 degrees at the end. If instead of using the common cross and elbow I use a 45-degree Y and two 45-degree elbows, will the friction be as great? I have stopped using the bull-headed tee on the ends of lines, and instead I use a Y with a 45-degree elbow. Is this advisable?"

[The friction of two 45-degree elbows is exactly equal to one of 90 degrees, the radius being the same in both cases. This is for pipe smooth on the inside, where the diameter of the elbows and the diameter of the pipe is the same. With common fittings the resistance of two 45-degree elbows is greater than one 90-degree. There is probably an advantage, however, in using a Y with a 45-degree elbow on straight lines, and also on the ends of lines as you are using them. It is only the substitution of two 45-degree elbows with their attendant nipples for one 90-degree elbow to which objection can be made.]

CLEANING OUT A HOT-WATER HEATER.

E. P. W., Xenia, O., writes:

"I have a hot-water heating boiler in my residence. It has not been cleaned out this winter, and I imagine it would heat better if it was cleaned. The person who put it in says there is no need of cleaning it until the end of the season, while the engineer at my factory 'blows off' his boiler every two weeks. The water is then found to be very dirty and the boiler works better after blowing off. Which is right?"

[Both are right, as the conditions are very different. Your factory boiler may evaporate a thousand gallons of water, more or less, during each day with its proportionate amount of dirt left to settle in the boiler. Your hot-water boiler is not intended to evaporate its water into steam. Its duty is to heat the water and pass it off to the various heater coils, radiators, or stacks through your house. This water is then returned to the boiler to be reheated and this action is kept up continuously so long as required. As there should be no escape of steam, the only loss from the original charging of the apparatus should be from imperceptible leaks, evaporation from the expansion tank, overflowing, or from water being drawn off. All of these sources with proper care should show a loss of only a small amount during the season. The probable amount of deposit in your boiler will not be enough under ordinary conditions to require the trouble of removing it until the heating season is over.]

TO PREVENT HOT-WATER RADIATORS FROM FREEZING WHEN NOT IN USE.

S. P. J., of Boston, inquires "whether, in a system of direct water heating, there is any way provided for relieving the pipes and radiators of water when shut off.

"As for instance, you have two spare rooms on the north side of your house which are rarely used. It follows that if connections are closed, leaving the pipes and radiators full of water, they would freeze up. Therefore it would seem as though a continuous circulation must be kept up in every room, in direct water heating, unless there is some way of relieving the pipes and radiators of water in rooms not used. How is that done?"

[It would ordinarily be impracticable to provide means for drawing off the water from every radiator of a hot-water system when not in use. The usual method of preventing freezing is to have the valves sufficiently off the seat; that is, just a little open, so as to keep up a slow circulation, just sufficient to prevent the temperature from falling below 40 to 50 degrees in cold weather.

In a single circuit apparatus in which there is a separate circuit from the boiler direct to each radiator or small group of radiators, with stop valves at the boiler and a drip valve in each pipe (such as is sometimes used in hot-water circuits in fine work, to draw off for repairs), each circuit may be emptied separately, if desired, but we do not advise the use of such valves in a hot-water heating apparatus merely for the purpose of preventing freezing. The best way to prevent the freezing of a radiator is to let it defend itself from frost.

An example of single-circuit work, where draw-off pipes are used as above described, will be found on page 173 of Baldwin's book on "Hot-Water Heating and Fitting."]

HEATING BY STEAM FROM AN ELECTRIC LIGHT PLANT.

J. H. F., Middletown, Pa., writes:

"I have a problem in relation to heating several buildings by steam from an electric plant. The plant in question is an electric light plant containing two horizontal tubular boilers of 150 horse-power each. In the summer-time these boilers supply steam for operating an electric street railway and an electric lighting plant, the latter requiring but one-fourth of my boiler power. As the railroad is not used in the winter-time I thought of supplying steam to heat several buildings in the vicinity. Two plans have been suggested, and I would like to have your opinion upon them. We thought of running one boiler to its full capacity and use the excess steam for heating, using a reducing valve to bring the pressure down. If we did this we would be unable to return the water by gravity to the boiler, and would have to use a pump unless we let the condensation from the radiators go to waste, and thus lose the heat it contained. We would have to lay out one pipe, as we are not returning the condensation from the radiators. The other plan is to disconnect the two boilers and use one at high pressure to supply the engine and one at low pressure for heating. The only point I can see in favor of this is, we do not lose the heat contained in the return water. Do you think this water could be returned through the main pipe similar to a one-pipe system? Do you think either of the two plans suggested practicable, and if so, which do you think would be the most economical, and what would be a fair price to charge per 100 square feet of radiation delivered, coal costing us $2.60 per gross ton?"

[You have a total of 300 horse-power in boilers, and of this 225 horse-power is available for heating buildings. We would not advise your using low-pressure steam, as you would probably have to carry it some distance, and this would require very large pipes. If you use high-pressure steam the pipes would be much smaller and there would probably be less loss due to condensation, as the radiating surface of the pipes would be much less.

If you use a high-pressure system you can put a reducing valve on the supply pipe as it enters each building. The fact that the steam expands in passing through a reducing valve, a portion of the condensation or moisture in the pipes would be re-evaporated. Generally it is better to let the condensation go to waste in each building than to use a return system for long lines, as the pressure would vary between the different buildings, each building being under the control of a different person. This, you can readily see, would cause trouble in the common return pipe, as the water would tend to back up in the buildings supplied with the lower pressures. The New York Steam Company abandoned the use of returns some years ago as they could not be made to work satisfactorily. You will not, in most instances, be obliged to throw away the heat in the condensed water, as it can be used in some of the buildings to warm the water for the house supply. In a hotel it can be used in a number of such ways, and in a private house for warming a hot-water indirect stack.

In estimating the steam used for heating in a low-pressure system a rough way is to allow that 1 square foot of radiating surface will condense from one-fourth to one-third of a pound of steam per square foot per hour if the radiating surface is properly proportioned. A more accurate way, as each system would probably be drained by a trap, would be to weigh the discharge from the trap and find out what each tenant is using. In estimating the running expenses of your plant you should add in the wages of your fireman, and 12½ per cent of the cost of your plant to cover the interest and depreciation in its value, as well as your coal bill. You could then charge accordingly, as you knew approximately what each tenant is using. Steam charges in New York vary from $50 to $100 per horse-power per year of 300 10-hour days, depending on the amount of steam taken by the customer. A horse-power in this instance means 30 pounds weight of steam per hour.

We believe it would pay you to engage the services of a competent expert to look over your plant and report upon the most advantageous system to adopt.]

HEATING SURFACE.

EFFICIENCY OF HOT-WATER RADIATORS.

W. B. L., Cleveland, O., writes:

"Can you give me a rule for determining the heat given off by hot-water radiators?"

[The rule as given by Baldwin's book on "Hot-Water Heating and Fitting" (Chapter X.) is, that a square foot of heating surfaces radiates twice as many heat units per hour as there is difference in degrees between the temperature of the hot water in the pipes and temperature in the room. For instance, if your water is at 170 degrees and the room at 70 degrees, the difference of these, 100, multiplied by twice the number of square feet of radiation, will be the number of heat units radiated per hour.]

PIPE SURFACE FOR GREENHOUSE WARMING.

Referring to the article on "Heating a Greenhouse," published in The Engineering RECORD March 29, 1890, John A. Scollay, of Brooklyn, sends the appended table of pipe surface required for different temperatures. The table is somewhat simpler than the one previously published, being arranged for 100 square feet of glass surface.

Table Showing the Amount of Pipe 4 Inches in Diameter that will Heat 100 Square Feet of Glass Exposure any Required Number of Degrees, the Temperature of the Pipes being 200 Degrees Fahrenheit:

Temperature of External Air.	Temperature at which the House is Required to be Kept.								
	40	45	50	55	60	65	70	75	80
20	26	30	33	36	40	45	48	51	59
18	25	29	32	35	39	43	47	52	58
16	25	28	31	34	38	42	46	51	56
14	24	27	30	33	37	41	45	50	55
12	23	26	29	32	36	40	44	49	54
10	22	25	28	31	35	39	43	48	53
8	21	24	27	30	34	38	42	47	52
6	20	23	26	29	33	37	41	46	50
4	19	22	25	29	32	36	40	45	49
2	19	21	24	28	31	35	39	44	48
Zero	18	20	22	27	30	34	38	43	47
2 above	17	19	22	26	29	33	37	41	46
4	16	18	21	25	28	32	36	40	44
6	15	17	20	24	27	31	35	39	43
8	14	17	19	23	26	30	34	38	42
10	13	16	18	22	25	29	33	37	41
12	13	15	18	21	24	28	32	35	39
14	12	14	17	19	23	27	31	34	38
16	11	13	16	19	22	26	30	33	37
18	10	12	15	18	21	25	29	32	36
20	9	11	14	17	20	23	28	31	35

VENTILATION—NOTES AND QUERIES.

LOUVERS.

LOUVER IN VENTILATOR FOR TRAIN-SHED ROOF TO LET OUT SMOKE AND EXCLUDE SNOW.

THE CANADIAN PACIFIC RAILWAY COMPANY,
ENGINEER'S DEPARTMENT,
MONTREAL, November 1, 1888.

SIR: Can you tell me where I can find a drawing or description, or where I can see a louver in a ventilator of a trainshed roof so arranged that it will properly carry out smoke and at the same time keep fine snow from drifting in?

ALLEN PETERSON, Engineer.

[Having heard that Mr. C. L. Strobel, C. E., of Chicago, had designed something for this purpose, we wrote him regarding it, and are indebted to him for the following description and accompanying sketch. Mr. Strobel says:

"I send you some sketches showing the arrange. of louver in trainshed of the Chicago, Milwaukee, and St. Paul Railroad at Milwaukee, and on the Chicago and Great Western trainshed at Chicago (Wisconsin Central Line). The latter is now under construction by the Keystone Bridge Company, S. S. Beman, architect. The sketches also show the smoke jack used on the Milwaukee trainshed. The arrangement adopted at Milwaukee has been used in a number of instances, and I think accomplishes the object fairly well of providing ventilation, keeping out the rain and, to some extent, snow. I do not know of any plan by which snow can be entirely excluded, *so long as there is ventilation*. Fine, drifting snow will find its way through the double windows of a Pullman car in a strong wind, and it will of course pass through the louver openings and into the trainshed, to some extent."]

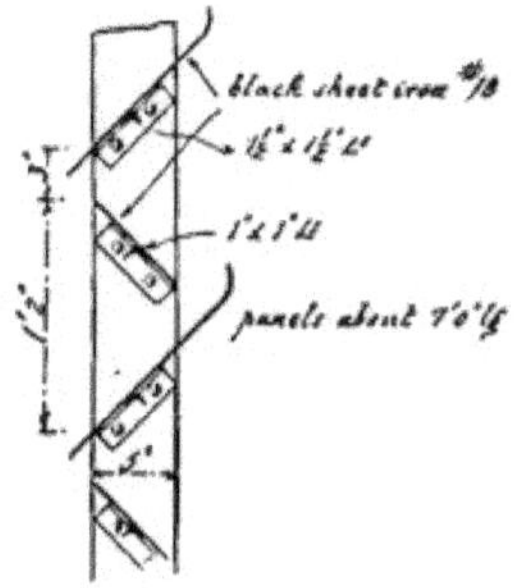

CHICAGO AND GREAT WESTERN TRAINSHED, CHICAGO.

We are indebted to Charles M. Jarvis, President of the Berlin Iron Bridge Company, New Berlin, Conn., for the shop drawings from which we have been enabled to prepare the accompanying illustrations. It is sent in response to the foregoing inquiry. Mr. Jarvis says:

"EAST BERLIN, CONN., November 19, 1888.

"SIR: Referring to the communication of Mr. Allen Peterson, Engineer of the Canada Pacific Railway Company, in reference to a louver in the ventilator of a trainshed roof, we inclose herewith a blue-print drawing of a louver designed by us for a trainshed which we built some two years ago for the N. Y. N. H. & H. R. R. Co., at New Haven, Conn.

"There can be no louver designed which will ventilate a trainshed and at the same time keep fine snow from blowing in. We believe, however, the

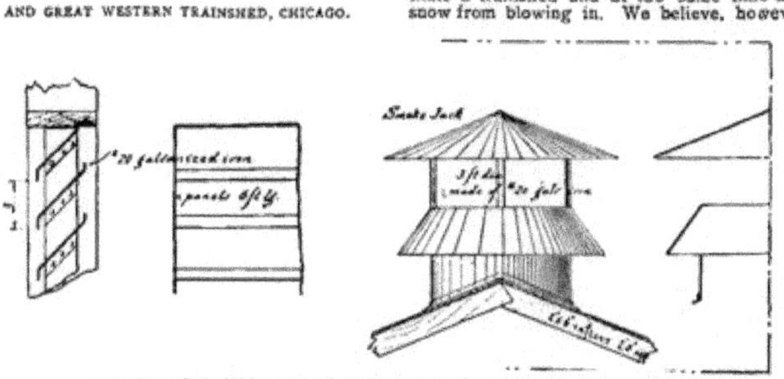

CHICAGO, MILWAUKEE, AND ST. PAUL RAILROAD TRAINSHED, MILWAUKEE.

inclosed to be the best thing of the kind that can be used, as the snow must have an upward motion in order to get through the slats of the louver."

[Figure 1 is a section of louver showing a main post attached to the roof truss.

Figure 4 is a section showing an intermediate post; B B is the line of the skylight, and A A that of the other purlins.

Figures 2 and 4 are side elevations of part of the louver, the former at an intermediate point and the latter at one end, at X X X X, Fig. 1.

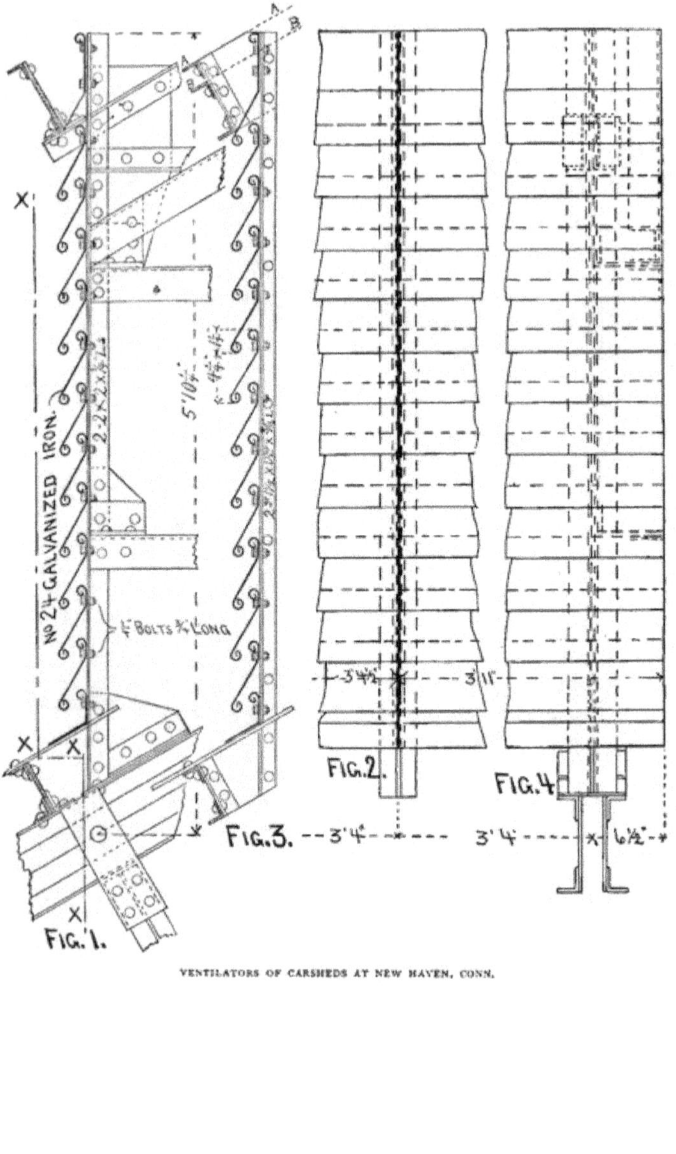

VENTILATORS OF CARSHEDS AT NEW HAVEN, CONN.

The louver boards overlap half an inch at the ends, as shown by the dimensions and the vertical lines at center of Fig. 2, which are both drawn full for clearness.]

In further response to Mr. Peterson's letter the arrangement in the New York Metropolitan Opera House is shown.

Figure 1 is a section of roof and louver over the auditorium. F H, F H is the roof line; H K, K H is a polygonal louver; D E, D E is a water-tight ceiling; C D, D C is a shaft that may be closed by the cap B when lowered by windlass A to position C C. When B is raised the foul air escapes as shown by the arrows.

Figures 3 and 4 are details in section and elevation respectively of the galvanized-iron outlet conduits T T, etc., through the louver walls.

Figure 2 is a partial elevation of the louver, where, as in Fig. 1, the conduits T T, etc., are shown by

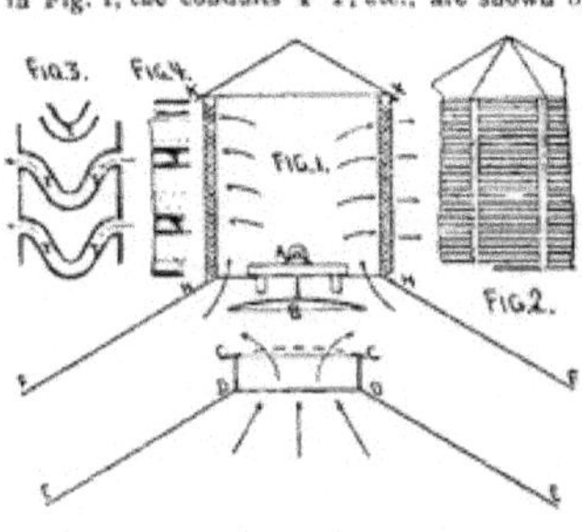

VENTILATING LOUVER IN THE METROPOLITAN OPERA HOUSE.

single heavy lines. The air in the building is generally under a slight pressure from the ventilation fans. It is said that the arrangement is satisfactory, and that no snow or rain penetrates the louver.

Our illustrations are prepared from sketches furnished through the kindness of Mr. S. J. McKay, the Chief Engineer of the steam plant, etc., of the Opera House.

DAMPER TO PREVENT BACK DRAFTS.

The two appended sketches show a device for a damper put upon the discharge of a ventilating duct

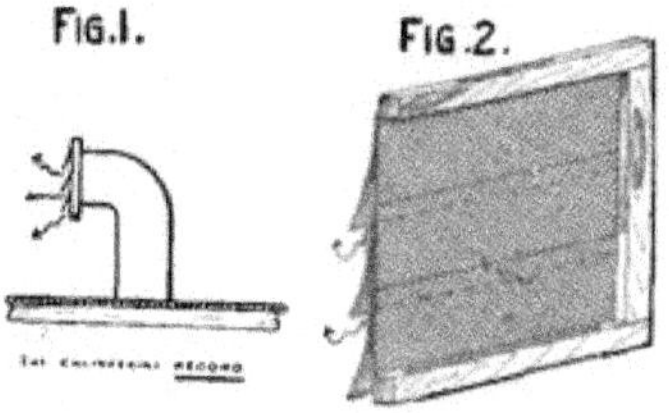

DAMPER TO PREVENT BACK DRAFTS.

to prevent a back draft by Messrs. Reed & Stern, architects, of St. Paul, Minn. It is a detail of the heating plant of the Medical College of the University of Minnesota. Figure 1 shows the end of the ventilating duct discharging into a space between the attic ceiling and the roof. The damper is shown in detail by Fig. 2. It consists of a wooden frame, which is fastened to the opening of the discharge duct. A fine wire screen is fastened over the frame. Small sheets of oil silk are then fastened on in the manner shown, the lower edge of each overlapping the upper edge of the sheet below in the manner of shingles on a roof. The slightest outward pressure causes this damper to open and allow the exit of air, while a slight back pressure will close the flaps, thus preventing the foul air in the chamber from being forced back into the living-rooms of the building.

SIZE OF FLUES.

EXHAUST VENTILATION UNUSED.

Inquirer, Bowdoinham, Me., writes:

"1. In ventilating a building by the exhaust-fan system, the ducts being small and velocity high, what consideration must be given to fireplace flues, built for ornament and not for practical use? If such flues are run up an inner wall the temperature of the air within them might be nearly that of the rooms. The question is, which way would the current be—down the flues into the room, being drawn by the fan, or up the flues notwithstanding the exhaust current?

"2. In ventilation is high velocity considered good practice?

"3. What advantage is gained by extending hot-air ducts nearly to ceiling?"

[The answer to the first question is that it depends upon the presence and sufficiency of special inlets for fresh air. If an exhaust fan is connected with a room to which special fresh-air inlets have not been provided—a not uncommon blunder—there will be a suction into the room through all available openings, including fireplace flues, and the result will depend on the number, size, and position of these openings. It will come from the point of least resistance, where there is the least friction and the least distance to travel. If the window is open it will pour in through that. If the windows fit tightly and there are few cracks in the walls or floors, it will come down the fireplace flue. If an exhaust fan is connected with a room having a fireplace flue, fresh-air inlets should be provided of sufficient size to give a free supply to both the fan and the flue in order to prevent them from pulling against each other.

In reply to the second question it may be said that high velocities are now in general use in the mechanical ventilation of buildings, and especially in what are known as hot-blast systems. The reason for this is that the pipes and ducts can be made comparatively small, thus saving in space and cost. It requires more power to force a given quantity of air in a given time through a system of small flues than it does to force the same quantity of air through large flues in the same time, owing to the rapid increase of friction with increase of velocity. Hence, high velocity plants are cheaper in original cost and take up less space, but are more expensive to run. They are preferred by contractors, but it is usually not to the owner's interest to have flues so small that a velocity of more than 360 feet per minute must be maintained in them to give the quantity of air required.

With regard to the third question, the chief advantage of extending hot-air ducts up so that their openings shall be well above the heads of those occupying the room, is that it permits of forcing the air in with considerable velocity without producing the unpleasant drafts which would be caused if the openings were near the floor, and therefore permits of the use of smaller flues and registers.]

SIZE OF CHIMNEY FLUE FOR BOILER.

W. W. Wood, Honesdale, Pa., writes:

"We have a house which calls for 300 square feet of direct and 190 square feet of indirect radiating surface. The boiler used is capable of supplying 650 square feet of direct radiating surface. The chimney flue is 8x9 inches and has a good draft, but fire in boiler burns very poorly; we claim the chimney flue should be at least 12x12 inches."

[A chimney flue of half a square foot of cross-section is rarely enough for a house-heating boiler even of the smallest size.

In very cold weather, with the apparatus you describe, the consumption of coal may be reasonably set at 20 pounds per hour. Each pound of fuel will require and liberate gases to the amount of about 600 cubic feet at the temperature which they pass into the chimney, or 12,000 cubic feet in all. To pass this through a flue 8x9 inches or half a square foot will require an effective velocity of 6.6 feet per second over its whole area. This velocity, though easily obtained theoretically in flues of any practical magnitude, is rarely obtained in house flues on account of the amount of resistance caused by rough brickwork and short or square turns. The leakage of air also through the comparatively thin walls of the flue is considerable; and when boilers have quick setting there is also a considerable infiltration or leakage of air through the walls of the setting, which has to pass off by the chimney flue. An 8x12-inch flue might be ample, though if we were building a chimney we would make it 12x12 inches, if of brick. A 12-inch circular pipe built into the walls we consider equally as good as a 12-inch square when made of brick, as they are usually built.]

SIZE OF VENTILATING FLUE.

William Jennings, Harrisburg, Pa., writes:

"We have an opera-house which we desire to ventilate. It has a ground floor, first and second galleries. Our idea is to put in a flue and take the foul air out at each of the three floors. What we desire to know is, shall the flue be of uniform size, or increase at each floor? If so, in what ratio and in what part of the flue shall the coil of steam pipes be placed?"

[Generally speaking, a ventilating flue or shaft, as in this case, must be larger than might at first be supposed in order to satisfactorily meet the requirements of successful operation. It is usually made of practically uniform size from the bottom up.

As to the location of the steam pipe coil in the shaft, it should be borne in mind that the aspirating power of the shaft depends on the height of the column of heated air as well as on the difference between the temperature in the shaft and that of the external air. It is therefore evident that the nearer the bottom of the shaft the heat is applied the greater will be the efficiency. The lowest point at which the coil is placed should, however, be above the entrance of the highest foul-air flue.]

SIZE OF REGISTERS.

RATIO OF REGISTER AREA TO RADIATING SURFACE.

J. Murchison, New York, writes:

"I have had a discussion with a friend as to the proper ratio between the area of a hot-air register and the surface of indirect stacks of a hot-water job. As we could not agree I thought I would write to you for your opinion in the matter."

[A common rule for this proportion is to allow 1 square inch of flue area for every square foot of radiating surface in your indirect stack. You must never have less flue area than this, and sometimes more, depending whether or not there are any sharp bends in flue or long horizontal distances to make an excessive amount of friction. Vertical distance does not enter into the problem to any great extent, as the increase in velocity due to a greater height of flue tends to balance the increase of friction due to the increase of length. You should have twice as much register area as flue area to allow for the fretwork in the design of the register.]

ALLOWANCE FOR FRICTION IN REGISTER OPENINGS.

E. F. KITTRO, Chicago, Ill., writes :

"In your issue giving an account of the warming and ventilation of the Honesdale School, I notice in the table of size of registers that a 26x20-inch is taken of the capacity of 3.5 square feet, whereas the actual area of opening of 'fretwork' is usually taken as one-third inch less 26x20 inches = 520 inches, less 33⅓ inches = 346.32 inches = 2.4 feet. The line

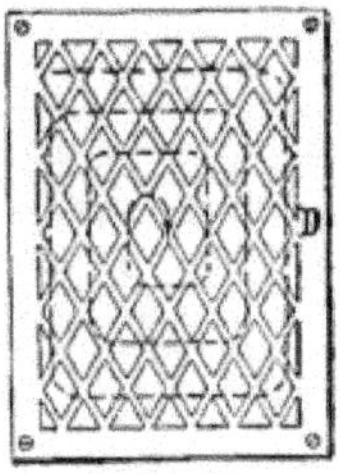

REGISTER

velocity through this register is given as 305 feet; 305 + 2.4 gives 43,920 cubic feet per hour, as against 64,020. Please explain if the proper allowance was made for friction or obstructions in the table."

[The engineer who made the measurements of the air in the Honesdale School experiments informs us that it would not be proper to deduct one-third from the area of the registers, as the air velocity was not measured in the holes of the fretwork, but 2 inches in front of the register face. His method of measuring air currents at register faces, he explains, is to commence at one corner, and pass the anemometer over the face of the register in a manner as shown by the dotted lines, reaching the center in about half a minute, then returning over the same course, reaching the corner again at the end of the minute as nearly as possible. Should the corner be reached before the minute was called, he would make a few oblique movements across the register face.

The registers used were Persian pattern, and consequently very open in the face; but even if they were asylum patterns—strong and close—it would be improper to deduct one-third from the flow of air when measured 2 or 3 inches from the face of the register.

When register faces are removed from the frames, and the velocity measured in the opening, then a large allowance must be made for the obstruction of the fretwork, but as this allowance would be different with every different pattern of register face, the error will be much less if the air is measured with the face on.

The distance of the anemometer from the face of the register is not very important, except that it must not be so far from it that air currents can spread and lose their velocity before they reach it. If the anemometer was not kept in motion as explained before, then it might be possible to get too high a reading by having a jet of air through a hole in the fretwork strike the blades; but while it is in motion it is the average effect only which reaches it, and by keeping it 2 or 3 inches from the face the stream of air 20x26 inches at the recorded velocity is registered.]

MISCELLANEOUS.

PRISON VENTILATION.

W. M. DUNLAP, City Engineer of Roanoke, Va., writes:

"Can you refer me to any article that you may have published concerning the ventilation of jails? I have heard that there is a recent device for this, the general plan being to have a high chimney for draft, into which the air from the cells passes, being led to it, in a downward direction, by a pipe through which also the sewage passes, and is either dried or burnt. I will be glad to get any information on this subject."

[We are unable to refer you to any treatise on jail or prison ventilation. It is quite common, however, to draw the air from prison corridors through the grated doors and thence to a cast-iron niche-shaped chamber in the wall at the floor level, the upper end of which connects with a flue, the dimensions of which are from 4 to 6 inches square. This flue runs to the head of the party wall between any two tiers of cells, and is sometimes connected with a vent shaft or large chimney; often, however, it simply opens into the roof space and a ventilator is placed on the roof. Many cells being treated this way makes the center wall a stack of small flues, and it is customary to make this wall thick and heavy for this reason. The niche-shaped receptacle is sufficiently wide and deep to let the prisoner's night bucket set within it. It is made of cast iron with a flange on the face of the wall and a short iron collar extending some little distance up the flue. The object of the casting is to make a neat finish and to prevent prisoners from cutting holes at this point through to the next cell. The castings are also anchored in, so that a prisoner cannot readily take it from the wall and use it as a shield to digging operations.

In some cases a second flue is provided, starting at the ceiling of the cell and running parallel with the first one and terminating as it does. These flues also have a strong casting at their lower ends, near the ceiling, and the hole measures usually 4x4 inches.

In this way the upper hole carries off the products of combustion from lights and other light vapors, and the lower opening ventilates the cell at the floor level, causing the air current to pass over the night bucket as it ascends into the flue. The latter probably gives rise to your surmise of using the vent hole for sewage. There is, however, a scheme for ventilating prison cells through water-closet hoppers without traps, through which the air of the cell is supposed to be drawn down into the hopper, and allowed to escape through a pipe that extends from the soil pipe to the roof. We would be opposed to any such system of ventilation for prisons, although it has several advocates both in prison and asylum construction. When used in asylums there is generally this difference, which may be in its favor—namely, plenum ventilation is generally used with it, helping to force the air out through the hopper. We consider such ventilation, however, both inadequate and dangerous, because of liability to reverse currents.]

HEATING AND VENTILATING AN HOSPITAL.

ENGINEER, Boston, Mass., writes:

"An architect from 'away down East,' with a little knowledge of steam heating and ventilating, has laid out a job in this section, or, rather, has partially laid it out, trying to separate the ventilating from the heating part. He thinks he knows all about ventilating, but admits that steam is 'beyond his realm.' I would like to have your opinion on a few points.

"The building is an hospital, each floor (first and second) having 300 beds, and containing 180,000 cubic feet, not including corridors, small rooms, etc. The building is of wood, exposed on all sides, and has fully the average of glass.

"The architect has planned for 36 hot-air flues, each 12 inches in diameter (18 to the first and 18 to the second floor), and for 44 ventilating flues, each 12 inches in diameter (22 to each floor). He proposes to draw the air down to the cellar by a ventilating fan located 250 feet from the extremes, about as shown in the sketch. He proposes to make one 15-inch pipe take the foul air from four 12-inch pipes, thus making 11 15-inch ducts, each to enter a 24-inch duct, or, say, five in one branch and six in the other. Where the two 24-inch main branches come together,

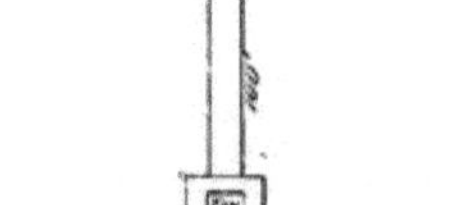

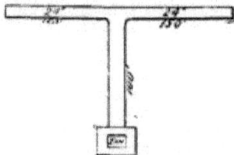

the size of the duct running to the fan has not yet been determined upon, except that the architect says that a 36 inch wheel will do the work easily. I tell him that he ought to use a 72-inch wheel. I do not fancy the idea of taking 22 12-inch ducts into one 24-inch duct, reducing the area 5½ times.

"I asked him how much indirect heating surface he wanted in each stack. He replied that each stack would supply two flues (12 inches), but he did not know how much surface was wanted. Then I asked how often he was going to change the air, and he said 'every 15 minutes.' I suggested to him that heating and ventilating went together, and that the job should be laid out as a whole, but he evidently has just about knowledge enough of the business to make the old saying good.

"His chimney for the boilers is only 20 inches in diameter, 50 feet high, and extends to 3 feet below the main ridge of the building. I asked why they were not carried up higher, and made 24 inches in diameter, at least. 'Oh, you will get a good draft,' he replied. The top floor is to be heated by direct radiation; total cubic feet of space, 362,000; hot water is to be used. I estimate that 75 horse-power of boiler will be required to do the work easily, and a 6 or 8-inch main at the boiler. What do you think of the matter?"

[You can safely predict a failure with such an arrangement of flues. An hospital of 300 beds will want from 1,000,000 to 2,000,000 cubic feet of air per hour, and this much air cannot be passed through two 24-inch round pipes by such fans or blowers as are generally used in hospitals, and certainly not by a 36-inch propeller or air-wheel. A system of exhaust flues should be designed so that the trunk flues aggregated the area of all the branches. This rule, you are aware, may be contradicted on the ground that for pipes of equal short lengths the increase of diameters will be in the ratios of the fifth root of the square of the branches, pressure remaining constant; but even this rule will give you pipes greatly in excess of those proposed by the architect. In practice, however, you have varying lengths, and the lengths of the branches, with their short bends and turns in the wall, are relatively much greater than the lengths of the trunks; therefore the only safe rule is to have the trunks aggregate the area of the branches.

To warm 2,000,000 cubic feet of air 100 degrees each hour is the equivalent of 4,000 pounds of water converted into steam in the same time—the equivalent of about 133 horse-power. Of course, if the hospital authorities are satisfied with only 1,000,000 cubic feet of air in an hour, a 67 horse-power boiler will be just ample, making no allowance for contingencies in other work. Assuming therefore that you pass 2,000,000 cubic feet of air through the hospital every hour, it will require a chimney at least 2 feet square or 4 square feet; but if we were building the same we would not risk our reputation by making it so small, and would certainly have it 30 inches on the side, or 6¼ square feet in area.]

VENTILATING A VAULT.

D., Washington, D. C., writes:

"If you will kindly answer the following without much trouble to yourself, I should be very glad to have you do so. I cannot ask you to consider it at length, but would like a brief explanation. I found the problem the other day in a country house near here and wondered if it could have any simple, inexpensive solution.

"A storage vault, 8x15 feet and 10 feet deep, below ground surface, with no opening when the door is closed, except a 1-foot square hole in the brick arched roof, opening into an air chamber 3 or 4 feet high over the whole vault, the latter having two windows to open air. How can the vault be sufficiently ventilated in a cheap way? Would a duct B (Fig. 1) from the side of the house above the windows to near the bottom convey any air?"

[Without some mechanical power, or its equivalent *in heat*, you cannot move air enough to give satisfactory results. You want, if possible, to dry the air, lessen its humidity, and cleanse it sufficient to prevent the growth of fungus. Cold air in a hole beneath the ground is just like water in a hole in one

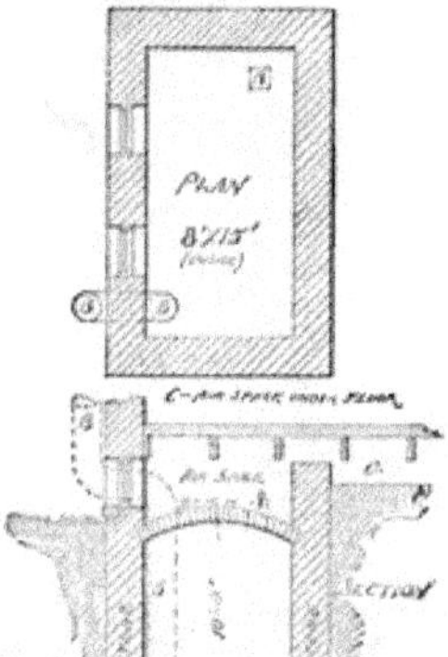

FIG-I

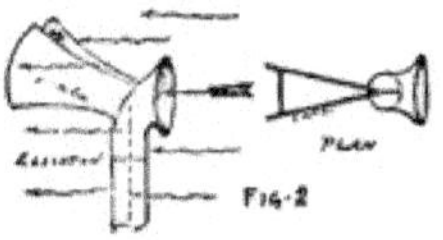

FIG-2

respect. It is heavier than the surrounding and superimposed atmosphere and must be pumped out; in other words, every pound of it that is lifted will represent work done.

Above the ground level it also takes power to move air, but generally the air is warmer than the outside atmosphere, and thus the draft in chimneys, etc., goes on, as the pressure outside is greater, but in the case of the vault there is light, warm air above it that cannot descend to press the cold and heavy air out, and there it will stay unless forced or drawn out, which means the same thing.

If you can connect the pipe B, which you propose, to a warm chimney, it will draw the air out from the bottom, admitting fresh air at A. But as you show it, against the wall of the house, it will do no good unless the sun shines against it all the time.

If there is no warm flue in the house to spare, put a little heat—steam coil—into B above the vault, or so that it will not not warm the vault, and that will do. If you can tap a fan or air pipe, which we presume improbable, a small inlet 4 to 6 inches will do, near the floor. Then if you can take advantage of prevailing winds, make a cowl C with a large vane that will always hold it into the wind; when placed on B it will do for a great part of the time if you extend the duct above the housetop. The double vane will keep the cowl from swinging about.]

HOW MUCH COLD AIR TO ADMIT AND HOW TO RETAIN IT WHEN WARMED.

SUBSCRIBER, of Buffalo, N. Y., asks:

"Have you any rule governing the admission of cold air to either hot-water or steam indirect stacks?

"What is your remedy for overcoming the warm air passing out through cold-air boxes, there being fireplaces with good draft for ventilation in each room? The house is isolated and very much exposed.

"What is the best mode of constructing cold-air boxes to such stacks?"

[The amount of cold fresh air that should be admitted to a stack of pipes or radiators depends on the requirements of the room to be warmed. Each occupant of a room should have at least 2,000 cubic feet of fresh air every hour and as much more as it is possible to supply within the bounds of economy.

If four persons sleep or work in a room containing 4,000 cubic feet of space, a sanitarian would not be likely to say it was unhealthy as long as the air was changed once every 30 minutes. If it was changed every 15 minutes, however, he would feel better satisfied and would probably have no hesitation in pronouncing it ample.

From such reasoning as the above, and on the assumption that there is about one person to the 1,000 cubic feet of occupied space in an ordinary private residence or well-arranged office, it has been generally agreed that to change all the air of a building once in 30 minutes will give fair ventilation and once in 15 minutes good ventilation.

It has been found by experiment that when the air of a room is changed once in 30 minutes it has to enter the room at a temperature about as much above 70° Fahr. as the outside air is below 70 degrees, and that when the air is changed every 15 minutes its excess of temperature above 70 degrees has only to be half as much as in the former case. The reason is that the heat lost by the air through the walls and windows depends only on the difference between the inside and outside temperatures, whereas the temperature of the incoming air required to maintain the room at 70° Fahr. varies with the quantity admitted, and is of course less the faster it comes in.

Assume, then, that you are going to change the air every 15 minutes in a room containing 4,000 cubic feet, or in other words, to supply 16,000 cubic feet per hour, and that it is to be warmed from zero to $70° + \frac{70°}{2} = 105°$, the temperature at which the air should enter the room, according to the principle we have just stated.

Then, as 1 heat unit, or the amount of heat that will raise the temperature of a pound of water 1° Fahr., will do the same for 50 cubic feet of air at atmosphere pressure, and as steam in condensing gives off about 1,000 heat units per pound of condensed water, you have $\frac{16{,}000 \text{ c. f.} \times 105^\circ}{50} = 33{,}600$ heat units, which, if divided by 1,000 = 33.6 pounds of steam that must be condensed in an hour to do this work, or in the case of hot water, 3,360 pounds of water must pass through the coil in an hour and be cooled 10 degrees, or half the amount cooled 20 degrees to do the work of the steam.

One hundred square feet of good indirect steam radiation should condense 33.6 pounds of water, and about 150 square feet of hot-water surface will do the same work in heating the entering air.

The reason why the warm air passes out through the cold-air boxes is because there is more pressure in the room than there is at the mouth of the cold-air box. How this can be the case with open fireplaces in the rooms we cannot understand without fuller particulars. The air will always press inwards unless there is some greater power resisting it. Unless you are sure by building a fire or otherwise that there is a good draft in your fireplace flues, you had better make sure that they are all clean, as sometimes obstructions are left in while building or the holes in the coping stones are forgotten.

The wind from some quarters may make a very strong eddy at the mouth of your cold-air box sufficient to overcome for a time at least the draft of your chimneys. Your difficulty could probably be overcome by having a cold-air opening on each side of the house and taking care to close the one on the lee side. A number of light canvas flaps about 4 inches deep, opening inwards in the cold-air box, and shutting against a piece of coarse wire netting, would do this automatically when necessary.

As to constructing cold-air boxes to heating stacks conditions vary too much to allow any one method to be pronounced best for all cases; the simple and ordinary way is to bring a wooden or sheet-metal cold-air duct to the bottom of the coil chambers from some convenient basement or cellar window or from a hole in the wall.]

A SIMPLE DAMPER REGULATOR.

A CONVENIENT device for controlling a hot or cold flue damper for indirect heating systems has been used by W. F. Porter & Co., Minneapolis, Minn. The damper is attached to the weighted chain C

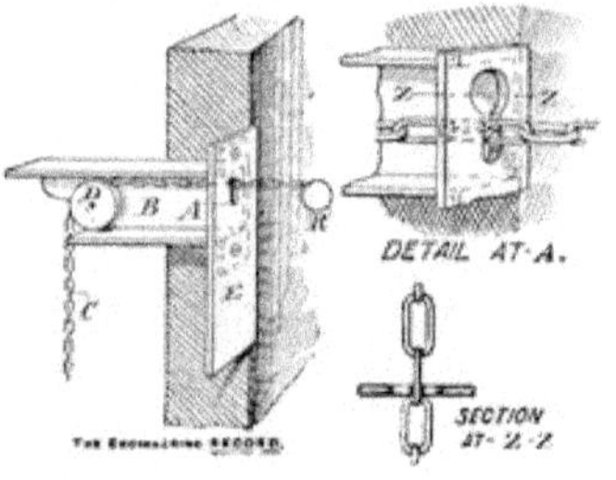

which passes over a sheave D which is supported on a cast bracket B, built into the wall and projecting into the flue. From sheave D the chain passes through a slot A in the escutcheon plate E and terminates in a handle ring R by which it is commanded to set the damper at any required position, where it is maintained by setting one link vertically in the narrow bottom of the slot A, which will not allow the next link to enter crosswise. The upper part of the hole is made circular and large enough to let the chain through freely, but not to admit the ring R.

UNWISE HEATING CONTRACTS.

THE following discussion upon contracts guaranteeing the heating of buildings to 70° Fahr. with the outside temperature at zero, and permitting the withholding of final payment until meteorological conditions allow practical demonstration, was contributed to THE ENGINEERING RECORD by a number of prominent heating engineers. As the arguments advanced are of permanent value to architects, owners, and heating contractors, the discussion is reprinted in chronological order, together with two editorial articles, one introducing the subject and the concluding article expressing the views of THE ENGINEERING RECORD.

[From THE ENGINEERING RECORD of January 20, 1894.]

HEATING GUARANTEE AND ZERO WEATHER.

THE letter of Mr. W. H. Francis, printed upon another page of this issue, brings up a question of interest to all who have to do with building engineering. For the provision of the heating of buildings the practice is, unfortunately, too general among architects to throw the arrangement of all the details upon the bidders, specifying only that the boilers and radiators shall heat the building to 70° Fahr. in zero weather. The requirement of this guarantee is not fair either to the owner or the heating contractor. The architect should determine definitely what he wants; it should be as much his duty to specify the square feet of heating surface required and the sizes of boiler and piping as to specify the thickness of the walls. When the terms of such specifications are complied with the contract should be considered completed, and the contractor should be entitled to receive his pay in the ordinary course of business the same as in any other branch of building engineering. We have also known of cases, as instanced by Mr. Francis, where heating contractors have been forced to wait for two years for their pay by a strict enforcement of the obnoxious clause, and do not doubt that its operation has been attended with serious injustice and hardship. We should advise heating engineers to refuse to accept contracts including this provision and to demand that architects make their specifications explicit.

In this connection there is another practice current which is not in the interest of an owner who desires first-class work nor that of a conservative contractor who aims to do what is right and to make a reasonable profit in his business. This is the requiring of a bond that the plant will perform a certain duty and at the same time the exaction of the right to withhold the final payment until it has been demonstrated that there is no occasion to call upon the bondsmen. If a guarantee or a bond is required payment should be made when the material is put in place, for it is unjust to demand a guarantee or bond and withhold the payment too. If a man puts in the plant he agrees to furnish he should be paid for it; if it fails to perform the duties guaranteed, the man, if responsible, can be made to make it good, or the bond is the recourse. Experience, however, shows that it is better for owners or building committees to deal with responsible parties whose guarantee is good than to trust to being recouped by bondsmen in cases of default.

THE REQUIRED HEATING OF BUILDINGS TO "SEVENTY DEGREES IN ZERO WEATHER."

KENSINGTON ENGINE WORKS, LTD.,
PHILADELPHIA, PA., January 15, 1894.

To the Editor of THE ENGINEERING RECORD.

SIR: I write in protest of the terms in which the guarantee of a steam-heating system is usually expressed in the specifications of architects and builders, "that the building shall be heated to a temperature of 70 degrees in zero weather." This has become the standard requirement and has led to endless trouble and misinterpretation, and caused bitter experience to the contractor in every case where be has met a principle governed only by the "letter of his bond." In one case I recall, the contractors were obliged to wait two years after finishing the work, for a literal zero test, when, as the result proved, the system was ample in every respect to maintain the temperature under the guarantee. It is very unjust to require the contractor to wait until the outside temperature is zero to test his apparatus.

In our own city of Philadelphia in the winter of 1892-93 the temperature did fall to zero on several occasions, but previous to that there had not been so low a temperature for 11 years. The expression, "70 degrees in zero weather," should be considered, as it is, in justice to the contractor, a commercial one, and be understood universally to mean when the system is capable of warming a building to 70 degrees under wind pressure, with outside temperature of 15 to 20 degrees above zero, it will be capable of warming it in zero weather, providing it is doing it easily under the above conditions. It is not so understood with people not familiar with work of this kind, and hence the trouble arises as to the terms of the contract. Why not abandon the phrase entirely and substitute its just and fair meaning, "the system shall be capable of maintaining a temperature of 70 degrees in zero weather."

Many users of steam-heating systems expect to reach a temperature of 70 degrees within an hour of steaming, in a very cold building, which is simply impossible with an ordinary well-designed heating system, and yet the system may be fully equal to attaining and maintaining the temperature with continuous warming. The guarantee to warm to "70 degrees in zero weather," as demonstrated by actual experience, is subject to too arbitrary interpretation, and all contractors for steam work should uniformly decline to sign a contract under such conditions, but at the same time no reputable party should object to guaranteeing their system "to be equal to maintaining the temperature of 70 degrees on a basis of zero weather," with not over a given steam pressure.

W. H. FRANCIS.

[This question is discussed editorially in this issue.]

PHILADELPHIA, January 18, 1894.

To the Editor of THE ENGINEERING RECORD.

SIR: A rule by which one could estimate what a given heater and radiator will do in zero weather, knowing what they actually do in 20 degrees above zero, would be very acceptable to many readers, no doubt. We bid on heaters, low-pressure steam, say five pounds pressure on the gauge, or on open system water radiators guaranteed to heat to 70 degrees in zero weather; we can show 80 to 90 degrees at 10 degrees above zero outside, but cannot collect our bills until zero weather allows a practical demonstration.

J. G.

BRANFORD LOCK WORKS,
BRANFORD, CONN., January 23, 1894.

To the Editor of THE ENGINEERING RECORD.

SIR: Your correspondent, Mr. W. H. Francis, has, in your issue of January 20, raised a question of great interest to all, especially to the heating and ventilating contractor. The injustice of this almost universal clause in the contract, together with the withholding of a portion of the contract price until a test in zero weather is made, which practice obtains with nearly all architects, cannot fail to be appreciated by all engineers.

Mr. Francis further takes the position that with the outside temperature from 15 to 20 degrees with high wind the conditions of heating are more difficult than at usual zero weather; this position, though seeming paradoxical, can readily be confirmed by engineers of experience in this particular branch, and further, it is well known to meteorologists that zero weather is usually unaccompanied by wind, while at temperatures from 10 to 30 degrees the most violent winds usually occur. Hence, if the heating to 70 degrees in zero weather is material, it would be perfectly competent to introduce a clause bearing upon what is, to the writer's mind, a far more important factor—viz., the direction and velocity of the wind.

A year or two ago (1892) the writer undertook a proximate theoretical solution of the problem. An apparatus is guaranteed to heat a building to 70 degrees in zero weather; to what temperature will it heat the building at 30 degrees, 40 degrees, 50 degrees, and 60 degrees outside temperature in calm weather? The proximate solution, which may be of interest to your readers, was effected as follows:

Preliminarily, the loss of heat between bodies of small differences of temperature, say about 100° Fahr., is about proportional to the difference in temperature, hence the loss of temperature between the building and the external air may, without sensible error, be said to be directly proportional to the difference. The loss of heat from the radiating surface to the air and walls of the room, where the difference in temperature is much larger, follows a complicated law for which Dulong has developed an approximate formula. Again, if we heat a room to 70 degrees in zero weather and apply the same apparatus under similar conditions, varying the external temperature only, we find we have a variation of a number of the factors resultant upon the change of the external temperature. For example, suppose an apparatus, having a given number of square feet, heats a room or building to 70 degrees in zero weather with a given pressure of steam. Suppose the external temperature is raised to 30 degrees, the loss of heat by the walls and windows by radiation, conduction, and contact with air, and by ventilation, is lessened, the temperature of the room rises above 70 degrees, the difference in temperature between the air and walls of the room becomes less, and the radiator thus furnishes less heat. Hence, the rise in the temperature of the room will not vary directly with the increase in external temperature, because:

1. The law of transmission of heat from the radiators to the air and walls of the room is not directly proportional to the temperature, and

2. The radiating surface in the room becomes less efficient as the temperature (external) rises.

The approximate solution of this question becomes therefore:

I. A determination of the loss of heat by the building at the various external temperatures.

II. A determination of the ratio of rise in temperature, heating surface and pressure constant, assuming the loss of heat to be that determined by I.

III. A determination of the rate of the loss of heat, heating surface and pressure constant, the loss of heat being occasioned by the temperature, determined by II.

IV. The final determination of the resultant temperature from the results of III.

I. *A Determination of the Loss of Heat by the Building at Various External Temperatures.*

The loss by radiation, conduction, and contact with air, is determined best by a formula given by Box.

(A practical Treatise on Heat, by Thomas Box, edition 1883, p. 216.)

$$U = \frac{(A \times C \times Q)(T - T')}{C(2A + R) + (E \times A \times Q)}$$

Where
U = units of heat lost per hour per square foot of outside surface of building.
A = loss by contact of air.
R = radiant power of the material of building.
$Q = R + A$.
C = conducting power of material of building.
T = internal temperature, 70 degrees.
T' = external temperature, and
E = thickness of the wall (inches).

From the appearance of the formula it is readily seen that the loss of heat incurred by the three factors just mentioned is directly proportioned to the difference in temperature. Similarly, within the limits spoken of above, the ventilation, if natural, may be taken without sensible error, to vary in the same way. Hence, if we take a square, Diagram 1, and draw its diagonal, the abscissas, or horizontal lines, may be made to represent the external temperatures; the ordinates, or the vertical lines, the ratios of heat losses. (It should be understood here that we may, under the conditions assumed and when dealing with the same building, use ratios of the quantities of heat the same as the actual thermal units themselves.)

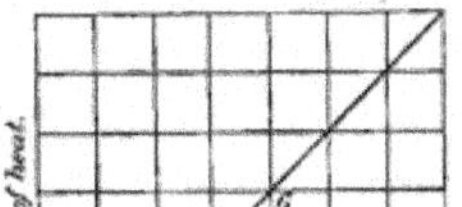

Diagram 1.

II. *A Determination of the Ratio of Rise in Temperature, Heating Surface and Pressure Constant, the Loss of Heat being Occasioned by the Temperature as Determined from* I.

The first step under this heading is to determine the ratio of heat required per unit of radiating surface, steam pressure constant, the external temperature varying from zero to 200 degrees. This is accomplished by the formulas for ratios for the loss of heat at high temperatures as modified by Box (pp. 226 and 228), first for radiation, and secondly for contact with air.

These formulas are as follows:

For radiation

$$R' = \frac{124.72 \times 1.0077^{t} \times (1.0077^{T} - 1)}{T}$$

Where
T = the difference in temperature between the steam and the temperature of the room (Cent.).
t = the temperature of the room (Cent.), and
R' = the factor by which the radiation at ordinary temperature is to be multiplied to allow for the large difference in temperature.

For loss of heat by contact with air

$$R'' = \frac{.552 \times t^{1.233}}{t}$$

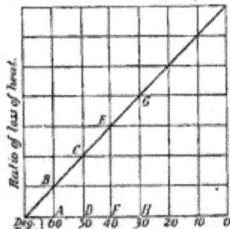

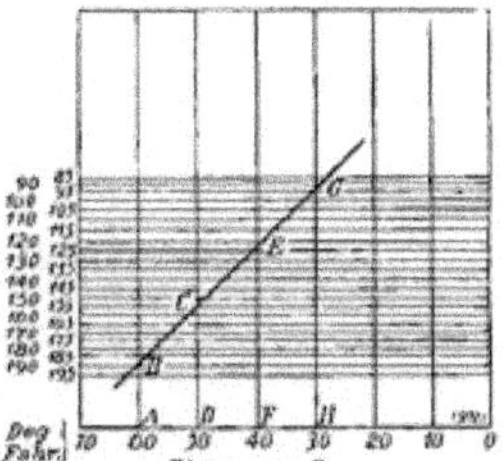

Diagram 2.

Where
R'' = the factor by which the loss of heat by contact with air at ordinary temperature is to be multiplied to allow for the large difference in temperatures, and
t = the difference in temperature between the steam and the air and walls of the room.

Solving this formula for 12 different cases between the above limits and plotting the results, using the same scale and nomenclature as in Diagram 1, we obtain Diagram 2. The results of the formula give ratios of thermal units corresponding to certain differences in temperature, changed to Fahr., given on the left; the figures given at the bottom represent external temperatures. Now, to find the temperature to which the building would be raised, heating surface and pressure constant, the loss of heat being assumed as due not to the internal but as from a constant temperature (70 degrees) to a varying external; for example, say 60 degrees, measure the height of the ordinate, or vertical line between the points A and B on Diagram 1, and on the similar vertical line starting from 60 degrees on Diagram 2, lay off the same distance A B and note from the left the abscissa, or horizontal with which the point B corresponds. This for the case of 60 degrees, external temperature, we find to be about 194 degrees, similarly for 50 degrees we find 158, for 40, 122, and for 30, 86. These results indicate the internal corresponding to the various external temperatures, assuming the loss of heat by the building was that due to an internal temperature of 70 degrees. But as the loss of heat by the building increases as the rise of internal temperature, if we obtain the mean temperature between the above figure and 70 degrees we get a figure which will approximately represent the temperature at which the building would be losing heat under such conditions. Hence, for 30 external we lose, under assumed conditions, as though the internal temperature was 78 degrees:

40 degrees.	96 degrees.
50 "	114 "
60 "	178 "

III. *A Determination of the Ratio of the Loss of Heat, Heating Surface and Pressure Constant, the Loss of Heat being Occasioned by the Temperatures Determined by* II.

If the internal temperature was as high as that given above, the efficiency of the radiating surface would be considerably reduced, and applying the formula under II. the reduction would be found to be as in the ratio of the following numbers:

0 degrees	307
30 "	202
40 "	164
50 "	[illegible]
60 "	[illegible]

IV. *The Final Determination of the Resultant Temperature from the Results of* III.

Finally expressing the results of III. in a proportion, and remembering that the apparatus has heated the building to 70 degrees in zero weather, we have

$$307 : 195 :: x : 70 \ldots\ldots \text{hence } x = 110 \text{ degrees}$$
$$307 : 226 :: x : 70 \ldots\ldots \text{hence } x = 95 \quad "$$
$$307 : 262 :: x : 70 \ldots\ldots \text{hence } x = 82 \quad "$$
$$307 : 290 :: x : 70 \ldots\ldots \text{hence } x = 74 \quad "$$

or, a heating apparatus sufficient to heat a given building to 70 degrees in zero weather with a given pressure of steam will be found to heat the same building, steam pressure constant, to 110 degrees at 60, 95 at 50, 82 at 40, and 74 at 32.

EDWARD E. MAGOVERN.

DETROIT, MICH., February 8, 1894.

To the Editor of THE ENGINEERING RECORD.

SIR: We have before us the issues of THE ENGINEERING RECORD for January 20 and February 3, and are much interested in the discussion of the question raised in regard to the requirement that buildings must be heated to a temperature of 70 degrees in zero weather. We hope that the agitation of this question may continue until some definite results are secured. We have had large amounts tied up and carried over from year to year simply for the lack of some means of forcing settlement without waiting for zero weather. We recall an instance three years ago where we had over $7,000 tied up, and our experience in this case may be of interest.

It was a mild winter and there was no prospect of zero weather. We had guaranteed to heat the building to 60 degrees with zero outside. We finally insisted upon making a test, but the temperature was about 20 degrees above. We were dealing with a mechanical engineer of high standing, and he insisted on our raising the temperature from 20 degrees to 80 degrees. We argued that that was not fair, and that it was much easier to raise the temperature from zero to 60 degrees than from 20 degrees to 80 degrees. Our representative stated at the time that he had good authority to prove this. He tried to find the table he had in mind in Haswell's, but for some reason could not locate it. It was finally agreed that if upon returning home our representative could furnish such evidence the engineer's clients would accept the plant and settle for it. Returning to Detroit the table was found, which is on page 526 of Haswell's 1893 edition. We expressed a copy of this to them, and as soon as it had time to reach its destination we received a dispatch stating that a New York draft had been mailed us. The account was in that way settled and everybody apparently satisfied. In our opinion neither that table nor the formula given by Mr. Magovern is fully satisfactory. We believe that a rule which could be given a more general application could be formulated, and would suggest that Mr. Baldwin or some one of his prominence as a heating and ventilating engineer should take this up and work out a rule that would be acceptable to all heating contractors.

We would then suggest that in drawing contracts it be stated in some such way as this: "Guaranteed to maintain a temperature of 70 degrees with outside temperature zero, or other temperatures, as per—— rule." Then it might be explained parenthetically, or in a foot-note of each contract, what was the nature of this rule and the reason of its insertion.

We further suggest that this rule, if satisfactorily formulated, be adopted by the Society of Mechanical Engineers, the organization of steamfitters, and it might be advisable for the manufacturers of steam-fitting appliances to form an organization and the association act upon this. We do not know that the latter class have ever formed any association, but we can see that much might be accomplished by such an organization.

Referring again to the paper of Mr. Magovern in your issue of the 3d inst., we wish to say that we think it a most excellent paper, and one well worth the careful consideration of your readers.

In the foregoing we have considered the subject from a business rather than a technical standpoint; but in any event we hope our suggestions may have some consideration, or that the results aimed at may be secured in some other way.

HUYETT & SMITH MANUFACTURING COMPANY.
By JAMES INGLIS, Secretary-Treasurer.

MINNEAPOLIS, MINN., February 19, 1894.

To the Editor of THE ENGINEERING RECORD.

SIR: I have noted with interest the discussion in your columns of the question of the requirement in heating contracts that buildings shall be heated to 70° Fahr. in zero weather as precedent to final payment. The similar clause in our contracts reads 70 degrees with outside temperature 40 degrees below zero. The experience of our firm (the Porter Radiator and Iron Company) has been varied, both sad and pleasant. In our form of contract it is distinctly stated that lack of cold weather shall not be made an excuse for delay of payments. The full text of the acceptance clause and also our guarantee to which it refers are as follows:

ACCEPTANCE.

The apparatus, in so far as the mechanical work thereof and construction of the same are concerned, shall be considered as accepted immediately upon completion. If it be found that the same does not comply with said specifications, notice thereof specifying the defects shall be given in writing immediately to the heating contractor.

It is distinctly understood that no payments or part thereof are to be delayed on account of lack of cold weather in which to test the heating apparatus, as the guarantee herein contained is binding upon the contractor as to fulfillment of contract. It is further understood that such acceptance shall not be deemed a waiver of our guarantee as to efficiency of the heating apparatus.

GUARANTEE.

When the apparatus is completed in accordance with the conditions hereof, we guarantee that when properly operated it will be capable of continuously warming all apartments of said building enumerated in Exhibit 1 to the "inside temperature" therein mentioned when the outside temperature is that specified in said Exhibit 1.

Further, the building and apparatus being kept in repair, and the apparatus properly operated, there shall be no snapping, cracking, or pounding whatsoever in any of the pipes and radiators, nor shall there be appreciable loss of steam or hot water from any part of said apparatus.

All material furnished by us shall be first-class in every particular, and shall be erected by competent workmen. There shall be no leaks, flaws, or other imperfections, and when completed the job shall present a finished appearance.

Our estimate for the capacity of the apparatus required for heating the building is based upon information furnished us by owner of building or his representative. In case of any change in the same, such change shall release us from the guarantee as to specified capacity of the heating apparatus, so far as such change shall affect it.

The chimney, to be furnished by the owner of the building, must be large enough and capable of passing sufficient air for the rapid consumption of fuel.

All necessary excavating for the setting of the boiler and running of mains is to be done by the owner of the building. Proper opening for admission of boiler, water supply and connection, and fuel for testing, are also to be furnished by him.

This guarantee to be good only for the time from the date of completion, as specified in Exhibit 1.

The temperatures referred to in the guarantee are to be filled in on blanks in Exhibit 1, which is a kind of general summary and acceptable of the contract, and is signed by contractor and owner.

For heating contracts we use a set of forms, numbered as Exhibits 1 to 14 inclusive, in which the items are enumerated and described in detail. The first sheet, as previously stated, embodies the essence of the contract, and by reference to the other sheets for details we are enabled to get out proposals covering almost any kind of a steam or hot-water heating job. The exhibits forming a part of the specifications are enumerated upon the first sheet, and the letter-press copies required are therefore reduced to a minimum. I think an error is made in making specifications in full detail. Our sheets in the main describe the systems in a general way, but do not give the sizes of mains or risers. Our experience seems to indicate that this is the better course, unless one is warranted in making special complete detailed drawings. In that event it is not desirable to let the plans go out of your hands, for they may be unfairly criticised by competitors, or the owner may be influenced by criticism, of the value or disinterestedness of which he is unable to judge. Our experience has led to the belief that architects' heating specifications have done more harm than good, and that they are not desirable unless drawn up by some person who is familiar with all the technical points, and who is also allowed to direct the work of erection from day to day. In such cases the heating contractor should only be required to furnish labor and material, and he should be entitled to his pay when the material is properly erected, or after a circulation test. We have done several jobs of this character, but always for a much higher price than we would have required to accomplish the same results under different conditions.

I believe it would be for the general benefit of all heating contractors if an agreement could be reached upon a general form of contract which should specify on one sheet the temperature, rooms, size of boiler, amount of radiation, chimneys, kind of valve, method of payment, and duration of guarantee, together with a short statement regarding the system and the manner in which work was to be done. This would cover all that the owner needs or cares to know, and with such a proposal the owner should look to the intelligence, capital, skill, experience, and business ability of the contractor for his assurance that he will get honest work and an effective plant, precisely as he depends upon those same qualities in a carriage-builder when he goes to buy a vehicle. Our experience has been that the more in detail heating plans have been the more they have harmed the owner. Where a contractor has a reputation to sustain and intends to put in good work, he must either follow the specifications blindly and neglect to avail himself of points which may suggest themselves to him in the work, or in varying from them he will find it well-nigh impossible to convince the owner that the changes were not made for the contractor's benefit.

The owner of a building is a copartner with the heating contractor to the extent that he is to furnish the variable quantity in the problem, in the construction of the walls and windows and chimney. It is impossible to tell from their appearance how much heat they will transmit, and it is a question if the heating contractor should undertake to do more than furnish a given amount of radiation and to guarantee the circulation and workmanship. In such instances if it should be found that the building transmitted more heat than was estimated, the additional radiation should be provided at the expense of the owner, as it was the building and not the radiation which was at fault. It is difficult to see why a heating contractor should be required to guarantee the temperature to which he will warm a building if he has specified and furnished a given amount of radiation and a boiler of agreed dimensions, for it is well known that the heat given off by radiators is, under ordinary conditions, practically a constant and outside the control of the contractor, and also that the efficiency of the boiler depends entirely upon the draft and the manner of firing. The province of the contractor is then to guarantee the workmanship and the circulation and not the variable features, the house and chimney, which the owner furnishes; and the sooner this is appreciated, the better for all concerned.

We have found a letter of advice as a contract outlining the work about as satisfactory a manner of arranging for steam and hot-water heating work as any other, and we prefer that the owner should look to the reputation of the contractor rather than to the letter of the specifications for his protection. You have to satisfy your customer in any event if you are to collect your pay, and if you do that you seldom have occasion to refer to the details of the specifications. Most of the points raised upon specifications have been made by persons who did not intend to pay if they could avoid doing so, and the very points which might be really in the contractor's favor would, if brought into court, have been used to his disadvantage in causing delay in settlement of the claim. I believe material and labor should be paid for as delivered and placed in position, and when the work is tested for tightness the whole amount should be due. The guarantee is something given, and the reputation of the contractor should be sufficient to assure the owner that it will be lived up to if a defect should occur during the time specified therein.

GEORGE C. ANDREWS.

NEW YORK, March 28, 1894.

To the Editor of THE ENGINEERING RECORD.

SIR: I have been somewhat interested in the correspondence published in recent issues of your journal regarding the term used in heating contracts, "That the building shall be heated to 70° Fahr. in zero weather," and the practice of architects and owners in withholding the final payment (sometimes a third of the entire amount of the contract) from the heating contractor until such time as they have had zero weather in which to test the apparatus, which, as your readers are aware, we sometimes do not get during an entire winter. I have always considered this more of a hardship on the heating contractor than it would be for an owner to retain a portion of the architect's fees for a number of years, or until he had fully satisfied himself that the building as designed would fully answer the requirements and prove suitable in all weathers.

In many cases a heating contractor has very little to say as to the make or general arrangement of a heating apparatus. The make, size, and location of heater or boiler are specified. The make, location, and sometimes the size of the radiating surface are specified, and while the heating contractor has the privilege of increasing the surface, if in his judgment he considers it necessary, in many cases the building is not erected and the heating contractor has no way of knowing whether it will be well or poorly built, and consequently can only judge from the exposure shown and his general experience as to whether the radiating surface and heater power will be ample or not.

The labor which the heating contractor employs in installing the apparatus has to be paid for in cash, the materials used have to be paid for in from 30 to 60 days from the time of delivery, and as the materials and labor on the average heating contract usually amounts to over 80 per cent. of the contract, unless some specified time for making the final payment is mentioned in the contract, the contractor in addition to being a mechanic has to be somewhat of a banker or else has to purchase goods in the hope of getting zero weather in which to test them before the time of payment arrives, and in the event of continued mild weather and consequent non-payment has to disappoint his supply man or the manufacturers.

I think a good deal of the trouble is brought on by an over-anxiety on the part of some heating contractors to secure work and a consequent willingness on their part to agree to almost any clause which may be submitted to them, with the hope that it will not be enforced, or that, if enforced, they will get zero weather in which to test the apparatus before the time for the final payment arrives. I have also found that another reason why the term has been allowed to exist and grow in use has been that manufacturers of new constructions of boiler heaters and other heating appliances, who could not show similar buildings heated with their apparatus, have been willing, so as to get their goods started, to allow an owner to use their appliances for an entire winter without payment, so as to give him ample time to demonstrate that the claims made for them were fully carried out in their operation during cold weather. My opinion is that for the above-mentioned reasons more than from any other cause steam and hot-water heating apparatus, although largely and successfully used for the past 25 to 50 years for heating all classes of buildings, have been looked upon by many as an experiment and treated as such, in requiring a contractor to demonstrate that each apparatus shall successfully heat the building in zero weather before final payment is made, while new materials in all other lines in connection with the building trades are being considered, used, and treated as commercial articles and paid for when placed or within a reasonable time after being placed.

In my opinion the term should be done away with or modified so as to name a date in the contract when final payment shall be made, provided the apparatus has proved ample for the requirements up to that time. Architects or owners should decide what is wanted in the way of a heating apparatus, the plans and specifications should be clearly drawn either by the architect or a competent heating engineer. Only responsible and reliable persons should be allowed to estimate on the work, and when completed, whether in the summer or winter, it should be considered as a commercial article, and there should be a specified time for making the final payment, which should be within a reasonable time after the completion of the work, and should not be delayed longer than from 30 to 60 days after that time.

Some years ago, before the hot-water heating system was as largely used as it is at the present time, I made plans and specifications for the heating of a large building by that system. It was looked upon somewhat as an experiment, although larger buildings had been successfully heated in the same way for years; the question came up of leaving the final payment until the apparatus had heated the building in zero weather. It was finally decided that the apparatus should be paid for in full by February 1 (the work having been completed November 1), provided the apparatus proved efficient up to that time. It was paid for and has proved satisfactory in all weathers since that time.

At another time I designed a heating apparatus for a school building which had formerly been heated by a hot-air system, and in arranging the details the commissioners wanted to hold one-half the amount of the contract until the building had been satisfactorily heated and ventilated in zero weather. I explained the unreasonableness of such an arrangement unless they could guarantee to furnish the contractor with zero weather within a reasonable time, when they explained that they had held back $1,000 from the former contractor for three years, and gave that as an excuse for doing the same in this case. It was finally arranged that the final payment should be made at a stated time; the apparatus was paid for and has proved satisfactory ever since.

I could cite hundreds of similar cases to prove that architects and owners have not suffered by accepting the guarantee of a responsible heating contractor, and by making payment for an apparatus in which they had already received the equivalent in labor and material for the amount paid.

My opinion is that it is just as necessary to mention the time of payment as it is to mention the amount in any heating contract, and while it might be possible to formulate a rule, such as Mr. Magovern has mentioned, so as to determine in moderate weather what an apparatus would do in zero weather, my experience has been (particularly in hot-water heating) that you do not get a proportionate result in moderate weather of what you do in extreme weather, on account of a poorer draft and consequent lower efficiency of boiler surface. This of course would not apply to a steam heater after steam was generated, and while in steam heating such a rule could be made to apply to steam at a certain pressure, say two pounds, in a hot-water heating apparatus the temperature at which the water should have to be carried would have to be determined—viz., whether it would have to be at 150 degrees, 180 degrees, or higher. Then, from the present tendency, some would want to experiment with such a rule in zero weather before they would decide to adopt and be governed by it, so that I think that the proper and business way out of it would be to have the plans and specifications properly drawn to suit the requirements, and to insist on the time of payment being specified in all heating contracts. I also advise putting the terms of payment in the estimate somewhat after the following order:

We hereby offer and agree to erect a low-pressure, steam-heating apparatus in your building located at , according to plans and specifications prepared by

Apparatus guaranteed to be noiseless in operation, and of ample capacity to thoroughly heat the building to the specified temperature in the coldest weather, and to be completed in a good and workmanlike manner for the sum of $, in payments to be arranged as follows:

$ when heater and mains are placed in the building.

$ when radiators are placed and connected.

Balance within days after completion of the work.

Another objectionable term that I find is taken advantage of at times in heating contracts are the words "to the satisfaction of the architect or owner."

If an architect or owner decides not to be satisfied with a heating apparatus he can keep a contractor out of his payments for a considerable time, even if the apparatus is placed exactly in accordance with the plans and specifications on which the contract is based. The use of the term does not increase the possibility of the contract being carried out, while it often makes a delay in the payments. It does not insure the placing of the apparatus in any better shape than if it was erected under the supervision of the architect and according to certain plans and specifications.

I have always looked upon this "satisfactory clause" as a meaningless term, in which there was no equity on which to base a contract, and in my opinion it should be dropped. W. M. MACKAY.

OFFICE OF EUREKA
HEATING AND VENTILATING COMPANY,
SAGINAW, MICH., April 20, 1894.

To the Editor of THE ENGINEERING RECORD

SIR: I have read with interest the letter of Mr. W. H. Francis on the requirement of 70 degrees in zero weather, and I think the matter should receive the due consideration of those in a position to hammer out a correction of the abuses.

I had made up several years ago a blank form of specification and contract, more especially for residence and smaller work. For our larger jobs, public buildings, etc., conditions vary so that we invariably make a separate specification to meet their special wants. The form in general use I find answers every requirement very fully for the purpose it was designed, special cases of course making it desirable to alter. On work this form of specification and contract is not best adapted to, and which requires a separate and distinct specification, say for special boiler setting, special forms of radiation, etc., we make a written or typewriter copy, but usually for the purpose of showing our customer that the condition is not special with him, but general with all, we put on the last two sheets in the printed form.

In Section 16 the guarantee contains the expression, "The said apparatus shall be capable," etc.; also in speaking about snapping, cracking, etc., we state that the building and apparatus is to be kept in repair and properly operated, etc. Our contract proper is the last page, and is self-explanatory. We have found such a form of contract fair to all, and absolutely necessary for safety in dealing with many.

I have known of a number of instances where contractors have been kept from the use of their money until zero weather or colder was had. While I believe our form of specification and contract protects us, yet it has been my intention when next it was necessary to print to either change the form or add an explanatory note. The guarantee, while not explicit, is enough to compel payments of itself without possibly a question, yet I think that the contract proper, specifying the amounts to be paid and when they are to be paid, settles that question.

It is a matter of importance to every fitter to exert his influence, not only for his own but for the benefit of the craft, to see that the right interpretation is given to specifications, and unitedly to oppose the careless specified exactions of many architects. Let me say for one, with all due respect for the architects, many of whom are very capable of writing heating specifications, that as a rule I do not believe it is right that an architect should specify in detail the heating work. There are in reality very few architects who are competent to properly apportion the amount of radiation, boiler surface, piping, etc. I am strictly in favor, on a building of any particular size, of the owner furnishing to those he wishes figures from a complete set of plans and specifications covering his work in detail. Such specifications should be prepared by a disinterested and competent heating engineer, to be found in all of our larger cities, and who has made a careful study of the different systems, and from practical knowledge and experience knows what is wanted.

Another point I am utterly opposed to, not only for the good of the craft but the general good of all concerned, and that is the common practice in many sections of giving a carpenter or mason the contract for the building, etc. There is practically nothing saved the owner, but it is frequently the case that the heating and plumbing work is cut down to enlarge the contractor's profit and work substituted to meet the condition of low price. Besides, one of the greatest evils is that the subcontractor who finished the work last has the poorest show to get his money.

I am glad to see this matter brought to notice in THE ENGINEERING RECORD. It should lead to a friendly exchange of ideas. C. W. LIGHT.

[Mr. Light has also sent a copy of the specifications and contract issued by his company. The guarantee reads as follows:

When the apparatus herein proposed to be furnished is completed in accordance with the conditions hereof, we guarantee that it will be so constructed as to permit steam to circulate in all its parts with.........pound pressure thereon, or any higher pressure; and that the said apparatus shall be capable of continuously warming all parts of said building that are enumerated in Section 5 of this proposal, to the temperature mentioned therein when the outside temperature is .....degrees below zero; further, the buildings and apparatus being kept in repair, and the apparatus properly operated, there shall be no snapping, cracking, or pounding whatsoever in the pipes or radiators, nor shall there be any appreciable loss of steam or water from any part of the said apparatus; we further guarantee that all work and materials furnished by us shall be free from any mechanical defects for a period of..........years from ....189.....

The variable quantity consisted in the construction of the building is thus provided for in another clause of the agreement.

Our estimate for the capacity of the apparatus required to heat said building, and our specifications and proposals are all based upon dimensions and plans of the building, and information respecting its construction received from you or your representatives. If it shall hereafter be found that such dimensions, plans, or information are erroneous, or if changes shall be made in the plans of construction of the building, then, so far as such error or changes shall affect the sufficiency of the heating apparatus, the foregoing guarantee as to the sufficiency thereof must be deemed canceled.

The terms of payment called for by the agreement are: "Fifty per cent. of the contract price when the boiler is set and main supply pipes run, 25 per cent. more when the radiation is delivered at the building, and the balance on completion of the work herein specified to be performed by us." There is also included provision for payment upon the installment plan, the apparatus to remain the personal property of the installing company until paid for, and the contractors being entitled to remove the apparatus without process of law in the event of default of payments, the moneys already paid to be held as liquidated damages for the use of the property.]

H. D. Crane, 234 North Pearl Street, Cincinnati, O., writes that he accompanies all his heating contracts with a guarantee, of which the following is a copy.

I herewith guarantee the above apparatus for one year from date against any defects in materials or workmanship, and that it will fulfill the guarantees called for in our specifications. Should any defect or deficiencies develop, we will, upon notice, make good the same at our own cost.

BOSTON, MASS., May 7, 1894.

To the Editor of THE ENGINEERING RECORD.

SIR: The discussion of this subject which has already appeared in these columns has been of marked interest to the writer and has brought out clearly the two aspects of the case—the commercial and the engineering. The point of view of the purchaser has not been directly presented, but it is evident that he is, after all, the most important factor, and that he may justly demand absolute protection against an inadequate heating system. Having the power to give or refuse his order, and within reasonable limits to pay or not to pay unless satisfied, he is, perforce, the one who eventually must decide upon the form of guarantee that will be acceptable.

Therefore, while the question is primarily one of engineering, the contractor, in point of fact, seeks the aid of the engineer merely to assist him in devising some scheme by which he may legitimately secure full payment for his work before it has been subjected to test under the most adverse conditions.

The matter is certainly a serious one, and yet it might be made far less so were all purchasers and contractors alike honorable men and financially responsible. It is well enough to suggest that one deal only with reputable houses, but circumstances will continually arise where something more substantial than a man's written promise as to the efficiency of a heating system is necessary to the protection of the purchaser. Nevertheless, it is evident from some of the letters presented in this discussion, that it is possible, under certain conditions, to enter into an agreement between owner and contractor that at once protects the one and secures payment to the other when his work is completed. Whether these conditions can be made general must, and can only, be determined by trial.

The injustice of the present form of guarantee must be evident to all; most assuredly so to those contractors who have suffered by it. Perhaps an additional instance of the effect may serve to still further emphasize the need of reform. Some years ago a contract was accepted by a certain heating concern guaranteeing to heat a given building to 70 degrees when the thermometer stood at 0 degree outside and the wind was blowing 21 miles an hour, the final payment (a large proportion of the total) being withheld until a test should be successfully made under these conditions. For three winters the contractor anxiously awaited the proper opportunity for a test, but in vain. The thermometer ran well below zero, and the wind rose far above 20 miles an hour, but on no day did they combine to give the desired conditions. The matter was finally compromised to the practical satisfaction of both parties, but the contractor lost three years' interest on his accounts due.

It is the experience of all contractors that so long as there exists any unsettled accounts the purchaser holds them as a financial lever which he is not slow to utilize. But once settled, the system is his for better or worse. Then, and not till then, does it appear that what has been pronounced a dismal failure may after all prove a reasonable success. The custom of deferred payments, subject to satisfactory test under circumstances of wind and weather of exceedingly rare occurrence, is responsible for this condition under which the contractor may be put to considerable inconvenience without just cause, or at least without a corresponding benefit to the purchaser.

It appears at first glance a simple matter of engineering to determine once for all a definite relation between the temperatures that may be maintained by a given system under varied outside temperatures. It such a relation could be established upon a basis so simple that its *raison d'être* might be comprehended by all, it would at once and for all time settle the difficulty.

Mr. Magovern's letter has made it evident that at least the matter is not a simple one; in fact, that it embodies so many elements that to a person not versed in the subject the deductions would make no logical appeal whatever. And yet he has presented the simplest of all cases, that of a room heated by direct radiation. Although not desiring to criticise his methods or results, it is at least to be noted that, notwithstanding he has clearly stated the well-known fact that the loss of heat between bodies of small difference of temperature is about proportional to that difference, nevertheless his final ratios of loss of heat are far from proportional to the differences between the internal and external temperatures for which they were calculated. In fact, his deduction, for instance, that an apparatus capable of heating a given room to 70 degrees with the thermometer at 0 degree outside will only heat it to 74 degrees with an outside temperature of 30 degrees, bears upon its face the evidence of error, which it will be found lies in the method of reasoning.

The formulas quoted from Box apply only to a theoretical, but a practically impossible, case; that is, where there is no loss of heat by leakage of air. No apartment devoted to the ordinary uses of man can be found into and from which there is not a continual natural leakage, while ventilation itself is but the provision for, and control of, such leakage on a more generous scale. The very variableness of this factor in heat loss is the strongest evidence of the difficulty, if not impossibility, of determining any absolute general basis of calculation to meet the requirements of all cases that may come under consideration.

We have only to appreciate the effect on the temperature of a room produced by a change in the air volume admitted, and to consider the difficulty of determining this volume when it enters and leaves by cracks, crevices, and porous walls, to comprehend the obstacles to the development of a generally applicable formula. Again, this natural ventilation will vary with the relation of internal and external temperatures, but not in the same ratio. Differences in wind velocity hardly perceptible to the senses will have an appreciable effect upon the results, while the humidity of the atmosphere should be taken into consideration in any complete formula.

Still further, the formula that would apply to one form and location of direct radiation would not hold with other designs and arrangements, and would be absolutely inapplicable with any method of indirect heating.

Such, then, appear to be some of the difficulties in the way of a solution of this question from an engineering standpoint; difficulties so nearly insuperable that we are forced back to the commercial side with the conviction that the reform must arise there.

At all events, in the writer's opinion, the owner must make it a part of his business, if not directly, at least by expert advice, to ascertain whether the system for which he contracts is so designed and proportioned as to effect the result desired. In the case of large buildings, at least, the specifications should cover everything necessary to a successful system, and the contractor be held responsible only for workmanship and material.

With smaller buildings, manufactories, and the like, where no regular architect is employed, the reform must come, more slowly, to be sure, along the lines of better knowledge of the subject of steam heating by the public at large, the employment of heating engineers to design and specify the systems, a more general dealing with reputable concerns, and the possible development of an approximate formula that shall relieve the contractor of a part of his responsibility for an actual test under the stated conditions of guarantee, if such be made.

WALTER B. SNOW,

Engineer, B. F. Sturtevant Company.

[From THE ENGINEERING RECORD of June 2, 1894.]

HEATING GUARANTEE AND ZERO WEATHER.

IN our issue of January 20 we published a communication from W. H. Francis, of Philadelphia, which was a well-considered protest against the usually expressed terms of the guarantee of a steam or hot-water heating contract—namely, that the building shall be heated to a temperature of 70 degrees when the external temperature is at zero, the final payment being dependent on a practical demonstration. The impropriety and unfairness of this form of guarantee was then pointed out, cases being cited where the final payment was withheld for several years until zero weather was obtained. Edward E. Magovern, M. Am. Soc. C. E.; the Huyett & Smith Manufacturing Company, of Detroit; George C. Andrews, of Minneapolis; W. M. Mackay, of New York; C. W. Light, of Saginaw, and Walter B. Snow, of Boston, have, in communications that are interesting and instructive, given some good reasons why the architect should not incorporate such a clause in his specifications, and why every prudent and reputable contractor should refuse to accept it in any contract he may sign.

A perusal of this correspondence, which we have reprinted in pamphlet form for convenience of reference, and for the future use of architects, contractors, and building committees, will further strengthen the position taken editorially by THE ENGINEERING RECORD in its issue of January 20, and at intervals during the past twelve years, when it has been maintained that the practice is unfortunately too general among architects to throw the arrangement of all the details upon the bidders, specifying only that the building "shall be heated to 70° Fahr. in zero weather." This disposition to shirk responsibility on the part of the architects, and the readiness of bidders to take contracts on any terms, and to gamble on the chances of being held to a strict construction of their requirements, is clearly responsible for this unfair and unwise provision. The difficulty of formulating a proportional equivalent as suggested by P. G., and discussed by Mr. Magovern and others, is made sufficiently clear in the correspondence, the conditions varying so widely in the various cases. With this correspondence before him, it would seem that any contractor who in future accepts a contract under which any portion of the contract price may be retained until Providence sees fit to furnish zero weather, so that a practical demonstration may be made, deserves no sympathy if his money is withheld for a period of several years.

If a heating contract is to be made with a contractor to whom the client is willing to give the work without competition, and the sufficiency of the plant is to be left to the contractor's judgment, then the guarantee for results may be in the following terms: "The system shall be capable of maintaining a temperature of 70° Fahr. with the external temperature at ——," but the payments should be made when the work is installed and the guarantee relied on. When, however, as is more often the case, competition is desired, detailed plans and specifications should be submitted alike for all to bid on. If the architect is not competent to prepare these, experts should be employed. These plans and specifications should state the kind, character, and size of boiler, size of chimney, and size of grate. If the work is to be on the indirect system they should include, size of both heat and vent flues, the amount of indirect radiating surface, whether in coils or radiators, how inclosed and in what manner—i. e., wood or iron, the size of mains and return pipes, the number and make of valves, method of supporting, size and description of fan, make and size of engine, and the minor details that go to make up a properly complete specification upon which each contractor can fairly and intelligently bid.

The assumption recently made by the *American Architect* in commenting on our former editorial, that the determination of the necessary boiler capacity and radiating surface is necessarily more a matter of inference than are the other details of building construction, is not justified. Architects who have not the special technical knowledge requisite to prepare a proper heating specification know perfectly well that such specifications can be secured the same as any other professional service. A heating and ventilating system must be designed to secure certain results under given conditions. These are best known by the designer of the building. The margin or factor of safety for which the owner must pay should be determined by the owner's professional adviser, just as it is done in giving the thickness of walls, size of beams, and details of a truss, and there is no more uncertainty in determining heating apparatus requirements than other construction details.

With regard to competitive plans from contractors, as before said, this is shirking responsibility, and results in the contractor gambling on how poor a job will be accepted by the building committee. Usually there is no expert to advise the committee which proposition is best for them to accept, so they take the cheapest, and while they may get all they pay for, they rarely get what they need, and they find it out when too late. Moreover, contractors in furnishing professional services in the shape of plans without compensation in the hope of getting the work, have largely increased their cost of doing business and reduced their legitimate profits, since competitive schemes are apt to be cut down to the lowest possible limit in order to bring the bids down. Thus, only a partial job can be done, the character of which is not realized by the building committee until too late.

VENTILATION AND HEATING.

By JOHN S. BILLINGS, A. M., M. D.,

LL.D. Edinb. and Harvard. D. C. L. Oxon. Member of the National Academy of Sciences. Surgeon, U. S. Army, etc.

"Dr. John S. Billings, the Director of the Department of Hygiene of the University of Pennsylvania, and probably the greatest authority on all hygienic matters in the world, published several years ago a brief work on Heating and Ventilation. The demand for this work has been constant, and its publication has greatly increased the general knowledge of this important subject. As the discussion of the matter has proceeded, many facts have been discovered and new inventions made, and Dr. Billings has therefore decided to recast and enlarge the work. His principal assistant, Prof. A. C. Abbott, of the University of Pennsylvania, has aided him in the compilation of a work which will undoubtedly rank as the chief authority on ventilation. The new volume contains 500 pages, and is fully illustrated with working drawings. Many of these were originally used in THE ENGINEERING RECORD, and are models of clearness and accuracy, though necessarily printed on a small scale. The work will be almost invaluable to all whose business requires them to study the difficult subject of ventilation, as well as to those who are interested in the construction of hospitals or other public institutions."—*Philadelphia Ledger, April 21, 1893.*

DR. BILLINGS' work is in all respects excellent, and furnishes a most convenient and complete reference volume for the very important subjects of which it treats. It is plainly and simply written, so as to be available for non-professional as well as for professional men, and leaves little to be desired as regards fullness. The opening chapter on the utility of ventilation may be studied with advantage by young people of both sexes, the subject being at least as important and as generally neglected as those of food, clothing and exercise. In this country, certainly, the average student, after completing a college course, goes out into the world with but the slightest knowledge of how to take care of his bodily and mental health. In his third chapter the author treats of the composition and physical properties of the atmosphere, and in the fourth of carbonic dioxide, familiarly known as carbonic acid. Then comes a chapter on the conditions which make ventilation necessary and the physiology of respiration. Moisture in the air, the quantity of air required for ventilation, the forces concerned in the process, and the methods of testing come next in order. All these subjects may be regarded as introductory. We come then to special modes of heating, and these are discussed with much care and thoroughness. Sources and methods of air supply follow, and then ventilating shafts with their accessories. Finally, in eight chapters, we have the various methods of ventilating mines and hospitals, halls and public buildings of all kinds, schools and dwellings, and lastly, miscellaneous applications of the now generally received principles.

The illustrations are good and very numerous, and we can safely venture to predict for the work a wide sphere of usefulness.—*The Nation, June 1, 1893.*

"In looking through the pages of the recent work on 'Ventilation and Heating,' by Dr. Billings, and recollecting that the subjects treated in the bulky five-hundred-paged volume relate to what is after all only a single department of architectural practice, and recollecting also that the architect is supposed to be well posted about the subject even if not a past-master, one cannot but feel the immensity of the professional matters which nowadays are added to the requirements of that busiest of professional men, the architect; but after reading the volume attentively and considering the vast amount of detail which enters into it, the extent of exact scientific knowledge which it implies, one appreciates that however ardently an architect might wish to know it all, in these days of complicated life he can hope at best to be only a leader among specialists in so far as relates to the so-called practical details of his profession. It is hopeless to expect that in the rush of business life a single man could master fully and hold available for daily practice the amount of knowledge involved in the mastery of ventilation and heating, and it is still more vain to hope that, having once mastered such a subject, an architect would have the time to keep abreast with the changing views, the fresh data and the more recent researches which are constantly being put forward in these special lines. Certain portions of Dr. Billings's work have appeared in a previous work, the 'Principles of Heating and Ventilation,' issued in 1884; but the present volume is substantially new, with numerous illustrations of recent practice, many of them drawn from the pages of THE ENGINEERING RECORD in which the descriptions first appeared. As to Dr. Billings's ability to speak on the subject there is no question. He has made it a long study and is one of the best authorities in every sense, and while it is perhaps to be regretted that the volume is not more condensed, it is very difficult to draw the line and say what could be omitted without sacrificing the perfect illustration of the subject. * * *

"Perhaps the most practically available portions of the work are the very numerous and complete illustrations, including nearly all the best examples of heating and ventilation throughout the world. The plans and other cuts are specially to be commended for their clearness, and the thorough manner in which particular systems are illustrated so as to be made perfectly manifest. The illustrations are by no means confined to the stock examples which are found in so many of the older works on ventilation and heating. The New Sorbonne at Paris, the Music Hall and the Metropolitan Opera House at New York, the Pueblo Opera House, Empire Theater, Philadelphia, as well as a number of very thorough instances of domestic work, are fully illustrated.

"Dr. Billings states in the preface that his object has been to produce a book which should be useful to students of architecture and engineering, as well as of interest to non-professional men who may be interested in the most important subjects which he treats. While his volume perfectly elucidates the points which he discusses, and is thoroughly admirable in every respect as a work on heating and ventilation, we fancy that any one except a heating engineer who would attempt to peruse the volume would feel so overpowered with the vastness of the subject that the first effect of the book would be to send the architect or the non-professional person immediately into the arms of the specialist, rather than to lead him to avail himself of the very complete information which the book affords. This can hardly be considered a defect; indeed, if the book should produce no other result than to convince the average architect that he should leave the matters of heating and ventilation as a rule entirely to specialists, it will certainly have accomplished a very important mission."—*American Architect.*

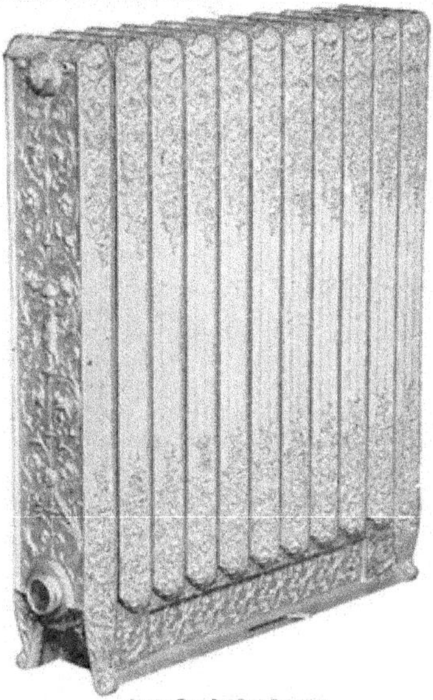

THE "ORNATE" RADIATOR.

STEAM OR WATER.

WE CLAIM—Superior Strength of Design,
Proper Proportion between Surface and Contents
Beauty of Finish.

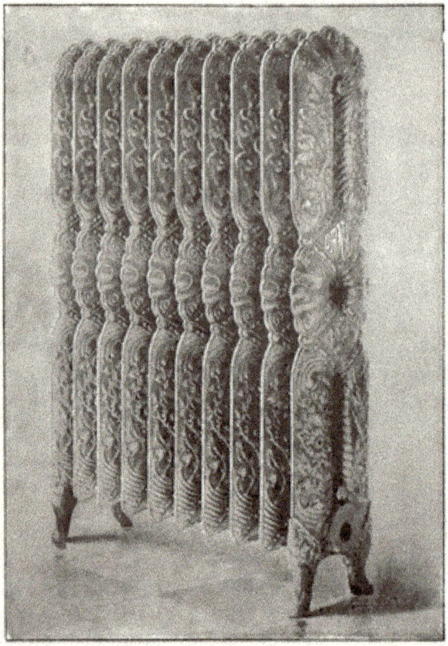

ALL PRIME SURFACE.
Made in 45, 38, 32, and 23-Inch Heights.
SEND FOR CIRCULARS AND PRICES.

MANUFACTURED BY THE

United States Radiator Company.

Works at Saltsburg, Pa.

PITTSBURG OFFICE, No. 102 WOOD STREET.

THE DUNNING BOILERS...

For Steam and Hot-Water ...Heating...

—ALSO—

Steam Engines, Boilers, and Machinery OF ALL KINDS.

Repair work done promptly, Satisfaction guaranteed.

PIPE AND FITTINGS, VALVES, IRON AND BRASS CASTINGS.

NEW YORK CENTRAL IRON WORKS CO.

52 EXCHANGE STREET, GENEVA, N. Y., U. S. A.

ESTABLISHED 1855

CRANE
ELEVATOR
COMPANY

Nine First Awards . . .
World's Columbian Exposition

CHICAGO

STEAM-HEATING PROBLEMS.

Warming churches (plan of placing a coil in each pew).
Warming churches.

PIPE AND FITTING.

Steam-heating work—good and indifferent.
Piping adjacent buildings; pumps *vs.* steam-traps.
True diameters and weights of standard pipes.
Expansion of pipes of various metals.
Expansion of steam-pipes.
Advantages claimed for overhead piping.
Position of valves on steam-riser connection.
Cause of noise in steam-pipes.
One-pipe system of steam-heating.
How to heat several adjacent buildings with a single apparatus.
Patents on Mills' system of steam-heating.
Air-binding in return steam-pipes.
Air-binding in return steam-pipes, and methods to overcome it.

VENTILATION.

Size of registers to heat certain rooms.
Determining the size of hot-air flues.
Window ventilation.
Removing vapor from dye-house.
Ventilation of Cunard steamer "Umbria."
Calculating sizes of flues and registers.
On methods of removing air from between ceiling and roof of a church.

STEAM.

Economy of using exhaust steam for heating.
Heat of steam for different conditions.
Superheating steam by the use of coils.
Effect of using a small pipe for exhaust steam-heating.
Explosion of a steam-table.

CUTTING NIPPLES AND BENDING PIPES.

Cutting large nipples—large in diameter and short in length.
Cutting crooked threads.
Cutting a close nipple out of a coupling after a thread is cut.
Bending pipe.
Cutting large nipples.
Cutting various sizes of thread with a solid die.

RAISING WATER AUTOMATICALLY.

Contrivance for raising water in high buildings.
Criticism of the foregoing and description of another device for a similar purpose.

MOISTURE ON WALLS, ETC.

Cause and prevention of moisture on walls.
Effect of moisture on sensible temperature.

MISCELLANEOUS.

Heating water in large tanks.
Heating water for large institutions and high city buildings.
Questions relating to water-tanks.
Faulty elevator-pump connections.
On heating several buildings from one source.
Coal-tar coating for water-pipe.
Filters for feeding house-boilers. Other means of clarifying water.
Testing gas-pipes for leaks and making pipe-joints.
Will boiling drinking-water purify it?
Differential rams for testing fittings and valves.
Percentage of ashes in coal.
Automatic pump-governor.
Cast-iron safe for steam-radiators.
Methods of graduating radiator service according to the weather.
Preventing fall of spray from steam-exhaust pipes.
Exhaust-condenser for preventing fall of spray from steam-exhaust pipes.
Steam-heating apparatus and plenum (ventilation), system in Kalamazoo Insane Asylum.
Heating and ventilation of a prison.
Amount of heat due to condensation of water.
Expansion-joints.
Resetting of house-heating boilers—a possible saving of fuel.
How to find the water-line of boilers and position of try-cocks.
Low-pressure hot-water system for heating buildings in England (comments by *The Sanitary Engineer*).
Steam-heating apparatus in Manhattan Company's and Merchants' Bank Building, New York.
Boilers in Manhattan Company's and Merchants' Bank Building, with extracts from specifications.
Steam-heating apparatus in Mutual Life Insurance Building on Broadway.
The setting of boilers in Tribune Building, New York.
Warming and ventilation of West Presbyterian Church, New York City.
Principles of heating-apparatus, Fine Arts Exhibition Building, Copenhagen.
Warming and ventilation of Opera-House at Ogdensburg, N. Y.
Systems of heating houses in Germany and Austria.
Steam-pipes under New York streets—difference between two systems adopted.
Some details of steam and ventilating apparatus used on the continent of Europe.

MISCELLANEOUS QUESTIONS.

Applying traps to gravity steam-apparatus.
Expansion of brass and iron pipe.
Connecting steam and return risers at their tops.
Power used in running hydraulic elevators.
On melting snow in the streets by steam.
Action of asher street fillings on iron pipes.
Arrangement of steam-coils for heating oil-stills.
Converting a steam-apparatus into a hot-water apparatus and back again.
Condensation per foot of steam-main when laid under ground.
Oil in boilers from exhaust steam, and methods of prevention.

PLUMBING PROBLEMS;

OR,

Questions, Answers and Descriptions,

FROM

THE ENGINEERING RECORD,

ESTABLISHED 1877.

(Prior to 1887, THE SANITARY ENGINEER.)

With 142 Illustrations.

"A feature of THE ENGINEERING RECORD (prior to 1887, *The Sanitary Engineer*), is its replies to questions on topics that come within its scope, included in which are Water-Supply, Sewage Disposal, Ventilation, Heating, Lighting, House-Drainage and Plumbing. Repeated inquiries concerning matters often explained in its columns, suggested the desirability of putting in a convenient form for reference a selection from its pages of questions and comments on various problems met with in house-drainage and plumbing, improper work being illustrated and explained as well as correct methods. It is, therefore, hoped that this book will be useful to those interested in this branch of Sanitary Engineering."

TABLE OF CONTENTS:

www.ingramcontent.com/pod-product-compliance
Lightning Source LLC
Chambersburg PA
CBHW021401210326
41599CB00011B/965
* 9 7 8 3 7 4 3 4 6 5 2 2 0 *